HITLER'S WAR ON RUSSIA

By the same author

THE FOXES OF THE DESERT
INVASION—THEY'RE COMING!

PAUL CARELL

HITLER'S WAR ON RUSSIA

The Story of the
German Defeat in the East

Translated from the German by
EWALD OSERS

ABERDEEN BOOKS
Sheridan, Colorado
2009

Hitler's War on Russia:
The Story of the German Defeat in the East
By Paul Carell

First published in Great Britain in 1964
By George G. Harrap & Co. Ltd.
(now an imprint of Chambers Harrap Publishers Limited)
English translation copyright George G. Harrap & Co. Ltd. 1964

This limited edition published in 2009 by
ABERDEEN BOOKS
3890 South Federal Boulevard
Sheridan, Colorado 80110
USA
303-795-1890
aberdeentp@earthlink.net · www.aberdeenbookstore.com

© 2009 Aberdeen Books
All rights reserved.
Printed in the United States of America.
No part of this book may be reproduced without
the written permission from the publisher except
in the case of brief quotations embodied
in critical articles and reviews.

ISBN 978-0-9713852-3-8

In this reprint edition, photos have been put into
two portfolios near the middle of the book.

Every reasonable effort has been made by the publishers
to trace the copyright holders of material quoted or
of illustrations used in this book. Any errors
or omissions should be notified in writing
to the publishers, who will endeavor
to rectify the situation for any
reprints and future editions.

CONTENTS

PART ONE: *Moscow*

1. Taken by Surprise — 11
   The forest of Pratulin—The white 'G'—0315 hours—Across the Bug, the San, and the Memel—Raseiniai and Liepaja—Surprise attack against Daugavpils—Manstein is made to halt—Rundstedt encounters difficulties—The citadel of Brest

2. Stalin looks for a Saviour — 45
   The first battles of encirclement—Why were the Soviets taken by surprise?—Stalin knew the date of the attack—The "Red Chapel" and Dr Sorge—Precursors of the U-2—Stalin and Hitler at poker—General Potaturchev is taken prisoner and is interrogated

3. Objective Smolensk — 68
   The forest of Bialowieza—The bridges over the Berezina—Soviet counter-attacks—The T-34, the great surprise—Fierce fighting at Rogachev and Vitebsk—Molotov cocktails—Across the Dnieper—Hoth's tanks cut the highway to Moscow—A Thuringian infantry regiment storms Smolensk—Potsdam Grenadiers against Mogilev

4. Moscow or Kiev? — 89
   Inferno in the Yelnya bend—A visit from the Mauerwald—Hitler does not want to make for Moscow—Guderian flies to see Hitler—Dramatic wranglins at Hitler's Headquarters—"My generals do not understand wartime economics"

5. Stalin's Great Mistake — 103
   Battles of annihilation at Roslavl and Klintsy—Stalin trusts his secret service—Armoured thrust to the south—Yeremenko expects an attack on Moscow

6. The Battle of Kiev — 116
   Rundstedt involved in heavy fighting on the southern wing—Kleist's tank victory at Uman—Marshal Budennyy tries to slip through the noose—Stalin's orders: Not a step back!—Guderian and Kleist close the trap: 665,000 prisoners

7. Code Name "Typhoon" — 129
   Caviare for Churchill—The mysterious town of Bryansk—Moscow's first line of defence over-run—Looting in Sadovaya Street—Stopped

6 Contents

by the mud—Fighting for Tula and Kalinin—The diary of a Russian lieutenant—Secret conference at Orsha—Marshal Zhukov reveals a Soviet bluff

8. Final Spurt towards Moscow 167
"The days of waiting are over"—Cavalry charge at Musino—On the Volga Canal—Within five miles of Moscow—Panic in the Kremlin—Stalin telephones the front—40 degrees below zero Centigrade—Battle for the motor highway—Men, horses, and tanks in ice and snow—Everything stop

9. Why couldn't Moscow be taken? 191
Cold weather and Siberian troops—The miracle of Moscow was no miracle—A chapter from the history of German-Soviet collaboration after the First World War—The unknown army—Tukhachevskiy's alliance with the Reichswehr—Himmler's grand intrigue—Stalin beheaded the Red Army

PART TWO: *Leningrad*

1. Chase through the Baltic Countries 219
Ostrov and Pskov—Artillery against KV-1 and KV-2 monsters—Hoepner is held back by the High Command—The swamp of Chudovo—Manstein's Corps cut off—The road to Leningrad is clear—Unsuccessful bathing party at Lake Samro

2. Break-through on the Luga Front 234
Critical situation at Staraya Russa—The battle of Novgorod—A Karelian supplies Russian maps—German 21st Infantry Division against Soviet 21st Armoured Division—Through the forests near Luga—On the Oredesh—The Luga pocket—On top of the Duderhof Hills—Radio signal from Second Lieutenant Darius: I can see St Petersburg and the sea

3. In the Suburbs of Leningrad 255
"All change—end of the line!"—In the gardens of Slutsk—Harry Hoppe takes Schlüsselburg—Order from the Fuehrer's Headquarters: Leningrad must not be taken—Hitler's great mistake

PART THREE: *Rostov*

1. Through the Nogay Steppe 271
New objectives for the Southern Front—The bridge of Berislav—Sappers tackle the lower Dnieper—Mölders's fighter aircraft intervene—The road to the Crimea is barred—Battle at the Tartar Ditch—Roundabout in the Nogay Steppe—Between Berdyansk and Mariupol

2. The Battle of the Crimea   284
Ghost fleet between Odessa and Sevastopol—Eight-day battle for the isthmus—the Askaniya Nova collective fruit farm—Pursuit across the Crimea—"Eight girls without baskets"—First assault on Sevastopol—In the communication trenches of Fort Stalin—Russian landing at Feodosiya—Disobedience of a general—Manstein suspends the attack on Sevastopol—The Sponeck affair

3. In the Industrial Region of the Soviet Union   302
Kleist's Panzer Army takes Stalino—Sixth Army captures Kharkov—First round in the battle for Rostov—Obersturmführer Olboeter and 30 men—Rundstedt is dismissed—Ringing of the alarm bells

PART FOUR: *Winter Battle*

1. The Siberians are coming   309
5th December 1941—No winter clothing—Fighting for Klin—3rd Panzer Group fights its way back—Second Panzer Army has to give ground—Drama on the ice of the Ruza—Brauchitsch leaves—Historic conversation at the Fuehrer's Headquarters—Hold on at all costs—Break-through at Ninth Army—The tragedy of XXIII Corps—Timetable for "Giessen"—Guderian is dismissed

2. South of Lake Ilmen   342
The fishing village of Vzvad—Charge across the frozen lake—Four Soviet Armies over-run one German Division—Staraya Russa—The Valday Hills—Yeremenko has a conversation with Stalin in the Kremlin shelter—The Guards are starving—Toropets and Andreapol—The tragedy of 189th Infantry Regiment

3. Model takes over   366
The supply dumps of Sychevka—"What have you brought with you, Herr General?"—A regiment holds the Volga bend—"I am the only one left from my company"—Stalin's offensive gets stuck—Sukhinichi or the mouse in the elephant's trunk—A padre and a cavalry sergeant—Spotlight on the other side: two Russian diaries and a farewell letter

4. Assault on the Valday Hills   390
The Soviet 57th Striking Brigade charges across the Volkhov—Rendezvous: Clearing Erika—Two Soviet Armies in the bag—Demyansk: 100,000 Germans surrounded—An unusual Order of the Day by Count Brockdorff-Ahlefeldt—A pocket is supplied from the air—Operation "Bridge-building"—Kholm, a fortress without guns

5. General Vlasov   409
A crack Soviet Army in the swamp—The wooden road across the clearing Erika—A merciless battle—Sapper Battalion 158—Break-out from hell—The disaster to the Soviet Second Striking Army—"Don't shoot, I am General Vlasov"—Maps buried under a river

PART FIVE: *The Ports on the Artic Ocean*

1. "Operation Platinum Fox" — 413
   The Murmansk railway—Offensive on the edge of the world—General Dietl reaches out for Murmansk—Across the Titovka and Litsa—No roads in the tundra—An error costs the Finns their victory—Mountain Jägers in the Litsa bridgehead

2. Battle in the Arctic Night — 425
   From Athens to Lapland—1400 horses must die—The Petsamojoki river—Supply crisis—Nightmare trek along the Arctic Ocean road—The Soviet 10th Rifle Division celebrates the October Revolution—An anniversary attack—Fighting at Hand Grenade Rock—Convoy PQ 17—The Soviet 155th Rifle Division freezes to death—The front in the Far North becomes icebound

PART SIX: *The Caucasus and the Oilfields*

1. Prelude to Stalingrad — 439
   Halder drives to Hitler's Headquarters—Anxieties of the Chief of the General Staff—The Izyum bend—Balakleya and Slavyansk—Fuehrer Directive No. 41—"Case Blue"—Curtain up over the Crimea—Failure of a Russian Dunkirk—Mid-May south of Kharkov—"Fridericus" will not take place—Kleist's one-pronged armoured pincers—The road to death—239,000 prisoners

2. Sevastopol — 464
   A grave in the cemetery of Yalta—Between Belbek Valley and Rose Hill—324 shells per second—"Karl" and "Dora," the giant mortars—A fire-belching fortress—The "Maxim Gorky" battery is blown up—"There are 22 of us left... Farewell"—Fighting for Rose Hill—Komsomols and commissars

3. A Plan betrayed to the Enemy — 476
   Venison and Crimean champagne—An interrupted feast—Major Reichel has disappeared—A disastrous flight—Two mysterious graves—The Russians know the plan for the offensive—The attack is mounted nevertheless—Birth of a tragedy

4. New Soviet Tactics — 487
   Fatal mistake at Voronezh—Timoshenko refuses battle—Hitler again changes his plan—Council of War at the Kremlin—The battle moves to the southern Don—Fighting for Rostov—Street fighting against NKVD units—The bridge of Bataysk

5. Action among the High Mountains 502
A blockhouse near Vinnitsa—Fuehrer Directive No. 45—By assault boat to Asia—Manychstroy and Martinovka—The approaches to the Caucasus—Chase through the Kuban—Mackensen takes Maykop—In the land of the Circassians

6. Between Novorossiysk and the Klukhor Pass 519
"The sea, the sea!"—The mountain passes of the Caucasus—Fighting for the old Military Highways—Expedition to the Summit of Mount Elbrus—Only 20 miles to the Black Sea coast—For the lack of the last battalion

7. Long-range Reconnaissance to Astrakhan 525
By armoured scout-car through 80 miles of enemy country—The unknown oil railway—Second Lieutenant Schliep telephones the stationmaster of Astrakhan—Captain Zagorodnyy's Cossacks

8. The Terek marks the Limit of the German Advance 535
Hitler's clash with Jodl—The Chief of the General Staff and Field-Marshal List are dismissed—An obsession with oil—Panzer Grenadiers on the Ossetian Military Highway—The Caucasus front freezes

PART SEVEN: *Stalingrad*

1. Between Don and Volga 541
Kalach, the bridge of destiny over the Don—Tank battle in the sands of the steppe—General Hube's armoured thrust to the Volga—"On the right the towers of Stalingrad"—Heavy anti-aircraft guns manned by women—The first engagement outside Stalin's city

2. Battle in the Approaches 552
The Tartar Ditch—T-34s straight off the assembly line—Counter-attack by the Soviet 35th Division—Seydlitz's Corps moves up—Insuperable Beketovka—Bold manœuvre by Hoth—Stalingrad's defences are torn open

3. The Drive into the City 561
General Lopatin wants to abandon Stalingrad—General Chuykov is sworn in by Khrushchev—The regiments of 71st Infantry Division storm Stalingrad Centre—Grenadiers of 24th Panzer Division at the main railway station—Chuykov's last brigade—Ten crucial hours—Rodimtsev's Guards

4. Last Front Line along the Cliff 568
Chuykov's escape from the underground passage near the Tsaritsa—The southern city in German hands—The secret of Stalingrad: the

steep river bank—The grain elevator—The bread factory—The "tennis racket"—Nine-tenths of the city in German hands

5. **Disaster on the Don**   575
   Danger signals along the flank of Sixth Army—Tanks knocked out by mice—November, a month of disaster—Renewed assault on the Volga bank—The Rumanian-held front collapses—Battle in the rear of Sixth Army—Break-through also south of Stalingrad—The 29th Motorized Infantry Division strikes—The Russians at Kalach—Paulus flies into the pocket

6. **Sixth Army in the Pocket**   587
   Get the hell out of here—"My Fuehrer, I request freedom of action"—Goering and supplies by air—The Army High Command sends a representative into the pocket—General von Seydlitz calls for disobedience—Manstein takes over—Wenck saves the situation on the Chir

7. **Hoth launches a Relief Attack**   604
   "Winter Storm" and "Thunder-clap"—The 19th December—Another 30 miles—Argument about "Thunder-clap"—Rokossovskiy offers honourable capitulation

8. **The End**   617
   The Soviets' final attack—The road to Pitomnik—The end in the Southern pocket—Paulus goes into captivity—Strecker continues to fight—Last flight over the city—The last bread for Stalingrad

Appendix   624

   Acknowledgment   624

   Bibliography   625

   Index of Names   630

PART ONE: *Moscow*

1. Taken by Surprise

The forest of Pratulin–The white 'G'–0315 hours–Across the Bug, the San, and the Memel–Raseiniai and Liepaja–Surprise attack against Daugavpils–Manstein is made to halt–Rundstedt encounters difficulties–The citadel of Brest.

FOR two days they had been lying in the dark pinewoods with their tanks and vehicles. They had arrived, driving with masked headlights, during the night of 19th/20th June. During the day they lay silent. They must not make a sound. At the mere rattle of a hatch-cover the troop commanders would have fits. Only when dusk fell were they allowed to go to the stream in the clearing to wash themselves, a troop at a time.

Second Lieutenant Weidner, the troop commander, was standing outside the company tent when Sergeant Sarge trotted past with his men of No. 2 Troop. "Nice spot for a holiday, Oberfeldwebel," he said with a chuckle. Sergeant Sarge stopped and grimaced. "I don't believe in holidays, Herr Leutnant." And more softly he added: "What's it all about, Herr Leutnant? Are we having a go at Ivan? Or is it true that we are only waiting for Stalin's leave to drive through Russia in order to get at the Tommies through their Persian back door and let the air out of their beautiful Empire?"

The question did not surprise Weidner. He knew as well as Sarge the many rumours and stories which had been going the rounds ever since their tank training battalion had been reorganized as the 3rd Battalion, 39th Panzer Regiment, which formed part of 17th Panzer Division, and had been moved first to Central Poland and then brought here into the woods of Pratulin. Here they were, less than three miles from the river Bug, which formed the frontier, almost exactly opposite the huge old fortress of Brest-Litovsk, occupied by the Russians since the partition of Poland in the autumn of 1939.

The regiment was bivouacking in the forest in full battle order. Each tank, moreover, carried ten jerricans of petrol strapped to its turret and had a trailer in tow with a further three drums. These were the preparations for a long journey, not for swift battle. "You don't go into battle with jerricans on your tank," the experienced tankmen were saying.

This was an argument against the stubborn ones who kept talking about imminent war with Russia. "Russia? Nonsense! We've got quite enough on our hands already. Why start another war? Ivan isn't doing us any harm, he's our ally, he's sending us grain, and he's against the British." That was what most of them were saying. It therefore followed that if they were not going into battle and not driving into Persia either, then the whole thing must be one huge diversionary manœuvre.

A diversionary manœuvre—but against whom? Surely, to bluff the British. All this build-up in the East might well be a blind for the invasion of Britain on the other side of Europe. This was an argument that was passed on in a whisper and with a knowing wink nearly everywhere. Those who spread it, and who believed in it, could not know of an entry in the diary of the Naval High Command, dated 18th February: "The build-up against Russia is to be presented as the greatest camouflage operation in military history, allegedly designed to divert attention from final preparations for the invasion of Britain."

Yet another story, breathtaking in its beauty and simplicity, was confidently bandied about by the old corporals, those old soldiers who, as is only too well known, can hear the grass growing, know all the secrets of the company office, and represent not only the soul but also the eyes and ears of each unit: Stalin, they patiently explained while washing up their mess-tins, playing "Skat," or polishing their boots, Stalin had leased the Ukraine to Hitler, and they were moving in merely as an army of occupation. In a war people will believe anything. And Sergeant Sarge was only too glad to believe in peace. He believed in the pact which Hitler had concluded with Stalin in August 1939. He believed in it, together with the rest of the German people, who regarded this pact as Hitler's greatest diplomatic achievement.

Second Lieutenant Weidner stepped up close to Sarge. "Do you believe in fairy-tales, Oberfeldwebel?" he asked. Sarge looked puzzled. The lieutenant glanced at his watch. "Be patient for another hour," he said significantly, and walked back to his tent.

While Sergeant Sarge and his second lieutenant had this conversation in the forest of Pratulin, a different conversation was taking place in the Wilhelmstrasse in Berlin, in the former residence of the Reich President. But here there was less mystery. Ribbentrop revealed the great secret. He informed his closest collaborators: Early tomorrow the Wehrmacht moves against Russia.

So that was it, after all. They had been suspecting it; now they knew. They had hoped it would remain merely a plan on paper; but now the die was cast. The time for politics and diplomacy, which were their concern, was over; now the weapons would speak. At that moment the ambassadors, the envoys, and the ministerial officials all asked themselves the same question: In view of this development, would Foreign Minister von Ribbentrop stay in office? Could he stay in office? Did not the rules demand his resignation?

Twenty-one months previously he had returned from Moscow with the German–Soviet Treaty of Friendship and explained to them: "Our treaty with Stalin keeps our rear covered and insures us against a war on two fronts such as brought disaster to Germany once before. I regard this alliance as the crowning achievement of my foreign policy."

And now it was to be war. The crowning achievement lay in the dust.

Ribbentrop sensed the wall of silence around him. He walked to the window overlooking the park where an earlier Chancellor, Prince Bismarck, used to take his constitutional—another man who had regarded the German–Russian alliance as the crowning achievement of his foreign policy. Was Ribbentrop reminded of his predecessor? He turned on his heel and said loudly and with emphasis, "The Fuehrer has information that Stalin has built up his forces against us in order to strike at us at a favourable moment. And the Fuehrer has always been right so far. He has assured me that the Wehrmacht will defeat the Soviet Union within eight weeks. Our rear will then be safe without having to depend solely on Stalin's goodwill."

Eight weeks. And supposing it took longer? It could not take longer. The Fuehrer had always been right before. For eight weeks one could, if necessary, fight on two fronts.

That was the position. And presently the troops would be told. In the dense pinewoods of Pratulin the hot day was drawing to its end. The pleasant smell of resin and the stench of petrol hung in the air. At 2110 hours an order was shouted softly from the company headquarters tent to tank No. 924: "Companies will fall in at 2200 hours. 4th Company, Panzer Lehr Regiment, in the large clearing." Wireless Operator Westphal called the order across to No. 925, and from there it was passed on from tank to tank.

Dusk had fallen by the time the company was lined up. First Lieutenant von Abendroth reported to the captain. The captain's eyes swept along the ranks of his men. Their faces beneath their field caps were unrecognizable in the twilight. The men were a grey-black wall, a tank company—without faces.

"4th Company!" Captain Streit shouted. "I shall read to you an order of the Fuehrer." There was dead silence in the forest near Brest-Litovsk. The captain switched on the flashlight he had hanging from the second button of his tunic. The sheet of paper in his hand shone white. With his voice slightly hoarse with excitement he began to read: "Soldiers of the Eastern Front!"

Eastern Front? Did he say Eastern Front? This was the first time the term had been used. So this was it, after all.

The captain read on. "Weighed down for many months by grave anxieties, compelled to keep silent, I can at last speak openly to you, my soldiers...." Eagerly the men listened to what had been worrying the Fuehrer for many months: "About 160 Russian divisions are lined up along our frontier. For weeks this frontier has been violated continually—not only the frontier of Germany but also that in the far north and in Rumania."

The men hear of Russian patrols penetrating into Reich territory and being driven back only after prolonged exchanges of fire. And they hear the conclusion: "At this moment, soldiers of the Eastern Front, a build-up is in progress which has no equal in world history, either in extent or in number. Allied with Finnish divisions, our comrades are standing side by side with the victor of Narvik on the Arctic Sea in the North....

"You are standing on the Eastern Front. In Rumania, on the banks of the Prut, on the Danube, down to the shores of the Black Sea, German and Rumanian troops are standing side by side, united under Head of State Antonescu. If this greatest front in world history is now going into action, then it does so not only in order to create the necessary conditions for the final conclusion of this great war, or to protect the countries threatened at this moment, but in order to save the whole of European civilization and culture.

"German soldiers! You are about to join battle, a hard and crucial battle. The destiny of Europe, the future of the German Reich, the existence of our nation, now lie in your hands alone."

For a moment the captain stood silent. The beam of his flashlight flickered over the paper in his hand. Then he added softly, almost as if these were his own words and not the conclusion of the Order of the Day: "May the Almighty help us all in this struggle."

After the men were dismissed there was a buzzing as of a swarm of bees. So they were going to fight Russia after all. First thing tomorrow morning. It was quite a thought. The men ran back to their tanks on the double.

Sergeant Fritz Ebert passed Sarge. "Extra comforts to be issued at once for each vehicle," he announced. He let down the tailboard of his lorry and opened a large crate: spirits, cigarettes, and chocolate. Thirty cigarettes per head. One bottle of brandy gratis for every four men. Drink and tobacco were the troops' traditional requirements.

There was feverish activity everywhere: tents were being taken down, tanks were being made ready. After that the men waited. They smoked. Very few touched the brandy. The spectre of a stomach wound still terrified them—in spite of sulfonamide. Only the very toughest slept that night.

It was a night of clock-watching. Slowly the hours ticked away, like eternity. It was the same all along the long frontier between Germany and the Soviet Union. Everywhere, strung out across an entire continent, the troops lay awake, from the Baltic to the Black Sea, a distance of 930 miles. And along these 930 miles three million troops were waiting. Hidden in forests, pastures, and cornfields. Shrouded by the night, waiting.

The German offensive front was divided into three sectors—North, Centre, South.

Army Group North, under Field-Marshal Ritter von Leeb, was to advance with two Armies and one Panzer Group from East Prussia across the Memel. Its object was the annihilation of the Soviet forces in the Baltic and the capture of Leningrad. The armoured spearhead of von Leeb's forces was Fourth Panzer Group, under Colonel-General Hoepner. Its two mobile corps were commanded by Generals von Manstein and Reinhardt. First Air Fleet, attached to this Army Group, was under Colonel-General Keller.

Army Group Centre was commanded by Field-Marshal von Bock. Its area of operations extended from Romintener Heide to south of Brest-Litovsk—a line of 250 miles. This was the strongest of the three Army Groups, and comprised two Armies as well as Second Panzer Group, under Colonel-General Guderian, and Third Panzer Group, under Colonel-General Hoth. Field-Marshal Kesselring's Second Air Fleet, with numerous Stuka wings, lent additional striking power to this tremendous armoured force. The object of Army Group Centre was the annihilation of the strong Soviet forces, with their many armoured and motorized units, in the triangle Brest–Vilna (Vilnius)–Smolensk. Once Smolensk had been taken by the mobile forces in a bold armoured thrust the decision would be made whether to wheel to the north or drive on towards Moscow.

In the southern sector, between the Pripet Marshes and the Carpathians, Army Group South, under Field-Marshal von Rundstedt, with its three Armies and one Panzer Group, was to engage and destroy Colonel-General Kirponos's Russian forces in Galicia and the Western Ukraine on the near side of the Dnieper, secure the Dnieper crossing, and finally take Kiev. Complete air cover was to be provided by Fourth Air Fleet under Colonel-General Löhr. The Rumanians and the German Eleventh Army, who came under Rundstedt's sphere of command, were to stand by as reinforcements. In the north Germany's other ally, Finland, was to stand by, ready for attack, until 11th July, the date for the German thrust against Leningrad.

This grouping of the German offensive line-up clearly shows its concentration of strength at Army Group Centre. In spite of unfavourable terrain, with river-courses and swamps, this sector was equipped with two Panzer Groups in order to bring about a rapid decision to the campaign.

Soviet intelligence evidently failed to spot this disposition, for the focus of the Soviet defensive system was in the south, opposite Rundstedt's Army Group. There Stalin had concentrated 64 divisions and 14 armoured brigades, while on the Central Front he had only 45 divisions and 15 armoured brigades, and on the northern front 30 divisions and 8 armoured brigades.

The Soviet High Command clearly expected the main German attack in the south, against Russia's principal agricultural and industrial areas. That was why it had concentrated the bulk of its armoured units on this sector for an elastic defence. But since the tank is primarily an offensive weapon, this concentration of powerful armoured forces on the southern wing simultaneously allowed for a Soviet offensive against Rumania, Germany's vital source of oil.

Hitler's plan of attack, therefore, was a gamble. It followed the recipe which had proved successful in the West, when, to the complete surprise of the French, he had broken rapidly through the unfavourable Ardennes terrain, piercing the Maginot Line, which was weak there, and thus bringing the campaign to a rapid conclusion. Hitler intended to apply the same plan to the Soviet Union: he would attack with all available forces in an unexpected place, tear open the enemy front, break through, utterly defeat the enemy, and seize his vital centres— Moscow, Leningrad, and Rostov—still carried by the momentum of the first great sweep. The second wave was then to advance to the line he had mapped out for himself—the line from Astrakhan to Archangel.

That was Operation Barbarossa on paper.

The time was 0300 hours. It was still dark. The summer night lay heavy over the banks of the Bug. Silence, only occasionally broken by the clank of a gas-mask case. From down by the river came the croaking of the frogs. No man who lay in the deep grass by the Bug that night of 21st/22nd June, with an assault troop or some advance detachment, will ever forget the plaintive croaking mating-call of the frogs on the Bug.

Nine miles on the near side of the Bug, outside the village of Volka Dobrynska, on Hill 158, stood one of those wooden observation towers which had sprung up on both sides of the frontier during the past few months. At the foot of Hill 158, in a patch of wood, was the advanced command post of Second Panzer Group, the brain of Guderian's tank force. "The white G's," the men called the group, because of the large white letter 'G' which all vehicles bore as their tactical identification sign. 'G' stood for Guderian. At a glance a vehicle was recognized as "one of ours." Guderian had introduced this idea during the campaign in France. It had proved so successful that Kleist had adopted it and had ordered the vehicles of his Panzer Group to be painted with a white 'K.'

During the preceding night, the night of 20th/21st June, the staff officers had arrived in greatest secrecy. They were now sitting in their tents or office buses, bending over maps and written orders. No signals came from the aerials: strict radio silence had been ordered, lest the monitoring posts of the Russians became suspicious. Use of the telephone was permitted only if strictly necessary. Guderian's personal command transport—two radio-vans, some jeeps, and several motor-cycles—stood parked behind the tents and buses, well camouflaged. The command armoured car approached. Guderian jumped out. "Morning, gentlemen."

The time was exactly 0310. A few words, then Guderian drove up the hill with his command transport to the observation tower. The luminous minute-hands of their wrist-watches crept round the dials.

0311 hours. In the tent of the operations staff the telephone jangled. Lieutenant-Colonel Bayerlein, the 1A, or chief of operations, picked up the receiver. Lieutenant-Colonel Brücker, the chief of operations of XXIV Panzer Corps—or XXIV Motorized Army Corps, as it then was—was on the line. Without greetings or formality he said, "Bayerlein, the Koden bridge was all right."

Bayerlein glanced across to Freiherr von Liebenstein, the chief of staff, and nodded. Then he said, "That's fine, Brücker. So long. Good luck." He replaced the receiver.

The bridge at Koden was the kingpin in the rapid tank thrust across the Bug to Brest. An assault troop of 3rd Panzer Division had orders to capture it by surprise a few minutes before the start of operations, to eliminate the Russian bridge guard on the far side, and to remove the explosive charges. The coup had succeeded.

A sigh of relief was heaved at Guderian's headquarters—even though provision had been made for the event that the surprise would not come off. Fourth Army had made preparations for bridging the Bug both above and below Brest. About fifty miles north of Brest, at Drohiczyn, Engineers Battalion 178 had crept up quietly to the intended spot, in laborious and lengthy secret marches, in order to build a pontoon bridge for the heavy weapons and equipment of 292nd and 78th Infantry Divisions.

It was 0312 hours. Everybody was watching the time. Everybody had a lump in his throat. Every one's heart was thumping. The silence was unbearable.

0313 hours. It was still not too late to change the course of events. Nothing irrevocable had yet happened. But as the minute-hands crept over the watch dials the war against the Soviet Union, which was lying ahead plunged in peaceful darkness, was drawing ineluctably nearer.

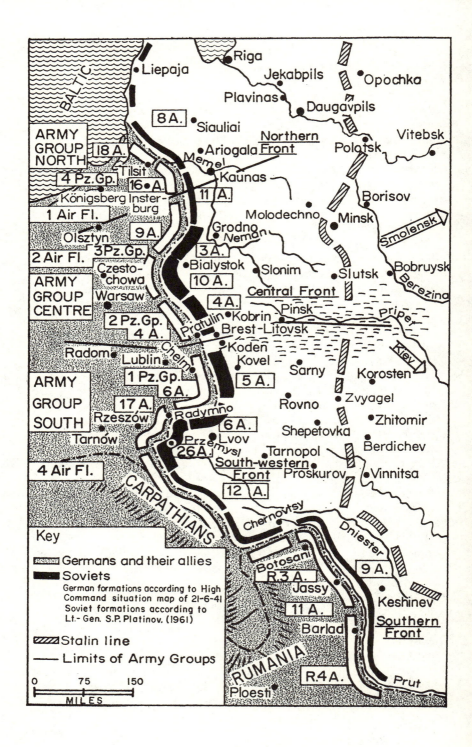

Appendix 3

The Fuehrer and Supreme Commander
       of the Wehrmacht
                           Fuehrer's Headquarters
                                      18.12.40

OKW/WFSt/Abt.L (I) No. 33 408/40 g.K. Chefs.

Directive No. 21
Case Barbarossa

The German Wehrmacht must be prepared, even before conclusion of the war against Britain, to overthrow Soviet Russia by a rapid campaign (Case Barbarossa).

Preparations are to be made by the High Command on the following basis:

(I) General intention:
    The bulk of the Russian Army in Western Russia is to be annihilated in bold operations by deeply penetrating Panzer wedges, and the withdrawal of combat-capable units into the wide-open spaces of Russia to be prevented.
    By means of a rapid pursuit a line must then be reached from which the Russian Air Force can no longer attack German Reich territory. The final objective of the operation is a screen against Russia-in-Asia from a general line Volga–Archangel. In this way the last industrial region left to Russia, in the Urals, can be eliminated by the Luftwaffe if necessary.

(II) Presumable allies and their tasks:
    . . . .

Map 1. The starting position for Operation Barbarossa. On 21st June the German forces in the East were poised for the attack with seven Armies, four Panzer Groups, and three Air Fleets—3,000,000 men, 600,000 vehicles, 750,000 horses, 3580 armoured fighting vehicles, 7184 guns, and 1830 aircraft. In the South, moreover, stood the Rumanian Third and Fourth Armies. The Soviets had ten Armies deployed in the frontier area, with 4,500,000 men.

(III) Conduct of operations:
(A) A r m y (in approval of intentions submitted to me):

In the operations area divided by the Pripet Marshes into a southern and a northern half the centre of gravity is to be formed north of this area. Here two Army Groups are to be envisaged.

The more southerly of these two Army Groups – Centre of the front as a whole – has the task of bursting forward with particularly strong Panzer and motorized formations from the area around and north of Warsaw and of smashing the enemy forces in Belorussia. In this way the prerequisite must be created for strong units of fast troops to wheel northward, where, in co-operation with the northern Army Group operating from East Prussia in the general direction of Leningrad, they will annihilate the enemy forces fighting in the Baltic area. Only after the accomplishment of this priority task, which must be followed by the occupation of Leningrad and Kronshtadt, are offensive operations to be continued with a view to the seizure of the important communications and armaments centre of Moscow.
Only an unexpectedly quick collapse of Russian resistance might justify the simultaneous pursuit of both these objectives.

The Army Group deployed south of the Pripet Marshes must form its centre of gravity in the Lublin area in the general direction of Kiev, in order to advance rapidly with strong Panzer forces into the deep flank and rear of the Russian forces and to roll it up along the Dnieper.

Once the battles south and north of the Pripet Marshes have been fought the following objectives are to be aimed at in the course of the pursuit:

In the south the early seizure of the Donets Basin, which is important for the war economy.

In the north the quick gaining of Moscow.

The capture of this city will represent a decisive success politically and economically, and will, moreover, mean the elimination of an important railway centre.

(Signed) Adolf Hitler

Bayerlein recalled September 1939. Then, too, he had been here at Brest with Guderian. That was a year and nine months back. Then, on 22nd September 1939, the Russians, in the shape of General Krivoshein's Armoured Brigade, arrived as allies. A demarcation line was drawn through their joint booty—defeated Poland. The Bug became the frontier. Under the treaty which Stalin had signed with Hitler the Germans had to withdraw behind the river and leave Brest with its citadel to the Soviets.

The arrangements had been meticulously observed, a joint parade had been organized, and colours had been exchanged. Finally, toasts had been proposed. For without vodka and toasts no treaty is considered valid by a Russian.

General Krivoshein had scraped together what little German he remembered from school and proposed his toast in German. In doing so he made a curious little mistake. He said, "I drink to eternal fiendship"—but instantly corrected himself with a smile: "eternal friendship between our nations."

Every one had raised his glass in high spirits. That was twenty-one months ago. Now the last few minutes of this "friendship" were ticking away. The letter 'r,' hurriedly inserted by General Krivoshein, was once more being deleted. "Fiendship" would break out with the first grey light of 22nd June.

The time was 0314. Like a spectre the wooden tower of Volka Dobrynska stood out against the sky. The first pale daylight appeared on the horizon. Deathly silence still reigned throughout the area of Army Group Centre. The forests were sleeping. The fields were quiet. Had the Russians not noticed that the woods and villages were bristling with assembled armies? Armies ready to spring? Division after division —all along the endless frontier.

The hands of the carefully synchronized watches jumped to 0315.

As though a switch had been thrown a gigantic flash of lightning rent the night. Guns of all calibres simultaneously belched fire. The tracks of tracer shells streaked across the sky. As far as the eye could see the front on the Bug was a sea of flames and flashes. A moment later the deep thunder of the guns swept over the tower of Volka Dobrynska like a steamroller. The whine of the mortar batteries mingled eerily with the rumble of the guns. Beyond the Bug a sea of fire and smoke was raging. The narrow sickle of the moon was hidden by a veil of cloud.

Peace was dead. War was drawing its first terrible breath.

Directly opposite the citadel of Brest stood the 45th Infantry Division—formerly the Austrian 4th Division—under Major-General Schlieper. The 130th and 135th Infantry Regiments were to mount the first assault against the bridges and the citadel. Still under cover of darkness the units composing the first wave had carefully worked up to the Bug. Like a black phantom, the railway bridge straddled the river. At 0200 a goods train chugged over the bridge, puffing, lamps brightly lit. It was the last grain train which Stalin sent his ally Hitler.

An ingenious ruse or an incredible degree of unsuspecting confidence? That was the question the officers and men of the assault battalions and shock companies were asking themselves as they lay in the crops and in the grass, by the railway embankment and opposite the Western Island. They did not know how many trains had crossed the bridge during the past few weeks. They did not know how conscientiously Stalin had implemented the German–Soviet trade agreement. Since 10th February 1940 up to that hour on 22nd June 1941 Stalin had delivered to Hitler 1,500,000 tons of grain. The Soviet Union was thus Germany's principal supplier of grain. But not only rye, oats, and wheat had been sent across the bridges over the Bug. During the sixteen months of friendship Stalin had sent Germany, strictly according to their contract, nearly 1,000,000 tons of mineral oil, 2700 kilograms of platinum, and large quantities of manganese ore, chrome, and cotton.

In contrast to the painstaking discharge of their obligations by the Russians, Germany had been a dilatory supplier right from the start. Even so, goods worth 467,000,000 marks had been supplied to the Soviet Union, including the half-finished heavy cruiser *Lützow*. When the last grain train from the east crossed the Bug at 0200 hours on 22nd June, Hitler was in debt to Stalin to the tune of 239,000,000 marks. Nothing of this was known to the officers and men by the railway bridge of Brest at first light on 22nd June. Above them, by the little wooden hut at the end of the bridge, reigned an atmosphere of peace and unsuspecting normality. The two German customs officials climbed on to the train. The sentry waved his hand to the Russian engine-driver. If any mistrustful eyes were watching from the far side they would see nothing suspicious or unusual. Slowly the engine puffed on towards the station of Terespol on the German side.

Then here too the minute-hand jumped to 0315.

"Fire!" And the hell dance began. The earth trembled.

The 4th Special Purpose Mortar Regiment with its nine heavy batteries lent a particular note to the inferno. Within half an hour 2880 mortar-bombs screamed with terrifying howls across the Bug into the town and the fortress. The heavy 60-cm. mortars and 21-cm. guns of 98th Artillery Regiment lobbed their shells across the river, into the ramparts of the fortress, and against pin-pointed Soviet gun positions. Was it possible for a single stone to be left standing after this? It was possible. That was to be the first of many nasty surprises.

Second Lieutenant Zumpe of 3rd Company, 135th Infantry Regiment, had watched the last few seconds tick away before 0315. At the very first crash of gunfire he leapt out of his ditch by the railway embankment. "Let's go!" he shouted to the men of his assault detachment. "Let's go!" Steel helmets rose up from the tall grass. Like sprinters the men tore across to the bridge—the second lieutenant in front. Past the abandoned German customs hut. The gunfire drowned the clatter of their boots on the heavy planks of the bridge. Ducking along the high parapet on both sides, they ran across. Always at the back of their minds was the fear: will the bridge go up or not? It did not go up. A single burst of fire from his sub-machine-gun was all the Soviet sentry had time for. Then he heeled over forward.

At that moment a machine-gun opened up from the dugout of the bridge guard. That had been expected. Lance-Corporal Holzer's light machine-gun sprayed the Russian position. Like a handful of fleeting shadows the obstacle-clearing detail of the 1st Company, Engineers Battalion 81, assigned to Zumpe's assault detachment, rushed to the spot. A dull thud; smoke and fumes. Finished.

Zumpe's men raced past the shattered dugout. They flung themselves down to the right and left of the bridge by the railway embankment, their machine-guns in position. The second lieutenant and the sappers ran back on to the bridge. The charge was fixed to the central pier. Out with it. Zumpe ran the beam of his flashlight over the pier, to make sure no other infernal machine was hidden anywhere. Nothing. He slid the green shield over the lens. The green light. Like a stationmaster he waved it above his head towards the German side: bridge clear! And already the first armoured scout-cars were racing across.

At Pratulin, where 17th and 18th Panzer Divisions were to cross the Bug, there was no bridge. At 0415 hours the advance detachments leaped into their rubber dinghies and assault boats, and swiftly crossed to the other side. The infantrymen and motor-cycle troops had with them light anti-tank guns and heavy machine-guns. The Russian pickets by the river opened up with automatic rifles and light machine-guns. They were quickly silenced. Units of the motor-cycle battalion dug in. Then everything that could be pumped into the bridgehead was ferried across. The sappers at once got down to building a pontoon bridge.

But what would happen if the Russians attacked the bridgehead with armour? How would the Germans oppose them? Tanks and heavy equipment could have been brought across only with the greatest difficulty in barges or over emergency bridges.

That was why an interesting new secret weapon was employed here for the first time—underwater tanks, also known as diving tanks. They were to cross the river under water, just like submarines. Then, on the far bank, they were to go into action as ordinary tanks, smashing enemy positions along the river and intercepting any counter-attacks.

It was an amazing plan. In fact, it was over a year old and had originally been intended for a different purpose—for Operation Sea Lion, the invasion of England. Soon after Hitler had decided to leap across to the British Isles, the project of using tanks under water was born. The idea was that they would be unloaded well off the south coast of England, in about twenty-five feet of water, to advance over the sea-floor to the flat beaches. There they were to have emerged from the waves, like Neptune, to have fought down the British coastal defences on both sides of Hastings, to have formed bridgeheads for the first German landing craft, and eventually to have advanced inland, causing havoc and panic in the coastal hinterland.

The idea was immediately put into effect. In July 1940 four diving-tank sections were formed from eight experienced Panzer regiments, and posted to Putlos on the German Baltic coast for special training. It was a strange course for the tank crews. In their Mark III and IV tanks they virtually turned into U-boat men.

The operational task required manœuvrability in water of twenty-five to thirty feet. That meant that the tanks had to withstand a water pressure of about two atmospheres and had to be appropriately sealed. This was achieved by a special adhesive. Sealing the joint between turret and tank body was done very simply by means of an extended bicycle inner tube which could be inflated by the gun-loader inside the tank. The gun itself was fitted with a rubber muzzle cap which could be blasted off from the turret within a second.

A special problem, however, was the supply of fresh air to the engine and the crew. Here the principle of the later U-boat snorkel was anticipated. A special hose about fifty feet long was fitted by a special suction device to a floating buoy, which, at the same time, carried an aerial. The tanks were steered with the aid of a gyro-compass.

Towards the end of July 1940 the four detachments practised in strictest secrecy at Hörnum on the Island of Sylt. An ancient ferry of the Rügen service would take them well out to sea; there they would slither down a hinged ramp to the sea-floor, and make their own way back to the coast. The unevenness of the seabed did not seem to worry the monsters. The experiments were highly successful. But then, in mid-October 1940, Operation Sea Lion was called off for good. The dream of the U-boat tanks had ended. Of the special detachments three were united into a plain tank regiment, 18th Panzer Regiment, while the remaining detachment was assigned to 6th Panzer Regiment, 3rd Panzer Division.

In the spring of 1941, when the High Command of the Army was discussing the crossing of the Bug north of Brest, in connection with the planning of Operation Barbarossa, somebody on the General Staff remembered the diving tanks. "Surely we had those things..." Inquiries were made. Questions were asked at 18th Panzer Regiment.

"Oh, yes, we still have those old diving tanks." An order came for diving basins to be built near Prague. 18th Panzer Regiment tested the diving capacity of the old tanks. Since they were no longer required to move under the sea, but merely to cross a river, the fifty-foot-long rubber snorkel was replaced by a ten-foot steel pipe. The exhaust pipes were fitted with one-way valves. Within a short time the U-boat tanks were again in perfect condition. On 22nd June 1941 they passed their ordeal by fire.

In the sector of 18th Panzer Division fifty batteries of all calibres opened fire at 0315 in order to clear the way to the other bank for the diving tanks. General Nehring, the divisional commander, has since described this as "a magnificent spectacle, but rather pointless since the Russians had been clever enough to withdraw their troops from the border area, leaving behind only weak frontier detachments, which subsequently fought very bravely."

At 0445 hours Sergeant Wierschin advanced into the Bug with diving tank No. 1. The infantrymen watched him in amazement. The water closed over the tank. "Playing at U-boats!" Only the slim steel tube which supplied fresh air to the crews and engine showed above the surface, indicating Wierschin's progress under water. There were also the exhaust bubbles, but these were quickly obliterated by the current.

Tank after tank—the whole of 1st Battalion, 18th Panzer Regiment, under the battalion commander, Manfred Graf Strachwitz—dived into the river.

And now the first ones were crawling up the far bank like mysterious amphibians. A soft plop and the rubber caps were blown off the gun muzzles. The gun-loaders let the air out of the bicycle inner tubes round the turrets. Turret hatches were flung open and the skippers wriggled out. An arm thrust into the air three times: the signal "Tanks forward."

Eighty tanks had crossed the frontier river under water. Eighty tanks were moving into action.

Their presence was more than welcome in the bridgehead. Enemy armoured scout-cars were approaching. At once came the firing orders for the leading tanks: "Turret—one o'clock—armour-piercing—800 yards—group of armoured scout-cars—fire at will."

The monsters fired. Several armoured scout-cars were burning. The rest retreated hurriedly. The armoured spearheads of Army Group Centre moved on in the direction of Minsk and Smolensk.

South of Brest too, at Koden, following the successful assault on the bridge, the surprise attack of XXIV Panzer Corps, under General Freiherr Geyr von Schweppenburg, had gone according to schedule. The tanks crossed the bridge, which had been captured intact. Advanced units of Lieutenant-General Model's 3rd Panzer Division crossed on rapidly built emergency bridges. The skippers stood in the turrets, scanning the landscape for the rearguards of the retreating Soviet frontier troops, overrunning the first anti-tank-gun positions, waving the first prisoners to the rear, moving nearer and nearer to their objective for the day—Kobrin on the Mukhavets.

North of Brest, near Drohiczyn, where Engineers Battalion 178 lay close to the Bug in the area of 292nd Infantry Division, in order to build a pontoon bridge as soon as possible for the heavy equipment of the divisions of IX Corps, everything went also according to plan. The reinforced 507th and 509th Infantry Regiments—with the 508th farther to the right—raced across the Bug in rubber dinghies and assault craft under heavy artillery cover. Within half an hour the Soviet pickets on the far bank were wiped out and a bridgehead was established. At the first artillery salvo the sappers had leaped to their feet and moved their first pontoons into the water. For a quarter of an hour the Russians fired from the far bank with machine-guns and rifles; then they fell silent. By 0900 exactly the bridge was finished—the first within the area of the Fourth Army. The heavy equipment rumbled over the swaying pontoons. The 78th Infantry Division was already lined up in close order to cross the river.

Of all the coups planned against the frontier bridges along the five-hundred-mile-long Bug not one went wrong. Similarly all the envisaged bridge-building operations across the frontier river succeeded, with the sole exception of one within the area of 62nd Infantry Division, which, belonging to the Sixth Army, was already part of the northern wing of Army Group South.

On 22nd June Field-Marshal von Rundstedt launched his offensive on his left wing with the Seventeenth and Sixth Armies, which stood to the north of the Carpathians. Farther to the south the Eleventh Army and the Rumanian Army were still standing by, both in order to deceive the Russians and to protect the Rumanian oil. The offensive in the Black Sea area was not to start till 1st July.

On the northern wing of Army Group South, at Reichenau's Sixth Army, on the Bug, good progress was made on the first day of the campaign, in spite of the difficulties which 62nd Infantry Division had in building its bridge.

Major-General von Oven's 56th Infantry Division crossed the river without any hitch with the very first wave of rubber dinghies. The artillery-fire lay so squarely on the well-reconnoitred enemy positions that the attackers suffered practically no casualties. Half-way through the morning a pontoon bridge was in position in the sector of 192nd Infantry Regiment at Chelm. The artillery crossed at high speed. On the very first day the regiments of XVII Corps pushed ahead nine miles right through the Russian frontier fortifications.

On the southern wing of the Army Group, where the frontier was formed by the river San, the divisions of General von Stülpnagel's Seventeenth Army found things more difficult. The bank of the San north of Przemysl was as flat as a pancake—without woods, without ravines, without any cover for whole regiments. That was why the assault battalions of 257th Infantry Division, from Berlin, could not move out of their deployment areas until the night of 21st/22nd June. "Not a sound" was the order of the regimental commander. Weapons were packed in blankets; bayonets and gas-mask cases were wrapped round with any soft material that was handy. "Thank God for the frogs," whispered Second Lieutenant Alicke. Their croaking drowned the creaking, rattling, and bumping of the companies making their way towards the river.

At 0315 precisely the assault detachments leaped to their feet on both sides of Radymno. The railway bridge was seized by a surprise stroke. But in front of the customs shed the Russians were already offering stubborn resistance. Second Lieutenant Alicke was killed. He was the division's first fatal casualty, the first of a long list. The men laid him beside the customs shed. The heavy weapons rolled on by him, over 'his' bridge.

In the south the Soviet alarm system functioned with surprising speed and precision. Only the most forward pickets were taken by surprise. The 457th Infantry Regiment had to battle all day long with the Soviet Non-commissioned Officers' Training School of Vysokoye, only a mile beyond the river. The 250 NCO cadets resisted stubbornly and skilfully. Not till the afternoon was their resistance broken by artillery-fire. The 466th Infantry Regiment fared even worse. No sooner were its battalions across the river than they were attacked from the flank by advanced detachments of the Soviet 199th Reserve Division.

In the fields of Stubienka the tall grain waved in the summer wind like the sea. Into this sea the troops now plunged. Both sides were lurking, invisible. Stalking each other. Hand-grenades, pistols, and machine carbines were the weapons of the day. Suddenly they would be facing one another amid the rye—the Russians and Germans. Eye to eye. Whose finger was quicker on the trigger? Whose spade would go up first? Over there a Russian machine pistol appeared from a foxhole. Would it score with its burst? Or would the hand-grenade do its work first? Only with the fall of dusk did this bloody fighting in the rye-fields come to an end. The enemy withdrew.

The sun set behind the horizon, large and red. And still from amid the grain came the voices, despairing, anguished, or softly dying away: "Stretcher! Stretcher!" The medical orderlies hurried into the fields with their stretchers. They gathered in the bloody harvest. The harvest of one day, of one regiment. It was a big harvest.

In the area of Army Group North concentrated artillery-fire preceded the attack on only a few sectors. For the most part the first wave of infantry, together with assault sappers, rose silently from their dugouts among the crops along the frontier of Soviet-occupied Lithuania shortly after 0300 hours. Shrouded in the morning mist, like phantoms, the tanks moved forward out of the woods.

The men of 30th Infantry Division, from Schleswig-Holstein, were in position south of the Memel. They had no water obstacles to overcome on their first day. The sapper platoon of their advanced detachment, under First Lieutenant Weiss, crept up to the barbed wire. For days they had been observing every detail. The Russians patrolled the wire only intermittently. Their defences were farther back, along some high ground.

Softly. Softly...

The wire-cutters clicked. A post rattled. Quiet—listen. But there was no movement on the other side. Keep going. Faster. Now the passages were clear. And already men of 6th Company were coming up on the double, ducking as they ran. Not a shot was fired. The two Soviet sentries stared terrified down the carbine-barrels and raised their hands.

Keep going.

The observation towers on Hills 71 and 67 stood out black against the sky. There the Russians were established in strong positions. The German troops were aware of it. And so were the gunners of the heavy group of 30th Artillery Regiment waiting in the frontier wood behind them. The Russian machine-guns opened up from the tower on Hill 71. These were the first shots fired between Memel and Dubysa. Immediately the reply came from the well-camouflaged heavy field howitzers of 2nd Battalion, 47th Artillery Regiment, in position behind the regiments of 30th Infantry Division on the road from Trappenen to Waldheide. Where their mortar-bombs burst there would be no grass growing for a long time.

Assault guns forward! Ducking behind the steel monsters, Weiss's advanced detachment was storming the high ground. Already they were inside the Soviet positions. The Russians were taken by surprise. Most of them were not even manning their newly built, though only partly finished, defences. They were still in their bivouacs. These were Mongolian construction battalions, employed here on building frontier defences. Wherever they were encountered, in groups or platoon strength, manning those defences, they fought stubbornly and fanatically.

The German troops were beginning to realize that this was not an opponent to be trifled with. These men were not only brave but also full of guile. They were masters at camouflage and ambush. They were first-rate riflemen. Fighting from an ambush had always been the great strength of the Russian infantry. Forward pickets, overrun and wounded, would wait for the first German wave to pass over them. Then they would resume fighting. Snipers would remain in their foxholes with their excellent automatic rifles with telescopic sights, waiting for their quarry. They would pick off the drivers of supply vehicles, officers, and orderlies on motor-cycles.

The 126th Infantry Division, from Rhine-Westphalia. fighting alongside the men from Schleswig-Holstein, also learned a bitter lesson from the tough Soviet frontier troops. The 2nd Battalion, 422nd Infantry Regiment, suffered heavy losses. Parts of a Soviet machine-gun picket had hidden themselves in a cornfield and allowed the first wave of the attack to pass by. In the afternoon, when Captain Lohmar unsuspectingly led his battalion from reserve positions to the front, the Russians in the crops suddenly opened up. Among those killed was the battalion commander, among the seriously wounded was his adjutant. It took an entire company three hours to flush the four Russians out of the field. They were still firing when the Germans had got within ten feet of them, and had to be silenced with hand-grenades.

On the northern flank, immediately on the Baltic coast, in the small corner of Memel territory, was General Herzog's Masurian 291st Infantry Division. Its tactical sign was an elk's head—in token of the division's Masurian home. At the moment when, 500 miles farther to the south, Second Lieutenant Zumpe stormed over the railway bridge of Brest, Colonel Lohmeyer, with an advanced detachment of 505th Infantry Regiment, pushed through the forward pillbox-line of an utterly surprised Soviet frontier position. Under cover of the morning mist the Russians withdrew quickly. But Lohmeyer gave them no respite: he pressed on hard, and by nightfall of the first day he had reached the Latvian–Lithuanian frontier. On the following morning the 505th took Priekule. After 34 hours Lohmeyer and his regiment were 44 miles deep in enemy territory.

In the area of General von Manstein's LVI Panzer Corps, in the wooded country north of the Memel, there was not much room for large-scale operations. That was why only the 8th Panzer Division and 290th Infantry Division were earmarked for the first thrust across the frontier. The forward line of pillboxes had to be pierced. And it had to be pierced quickly. The corps was scheduled to drive 50 miles right through the enemy on the first day, without stopping, without regard to anything else, with the object of capturing intact by a surprise stroke the big road viaduct across the Dubysa valley at Ariogala. If they failed in this the corps would be stuck in a deep and narrow river valley, and the enemy would have time to re-form. But most important of all, any idea of a surprise stroke against the important centre of Daugavpils (Dvinsk) would have to be dropped.

The companies of 290th Infantry Division suffered heavy casualties even while crossing the frontier stream—above all in officers. Second Lieutenant Weinrowski of the 7th Company, 501st Infantry Regiment, was probably the first soldier killed by the bullets of Soviet frontier guards up in the north during the first minute of this war. The burst came from a pillbox camouflaged as a farm cart. But the Russian frontier troops were unable to halt the German attack. The 11th Company of 501st Regiment led the assault ahead of the spearheads of 8th Panzer Division, clearing tree-trunk obstacles under Russian fire, sweeping through the wood, past a small village. First Lieutenant Hinkmann, the company commander, was killed. Second Lieutenant Silzer ran forward. "The company will take orders from me!" They reached the Mituva, a small river. They captured the bridge and, as instructed, established a bridgehead.

Presently General Brandenberger's 8th Panzer Division drove up. General von Manstein, the GOC, was accompanying the division in his command tank. "Keep going!" he urged them. "Keep going!" Never mind about your flanks. Never mind about cover. The Ariogala viaduct must be captured. And Daugavpils must be taken by surprise.

Manstein, a bold but coolly calculating strategist, knew very well that this gamble of a war called Operation Barbarossa could be won only if the Germans succeeded in knocking the Russians out during the very first weeks of the attack. He knew what Clausewitz knew before him: this vast country could not be conquered and occupied. At best it might be possible, by risky surprise strokes, by swift and hard blows at the military and political heart of the country, to overthrow the regime, to deprive the country of its leadership, and thus to paralyse its vast military potential. That was the only way in which it might be done—perhaps. Otherwise the war would be lost that very summer.

But unless it was to be lost during the very first eight weeks of the 1941 campaign, Leningrad had to fall quickly, Moscow had to fall quickly, and the bulk of the Russian forces in the Baltic and in Belorussia had to be outmanœuvred, smashed, and captured. And so that this could be done, the Panzer corps had to drive on regardless of everything, aiming their blows straight at the great nerve centres. And that, in the area of this particular Army Group, meant that Leningrad must fall. But to get to Leningrad the Daugava had to be crossed first, and it was against that river that Manstein's LVI Panzer Corps and, to the left of it, General Reinhardt's XLI Panzer Corps were pressing forward. And in order to get across this mighty river without a dangerous delay, the bridges across it at Daugavpils (Dvinsk) and Jekabpils had to be captured intact. But these bridges lay 220 miles behind the frontier. That was the situation.

At 1900 a signal was received at 8th Panzer Division headquarters from its advanced units: "Ariogala viaduct taken." Manstein nodded. All he said was: "Keep going."

The tanks were moving forward. The grenadiers were riding through clouds of hot dust. Keep going. Manstein was executing an armoured thrust such as no military tactician would have thought possible. Would his corps succeed in taking Daugavpils by surprise? Would he be able to drive straight through strongly held enemy territory for a distance of 230 miles and yet take the bridges across the Daugava by a surprise stroke?

That this tank war by the Baltic was not going to be a light-hearted adventure, no easy Blitzkrieg against an inferior enemy, was painfully clear after the first forty-eight hours. The Russians, too, had tanks—and what tanks! The XLI Panzer Corps, operating on the left wing of Fourth Panzer Group, was the first to make this discovery.

On 24th June, at 1330 hours, Reinhardt arrived at the command post of 1st Panzer Division with the news that 6th Panzer Division had encountered very strong enemy armour on its way to the Daugava, at a point east of Raseiniai on the Dubysa, and was involved in heavy fighting. Over 100 super-heavy Soviet tanks had come from the east to meet XLI Panzer Corps, and had clashed first of all with General Landgraf's 6th Panzer Division. No one suspected at that time that Raseiniai was to become a name in military history. It marked the first great crisis on the German northern front, a long way behind the spearhead of Manstein's Panzer corps.

1st Panzer Division therefore moved to relieve the 6th. Laboriously the tanks struggled forward along soft sandy or marshy tracks. The day was full of minor skirmishes, and the next morning began with an alarm. A Soviet tank attack with super-heavy armoured giants had overrun the 2nd Battalion, 113th Rifle Regiment. Neither the infantry's anti-tank guns, nor those of the Panzerjägers, nor the guns of the German tanks, were able to pierce the plating of these heavy enemy monsters. German artillery had to depress their barrels into the horizontal, and eventually halted the enemy attack by direct fire from open positions. Only because of their greater speed and their more skilful handling were the German tanks able to stand up to their heavy Soviet opponents. By using every trick in the book, especially good fire discipline and efficient radio communications, the tank companies succeeded in throwing the enemy back two miles.

The Soviet tanks which made this astonishing appearance were the as yet unknown types of the Klim Voroshilov series, the KV-1 and the KV-2, of 43 and 52 tons respectively.

An account by the Thuringian 1st Panzer Division describes this tank battle:

> The KV-1 and -2, which we first met here, were really something!

Our companies opened fire at about 800 yards, but it remained ineffective. We moved closer and closer to the enemy, who for his part continued to approach us unconcerned. Very soon we were facing each other at 50 to 100 yards. A fantastic exchange of fire took place without any visible German success. The Russian tanks continued to advance, and all armour-piercing shells simply bounced off them. Thus we were presently faced with the alarming situation of the Russian tanks driving through the ranks of 1st Panzer Regiment towards our own infantry and our hinterland. Our Panzer regiment therefore about-turned and rumbled back with the KV-1s and KV-2s, roughly in line with them. In the course of that operation we succeeded in immobilizing some of them with special-purpose shells at very close range—30 to 60 yards. A counter-attack was launched and the Russians were thrown back. A protective front was then established at Vosiliskis. Defensive fighting continued.

For several days a critical battle raged on the Dubysa between the German XLI Panzer Corps and the Soviet III Armoured Corps, which had thrown into battle 400 tanks, most of them super-heavy ones. Colonel-General Fedor Kuznetsov was employing his crack armoured units, including the 1st and 2nd Armoured Divisions.

These heavy Soviet tanks were protected by 80-mm. plating all round, reinforced in some places to 120-mm. They carried a 7·62- or 15·5-cm. long-barrel gun as well as four machine-guns. Their speed over open ground was about 25 miles per hour. The greatest headache at first was their armour-plating: one KV-2 bore the marks of over seventy hits, but not a single one had pierced the armour. Since the German anti-tank guns were useless against these tanks an attempt had to be made to immobilize these giants first by firing at their tracks, and then by tackling them with artillery and AA guns, or blowing them up at close range by high-explosive charges of the sticky-bomb type.

The battle was decided in the early morning of 26th June. The Russians attacked. German artillery had taken up position on high ground among the tank regiments and was firing point-blank at the Russian tanks. The German regiments then mounted a counter-attack. At 0838 hours the 1st Panzer Regiment linked up with advanced units of 6th Panzer Division. The Soviet III Armoured Corps was smashed.

These two German Panzer divisions, together with 36th Motorized Infantry Division and 269th Infantry Division, between them destroyed the bulk of the Soviet armoured forces in the Baltic countries. Two hundred Soviet tanks were wrecked. Twenty-nine super-heavy KV-1 and KV-2 monsters, built by the Kolpino works in Leningrad, were left gutted on the battlefield. The road to Jekabpils on the Daugava was now free also for XLI Panzer Corps.

And where was Lohmeyer? This question had become a daily routine at Eighteenth Army and at 291st Division headquarters.

In the evening of 24th June the colonel, with his 505th Infantry Regiment, was seven miles from Liepaja. On 25th June he tried to take the town by a surprise attack. The infantrymen and sailors of a naval assault detachment, under Lieutenant-Commander von Diest, subordinated to Lohmeyer, charged across the narrow neck of land against the fortifications. But they did not get through. A determined assault made by Lieutenant-Commander Schenke with men of Naval Artillery Detachment 530 likewise failed to achieve any success. Before Lohmeyer was able to regroup his forces and before the other two regiments of the division could be brought up, the garrison of Liepaja launched a counter-attack. Combat units supported by tanks mounted relief attacks, some of them right up to the German gun positions. On 27th June the Russians staged a massive sortie, actually tore a gap in the German encircling front, broke through to the coast road in the strength of several combat groups, and brought about a critical position on the German front. The gap was closed only with great difficulty. Eventually, about noon, the battalions of 505th Infantry Regiment and some infantry assault detachments succeeded in penetrating the southern part of the defences. On the following day assault units fought their way into the town.

Furious street fighting continued for the next forty-eight hours. The cunningly camouflaged Russian machine-gun posts in the barricaded buildings were wiped out only when heavy infantry guns, field howitzers, and mortars had been brought up.

The defence of the town was magnificently organized. The individual Soviet soldier was well trained and fought with fanatical bravery. The Russian troops regarded it as perfectly natural that they should be sacrificed in order to enable the higher command to gain time, or to provide the prerequisite for regroupings or break-outs. The ruthless sacrifice of detachments in order to save larger units was first revealed at Liepaja as a basic part of Soviet military thinking. Its application caused heavy losses to the German attackers: at Liepaja, for instance, the commanding officers of both naval units were killed.

At last, on 29th June, the naval fortress was conquered. The infantry of Eighteenth Army had scored its first great victory. But the victory also held a bitter lesson: Liepaja was the first demonstration of what the Red Army soldier was capable of in defending firm positions, provided he had a cool, resolute head to lead him, and provided the cumbersome Soviet chain of command had time enough to organize the defence.

In contrast to this self-sacrificing defence of Liepaja the resistance in Daugavpils was half-hearted, confused, and panicky.

In the early light of 26th June the spearhead of the Lusatian 8th Panzer Division was speeding along the great highroad which runs straight from Kaunas to Leningrad. The tracks clanked, the engines

B

34 Part One: Moscow

roared. The tank commanders stood propped up in the open turrets, their field-glasses at their eyes. For the past four days they had been rolling like this across hills and through marshy lowlands, breaking any resistance they encountered, keeping on the move, through forests, sand, swamps, and Russian lines. Right through two of Kuznetsov's Armies. A distance of 190 miles.

There were only 5 miles left to Daugavpils. Then only 4 miles. It was uncanny.

In the leading tank the commander's hand cut through the air and then came down to the right—the signal for "Close up on my right and halt." As the armoured spearhead came to a stop a strange column overtook it—four captured Soviet lorries, the drivers in Russian uniforms. The tank commanders in their turrets grinned knowingly. They knew what this mysterious convoy was: men of the Brandenburg Regiment, a special unit under Admiral Canaris, the head of German Military Intelligence.

Under the tarpaulins sat First Lieutenant Knaak with his men. Their task was as fantastic as it was simple—drive into the town, seize the bridges over the Daugava, prevent the Russians from blowing them up, and hold them until 8th Panzer Division had fought its way up to join them.

Knaak's lorries rolled past the armoured spearhead. They climbed the slight hill. Down there was the river-bend and the town. And down there were the bridges. Across the road bridge in the centre of Daugavpils traffic was flowing as in peacetime. Across the big railway bridge a locomotive chugged amid puffs of steam. The lorries bumped along towards the town. Past the Soviet outposts. The drivers in their Russian uniforms exchanged jokes with the pickets. "Where are the Germans?" the Russians asked. "Oh, a long way back!" On they moved. Into the suburb of Griva. The time was shortly before 0700. Threading their way through the local traffic, passing the tram-cars, Knaak's lorries rolled on. In front of them lay the great road bridge. Foot hard down on the accelerator! Forward!

The first lorry got across. But as the second approached the Russian sentry on the bridge tried to stop it. When it failed to halt it came under machine-gun fire. The platoon commander shouted, "Get out, boys, and let them have it!" The exchange of fire had aroused the guard at the far end. They now opened up with machine-guns on the leading lorry as it approached. But Knaak managed to get his men out. The Soviet bridge guard were forced to take cover. The second platoon managed to get on to the railway bridge, overcome the sentries, and cut the detonator wires. But through an accident part of the demolition charge went off nevertheless, wrecking a short stretch of the bridge.

On the high ground outside the town the observers of General Brandenberger's armoured spearhead had closely watched Knaak's operation. The moment the gun flashes were seen the commander of the leading tank slammed the hatch down. "We're off!" he shouted into his microphone, quite unmilitarily. "We're off!" his driver echoed. Secure hatches! Turret at 12 o'clock! HE shell! They raced into the town.

At 0800 hours General von Manstein received the signal, "Surprise of Daugavpils town and bridges successful. Road bridge intact. Railway bridge slightly damaged by demolition charge, but passable."

First Lieutenant Wolfram Knaak and five men had been killed, the remaining twenty under his command had all been wounded. The officer in charge of the Soviet guard party by the road bridge was taken prisoner. Under interrogation he said, "I had no order to blow up the bridge. Without such an order I could not take the responsibility. But there was no one about whom I could have asked."

Here we find revealed a decisive weakness in the lower echelons of the Soviet military command, a weakness we are to encounter many times yet. But in a war no one cares about reasons. The main thing was: Manstein had pulled it off. An armoured thrust without parallel had succeeded. True, there was some fighting in Daugavpils, but Daugavpils was no Liepaja. The commander of the Russian troops ordered a few demolitions and the burning of all stores, and then withdrew his forces. Soviet artillery bombarded the town. Soviet bomber squadrons appeared in the sky, persistently and stubbornly trying to destroy the bridges with bombs at this late stage. German army AA gunners and the fighter pilots of First Air Fleet had a field day and ensured the victory of the Daugavpils bridge.

But what use is a victory if it is not exploited? The wide Daugava had been crossed and the vital railway centre between Vilna and Leningrad was in German hands. The 8th Panzer Division and 3rd Motorized Infantry Division were on the far bank of the river. What must be done next?

What indeed? Was Manstein to push on? Was he to take advantage of the enemy's hopeless confusion and assume that he was unable to put in the field any superior or well-led forces against the phantom-like German tank thrust? Or should he adopt the textbook solution, the safety-first solution, and halt until the infantry came up? That was the question—the question which would decide the fate of Leningrad.

One would have thought that Hitler would have chosen the bold alternative. Indeed, on closer scrutiny, there was no real choice. The next move had to follow logically from the entire plan of campaign. And this campaign in the East was based on boldness and gamble. Hitler proposed to crush by rapid assault a gigantic empire which, to his certain knowledge, had over 200 combat-ready divisions in its western part alone. And behind these divisions? Beyond the Urals was unknown territory about which only vague reports were available—

reports of gigantic industrial plants, enormous armament industries, and inexhaustible human reserves. Hence this military gamble could be concluded successfully, if at all, only if the oak was felled by lightning. And that lightning had to be swift, powerful, surprise blows straight at the political and military heart of the Soviet empire. The enemy must not be allowed to collect himself or to deploy his strength. The very first days of this war had provided a lesson and a warning: wherever the enemy command was paralysed by surprise, victory was certain; wherever it was given time to resist, its troops would fight like the devil.

This realization and the whole logic of Operation Barbarossa therefore demanded that the bold advance should be maintained. Manstein realized this clearly. The enemy must not be given the opportunity to bring up his reserves against identified and stationary German spearheads. If he was allowed to do so, then—but only then—would the open flanks of numerically small armoured units deep in enemy territory be exposed to mortal danger. So long as the push was kept up Kuznetsov would have to throw into battle whatever he had to hand.

Long ago Guderian had formulated the basic commandment of armoured warfare: "Not driblets but mass." Manstein added a second commandment: "The safety of an armoured formation in the enemy's rear depends on its continued movement."

Of course, it was risky to have Manstein's corps operating alone north of the Daugava while Reinhardt's XLI Panzer Corps and the entire left wing of Colonel-General Busch's Sixteenth Army were still over sixty miles farther back—but without risks this campaign could not be waged at all, let alone won. The enemy had shown himself not too sensitive to the German armoured wedges—in other words, he had not taken back his other fronts but merely concentrated what units he could scrape together against Manstein's Daugava crossing. But this was not because the Soviet High Command was prepared to accept the swift armoured wedges driven into its lines, but because it was in total ignorance of the true position. Neither Kuznetsov nor the High Command in the Kremlin had a clear picture of the situation. This state of affairs should have been exploited.

However, the German High Command failed to understand the logic of its own strategy. Hitler suddenly became jittery—afraid of his own courage. It became clear that the man who based his plans so largely on boldness, recklessness, and luck was in practice the first to point an anxious finger at the exposed flanks on the situation map. He lacked confidence in the military skill of his generals. Against Hitler the German High Command could not win its point. Thus it was that Manstein received the orders: "Halt. Daugavpils bridgehead will be defended. Arrival of Sixteenth Army's left wing will be awaited."

The argument that supply considerations and enemy attacks made this halt unavoidable is, of course, quite correct in terms of a conservative general-staff assessment of the situation—but if that were to be made the yardstick, then surely Manstein should not have crossed the Daugava at all, nor, two weeks later, Guderian the Dnieper. No, Hitler's halt sprang from anxiety and even more from uncertainty whether he should first strike at Leningrad or at Moscow. It was this indecision that had halted Manstein. And this halt was Leningrad's first salvation. Like the rumbling of distant thunder the commanders in the field became aware of a crisis between Fuehrer and High Command, of the issue of Moscow versus Leningrad, of that crisis from which the great mistakes were to spring later, those mistakes which, one by one, were nails in the coffin of the German armies in the East.

For six days Manstein's Panzer Corps was made to stand still. For three of these days it was a long way in front of the Army Group. What was bound to happen happened. Kuznetsov scraped together what reserves he could lay hands on. From the Pskov area. From Moscow. From Minsk. He flung everything he had against Manstein's advanced positions. At long last, on 2nd July, when the green light was given for the resumption of the thrust, with Leningrad as the distant objective, valuable time had been lost. Time which the Soviet High Command had used to steady its panicking divisions and to prepare the defence of the Stalin Line, the old and often well-built defences along the former Russian–Estonian frontier, between Lake Peipus and Sebezh. The second round began.

And how did the operation go in the south during these first few days?

Field-Marshal von Rundstedt and the commander of his First Panzer Group, Colonel-General von Kleist, had drawn the most difficult position of the campaign. The Russian southern front, protecting the Ukrainian grain areas, had been organized in particular strength and with great care. Colonel-General Kirponos, who commanded the Soviet Army Group South-west Front, had deployed his four Armies in two groups in considerable depth. Well-camouflaged lines of pillboxes, heavy field-artillery positions, and cunning obstacles turned the first German leap across the frontier into a costly operation.

The divisions of Seventeenth Army under General of Infantry von Stülpnagel had to nibble their way through the lines of pillboxes before Lvov and Przemysl. Reichenau's Sixth Army crossed the Styr in the face of stubborn opposition. When von Kleist's Panzer Group had succeeded in breaking through east of Lvov and the vehicles with the white 'K' were about to mount their Blitzkrieg offensive, Kirponos instantly blocked the development of large-scale operations and the encirclement of Soviet forces. With armoured units rapidly brought up he launched strong counter-attacks and struck heavily at the spearheads of the advancing German divisions.

He sent his heavy KV-1 and KV-2 tanks into action, as well as the super-heavy Voroshilov model with its five revolving turrets. Against these the German Mark III with its 3·7- or 5-cm. gun was pretty helpless and forced to retreat. Anti-aircraft guns and artillery had to be brought up to fight the enemy armour. But the most dangerous of all was the Soviet T-34—an armoured giant of great speed and manœuvrability. It was 19 feet long, 10 feet wide, and 8 feet high. It had wide tracks, a massive turret with outward-sloping sides, it weighed 26 tons, and it carried a 7·62-cm. gun. It was near the Styr that the rifle brigade of 16th Panzer Division encountered the first of them.

The Panzerjäger unit of 16th Panzer Division hurriedly brought up its 3·7 anti-tank gun. Position! Range 100 yards. The Russian tank continued to advance. Fire! A hit. And another hit. And more hits. The men counted them: 21, 22, 23 times the 3·7-cm. shells smacked against the steel colossus. But the shells merely bounced off. The gunners were screaming with fury. The troop commander was pale with tension. The range was down to 20 yards. "Aim at the turret ring," the second lieutenant commanded.

Now they had got him. The tank wheeled about and moved off. The turret ring was damaged and the turret immobilized—but otherwise it was unscathed. The anti-tank gunners drew a deep breath. "Would you believe it?" they were saying. From then onward the T-34 was their great bogy. And their 3·7-c.m. anti-tank gun, which had rendered such good account of itself in the past, was henceforward nicknamed contemptuously "the army's door-knocker."

Major-General Hube, OC 16th Panzer Division, described the developments during the first few days of the campaign in the south as "slow but sure progress." But "slow and sure" was not provided for by Operation Barbarossa. Kirponos's forces in Galicia and the Western Ukraine were likewise to have been defeated speedily by means of crushing battles of encirclement.

On the Rumanian–Russian frontier, where the Eleventh Army under Colonel-General Ritter von Schobert stood, nothing much happened on 22nd June. There was no artillery bombardment and no assault was launched. Apart from slight patrol activity across the river Prut, which formed the frontier here, and a few Russian air-raids, things were fairly peaceful. Hitler's timetable purposely envisaged a delay on this sector; the Soviet forces here were to be driven, at the beginning of July, into a pocket that was being formed in the north.

On that fatal day, therefore, at 0315 hours, the Prut was flowing sluggishly to the south, under its usual blanket of light haze. Major-General Roettig, OC 198th Infantry Division, was lying by the river near the village of Sculeni, with his intelligence officer and an orderly officer, watching the opposite bank. The Russian frontier posts were keeping quiet, until suddenly an explosion rent the air. A patrol of 198th Infantry Division had paddled themselves across the Prut and blown up a Soviet guard tower. That was the only noisy incident on the southern flank of the Eastern Front.

Not until the evening of 22nd June did 198th Infantry Division carry out a reconnaissance in force across the Prut in order to occupy the village of Sculeni, through which the river and the frontier ran. The 305th Infantry Regiment occupied the village and formed a bridgehead. The bridgehead was held against strong enemy pressure during the following days.

Day after day passed. The delays on the northern wing of the Army Group in the area of the Sixth and Seventeenth Armies meant that Schobert's divisions had to wait also. At last, on 1st July, the green light was given. The 198th Infantry Division attacked from its bridgehead. Twenty-four hours later the remaining divisions of XXX Corps followed suit: the 170th Infantry Division, under Major-General Wittke, as well as the Rumanian 13th and 14th Divisions. The other two corps of the Army, LIV and XI Corps, crossed the Prut to the right and left of XXX Corps.

Even though one could hardly have expected the enemy to have been taken entirely by surprise eight days after the start of operations, 170th Division nevertheless succeeded in capturing intact the wooden bridge over the Prut near the village of Tutora. In a bold and cunning action Second Lieutenant Jordan led his platoon swiftly through the anti-tank defences along the Soviet frontier. The 800-yard-long causeway through the marsh was cleared of the enemy. Soviet posts were overcome in hand-to-hand fighting. In the morning 40 Russians lay dead by their machine-guns at the bridge and in the marsh. But Jordan's platoon paid a heavy price: 24 men killed or wounded.

The offensive of Eleventh Army was gaining momentum. Its direction was towards the north-east, towards the Dniester. But things did not go according to schedule; Schobert was not able to drive a retreating enemy into a trap, but had to content himself with slowly pushing back a strongly resisting enemy.

After ten days of very fierce fighting Rundstedt's armoured divisions had penetrated 60 miles into enemy territory. They were involved with superior forces, compelled to beat back counter-attacks from all sides, and defend themselves from the right and the left, from the front and the rear. A strong enemy was offering stubborn but elastic resistance. Colonel-General Kirponos succeeded in evading the planned German encirclement north of the Dniester and in taking his troops back, still in an unbroken front, to the strongly fortified Stalin Line to both sides of Mogilev. Rundstedt had therefore not succeeded in achieving the planned large-scale break-through. The timetable of Army Group South had been upset. Could the delay be made up?

On the Central Front, on the other hand, all went well. After a swift break-through the armoured and motorized divisions of Hoth's and Guderian's Panzer Groups on the wings of the Army Group advanced rapidly according to plan, right through the startled and badly led armies of the Russian Western Front, and got into position for their large-scale pincer movement. It was here on the Central Front that the decisive action of the entire campaign had been scheduled from the outset: it was to be prepared by some 1600 tanks and to be finally consummated—in collaboration with Fourth Panzer Group under Colonel-General Hoepner, then still operating in the area of Army Group North—by the capture of Moscow. The plan seemed to work. The Panzer divisions were once more giving a demonstration of Blitzkrieg—as in the old days, as in Poland and in the West. At least, that was how things looked from where the armoured spearheads stood. The infantry here, as on the northern sector, had a somewhat different experience. The fortress of Brest-Litovsk was a typical example.

On 22nd June, 45th Infantry Division did not suspect that it would suffer such heavy losses in this ancient frontier fortress. Captain Praxa had prepared his assault against the heart of the citadel of Brest with great caution. The 3rd Battalion, 135th Infantry Regiment, was to take the Western Island and the central area with the barracks block. They had studied it all thoroughly at the sand-table. They had built a model from aerial photographs and old plans from the days of the Polish campaign, when, until it had to be surrendered to the Russians, Brest was in German hands. Guderian's staff officers realized from the outset that the citadel could be taken only by infantry, since it was proof against tanks.

The circular fortress, occupying an area of nearly two square miles, was surrounded by moats and river branches, and sub-divided internally by canals and artificial water-courses into four small islands. Casemates, snipers' positions, armoured cupolas with anti-tank and anti-aircraft guns, were established, well camouflaged, behind shrubs and under trees. On 22nd June there were in all five Soviet regiments in Brest; these included two artillery regiments, one reconnaissance battalion, an independent anti-aircraft detachment, and supply and medical battalions.

General Karabichev, who was captured beyond the Berezina very early in the campaign, stated under interrogation that in May 1941 he had been instructed, as an expert in fortification engineering, to inspect the western defences. On 8th June he had set out on his trip.

On 3rd June the Soviet Fourth Army had staged a practice alarm. The report on this exercise, which was captured by German units, had this to say about the 204th (Heavy) Howitzer Regiment: "The batteries were not ready to fire until six hours after the alarm." About the 33rd Rifle Regiment it said this: "The duty officers were unacquainted with the alarm regulations. Field kitchens are not functioning. The regiment marches without cover...." About the 246th Anti-Aircraft Detachment it said: "When the alarm was given the duty officer was unable to make a decision." When one has read this report one is no longer surprised at the lack of organized resistance in the town of Brest. But in the citadel proper the Germans got a surprise after all.

When the artillery bombardment began at 0315 hours the 3rd Battalion, 135th Infantry Regiment, was 30 yards from the river Bug, directly opposite the Western Island. The earth trembled. The sky was plunged in fire and smoke. Everything had been arranged in minute detail with the artillery units which were softening up the citadel: every four minutes the hail of death was to be advanced by 100 yards. It was an accurately planned inferno.

No stone could be left standing after this lot. That, at least, was what the men thought as they lay pressed to the ground by the river. That was what they hoped. For if death did not reap its harvest inside the citadel, then it would surely get them.

After the first four minutes, which seemed like an eternity of thunder, at exactly 0319, the first wave leaped to their feet. They dragged their rubber dinghies down into the water. They jumped in. And like shadows, veiled by smoke and fumes, they paddled across. The second wave followed at 0323. The men reached the other bank just as if they were on an exercise. Swiftly they climbed the sloping ground. Then they crouched down in the tall grass. Hell above them and hell in front of them. At 0327 Second Lieutenant Wieltsch, commanding No. 1 Platoon, straightened up. The pistol in his right hand was secured by a lanyard so that, if necessary, he had both hands free for the hand-grenades he was carrying in his belt and in two linen bags slung over his shoulders. No word of command was needed. Bent double, they crossed a garden. They moved past fruit-trees and through old stables. They crossed the road which ran along the ramparts. And now they would enter the fortress through the shattered gate-house. But here they had their first surprise. The bombardment, even the heavy shells of the 60-cm. mortars, had done very little damage to the massive masonry of the citadel. All it had done was to waken the garrison and give the alarm. Half dressed, the Russians were scurrying to their posts.

Towards midday the battalions of 135th and 130th Infantry Regiments had forced their way deep into the fortress in one or two places. But at the eastern fort of the Northern Island, as well as by the officers' mess and the barracks block on the Central Island, they had not gained an inch. Soviet snipers and machine-guns in armoured cupolas barred their way. Because of the close interlocking of attacker and defender the German artillery could not intervene. In the afternoon the corps' reserve, 133rd Infantry Regiment, was thrown into the fighting. In vain. A battery of assault guns was brought forward. With their 7·5-cm. guns they blasted the bunkers directly. In vain.

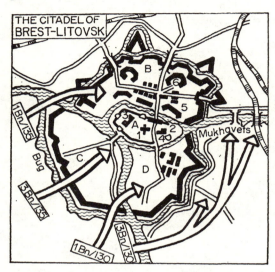

Map 2. The citadel of Brest-Litovsk. Attack by the battalions of 130th and 135th Infantry Regiments. A Central Island; B Northern Island; C Western Island; D Southern Island; 1 ancient fortress church; 2 officers' mess; 3 barracks; 4 barracks; 5 strongpoint Fomin; 6 Eastern fort.

By evening 21 officers and 290 NCOs and men had been killed. They included Captain Praxa, the battalion commander, and Captain Krauss, commander of 1st Battalion, 99th Artillery Regiment, as well as their combat staffs. Clearly, it could not be done that way. The combat units were pulled back from the fortress, and artillery and bombers had another go. Carefully they avoided the ancient fortress church: there seventy men of the 3rd Battalion sat surrounded, unable to move forward or back. Luckily for them they had a transmitter and had been able to report their position to Division.

The third day of Brest dawned.
As the sun's rays penetrated the smoke they fell upon an old and wrecked Russian anti-aircraft position. Amid the rubble was Lance-Corporal Teuchler's machine-gun party, belonging to Second Lieutenant Wieltsch's platoon. A painful rattle came from the gunner's throat. He had been shot through the lung and was dying. The machine-gun commander was sitting up stiffly, his back against the tripod. He had been dead for hours. Lance-Corporal Teuchler was lying shot through the chest, slumped over his ammunition-box. The sun on his face brought him round again. Cautiously he rolled over on his side. He could hear agonized voices. He saw a muzzle flash from a

Taken by Surprise 43

casemate some 300 yards away every time a wounded man sat up or tried to crawl behind cover. Snipers! It was they who wiped out Teuchler's party.

At noon a strong assault detachment of 1st Battalion, 133rd Infantry Regiment, broke through from the Western Island to the citadel church. The trapped German troops were freed; Lance-Corporal Teuchler was found. But the relieving units got no nearer to the officers' mess.

The eastern fort on the Northern Island was likewise still holding out. On 29th June Field-Marshal Kesselring sent in a Stuka Geschwader[1] against it. But the 1000-pound bombs had no effect. In the afternoon 4000-pounders were dropped. Now the masonry was shattered. Women and children came out of the fort, followed by 400 troops. But the officers' mess was still being stubbornly defended. The building had to be demolished piece by piece. Not one man surrendered.

On 30th June the operations report of 45th Infantry Division recorded the conclusion of the operation and the capture of the fortress. The division took 7000 prisoners, including 100 officers. German losses totalled 482 killed, including 40 officers, and over 1000 wounded, of whom many died subsequently. The magnitude of these losses can be judged by the fact that the total German losses on the entire Eastern Front up to 30th June amounted to 8886 killed. The citadel of Brest therefore accounted for over 5 per cent. of all fatal casualties.

A story such as the defence of the citadel of Brest would have received tremendous publicity in any other country. But the bravery and heroism of the Soviet defenders of Brest remained unsung. Up to Stalin's death the Soviets simply took no notice of the heroic defence of the fortress. The fortress had fallen and many soldiers had surrendered—that, in the eyes of the Stalinists, was a disgrace. Hence there were no heroes of Brest. The chapter was simply expunged from military history. The names of the commanders were erased.

But in 1956, three years after Stalin's death, an interesting attempt was made to rehabilitate the defenders of Brest. The publicist Sergey Smirnov published a little book entitled *In Search of the Heroes of Brest-Litovsk*. The reader discovers that the author had to go to a lot of trouble to track down the heroes who had survived the hell of Brest: they were all living inconspicuously, because fifteen years after the battle and ten years after the end of the war they were still being regarded as suspect and dishonoured. Smirnov writes:

> We have in Russia about 400 survivors of the battle of the citadel of Brest. Most of them were seriously wounded when the Germans took them prisoners. It must be admitted that we have not always

[1] Unit consisting of 3 Gruppen, usually 93 aircraft.

treated these men as we should have done. It is no secret that the people's enemy Beria and his henchmen encouraged an incorrect attitude to former prisoners of war, regardless of the manner in which these men became prisoners or how they bore themselves while in captivity. That is the reason why we have not so far been told the truth about Brest-Litovsk.

And what was that truth?

Smirnov found it on the walls of the casemates. There, scratched with a nail into the plaster, we read: "We are three men from Moscow —Ivanov, Stepanchikov, and Shuntyayev. We are defending this church, and we have sworn not to surrender. July 1941." And below we read: "I am alone now. Stepanchikov and Shuntyayev have been killed. The Germans are inside the church. I have one hand-grenade left. They shall not get me alive."

In another place we read: "Things are difficult, but we are not losing courage. We die confidently. July 1941."

In the basement of the barracks on the Western Island there is an inscription: "I will die but I will not surrender. Farewell, native country." There is no signature, but instead the date, 20.7.41. It appears therefore that individual groups in the dungeons of the citadel continued resisting until the end of July.

In 1956 the world was at last told who commanded the defence of the citadel. Smirnov writes: "From combat order No. 1, which has been found, we know the names of the unit commanders defending the central citadel: Troop Commissar Fomin, Captain Zubachev, First Lieutenant Semenenko, and Second Lieutenant Vinogradov." The 44th Rifle Regiment was commanded by Petr Mikhaylovich Gavrilov. Commissar Fomin, Captain Zubachev, and Second Lieutenant Vinogradov belonged to a combat group which broke out of the fortress on 25th June, but they were intercepted on the Warsaw highway and wiped out. The three officers were taken prisoners. Vinogradov survived the war. Smirnov found him in Vologda, where, still unrecognized in 1956, he worked as a blacksmith. According to his account, "Commissar Fomin, before the break-out, put on the uniform of a private soldier who had been killed; but he was identified in the POW camp by another soldier, denounced, and shot. Zubachev died in captivity. Major Gavrilov survived his captivity although, seriously wounded, he had resisted capture by throwing a hand-grenade and killing a German soldier."

It was a long time before the heroes of the citadel of Brest were recorded in Soviet history. They have earned their place there. The manner in which they fought, their perseverance, their devotion to duty, their bravery in the face of hopeless odds—all these were typical of the fighting morale and powers of resistance of the Soviet soldier. The German divisions were to encounter many more such instances.

The stubbornness and devotion of the defenders of Brest made a deep impression on the German troops. Military history has but few examples of similar disdain for death. When Colonel-General Guderian received the reports on the operations he said to Major von Below, the Army High Command's liaison officer with the Panzer Group, "These men deserve the highest admiration."

2. Stalin looks for a Saviour

The first battles of encirclement—Why were the Soviets taken by surprise?—Stalin knew the date of the attack—The "Red Chapel" and Dr Sorge—Precursors of the U-2—Stalin and Hitler at poker—General Potaturchev is taken prisoner and is interrogated.

"THE material and moral consequences of every major engagement," Field-Marshal Count Moltke wrote over eighty years ago, "are of such far-reaching character that as a rule they create an entirely changed situation."

Military experts agree that this dictum is valid to this day, and certainly applied in 1941. It is not known whether Stalin had read Moltke, but he acted in accordance with his thesis. He realized that on the Central Front disaster was staring him in the face because something decisive was lacking—a bold organizer, a tough, experienced commander in the field, a man who could by ruthless improvisation master the chaos caused by Guderian's and Hoth's advancing tanks.

Where was there such a man?

Stalin believed he had found him in the Far East. And he did not hesitate a minute to entrust to him the salvation of the Central Front.

At the moment when Second Lieutenant Wieltsch was bursting into the citadel of Brest, when Manstein was crossing the bridge of Daugavpils, and Hoth's tanks were racing towards the historic gap of Molodechno, from where Napoleon after his disastrous retreat from Moscow informed the world that the Grande Armée had been destroyed but the Emperor was in excellent health—at that moment, at the railway station of Novosibirsk, 900 miles east of the Urals, the stationmaster and the quartermaster of the Siberian Military District were running along the platform at which the trans-Siberian express stood. They were looking for a certain special compartment. At last they found it.

The stationmaster stepped up to the open window. Comrade General," he said to the broad-shouldered man in the compartment, "Comrade General, the Defence Minister requests you to leave the train and continue your journey by air."

"Very well, very well," said the general. The quartermaster dashed up the steps into the carriage to bring out the general's luggage.

The date was 27th June 1941. It was a hot afternoon. The platform was packed with milling uniformed crowds. Outside, in the station square, a loudspeaker was blaring. It was relaying a recruiting appeal from the Siberian Military District command.

The general, escorted by the quartermaster and the stationmaster, pushed his way through the crowd of men called up for service and now waiting for their trains to their respective garrisons. The general's name was Andrey Ivanovich Yeremenko. He was wearing the Order of the Red Banner of Labour. He had come from Khabarovsk, where until a week before he had commanded the First Far Eastern Army. In the Soviet High Command he enjoyed the reputation of being a tough commander of great personal courage, a brilliant tactician, and an absolutely reliable member of the Communist Party. He was a veteran of the Red Army, one of Trotsky's old guard, who had gone over to the Red Army as an NCO and gone through the entire campaign against the Whites. In this campaign he had earned his commission as an officer.

On 22nd June, the day war broke out, shortly after noon, General Smorodinov, the Chief of Staff of Army Group Far East, had rung up Yeremenko in great excitement: "Andrey Ivanovich, the Germans have been shelling our towns since early morning. The war has begun."

Yeremenko describes the scene in his memoirs:

> As a man who had dedicated his life to the military profession I had frequently thought about the possibility of war, in particular about the way in which it might start. I had been convinced that we would always be able to discern the enemy's intentions in good time, and would never be taken by surprise. But now, listening to Smorodinov, I realized instantly, we had been taken by surprise. We had been utterly unsuspecting. All of us—soldiers, officers, the Soviet people. What a *disastrous* failure of our intelligence service!

But Smorodinov did not give Yeremenko time for meditation. He was passing on to him definite orders. One: the First Far Eastern Army was to be put on full alert. "That means an attack is threatening here too—by the Japanese?" Yeremenko asked, startled.

Smorodinov put his mind at rest. The alert, he explained, was a precautionary measure. There were no indications that the Japanese intended to attack. Indeed, the High Command's assurance on this point was clear from the second order, which instructed Yeremenko to leave for Moscow at once to take up a new command.

Stalin Looks for a Saviour 47

Lieutenant-General Yeremenko did not know what awaited him. He did not know that from all the marshals and generals Stalin had chosen him, the lieutenant-general from the Far East, to save the Central Front. Stalin considered him to be the very man he needed—a master of improvisation, a Russian Rommel, familiar with the problems involved in commanding large formations. His First Far Eastern Army had been awarded the Order of the Red Banner of Labour for its excellent state of training. He seemed the obvious choice of a new iron broom for the wavering Western Front. If anyone could save the desperate situation there it was Yeremenko with his strong arm and unshakable belief in Stalin.

Certainly the situation on the Bialystok front was desperate enough. Three Soviet infantry divisions—the 12th, the 89th, and the 103rd—had not only offered no resistance to the Germans, but when their political commissars had tried, pistol in hand, to make the troops fight the troops had shot them and had then melted away. Most of them had been only too glad to go into captivity. It was this particular incident that had shaken Stalin. The situation required a very hard man.

Yeremenko had left Khabarovsk by the trans-Siberian express on the same day—22nd June. Anxiously he counted the hours he would have to spend en route. The man whom Moscow had chosen as the saviour of the Central Front was to make his journey by train! At last some one evidently thought better of it, and that was why he was snatched from the train at Novosibirsk.

Yeremenko drove straight to the headquarters of the Siberian Military District. But they had no news for him there from the front. As always in such circumstances, rumours were rife and were being spread even by senior officers. The Germans, they said, had been knocked on the head. General Pavlov's tanks had already moved forward from the famous Bialystok bend and were clearing the road to Warsaw for the infantry. Captain Gorobin, who had only recently been transferred to Novosibirsk from the staff of the First Cossack Army, said with a wink, "The maps we had there covered the territory all the way to the Rhine—and every single division was marked on them." There was optimism in Novosibirsk. On 26th June the communiqué announced: "The enemy has taken Brest," but no one took the news very seriously. Brest? *Nichevo*—surely Brest was somewhere in Poland!

Two hours later Yeremenko climbed into a twin-engined bomber and took off for Moscow. He had 1750 miles to cover. There were four intermediate stops for refuelling, overhaul, and rest. Russia is a big country. Some 2200 miles from Novosibirsk battles were raging on the Western Front. Yet Novosibirsk was only about half-way between Brest-Litovsk and Vladivostok.

While Yeremenko sat in his bomber on that 28th June, flying towards Omsk some 2600 feet above the dark *tayga*, over the huge Siberian plain with its boundless fields of wheat, over the cheerless industrial landscape around Sverdlovsk, towards the Urals, the man against whom he was to measure his skill was standing in his armoured command vehicle barely fifty miles south-west of Minsk, the Belorussian capital.

Colonel-General Heinz Guderian, commanding Second Panzer Group, had just sent a signal to Colonel Freiherr von Liebenstein, Chief of Staff of the Panzer Group: "The 29th Motorized Division, at present engaged on a broad front against Russian break-out attempts 110 miles south-west of Minsk in the Slonim–Zelva area, is to wheel round as soon as possible for a thrust towards Minsk–Smolensk."

As Guderian's order arrived at the headquarters of the Panzer Group in the ancient Radziwill château at Nieswiez, Bayerlein and Liebenstein, Guderian's Chief of Operations and Chief of Staff respectively, were bending over their map-tables, swiftly sketching in the latest situation. Their headquarters had been moved into the château only that morning. Two gutted Russian tanks were still lying by the bridge. Their story was being told throughout the Panzer Group.

During the night of 26th/27th June General Nehring, commanding 18th Panzer Division, was looking for the headquarters of his Panzer regiment. In his open armoured car he cautiously drove up to the château. A German Mark III tank was covering the approach to the bridge. Nehring ordered his driver to pull up, some forty yards from the tank. He hailed it. At that moment he heard the clank of tracks. Nehring stood up and shone his flashlight to the rear. He froze with shock. Two light Russian tanks of the old T-26 type stood close behind him, their machine-guns pointing forward.

"Break away half right!" Nehring hissed to his driver. He let in his clutch and roared off. But the German tank had noticed that there was something amiss. Within a second the first shell came from its 5-cm. gun. A second and a third followed at once. Not a single burst of machine-gun fire came from the Russian tanks.

Now the Soviet tanks were lying outside the château of Nieswiez, smoke-blackened witnesses to a general's strange adventure. Inside the Radziwill château, up on the third floor, another curious souvenir hung on the wall—a photograph of a hunting party in 1912. The guest of honour in the centre was Kaiser Wilhelm II.

Liebenstein and Bayerlein at once realized the idea behind Guderian's signal. The campaign on the Central Front had reached a decisive phase. The first major success was beginning to show in outline: the 17th Panzer Division, the spearhead of the units wheeling towards Minsk from the south, had reached the city. In the north Colonel-General Hoth with his Third Panzer Group had formed the

Stalin Looks for a Saviour   49

northern enveloping arc and, with General Stumpff's 20th Panzer Division, had penetrated into Minsk on 26th June. Hoth's and Guderian's Groups were therefore linking up. This meant that the huge pincers which the Fourth and Ninth Armies had formed round the Bialystok bend had now closed. The pocket, in which 4 Soviet Armies were caught, with 23 divisions and 6 independent brigades between Bialystok-Novogrodek and Minsk, was being sealed up. Four Armies—half a million men. The first gigantic battle of annihilation on the Eastern Front was unrolling, a battle of annihilation rarely equalled in military history. It was typical of Guderian's strategic grasp that he did not get intoxicated with the victory that was taking shape, that he was not yielding to the temptation to collect a few hundred thousand prisoners. He knew that it was no part of an armoured formation's duties to act as beaters, or to guard the sides of a pocket, or shepherd processions of prisoners. All these were tasks for the infantry. The fast troops had to keep moving, exploiting their opportunity. They must advance over the Berezina. And then over the Dnieper. Always moving, towards the first great strategic objective of the campaign—Smolensk.

That was why Guderian wanted to detach Major-General von Boltenstern's 29th Motorized Infantry Division from the defensive operations on the southern side of the pocket near the little river Zelvyanka and to both sides of the township of Zelva, where the Russians were trying to break out, and to employ it instead for the northward thrust towards Smolensk. But the 29th Motorized Infantry Division, known as "the Hawks" from its divisional tactical sign, was deeply involved in defensive fighting against desperate Soviet breakout attempts on a 40-mile sector along the edge of the great bulge. The Russians hoped to force a way through at this point, to tear open a gap to escape from the pocket. Again and again they had assembled in the thick forests and then, supported by tanks and artillery, charged against the thinly held lines of the German division.

South-west of the village of Yeziornitsa they made a cavalry charge straight into the machine-gun fire of the motor-cycle battalion and the machine-gun battalion of 5th Regiment, and, amid shouts of "Urra! Urra!" time and again re-formed in battalion and regiment strength. Near Zelva they penetrated into the forward positions of the reconnaissance detachments. The German 15th and 71st Infantry Regiments, from Kassel and Erfurt, were ceaselessly in action. The battalions of 15th Infantry Regiment, in particular, had a difficult time. The 5th Company was in position just over a mile outside the little town of Zelva, which was crammed full with Russians. Again and again they swept up against the German positions with their unnerving cries of "Urra!"—companies, battalions, regiments.

The picture was one that made the German troops' imagination boggle. The Russians were charging on a broad front, in an almost endless-seeming solid line, their arms linked. Behind them a second, a third, and a fourth line abreast.

"They must be mad," said the men of 29th Division. Mesmerized they stared at the earth-brown-uniformed wall of human bodies, of men pressed close together, approaching at a steady trot. Their long fixed bayonets were held rigidly in front of them—a wall bristling with lances.

"Urra! Urra!"

"This is murder," groaned Captain Schmidt, the commander of 1st Battalion. But then what else is war? If this gigantic storm was to be smashed and not just forced to the ground they would have to wait for the right moment. "Wait for my order to fire!" he commanded. The wall was still getting nearer. "Urra! Urra!"

The German troops behind their machine-guns could hear their hearts thumping. It was almost too much to bear. At last came the order: "Fire at will!" They squeezed the triggers. They knew that if they did not get the attackers the attackers would get them.

The machine-guns were rattling. "Fire!" Carbines barked. Sub-machine-guns spluttered. The first wave collapsed. The second collapsed on top of them. The third ebbed back. Brown mounds covered the vast field.

In the evening they came again. This time they came with an armoured train—a Soviet weapon that might have been useful in a civil war but was hardly suited to a modern battle of *matériel*. An armoured locomotive hauled behind it gun-platforms and armoured infantry wagons. Puffing heavily and firing from all guns, the monster approached from the direction of the little town of Zelva. Simultaneously, two squadrons of cavalry were charging on the left of the rail track, and on the right of it several T-26 tanks were rolling towards the headquarters of 2nd Battalion.

A 3·7-cm. anti-tank gun of the 14th Company, hurriedly brought up, set the armoured train on fire after the sappers had blown up the track and thus halted it. The cavalry charge collapsed in the machine-gun fire of 8th Company. It was the most terrible thing the men had experienced so far—the screaming of the horses. Horses howling with pain as their torn bodies twisted in agony. They rolled on top of one another, and, sitting up on their lacerated hindquarters, flailed the air with their forelegs like beasts demented.

"Fire!" Put them out of their misery. Make an end of it.

Things were easier for the men behind the anti-tank guns. Tanks do not scream. Besides, the Russian T-26 was not a match for the German 5-cm. anti-tank guns. Not one of them broke through the line.

However, the 29th Motorized Infantry Division could not be switched to the north, which had been Guderian's intention.

That same evening, on 28th June, Yeremenko's bomber landed at the military airfield of the Soviet capital. The general drove straight to the War Ministry. Marshal Timoshenko, the Defence Minister, met him with the words: "We've been waiting for you." There were no courtesies or polite phrases. The marshal came straight to the point. He went over to the situation map of the Central Front and—as Yeremenko reports in his memoirs—said, "Our failures on the Western Front are due to the fact that the commanders in the frontier zone did not show themselves equal to their tasks."

Yeremenko was amazed.

Timoshenko passed a shattering judgment on the Commander-in-Chief, Colonel-General Dmitriy Pavlov, who had been stationed in the Bialystok bend with the bulk of the Soviet motorized forces and who had been known in the Red Army before the war as "the Soviet Guderian."

Yeremenko was horrified when Timoshenko indicated on the map the territories which had already been lost during the first week of the war. Timoshenko's pencil moved over the map. "The Germans are now along the line Jelgava–Daugavpils–Minsk–Bobruysk. Belorussia is lost. Four Armies of the Western Front are cut off. The enemy is obviously aiming at Smolensk. And we have no forces left to protect it."

Timoshenko paused. According to Yeremenko's report, there was complete silence in the room. Then the marshal continued in a cold, angry voice: "The danger of the fascists lies in their tank strategy. They attack in large units. Unlike us, they have entire armoured corps operating independently, whereas our armoured brigades are no more than support for the infantry, and our tanks are employed piecemeal. And yet those German tanks are not invincible. They have no super-heavy types—at least they have not used any so far. I have released the T-34 for operational use. All those available will be supplied to the front as quickly as possible by the Moscow Tank Training Regiment."

The dramatic quality of the situation cannot be described more fittingly than in Yeremenko's own words:

> Marshal Timoshenko said, "Well, then, Comrade Yeremenko, now you have a clear picture." "It's a sad picture," I replied. After a while Timoshenko continued: "General Pavlov and his Chief of Staff are being relieved at once. By decree of the Government you have been appointed Commander-in-Chief of the Western Front."

"What is the task of that front?" Yeremenko asked concisely. Timoshenko replied: "To stop the enemy advance."

It was a clear order. A precise order. The fate of Moscow depended on its implementation.

The question inevitably arises: Why was not Stalin present at this conversation? What other Head of State and Supreme Commander-in-

Chief would have denied himself the opportunity, at such a crucial moment, of personally swearing in the general he had chosen as his country's military saviour? But not only Yeremenko—no one in Moscow heard anything of Stalin during the first two weeks of the war. It was Molotov who, in a nation-wide broadcast, had told the country of the German attack and called on the people to fight. Yet Stalin had been Chairman of the Council of People's Commissars—*i.e.*, Head of the Government—since the beginning of May.

"Where is he?" the Muscovites were asking. He did not speak. He made no public appearances. He had not even received the British Military Mission which arrived on 27th June to offer economic and military assistance. The wildest rumours were circulating. Had he been overthrown because he had trusted Hitler? It was even being said that he had fled the country. That he had gone to Turkey, or to Persia. At any rate, there was no sign of life from him. And on that night of 28th/29th June, Yeremenko had to embark on his difficult task without Stalin's blessing.

Meanwhile the German supply columns were moving in unbroken lines over the hot, dusty, rutted roads of the central sector. They were moving ceaselessly. 'Roads' was a misnomer for the deep, sandy tracks. Forward, ceaselessly forward, to where the armoured spearheads were waiting for fuel and their crews for cigarettes. Those cursed Russian roads! The arteries of the war! A Blitzkrieg was not just a question of fighting morale, but equally one of transport morale. The roads determined the pace of the war. And that pace decided the battles of the armoured corps. Only some one with experience of Russian roads can begin to suspect the amount of quiet planning that had to be done by the quartermasters.

Thus, in the operations area of Guderian's Panzer Group there were, once the Bug had been crossed, only two good roads of advance— from Brest to Bobruysk and to Minsk. Along these two roads some 27,000 vehicles of the Panzer Group and another 60,000 of the following infantry, headquarters personnel, and communications troops had to be ferried. To cope with this problem and to avoid complete chaos, Guderian had introduced three priority ratings. For any traffic with No. 1 priority the road had to be cleared. Anything with No. 2 priority had to yield precedence to anything with No. 1. Only when no formation with No. 1 or No. 2 priority was on the road could it be used by No. 3 priority traffic. Needless to say, this arrangement gave rise to fierce squabbles and rivalry. The Hermann Göring Luftwaffe Communications Regiment, for instance, had been assigned No. 3 priority since at that time it was concerned only with the transport and erection of telegraph posts. The Reich Marshal was very angry and ordered the regimental commander to see Guderian about it. Göring demanded No. 1 priority.

Guderian listened to the complaint and then asked, "Can telegraph posts shoot?" "Of course not, Herr Generaloberst," the regimental commander replied. "And that," Guderian said to him, "is why you'll keep No. 3 priority." That was the end of the matter. At least of the official side of it. On the human plane it ended more tragically. The regimental commander dared not report his failure to the Reich Marshal and shot himself.

Thus a mere handful of roads had to serve as the main arteries of the war against Russia. If only the Soviet Command had realized the significance of this fact in good time it could have inflicted upon the German supplies far more crises than in fact they had to face. There was the instance of 3rd Battalion, 39th Panzer Regiment. Those seasoned old warriors of the Panzer Lehr Regiment were lying in a patch of wood near Minsk on the evening of 28th June. They were waiting for motor-fuel. A tanker lorry pulled up. Lance-Corporal Piontek had to allow himself to be teased by Sergeant Willi Born: "Had a good journey, fuel-driver? Let's have thirty jerricans!" He unfastened the small trap in the armour-plating behind which the filler-cap was located.

But Piontek did not feel like joking. "Twelve jerricans—that's all," he decided. "That's hardly enough to fill my lighter," complained Born. But then he saw Piontek's face and fell silent. Piontek explained: "The Russian fighters—Ratas—got us. Five lorries burned out. All the drivers killed. And farther back the Russians have broken through, cut the road, and made a frightful mess of our entire supplies."

These were the drawbacks of the armoured thrusts through thickly held enemy territory, where entire Russian divisions were lurking in the woods. This was not the first time the regiment was in trouble. Things had been pretty bad at Slonim. They had driven on as far as the railway embankment of the Bialystok–Baranovichi line. Suddenly they heard the noise of battle coming from the town. The Russian infantry had lain low while the armoured spearheads passed, but now they pounced on the anti-aircraft gunners, the sappers, and the supply columns with everything they had.

No. 1 and No. 2 Troops of 9th Company, 39th Panzer Regiment, turned back. They went back into the town. "Clear them out!" That was easily said. For at the same moment the Russians attacked across the railway embankment. Slonim was in flames. The regiment was cut off and was being harassed from all sides. The troops dug in for all-round defence.

In the grey light of dawn Russian columns were made out through field-glasses on the far side of the railway embankment. The German tanks had all switched their radios to receiving. Time and again the battalion commander's stand-by signal came through for each tank in turn. The radio operator would push his key over to the right, so that his whole crew could hear the commander's orders: "Fire is not to be opened until you see the red Very light. Let the enemy get in close. Then concentrate your fire on the tanks." The engine noise was getting louder. "The old man must be asleep," the men in the tanks were saying. "They're practically on top of us!" The enemy column was headed by tanks. These were followed by lorries, horse-drawn carts, field-kitchens, and ammunition vehicles. The leading vehicles were now within fifty yards of the German line. At last the red Very light.

At a single blow a veritable wall of fire and smoke was thrown up by the German tanks. On the other side vehicle after vehicle went up in flames. The column was scattered. The tanks wheeled about and made for the fields with the tall crops. It was afternoon before Slonim was cleared and the Russian break-out attempt smashed. That had been three days ago—six days after the campaign had started.

And now General von Arnim's 17th Panzer Division was on the southern edge of Minsk. The troops saw the burning city. On the highway in the distance columns of traffic were moving in both directions. Radio Operator Westphal slung his machine pistol over his shoulder. He pushed his field-glasses back inside his tunic and climbed up on his tank. Three hours' watch lay in front of him. By the time he was relieved by the gun-loader it would be daylight. How far was it to Moscow? And how big was this country?

The distance from Moscow to Minsk is exactly 420 miles. And to Mogilev, where General Pavlov, C-in-C of the Bialystok sector, had his headquarters, it is 305 miles. Until the publication of Yeremenko's memoirs it had generally been believed that Pavlov shot himself after Marshal Kulik had deposed him on Stalin's orders and had put a pistol on his desk. Yeremenko tells a different story. According to his account, he arrived at Pavlov's headquarters early in the morning of 29th June, just as Pavlov was having breakfast in his tent. Pavlov was surprised to see him. He welcomed him rather glumly: "What brings you to this lousy spot?" Then he motioned towards the table. "Come on, sit down and have a spot of breakfast with me. Tell me the news."

But in mid-sentence Pavlov's voice tailed off. He sensed the icy chill which emanated from Yeremenko. Yeremenko said nothing. Instead of an answer he handed Pavlov his letter of dismissal. Pavlov ran his eyes over it. His face went rigid with shock. "And where am I to go?" he asked.

"The People's Commissar has ordered that you should go to Moscow."

Pavlov nodded. He bowed slightly. "Wouldn't you like a cup of tea after all?"

Yeremenko shook his head. "I consider it more important to acquaint myself with the situation at the front."

Pavlov sensed the reprimand. He justified himself: "The enemy's surprise attack found my units unprepared. We were not organized for action. A large part of the troops was in the garrisons or on the practice ranges. The troops were all set for a peaceful life. That is how the enemy found us. He simply drove straight through us, he smashed us, and he has now taken Bobruysk and Minsk. We had no warning. The order alerting the frontier units arrived much too late. We had no idea."

Caught napping—that was the great excuse. And Yeremenko, who otherwise has not a good word to say for Pavlov, writes: "On this point Pavlov was right. Today we know it. Had the order alerting the frontier units arrived sooner, everything would have turned out differently."

This raises a question of vital importance to military historians: Were the Russians really taken by surprise by the German attack, completely unsuspecting and engaged in harmless peaceful pursuits? Were they really so unprepared, and did they withdraw their allegedly inferior forces—as is still being maintained in many quarters to this day—to the Don and the Lower Volga in order to lure the German armies deep into Soviet territory and defeat them there? Was that what happened? It was not.

It is, of course, true that the Soviet frontier troops suffered a complete tactical surprise from the German attack on 22nd June. Only a few bridges along the 1000-mile frontier had been blown up by the Russians in time. The crucial bridges over the Memel, the Neman, the Bug, the San, and the Prut—and even that over the Daugava at Daugavpils, though 155 miles behind the frontier—were captured by German assault detachments by bold or cunning strokes. Does this prove that the Russians were unsuspecting?

Then how is it to be explained that on 22nd June the 146 attacking German divisions, with 3,000,000 men, were faced on the Russian side by 139 Soviet divisions and 29 independent brigades, with some 4,700,000 men? The Soviet Air Force had 6000 aircraft stationed in Belorussia alone. A large part of them, admittedly, were obsolete, but at least 1300 to 1500 were of the latest types. The German Luftwaffe, on the other hand, started the campaign with no more than 1800 operational machines.

This seems to suggest that the Russians were, after all, well prepared and equipped for defence. How then can the incredible mismanagement on the frontier be explained? What is the solution of the riddle?

On 23rd February 1941 Soviet Defence Minister Timoshenko had issued the following decree: "In spite of the successes of our policy of neutrality the entire Soviet people must remain in a constant state of readiness against the threat of an enemy attack."

On 10th April 1941 the Soviet War Council had decreed a secret alert for the so-called Western Front. Why? On the strength of what circumstances, what information, what news?

Well, the news which had been reaching Moscow since January 1941 had all been rather alarming. This information was supplied by the magnificently organized Soviet intelligence service. Between Paris and Berlin one Leopold Trepper, alias Gilbert, also known as "Grand Chef," was travelling freely, collecting information throughout Hitler's sphere of control, and this information he passed on to Moscow via the Soviet Embassy in Berlin.

In Brussels, Viktor Sokolov, alias Kent, a major in the Soviet intelligence service, maintained an office, and there he received his precious pieces of information from well-informed Communist contacts. His intelligence network was known as the "Rote Kapelle" (the Red Choir).

In Switzerland the most cunning of Soviet agents in Europe was operating—Rudolf Rössler, known as Lucy, a member of the "Red Choir," subordinated to Rado, the Soviet top agent.

But the best man of the Soviet military intelligence service was in Tokyo—Dr Richard Sorge, Press Assistant at the German Embassy there, a man who did more for the Soviet Fatherland War than a whole army. It was he who had given Stalin the certain information that Japan would not move against the Red Army in Manchuria. Sorge's report enabled him to withdraw the Siberian divisions from the Far Eastern Front, and it was these divisions which later turned the tide of the war at Moscow, Kursk, and Stalingrad.

All these agents supplied the Red Army's intelligence departments with mountains of information about Hitler's military plans against the Soviet Union. All of them predicted the attack. And such gaps as there may have been in their reports were filled in by the diplomatic representatives of the Western Powers from the inexhaustible fund of information of the British and American secret services.

Here is one piece of evidence that the German attack, including its exact date, cannot have been a surprise for the Russians. On 25th April 1941 the German Naval Attaché in Moscow, in a telegram routed to the Naval High Command via the Foreign Office in Berlin, reported: "Rumours about impending German-Russian war greatly increasing in scope. British Ambassador gives 22nd June as date of beginning of war."

This suggests that some two months before the outbreak of the war half Moscow was informed about Hitler's date of attack. And Stalin? Would he not have been told? Of course he was told, and he was fully aware of the importance of espionage and personally looked after this department.

In March 1937, addressing the Central Committee of the Communist Party about the tasks of secret intelligence work, he had said, "To win a battle in war one must have several corps of Red Army men. But to

Stalin Looks for a Saviour 57

prevent a victory at the front it is enough to have a few spies in an army staff, or even in a divisional staff, who would steal the plan of operations and pass it on to the opponent."

At the Eighteenth Party Congress, in 1939, Stalin had again broached the subject, in the following significant words: "Our Army and intelligence service have their sharp eyes no longer on the enemy within our country, but on the enemy abroad." In view of these remarks, is it credible that in 1941 Stalin would have taken no notice of the information supplied to him by his secret service about the German preparations for an attack? He must have been informed. After all, he had first-rate informants. From Berlin to Tokyo, from Paris to Geneva, his informants—many of them highly respected men beyond the breath of suspicion—sat in high positions and supplied valuable information.

The thoroughness of their work was revealed during the very first few weeks of the war. When the 221st Defence Division in Lomza cracked the safe left behind by the C-in-C of the First Cossack Army, they found in it maps for the whole of Germany, with the location of German Armies, Army Groups, and divisions accurately entered. The information was complete—nothing was lacking.

But this, by comparison, was peanuts. Some much more exciting discoveries were made.

The German radio monitoring service in the East Prussian seaside resort of Cranz had been intercepting the coded messages of countless unknown agents' transmitters since the beginning of the war. Attempts to crack the ingenious figure codes had been in vain. At last, in November 1942, German intelligence received the key. The Soviet chief agent Viktor Sokolov, alias Kent, had been captured in Marseilles. In order to save his mistress, Margarete Barcza, he offered to work for the Germans and betrayed the code.

What Admiral Canaris was shown after the decoding of the messages was far worse than the greatest pessimists had feared. There was a message of 2nd July 1941, for instance. Ten days after the outbreak of war Alexander Rado reported from Geneva to Moscow: "Rdo. To Director. KNR 34. Valid German plan of operations is Plan 1 with objective Moscow. Operations on wings merely diversions. Main thrust on Central Front. Rado."

About three weeks later, on 27th July, Rado amplified his message in reply to an inquiry from Moscow: "Rdo. To Director. KNR 92. Re RSK 1211. In case Plan 1 meets with difficulties Plan 2 will be used with main thrust on wings. Change of plan will be known to me within two days. Plan 3 with objective Caucasus not envisaged before November. Rado."

Needless to say, Berlin was flabbergasted to find a Soviet agent in Switzerland so accurately informed, and every effort was made to discover his source—a source which could discover a "change of plan" in the German High Command "within two days." But this source was never discovered. It has not been discovered to this day. Right through the war Alexander Rado continued to send his information to Moscow by radio. One thing is certain, however: Rado's main contact was Rudolf Rössler, alias Lucy, a Communist émigré from Bavaria who worked in Switzerland. In *The Soviet Army*, edited by the British military historian Liddell Hart, Dr Raymond L. Garthoff, who made a thorough study of the evidence, states that an anonymous source on the German General Staff informed this net of the German plans for the invasion of the USSR, and even provided the date of the invasion.

What more could Stalin or the Soviet General Staff want? Hitler's secrets were openly revealed to the Kremlin. Moscow, therefore, could have turned Operation Barbarossa, based as it was on surprise, into a crushing defeat for Hitler within the first twenty-four hours. Provided, of course, Stalin drew the correct military conclusions from his information. Why did he not do it?

To decide this key question of the German-Soviet war we have to turn our attention to a different one first. What was the state of German espionage against Russia? What did the German Command know about the military secrets of the Soviet Union? The question can be answered in two words—very little. The German secret service was very thinly established in Russia. It knew nothing of the vital military secrets of the Soviets, whereas they knew everything about Germany. They knew all about German weapons, about German garrisons, they knew where the German training areas were situated, and where the armament factories were. They knew the exact figures for German tank production. They had a clear picture about the number of German divisions. The German Command, by way of contrast, at the beginning of the war estimated the Red Army at 200 divisions. Within six weeks of the start of operations it discovered that there were at least 360. The German Command had no idea that the Russians had super-heavy KV tanks, or the T-34, or those terrifying multiple mortars, soon to be nicknamed "Stalin's organ-pipes."

Naturally, the German military secret service had tried, especially after 1933, to look behind the Soviet scenes. But the Soviets' mistrust of Hitler's Third Reich had been greater still than their suspicion of the Weimar Republic, and consequently the prospects of establishing secret agents within the Soviet Union were not promising. Besides, German intelligence was not overzealous in this direction and was unwilling to take risks. After all, no one in the German High Command envisaged a war between Germany and Russia.

Later, when Hitler demanded intensified intelligence work in Russia, it was found that this could not be organized at such short notice. The strict controls on the frontiers of the Communist empire, the close surveillance of every traveller, and indeed every stranger, made it virtually impossible to build up a network of agents. If, now and again, a spy was insinuated nevertheless from Finland, Turkey, or Iraq he would encounter almost insuperable difficulties in transmitting his information. A courier service was out of the question since no Soviet citizen was allowed to travel abroad. What few tourists there were were under strict supervision. That left only carrier pigeons from the frontier regions, and the radio. Both methods were enormously dangerous, and very few people were prepared to take the risk.

Nevertheless, in conjunction with the work of the German Military Attachés, some useful information was obtained in this way. Thus Guderian published a book entitled *Achtung—Panzer!* in which, on the grounds of reliable information, he put the number of Soviet tanks at 10,000. But in the German High Command the general was ridiculed. The then Chief of Army General Staff, Colonel-General Beck, accused Guderian of exaggerating, and even of creating alarm and despondency. Yet Guderian had deliberately erred on the cautious side and deducted a few thousand from the number reported to him. Quite unnecessarily, as it turned out, since the Russians at the outbreak of war possessed over 17,000 tanks.

In 1941 nobody would have thought that possible. The Finnish-Soviet winter war of 1939–40 had had a disastrous effect on the assessment of Soviet strength. The fact that little Finland offered such prolonged resistance to the Soviets was taken as evidence of Soviet weakness. To this day there are quite a few serious observers who maintain that Stalin deliberately conducted the Finnish war with outmoded weapons and inferior forces as a gigantic bluff, in order to deceive the world. Certainly the Soviet High Command did not employ the T-34 or the super-heavy KV tanks—even though these were manufactured right on Finland's doorstep, at Kolpino—nor yet the multiple mortars.

Finland's Marshal Mannerheim reports in his memoirs that Hitler told him in 1942 that the Russian armaments came as a colossal surprise to him. "If anyone had told me before the beginning of the war that the Russians could mobilize 35,000 fighting vehicles I would have had him declared insane. But up to date they have in fact thrown 35,000 of them into battle."

In order to get a peep behind Russia's walls after all, in spite of the well-nigh insuperable Soviet precautions against conventional forms of espionage, the German Command resorted to a method which was employed twenty years later, in our own time, by the Americans, and which when discovered gave rise to a serious crisis—secret aerial reconnaissance from great altitudes. The idea of spying inside Soviet territory by means of very fast and exceptionally high-ceiling aircraft was not an American invention. Hitler had practised the method successfully long before the Americans. This interesting chapter has so far not had the publicity it deserves. The evidence for it is in American secret archives. It may be assumed that it was the study of these papers which induced the Americans to experiment with their U-2s. The secret documents about German aerial reconnaissance bore the code name "Reconnaissance Group under the C-in-C Luftwaffe."

In October 1940 Lieutenant-Colonel Rowehl received a personal and top-secret order from Hitler: "You will organize long-range reconnaissance formations, capable of photographic reconnaissance of Western Russian territory from a great height. This height must be so exceptional that the Soviets will not notice anything. You must be ready by 15th June 1941."

Feverishly special machines were developed by various aircraft firms from the suitable available types. They were equipped with pressurized cabins, with engines specially tuned for high-altitude flying, with special photographic equipment and a wide angle of vision. In the late winter the "Rowehl Geschwader" began its secret flights. The first squadron operated from Seerappen in East Prussia and reconnoitred the area of Belorussia. The aircraft were He-111 machines with special high-altitude engines. The second squadron, operating from Insterburg, photographed the territory of the Baltic States as far as Lake Ilmen. They used the Do-215-B2, a special model made by the Dornier works. The machine had a ceiling of 30,000 feet. The area north of the Black Sea coast was photographed by the third squadron, operating from Bucharest with He-111 and Do-215-B2 machines. From Cracow and Budapest the special squadron of the Research Centre for High-altitude Flying covered the area between Minsk and Kiev. They employed special Junkers models, the Ju-88B and Ju-86P—magnificent machines capable of reaching 33,000 and 39,000 feet respectively. That was a sensational height for those days.

The plan worked smoothly. The Russians noticed nothing. Only one machine had engine trouble, and made a forced landing in the Minsk area on 20th June, two days before the outbreak of war. But the crew were able to set their secret machine on fire before they were captured. The outbreak of the war caused the incident to be forgotten.

The long-range reconnaissance flights of the Rowehl Geschwader were virtually the only source of really significant intelligence material for the first phase of the campaign. All the airfields in Western Russia, including the well-camouflaged fighter bases near the frontier, were photographed. What the human eye would never have seen was revealed by special films. Surprisingly numerous units were spotted on forward fields, and huge concentrations of armour were made out in the forests in the north.

This information enabled a resounding blow to be struck against the Soviet defensive capacity. For days Field-Marshal Kesselring and his

Air Corps commanders sat evaluating the aerial photographs and discussing operations.

There was just one problem that troubled them—the timing of their attack. Zero hour on 22nd June had been chosen to give the infantry enough light to make out their targets. That was why the artillery bombardment was scheduled to start at 0315 hours. On the Central Front, however, it was still dark at 0315, and air-force operations were therefore not yet possible. The Russian fighter and bomber formations, which would naturally be alerted by the artillery bombardment, would thus have thirty or forty minutes before the first German aircraft appeared over their fields. Needless to say, experienced pilots could have found their targets in the dark even twenty years ago, but the point was that no air forces should be spotted crossing the frontier too soon. For that would have warned the Russians and deprived the ground forces of their element of surprise. At last somebody thought of the solution—General Loerzer, General von Richthofen, or Colonel Mölders, nobody remembers for certain who it was. The idea was that the aircraft would approach the enemy airfields at great height in the dark, in the manner of long-range reconnaissance planes.

The plan was adopted. For each airfield from which Soviet fighters were operating three German bomber crews with experience of night flying set out. Flying at great height and taking advantage of uninhabited areas of marsh or forest, they crossed the frontier and sneaked up on their targets, so that they were over the fields exactly at first light, at 0315 hours on 22nd June.

At the same time as the bombers, but very much higher, flew Rowehl's long-range reconnaissance machines, carrying men of the "Brandenburg" Intelligence Regiment. They were to be dropped by parachute near railway junctions and road intersections, for sabotage actions or for work as undercover agents.

The plan went according to schedule. On the Russian fields the fighters were lined up in formation. Row by row they were bombed and shot up. Only from a single airfield did a fighter formation attempt to take off, just as the German bombers arrived. But the Russians were a few minutes too late. The bombs and shells burst right among the formation about to take off. Thus the pilots were written off as well as the machines. Right at the beginning of the war the Soviet fighter strength had been wiped out by a terrible "Pearl Harbor of the air." As a result, the German Stuka and bomber formations were able, on that first day of the offensive, to clear the way for the ground forces untroubled by enemy fighters. They penetrated some two hundred miles into Russian territory and destroyed Soviet bomber bases. Without this blow the Red Air Force would have been a dangerous enemy during the first crucial operations. Anyone questioning this assertion need only look at the losses suffered by the German Luftwaffe in the first four weeks of the war. Between 22nd June and 19th July the Luftwaffe lost, in spite of its shattering opening strikes, a total of 1284 aircraft shot down or damaged. The war in the air on the Eastern Front was therefore no walk-over. On 22nd June the three air fleets on the Eastern Front flew 2272 missions, with 1766 bombers and 506 fighters. Seven days later their operational strength had dropped to 960 aircraft. Not till 3rd July did it rise again above the thousand mark.

It is clear that the surprise blow at the Soviet Air Force was of decisive importance for the ground troops. This raises once more the question: How was this surprise possible if Moscow knew Hitler's attack to be imminent? What was the explanation of the curious fact that in the front line the Soviet ground troops and Air Force were blissfully and unconcernedly asleep, while in the hinterland all preparations for war had been carefully made? The blackout, to quote just one instance, had been planned so thoroughly that throughout Western Russia the blue blackout bulbs and blackout materials were available right from the start. Strips of gummed paper were found even in the smallest villages, protecting window-panes against being shattered by blast.

Mobilization, too, functioned smoothly. Altogether, military traffic throughout the hinterland worked exceedingly well. The switch-over to a total war footing was performed by industry without a hitch in accordance with prepared plans. The elimination of all possible "enemies of the State" in the border territories went like clockwork. As early as the night of 13th/14th June 1941—*i.e.*, eight days before the German attack—the Soviet State Security Service had all "suspect families" in the Baltic countries transported to the interior of Russia. Some 11,000 Estonians, 15,600 Latvians, and 34,260 Lithuanians were put aboard trains within a few hours and shipped wholesale to Siberia. Everything was functioning smoothly. Henry D. Cassidy, the Associated Press correspondent, in his first major report for the American papers sent from Moscow on 26th June, described his journey aboard a military train from the Black Sea to the Soviet capital. He said, "I got the impression from this journey that the Soviets are off to a good start."

Off to a good start! But why then were the forward lines on the Central Front off to such a bad start? So bad in fact that Colonel-General Guderian remarks in his memoirs: "Careful observation of the Russians convinced me that they knew nothing of our intentions." The enemy was taken by surprise all along the front of the Panzer Group.

How was this possible? A surprising but satisfactory answer is provided by Marshal Yeremenko in his memoirs, published in Moscow in 1956. Stalin alone was responsible, is Yeremenko's verdict.

I. V. Stalin, as the Head of State, believed that he could trust the agreement with Germany, and failed to pay the necessary attention to such symptoms as indicated a fascist attack against our country. He regarded information about an impending German attack as lies and provocations by the Western Powers, whom he suspected of wanting to wreck relations between Germany and the Soviet Union in order to implicate us in the war. That was why he failed to authorize all urgent or decisive defence measures along the frontier, for fear that this would serve the Hitlerites as a pretext for attacking our country.

It seems therefore that it was Stalin who, against the insistence of his General Staff, refused to authorize the alert for the frontier troops and forbade the organization of effective defence measures throughout the border regions. Stalin did not believe Richard Sorge, nor "Grand Chef" Gilbert, nor yet "Petit Chef" Kent. He did not believe Lucy, and least of all did he believe the British Ambassador.

Does that seem credible? It certainly does not seem incredible. The history of espionage and diplomacy is full of instances when excessively accurate reports supplied by agents about some great secret encountered not enthusiasm but mistrust. One such example is the story of Elyesa Bazna, the Armenian valet of the British Ambassador in Ankara, who from 1943 onward gained access to all the top-secret telegrams from the embassy safe and sold them to Hitler's espionage service. Cicero, as Sir Hughe Knatchbull-Hugessen's valet called himself, laid his hands on the secret documents in the simplest possible way. While at breakfast His Excellency usually left the key to his safe in his jacket in the bedroom. The Armenian would take it, go and dust the Ambassador's study, unlock the safe, photograph the documents, lock the safe again, and slip the key back into the jacket. It was as simple as that.

But Adolf Hitler refused to believe it. He regarded the whole thing as an elaborate plant by the British secret service, of which he was more terrified than the devil is of holy water. He would sweep the reports off his desk and refuse to draw any conclusions from the Allied plans which lay clearly revealed before him.

It seems that Stalin was filled with the same deep distrust of his informants, and that this suspicion grew stronger with every report confirming the imminence of a German attack. A master of casting suspicion on others, and a cunning tactician, he fell victim to his own conspiratorial mode of thinking. "The capitalist West is trying to manœuvre me into opposition to Hitler," he would speculate. With the stubbornness so often found in dictators he clung to his conviction that Hitler could not possibly be so foolish as to attack Russia until he had defeated Britain. He regarded the German concentrations along his military frontier in Poland as a bluff. Perhaps the Soviet dictator was himself influenced by the rumour deliberately spread by German intelligence that the concentration of forces in the East was intended to deceive Britain and divert attention from the planned invasion of the British Isles. Besides, a man like Stalin would hardly believe that an important secret such as a German war of aggression was so badly guarded that all the world seemed to be privy to it.

This view is confirmed by the greatest authority on the Kremlin backstage and the Red Army's secret intelligence activities, David J. Dallin. In his book *Soviet Espionage* he writes:

> In April 1941 a Czech agent named Škvor confirmed a report to the effect that the Germans were concentrating troops on the Soviet frontier and that the Skoda armament works in Bohemia had been instructed to stop fulfilling Soviet orders. Izmail Akhmedov confirms that Stalin wrote in the margin of this report with red ink: "This information is a British provocation. Find out where it comes from and punish the culprit."

Stalin's order was obeyed. Major Akhmedov of the Soviet secret service was sent to Berlin, disguised as a Tass correspondent, in order to find the culprit. There Akhmedov was caught by the war.

Quite clearly the reports of an intended attack by Hitler did not fit into Stalin's concept. His plan was to allow the capitalists and fascists to fight each other to exhaustion, and then he could do as he wished. That was what he was waiting for. That was why he was rearming. And that was also why he wanted to avoid making Hitler suspicious or provoke him into striking prematurely.

For that reason, according to Yeremenko, he prohibited all emergency mobilization or alerting of the frontier troops. In the hinterland, however, Stalin let the General Staff have its own way. And the General Staff, possessing the same secret information about the German offensive plans, set its mobilization in train and deployed its forces in the hinterland, not for an attack but, in the summer of 1941, for defence.

True, Field-Marshal von Manstein, when asked by the author whether in his opinion the Soviet deployment of forces had been offensive or defensive in character, expressed the view he had already stated in his memoirs: "Considering the numerical strength of the forces in the western areas of the Soviet Union, as well as the heavy concentration of tanks, both in the Bialystok area and near Lvov, it would have been quite easy for the Soviets to switch to the offensive. On the other hand, the deployment of Soviet forces on 22nd June did not suggest immediate offensive intentions.... One would probably get nearest to the truth by describing the Soviet concentrations as a 'deployment for all eventualities.' On 22nd June 1941 the Soviet forces were unquestionably organized in such depth that they could only be used for defensive operations. But that picture could have been changed very quickly. The Red Army could have regrouped for offensive operations within a very short tlme."

Colonel-General Hoth, when questioned by the author, repeated the conclusion he had drawn in his excellent study of armoured warfare on the northern wing of the Central Front: "The strategic surprise had come off. But one could not overlook the fact that in the Bialystok bend the Russians had concentrated strikingly large forces, especially mechanized ones, in greater numbers than would seem necessary for defensive operations."

Whichever view one inclines to, Stalin quite certainly did not intend to attack in the summer of 1941. The Red Army was in the middle of a complete change-over in equipment and a reorganization, especially in the armoured groups. New tanks and new aircraft were being supplied to the units. That, most probably, was the reason why Stalin did not want to provoke Hitler into action.

This attitude on the part of Stalin in turn confirmed Hitler in his intentions. Indeed, it might be said that this war and the cruel tragedy which sprang from it were the outcome of a sinister game of political poker played by the two dictators of the twentieth century.

An impartial witness in support of this theory of the political mechanism behind the German-Soviet war is Liddell Hart, the most searching of military historians in the West. In his essay "The Russo-German Campaign," in *The Soviet Army*, he has expounded it convincingly. He believes that it was Stalin's intention to extend his own positions in Central Europe in the course of the German-Allied war, and perhaps, at a suitable moment, extort further concessions from a hard-pressed Hitler.

Liddell Hart recalls that as early as 1940, while Hitler was still fighting in France, Stalin used the opportunity to occupy the three Baltic States—although under the German–Soviet secret treaty one of them, Lithuania, belonged to the German sphere of influence. Hitler must have then realized for the first time that Stalin was out to pull a fast one on him while his back was turned.

Shortly afterwards, when the Kremlin issued a 24-hour ultimatum to Rumania, extorting from her the cession of Bessarabia, and in this way drew closer to Rumania's oilfields, which were so vital to Germany, Hitler became nervous. He moved troops into Rumania and guaranteed the country's integrity.

Stalin saw this as an unfriendly act. Propaganda within the Red Army was being tuned more and more to an anti-fascist note. This was reported to Hitler, who promptly strengthened his troops on the Eastern Front. To this the Russians reacted by moving more of their troops to their western frontier.

Molotov was invited to Berlin. But the planned grand-scale understanding between the two dictators about the division of the world—Hitler was prepared to reward the Soviets with parts of the British Empire—did not come off. Hitler, in his egocentric way of looking at

C

things, took this as evidence of Stalin's ill-will. He saw the threat of a war on two fronts and went on record with the words: "I am now convinced that the Russians will not wait until I have defeated Britain." Three weeks later, on 21st December 1940, he signed "Directive No. 21 —Event Barbarossa." This contains the significant sentence: "All measures taken by the commanders-in-chief on the strength of this Directive must be unequivocally presented as precautionary measures in the event of Russia changing her attitude to us."

Stalin, on his part, had regarded the German offer to Molotov as a sign of weakness; he felt in a superior position and believed that Hitler, like himself, was merely out for political blackmail. In spite of his information he did not take Hitler's military plans seriously—or at least he did not believe that Hitler would consider he had any reason for striking already. That was why he avoided doing anything that might provide him with such a reason.

How strictly and meticulously—one might almost say, anxiously—Stalin's High Command was made to conform with this attitude is shown by the fact that General Karabichev, then Inspector of Engineers, was strictly forbidden during his tour of inspection in the Brest area at the beginning of June 1941 to visit the most forward frontier fortifications. Stalin did not wish to create a war atmosphere among the frontier troops; he wanted to avoid anything that looked like war preparations—either to his own troops or to Hitler's intelligence service. Therefore, in spite of the obvious German troop concentrations, the Soviet frontier troops were not on a proper combat-footing; no long-range artillery was in position for use against German reserves beyond the frontier, and no plans existed for heavy-artillery barrages. The consequences of Stalin's disastrous theory were terrible. One striking illustration was the action and destruction of the Soviet 4th Armoured Division.

Major-General Potaturchev, born in 1898—*i.e.*, forty-three years old in the summer of 1941—with his hair and moustache cut à la Stalin, was one of the first Soviet generals in the field to be taken prisoner. Potaturchev was in command of the Soviet 4th Armoured Division at Bialystok, the spearhead of the Soviet defences at a crucial point on the Central Front. The Soviet High Command thought highly of him. He was a member of the Party, the son of a small peasant from the Moscow area. As a lance-corporal under the Tsar he had gone over to the Red Army, and had advanced to the rank of general commanding a division. His story is of considerable interest:

"On 22nd June, at 0000 hours [Russian time—*i.e.*, 0100 German summer time], I was summoned to Major-General Khotskilevich, GOC VI Corps," Potaturchev wrote in the deposition he made on 30th August 1941 at the headquarters of the German 221st Defence

Division. "I was kept waiting because the General had himself been summoned to Major-General Golubyev, the C-in-C Tenth Army. At 0200 hours [*i.e.*, 0300 German time] he came back and said to me, 'Germany and Russia are at war.' 'And what are our orders?' I asked. He replied, 'We've got to wait.'"

An astonishing situation. War was imminent. The C-in-C of the Soviet Tenth Army knew it two hours before. But he would not, or could not, give any orders other than "Wait!"

They waited two hours—until 0500 German time. At last the first order came down from Tenth Army. "Alert! Occupy positions envisaged." Positions envisaged? What did that mean? Did it mean that the counter-attacks they had rehearsed in many manœuvres should now be launched? Nothing of the sort. The "positions envisaged" for the 4th Armoured Division were in the vast forest east of Bialystok. There the division should go into hiding—and wait.

"When the 10,900-strong division moved off, 500 men were missing. The medical detachment, with an establishment of 150 men, was 125 men short. Thirty per cent. of all tanks were not in working order, and of the rest several had to be left behind for lack of fuel."

That was how a key unit of the Soviet defensive line-up in the Bialystok area moved into action.

But no sooner had Potaturchev got his two tank regiments and his infantry brigade moving than a new order came down from Corps: the tank and infantry units were to be separated. The infantry was ordered to defend the Narev crossing, while the tank regiments were to hold up the German formations advancing from the direction of Grodno.

This order reveals the utter confusion in the Soviet Command. An armoured division was being torn apart and used piecemeal instead of being employed as a whole, frontally or from the flank, in a counter-attack. The fate of Potaturchev and his units was typical of the Soviet collapse in the border area. First they were battered by German Stukas. Admittedly, they did not lose many tanks, but the troops were badly shaken. Nevertheless Potaturchev reached his prescribed line. But then things began to go wrong for him. The advancing German armoured spearheads did not attack him, but thrust past him and cut him off. Potaturchev tried to evade encirclement. His companies got into a muddle, they were caught by German armoured forces, and smashed one by one. The infantry brigade suffered the same fate.

By 29th June Stalin's famous 4th Armoured Division was only a heap of wreckage. The password was "Every man for himself." They sought salvation in the big forest. In twos and threes, at most in handfuls of twenty or thirty men, infantry, artillery, and tank troops made for the woods. The few armoured cars of the 7th and 8th Tank Regiments which had escaped destruction hid out during the day and at night

68   Part One: Moscow

rolled towards the forest of Bialowieza. The vast forest was their only hope.

On 30th June General Potaturchev and a few officers broke away from their men. They intended to make their way on foot to Minsk and fight their way through from there to Smolensk. Potaturchev walked until his feet were sore, and, because he did not want to be seen on the roads as a shambling, bedraggled general, he got some civilian clothes from a farm. Nevertheless, he was intercepted by the Germans near Minsk and put in a POW cage. There he revealed his identity to the officer of the guard.

3. Objective Smolensk

The forest of Bialowieza–The bridges over the Berezina– Soviet counter-attacks–The T-34, the great surprise– Fierce fighting at Rogachev and Vitebsk–Molotov cocktails–Across the Dnieper–Hoth's tanks cut the highway to Moscow–A Thuringian infantry regiment storms Smolensk –Potsdam Grenadiers against Mogilev.

POTATURCHEV'S information surprised his captors: they had had no idea of the division's fire power. The Soviet 4th Armoured Division had 355 tanks and 30 armoured scout-cars; the tanks included 21 T-34s and 10 huge 68-ton KV models with 15·2-cm. guns. The artillery regiment was equipped with 24 guns of 12·2- and 15·2-cm. calibre. A bridge-building battalion had pontoon sections for bridges 60 yards long and capable of carrying 60-ton tanks.

Not a single German Panzer division in the East in the summer of 1941 was so well equipped. Guderian's entire Panzer Group with its five Panzer divisions and three and a half motorized divisions had only 850 tanks. But then, on the other hand, no German Panzer division was so badly led or so senselessly sacrificed as Potaturchev's 4th. It was against the remnants of this division that the German units were engaged in such fierce fighting in the forest of Bialowieza.

"That damned forest of Bialowieza!" the men grumbled. The whole of Germany made the acquaintance of this terrible virgin forest, the last surviving one in Europe. Bavarians and Austrians, men from Hesse, the Rhineland, Thuringia, and Pomerania, fought in this green hell.

The forest of Bialowieza meant ambush. It was a natural strongpoint in the rear and on the flank of the German forces. There was the village of Staryy Berezov, and, even better remembered, the village of Mokhnata.

Cossack squadrons were galloping across the open country, desperately anxious to gain the cover of the forest. The outposts of 508th Infantry Regiment were trampled down by them. Hooves pounded; sabres flashed. "Urra! Urra!" They got within a hundred yards of the village. Then the 2nd Battery, 292nd Artillery Regiment, smashed the attack with direct fire.

The 78th Infantry Division from Württemberg, the same which later received the title 78th Assault Division, was ordered to break into the green hell of Bialowieza, to comb the forest, and to drive the Russians out towards the intercepting line established by 17th Infantry Division along the northern edge of the huge forest.

The Russians were past masters of forest fighting. The German troops, by way of contrast, had little experience at that time of this difficult form of operation in the uninhabited, swampy forests of Eastern Poland and Western Russia. Forest fighting had been the poor relation of German Army training, for the German Forestry Commission kept a jealous eye on its woods and plantations. They could only be used with great care. As for virgin forests, the Wehrmacht had none at all for training purposes. The Russians, on the other hand, had practised this type of fighting extensively. Unlike the German infantry, they did not take up position in front of a wood, or on the wood's edge, but invariably right inside it, preferably behind swampy ground. Behind their all-round positions they kept their tactical reserves. In forest fighting, too, the Red Army men preferred the close combat in which they had been trained.

A particular feature of these Soviet defence positions were infantry foxholes which were unidentifiable from the front and provided a field of fire only to the rear; they were intended for picking off the enemy from behind after he had pushed past.

Whereas the German infantry would clear lanes of fire for themselves, if necessary by considerable felling of trees—which, of course, meant they were easily spotted from the air—the Russians worked like Red Indians. They would cut down the undergrowth only up to waist-height, creating tunnels of fire both forward and towards the sides. This gave them cover and a clear field of fire at the same time. The German divisions had to pay a heavy toll before they mastered this kind of fighting. Some of their costliest lessons they learned in the forest of Bialowieza.

On 29th June the 78th Infantry Division moved off in three columns of march—215th Infantry Regiment on the right, 195th Infantry Regiment on the left, and 238th Infantry Regiment in the rear, in echelon.

Contact was made with the enemy near the village of Popelevo. Here the last formations of General Potaturchev's scattered 4th Armoured Division, together with parts of three other divisions, brigades, and artillery detachments, had been re-formed into a new regiment, brilliantly led by Colonel Yashin. It was a case of hand-to-hand fighting—man stalking man with hand-grenade, pistol, and bayonet. The artillery was unable to intervene because friend and foe were too closely interlocked. Only the mortars were useful.

The afternoon of 29th June saw a massacre. The 3rd Battalion, 215th Infantry Regiment, succeeded in engaging the Russians in the flank and in the rear. Panic broke out. The Russians fled. Colonel Yashin lay dead by a road-block made of tree-trunks. Popelevo was again silent.

On the following day the division was more careful. The gunners pounded each patch of forest before the companies moved in. "Infantry will enter platoon by platoon!" A white Very light meant: Germans here. Red meant: Enemy attack. Green meant: Artillery fire to be moved forward. Blue meant: Enemy tanks. Yes, tanks—even in the forest the Russians employed individual tanks for infantry support.

By evening 78th Infantry Division was at last through that accursed forest of Bialowieza. The Russians had left behind 600 dead. The regiments had taken 1140 prisoners. Some 3000 Soviet troops were being pushed towards the interception line of 17th Infantry Division. In its two days of fighting in the forest of Bialowieza the 78th Infantry Division lost 114 killed and 125 wounded.

The 197th Infantry Division established its headquarters in the ancient Polish château of Bialowieza. Its regiments were instructed to clear the virgin forest of the last scattered remnants of the enemy, who were still established in several places and represented a permanent danger behind the line.

The 29th Motorized Infantry Division and the "Grossdeutschland" Infantry Regiment, who were keeping the big pocket around the Russian armies closed in the Slonim area, east of the forest, were involved on 29th June in fierce fighting against enemy forces attempting to break out. The infantry divisions of the Fourth and Ninth Armies had still not arrived to finish off the encircled Russians. True, they were hastening to the scene, in forced marches along terrible roads, covered in sweat and dust. But until they arrived the pocket had to be kept sealed by the 29th Motorized Infantry Division and Hoth's 18th Motorized Infantry Division, as well as by 19th Panzer Division. These units were itching to be relieved of their prison guard duties; they were anxious to move on, towards the east, towards their great strategic objective—Smolensk.

"We've got to strike at the root of these continuous Russian break-out attempts. We've got to ferret them out of their woods," Lieutenant-Colonel Franz, the Chief of Operations of 29th Motorized Infantry Division, suggested to his commander, Major-General von Boltenstern. The divisional commander agreed.

"Colonel Thomas to the commander!" The CO of the Thuringian 71st Infantry Regiment reported at headquarters. Maps were studied. A plan was worked out. And presently Thomas's combat group moved off into the wooded ground on the Zelvyanka sector, with parts of 10th Panzer Division, Panzerjägers (Panzer killers), two battalions of 71st Infantry Regiment, two artillery detachments, and sappers. They moved in two wedge-shaped formations. The divisional commander went along with them. Only then did they discover the kind of forces they had to deal with—considerable parts of the Soviet Fourth Army which, having rallied at Zelvyanka, were now trying to fight their way out of the pocket to the east. They intended to break through towards the Berezina. There they hoped they would be able to hold a new defensive position, the Yeremenko Line, which they had been told about in radio messages.

Numerically the German formations were greatly inferior. The Russians fought fanatically and were led by resolute officers and commissars who had not been affected by the panic which followed the first defeats. They broke through, cut off Thomas's combat group, moved their tanks against the rear of the 1st Battalion, 15th Infantry Regiment, and tried to recapture the railway-bridge to Zelva.

The divisional staff officers were lying in the infantry foxholes with carbines and machine pistols. Lieutenant-Colonel Franz commanded a hurriedly established road-block of anti-tank guns. The Russians were stopped. And at long last the German infantry divisions arrived. The 29th Motorized Infantry Division was able to move off to the north, towards new operations of decisive importance. A fortnight later the division's name would be on everybody's lips.

The Berezina, literally the "Birch River," a right-bank tributary of the Dnieper, enjoys a fame of its own in Russian history. It was here that in November 1812 Napoleon, retreating from Moscow, suffered those crushing losses which meant the final end of his Grande Armée. There is no doubt that Yeremenko too had this historic precedent in mind when, in the evening of 29th June 1941, upon assuming command of the Soviet Western Front in the Minsk area, he issued his first order. It ran: "The Berezina crossings are to be held at all costs. The Germans must be halted at the river."

When Yeremenko issued this order he was not yet aware of the full extent of the Soviet disaster on the Central Front. He supposed fighting divisions where none were left. He relied upon defences which had long been abandoned. He wanted to hold the Germans on the Berezina at a time when the marching orders of Guderian's Panzer divisions already mentioned the Dnieper. He placed his hopes in units which were already shuffling into captivity, such as General Potaturchev's 4th Armoured Division.

How Yeremenko's hopes came to naught was recounted to the author by General Nehring, commanding the German 18th Panzer Division. "In the evening of 29th June," Nehring recalled, "the spearheads of 18th Panzer Division had reached Minsk. Parts of Hoth's Panzer Group—the 20th Panzer Division—had taken the city on 28th June. The 18th Panzer Division was ordered to drive past Minsk in the south, along the motor highway, towards Borisov on the Berezina, and to form a bridgehead there. At the time the whole enterprise seemed like suicide, but it was nothing of the sort. However, that could hardly have been foreseen. The division, relying entirely upon itself, thrust some sixty miles into enemy-held territory."

Nehring moved off early on 30th June. Ahead lay excellent new roads. The tank commanders were delighted. But presently the division met Russian resistance from strongly fortified positions. The Russians fought desperately. It was clear that Yeremenko's orders had been: Hold out or die. He needed time to establish a new line of defences. Could the race against time be won? Nehring was determined to outrace Yeremenko. While the bulk of his division was engaged against the Russians he formed an advanced detachment under Major Teege—the 2nd Battalion, 18th Panzer Regiment, and with them, riding on the tanks, men of the regiment's motor-cycle battalion and parts of a reconnaissance detachment, as well as Major Teichert's artillery battalion.

By noon on 1st July Teege had reached Borisov. The Russians were taken by surprise but resisted furiously. They were officer-cadets and NCOs of the armoured forces training college in Borisov. They were crack troops. They realized the importance of the bridge over the Berezina. They defended it fanatically, but, strangely enough, did not blow it up. The German advanced detachment suffered heavy losses. Yeremenko threw into the battle whatever he could lay hands on in the Borisov area. But then the bulk of the German division came up. In the early afternoon two battalions of 52nd Rifle Regiment, supported by tanks, launched an assault on the Russian bridgehead on the western bank. 10th Company of this regiment worked its way through the Soviet defences. Sergeant Bukatschek led No. 1 Platoon. He reached the bridge. He fought down the two machine-gun posts on the ramp. He got a rifle bullet in his shoulder, but regardless he raced across the bridge with his men and captured the demolition squad on the other bank before the Soviet lieutenant could push down the plunger.

Teege's tanks and the motor-cycle troops, together with Laube's anti-aircraft battery, crossed the Berezina. The 8·8-cm. guns of the

Objective Smolensk 73

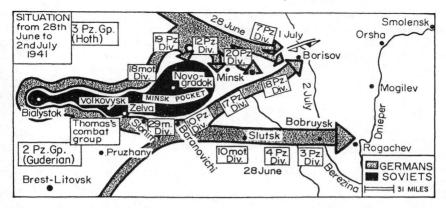

Map 3. The Bialystok-Minsk pocket. Between Bialystok and Minsk the first great battle of annihilation was fought on the Central Front. Four Soviet Armies were encircled by fast German divisions.

second battery secured the bridge against Soviet attacks. On the following morning, at first light, when Soviet crack battalions drove down the road on lorries from Borisov in order to eliminate the bridgehead, Second Lieutenant Döll with his 8·8-cm. battery blasted the column off the highroad and, at the cost of heavy losses, held the vital bridge against snipers, assault detachments, and tanks. The river, fateful since Napoleon's campaign in Russia, had been conquered. The road to the Dnieper was clear. Fifty miles farther south General Model's 3rd Panzer Division had already crossed the river at Bobruysk, and farther south still 4th Panzer Division, of General Freiherr Geyr von Schweppenburg's corps, was likewise across and driving towards Mogilev. Yeremenko had lost the round on the Berezina. The date was 2nd July 1941, the day when Alexander Rado in Geneva sent the radio signal to the Kremlin: "The object of the German operation is Moscow."

On the following day Marshal Timoshenko personally assumed supreme command of the Russian Western Front. Yeremenko became his second-in-command.

During the night of 2nd/3rd July, however, the Berezina was crossed yet again between the key points of Borisov and Bobruysk. Units of the 69th and 86th Rifle Regiments of 10th Panzer Division established a bridgehead at Berezino before daybreak and succeeded in holding it even though the wooden bridge behind them went up in flames.

On the same day—3rd July 1941, the twelfth day of the German campaign in the East—Colonel-General Halder, Chief of the German General Staff, wrote in his diary:

> Generally speaking, the enemy can now be regarded as written off in the Bialystok bend, with the exception of quite insignificant remnants. Along the front of Army Group North 12 to 15 enemy divisions can likewise be considered to have been completely wiped out. In front of Army Group South the enemy has also been battered by ceaseless heavy blows and is now largely smashed. Generally speaking, it is therefore already possible to say that the task of smashing the Soviet armies in front of the Western Dvina and Dnieper has been accomplished. It would probably be no exaggeration to say that the campaign against Russia has been won within the first fortnight. Naturally, this does not mean that it has been concluded. The vastness of the country and the stubborn resistance offered us in every possible way will keep our forces busy for many more weeks.

It is worth noting that these words were written not by Hitler but by the coolly calculating Chief of General Staff, Halder. He too was impressed by the headlong German advance and the breathtaking losses of the Red Army. To an officer thinking in Central European terms they were bound to spell the complete collapse of the enemy.

And, in fairness, what Field-Marshal von Bock, C-in-C Army Group Centre, wrote in his Order of the Day on 8th July was enough to go to anyone's head:

> The double battle of Bialystok and Minsk is over. The Army Group was engaged against four Russian Armies in the strength of about 32 infantry divisions, 8 armoured divisions, 6 motorized or mechanized brigades, and 3 cavalry divisions. Of these, 22 infantry divisions, 7 armoured divisions, 6 motorized or mechanized brigades, and 3 cavalry divisions were smashed.
> Even the formations which succeeded in evading encirclement have been weakened in their fighting strength. The enemy's casualties are exceedingly high. Counting of prisoners and booty up to yesterday has yielded the following totals: 287,704 prisoners, including several divisional and corps commanders; 2585 tanks captured or destroyed, including some super-heavy types; 1449 guns and 242 aircraft captured. To this must be added large quantities of small arms, ammunition, and vehicles of all kinds, as well as numerous stores of foodstuffs and fuel. We must now exploit the victory.

How could it possibly not be exploited?

But Stalin and his marshals saw matters differently. To them 300,000 men did not mean the earth. Russia was 46 times as big as the German Reich in its 1938 frontiers. The Soviet Union had 190,000,000 inhabitants. Some 16,000,000 men of military age could be mobilized. A huge armaments industry had been built up behind the Urals. Ten million soldiers could be called to the colours without difficulty even after the loss of Western Russia—provided the Soviets were given a little time.

Time was what the Soviet Command was fighting for in July 1941. "Gain time! Stop the eastward rush of the German tanks! Build up a

Objective Smolensk 75

line of defence whatever the cost!" That, in effect, was the order which Marshal Timoshenko gave his deputy Yeremenko.

Timoshenko realized clearly that unless the Germans, who had now crossed the Berezina, were held on the Dnieper and on the lower Western Dvina they would drive on from Borisov and Vitebsk to Smolensk. Once Smolensk had fallen, Moscow would be no more than 230 miles behind the fighting line. If Moscow also was lost, then the Soviet Union would be deprived of its political, spiritual, and economic heart. Would its separate parts continue in existence? Would they obey? Would they obey a Central Government in some remote provincial metropolis? Maybe they would. And maybe they would not. The fate of the Soviet Union would clearly be decided before Moscow. Victory or defeat would be determined outside the gates of the Soviet capital. Stalin realized this and acted accordingly.

There were surprised faces at 18th Panzer Division headquarters in the Borisov bridgehead when, on 3rd July, a signal was received from the division's air unit: "Strong enemy armoured columns with at least 100 heavy tanks advancing along both sides of Borisov–Orsha–Smolensk road in the area of Orsha. Among them very heavy, hitherto unobserved models."

"Where do they come from?" General Nehring asked in surprise. "These Russians seem to have nine lives."

It was, in fact, the 1st Moscow Motorized Rifle Division under Major-General I. G. Kreyzer, whom Yeremenko had sent into action against Guderian's armoured spearhead. It was a crack unit, the pride of the Soviet High Command.

The aerial reconnaissance report proved to have been entirely accurate. For in his memoirs Yeremenko writes: "The division had at its disposal about 100 tanks, including some T-34s not previously employed on the Central Front."

T-34s! Now it was the turn of the Central Front to experience that wonder-weapon which had made its appearance on the southern sector during the first forty-eight hours of the war, spreading terror and fear wherever it moved.

Six miles east of Borisov, near the village of Lipki, Nehring's and Kreyzer's armoured spearheads made contact. The 18th Panzer Division, from Chemnitz, to-day called Karl-Marx-Stadt, clashed with a crack unit from the centre of Karl Marx's world revolution.

When it first hove into sight the T-34 struck a good deal of terror among the German armoured spearheads and Panzerjägers. But abreast of it, at a distance of about 100 feet, came an even bigger monster—a KV-2, weighing 52 tons. The light T-26 and BT tanks between the two giants were soon set on fire by the German Mark IIIs. But their 5-cm. shells made no impression whatever on the two giants. The first Mark III received a direct hit and went up in flames. The other German tanks scuttled out of the way. The two Soviet monsters continued to advance.

Three German Mark IVs, nicknamed "the stubs," hastened to the scene, with their 7·5-cm. short-barrel cannon. But the heaviest German tanks then in existence were still some three tons lighter than the T-34, and the range of their guns was considerably less. However, the German commanders soon discovered that the crew of the T-34 were unsure of themselves and very slow in their fire. The German tanks underran its fire, weaved round it, and dodged its shells. They got the giant between them. They shot up its tracks. The Soviet crew got out and tried to escape, but ran straight into a burst of machine-gun fire from a Mark III.

Meanwhile the huge 52-ton KV-2 with its 15·2-cm. cannon was still shooting it out with two German Mark IIIs. The German shells penetrated into the Russian tank's plating as far as their driving bands, and then got stuck. Nevertheless the Russians suddenly abandoned their vehicle—probably because of engine trouble.

This incident reveals the cardinal mistake of the Russians. They employed their T-34s and super-heavy KVs not in formation, but individually among light and medium tanks, and as support for the infantry. Those were very outdated tank tactics. The result was that these vastly superior Soviet tanks were smashed up one by one by the German tank companies, in spite of the terror they originally struck among them. In this way General Kreyzer's counter-attack near Lipki collapsed.

Open-mouthed, Nehring's men inspected the Soviet armoured giants. The general himself stood thoughtfully in front of a KV, counting the tank shells lodged in its plating—11 hits and not a single penetration.

Colonel-General Guderian also saw his first T-34 on the Moscow highway, west of Borisov. Three of these giants had got stuck in marshy ground and had thus fallen undamaged into German hands. Guderian was full of admiration for the tank's excellent and purposeful design, and was particularly impressed by its powerful cannon.

The 1st Moscow Motorized Rifle Division continued to resist the German 18th Panzer Division with all the strength it possessed. The T-34 and the KV continued to be their most dangerous weapons. The German infantryman was faced with his first ordeal of the war in the East. This emerges clearly from the war diary of 101st Rifle Regiment, which contains the following accounts of engagements by its 2nd Battalion:

5th July. Russian tank attack on the near side of Tolochino. One of their tanks got stuck in the forest. Sergeant Findeisen with men of 6th and 7th Companies finished it off with close-combat weapons. Ten T-26s appeared in front of our lines, on the motor highway. Second

Lieutenant Isenbeck, leading a Panzerjäger platoon, blocked the road with a 5-cm. anti-tank gun. The Russian tanks were advancing well spaced out. Isenbeck knelt by his gun, firing shell after shell. The leading T-26 was on fire. The second slewed into the roadside ditch. The third one, its track shot to pieces, stood motionless by the side of the road, a sitting duck. Change of target. Fire! Five more tanks were knocked out. The ninth was hit just below its turret at 30 yards' range and was now blazing like a torch. The tenth, behind it, was able to turn and get away by zigzagging wildly.

7th July. Renewed Russian tank attack. Second Lieutenant Isenbeck's leading anti-tank gun was hit. The crew were killed or wounded. A 52-ton tank steam-rollered our anti-tank barrier. But presently it got stuck. Even so, it continued to paste the company's positions with its heavy gun.

Second Lieutenant Kreuter, leading the headquarters company of 101st Rifle Regiment, worked his way up to the colossus with a dozen men. A machine-gun, firing special hard-nosed anti-tank bullets, gave them cover. But the missiles bounced off like so many peas.

Sergeant Weber leaped to his feet. Corporal Kühne followed suit. They ran towards the Russian tank regardless of its machine-gun fire. Mud and earth spurted up in front of them. But they managed to get into the dead angle of the machine-gun. They had tied some hand-grenades together to make heavy explosive charges. Weber threw first, then Kühne. They flung themselves down. A flash, a burst, the clatter of fragments. The upper part of Kühne's arm was torn open. But the turret mechanism of the KV had been damaged. It could no longer traverse its gun.

Like hunters stalking some prehistoric beast, Kreuter's men lay on the ground around the giant with their machine-pistols and machine-guns. The second lieutenant jumped up on to the steel box. He ducked under the gun-barrel of the massive turret.

"Hand-grenade!" he called. Private Jedermann lobbed a stick hand-grenade up to him. The lieutenant caught it, pulled the pin, and thrust the grenade down the fat barrel of the cannon. He leapt down from the tank and rolled over. He was only just in time. Like a clap of thunder came the burst of the hand-grenade, and a moment later that of the shell in the breech. The explosion must have blown the breech-block into the turret, for a hatch was flung open. Corporal Klein, with great presence of mind, and even greater skill, chucked in an explosive charge at 25 feet range. A blinding flash and an explosion. The heavy turret was blown 15 feet into the field. For hours the giant blazed like a torch. It was still smouldering when, in the evening twilight, Captain Pepper, the battalion commander, came round the company's positions with Second Lieutenant Krauss.

"What a crate!" said Pepper. "Just look at..." He did not finish. A

Russian automatic rifle cracked twice. Pepper and Krauss flung themselves under cover. This time they were lucky. But on the following day the battalion commander was picked off on his way to regimental headquarters. The bullets came from a Russian tree-top sniper. Pepper was killed instantly, and Second Lieutenant Krauss, who was again accompanying him, was gravely wounded and died in hospital a few hours later. The sniper, a slightly wounded Russian who had hidden in a tree, survived the captain by only a quarter of an hour. He refused to surrender.

Thus far the war diary of 101st Rifle Regiment.

On the same day, 8th July 1941, the 17th Panzer Division also had its first encounter with a T-34—farther north, in the area of Senno, in the historic strip of land between the Western Dvina and the Dnieper. Yeremenko had brought up fresh units of the Soviet Twentieth Army and moved them into the strategically important strip of land between Orsha and Vitebsk in order to bar the road to Smolensk from this side also, the road which Hoth's and Guderian's Panzer divisions were trying to force.

At dawn the leading regiment of 17th Panzer Division moved into action. They went through waving grain crops, across potato-fields, and over shrub-grown heath. Towards 1100 hours Second Lieutenant von Ziegler's platoon made contact with the enemy. The Russians were in well-camouflaged positions and opened fire at close range. At the first shots the three battalions of 39th Panzer Regiment fanned out on a broad front. Troops of anti-tank artillery raced up to protect their flanks. A tank battle began, a battle which earned a place in military history—the battle of Senno. Fierce fighting raged from 1100 till nightfall. The Russians operated with considerable skill. They tried to take the Germans in the flank or in the rear. The sun was burning down upon them. The vast battlefield was dotted with blazing and smouldering tanks, German and Russian.

At 1700 hours the German tanks received a signal over their radios: "Ammunition must be used sparingly." At the same moment Radio Operator Westphal in his tank heard his commander's excited voice: "Heavy enemy tank! Turret 10 o'clock. Armour-piercing shell. Fire!"

"Direct hit!" Sergeant Sarge called out. But the Russian did not even seem to feel the shell. He simply drove on. He took no notice of it whatever. Two, three, and then four tanks of 9th Company were weaving around the Russian at 800–1000 yards' distance, firing. Nothing happened. Then he stopped. His turret swung round. With a bright flash his gun fired. A fountain of dirt shot up 40 yards in front of Sergeant Hornbogen's tank of 7th Company. Hornbogen swung out of the line of fire. The Russian continued to advance along a farm track. A German 3·7-cm. anti-tank gun was in position there.

"Fire!"

But the giant just seemed to shrug the shells off. Its broad tracks were full of tufts of grass and crushed haulms of grain. Its engine note rose. The Russian driver was engaging his top gear. That was not such an easy operation with their sturdily built vehicles. Nearly every driver therefore had a hammer lying by his feet; if the gear would not engage, striking the gear-lever with the hammer usually did the trick. A case of Soviet improvisation. Nevertheless, these things moved all right. This one was making straight for the anti-tank gun. The gunners fired furiously. Only twenty yards to go. Then ten, and then five.

Now it was on top of them. The men leaped out of its way, scattering. Like some huge monster the tank went straight over the gun. It then bore slightly to the right and drove on, through the German lines, towards the heavy artillery positions in the rear. Its journey did not end until nine miles behind the main fighting line, when it got stuck in marshy ground a short way in front of the German gun positions. A 10-cm. long-barrel gun of the divisional artillery finished it off.

The tank battle continued into the hours of darkness. Eerily the blazing tanks lay in the cornfield. Tank ammunition exploded, jerricans full of fuel blew up. Medical orderlies darted across the scene, looking for the screaming wounded and covering the dead with blankets or some tent canvas. The crew of the smouldering tank No. 925 laboriously pulled out their heavy skipper—Sergeant Sarge. He was dead. Many were dead who 17 days previously had stood in rank in the forest clearing near Pratulin, listening to the Fuehrer's orders. Many were wounded. But the 17th Panzer Division commanded the battlefield. And he who commands the battlefield is the victor.

There are two reasons why the T-34 did not become a decisive weapon in the summer of 1941. One was the wrong Soviet tank tactics, their practice of using the T-34 in driblets, in conjunction with lighter units or for infantry support, instead of—in line with German thinking—using them in bulk at selected points, tearing surprise gaps into the enemy's front, wrecking his rearward communications, and driving deep into his hinterland. The Russians disregarded this fundamental rule of modern tank warfare, a rule summed up by Guderian in a phrase valid to this day: "Not driblets but mass."

The second mistake of the Russians was in their combat technique. Here the T-34 suffered from one crucial weakness. Its crew of four—driver, gunner, gun-loader, and radio operator—lacked the fifth man, the commander. In a T-34 the gunner at the same time commanded the tank. This dual function—working the gun and looking out in between—interfered with efficient and rapid fire. By the time the T-34 got a shell out, a German Mark IV had fired three. In this way the German tanks underran the longer range of the T-34s and, in spite of the Russian tanks' massive 4·5-cm. plating, managed to score hits against their tracks and other 'soft spots.' Besides, each Soviet armoured unit had only one radio transmitter—in the company commander's tank. That made them far less mobile in action than their German opponents.

Even so, the T-34 remained a dangerous and much-feared weapon throughout the war. The effect which its mass employment might have had during the first few weeks of the campaign is difficult to imagine. The impression made on the Soviet infantry by the mass employment of German tanks, on the other hand, is described most impressively and frankly by Guderian's opponent, General Yeremenko. In his memoirs he says:

> The Germans attacked with large armoured formations, often with infantrymen riding on the tanks. Our infantry were not prepared for that. At the shout "Enemy tanks!" our companies, battalions, and even entire regiments scuttled to and fro, seeking cover behind anti-tank-gun or artillery positions, causing havoc to the whole combat order, and bunching up near anti-tank-gun positions. They lost their ability to manœuvre, their combat readiness was diminished, and all operational control, contact, and co-operation were rendered impossible.

Yeremenko understood clearly what made the German armour superior to his own. And he drew the necessary conclusions. He issued strict orders that the German tanks must be engaged. His recipe was concentrated artillery-fire, attack by aircraft with bombs and cannon, and, above all, engagement at close range with hand-grenades and with a new close-combat weapon which has to this day kept its German Army nickname—the Molotov cocktail. This weapon, still a great favourite in domestic revolutions, has an interesting history.

By chance Yeremenko learned that in Gomel there was a store of a highly inflammable liquid called KS—a petrol-and-phosphorus mixture with which the Red Army had experimented before the war, probably with a view to setting enemy stores and important installations on fire quickly. Yeremenko, ingenious as ever, immediately ordered 10,000 bottles of the liquid to be delivered to his sector of the front, and issued them to combat units for use against enemy tanks. The Molotov cocktail was no wonder weapon, but a piece of improvisation, a desperate makeshift. But quite often it was highly effective. The liquid burst into flames the moment it came into contact with air. A second bottle, filled with petrol, added to the effect. When only petrol was available an improvised fuse tied to the bottle and lit before throwing did the trick. Provided the bottles burst high up on a tank or on its side-wall, the burning mixture would run into the combat quarters or into the engine, setting the oil and fuel on fire immediately. These large boxes of steel and tin burned surprisingly readily—probably because the metal was usually covered with a film of oil, grease, and petrol.

Needless to say, however, tank armies could not be stopped with

Objective Smolensk 81

petrol bottles, especially once the German tanks—whose strength had always consisted in their close co-operation with the infantry—paid increased attention to enemy troops trying to engage them at close range. If the Russians wanted to halt the Germans, to prevent them from driving via Smolensk to Moscow, they would have to bring up large numbers of men and a lot of artillery.

The Soviet High Command therefore switched parts of its Nineteenth Army from Southern Russia to the Vitebsk area. The Russian regiments leaped out of their goods trucks and went straight into battle against Hoth's 7th and 12th Panzer Divisions. Yeremenko realized that he was slowly sacrificing a considerable force of six infantry divisions and a motorized corps. But what else could he do? He hoped that in this way he would at least delay the German spearheads. Time was what he needed.

But Yeremenko's hopes were in vain. The reconnaissance detachment of 7th Panzer Division captured a Soviet officer from an anti-aircraft unit. In his possession were found orders, dated 8th July, which revealed Yeremenko's plan to detrain divisions of the Nineteenth Army north of Vitebsk and employ them on the narrow strip of land between the rivers. Colonel-General Hoth took immediate counter-measures. He ordered Lieutenant-General Stumpff's 20th Panzer Division, which on 7th July had crossed over to the northern bank of the Western Dvina at Ulla, to advance on 9th July along that bank of the river in the direction of Vitebsk. On the neck of land south of the Western Dvina the 7th and 12th Panzer Divisions were meanwhile tying down Yeremenko's forces. Stumpff's tanks, together with the swiftly brought-up 20th Motorized Infantry Division, under Major-General Zorn, drove straight into the Russian rear and caused chaos to the enemy's detraining operations.

It was the early morning of 10th July—the nineteenth day of the campaign. It was to be a day of dramatic decisions. The German Blitzkrieg was still in full swing. Pskov, south of Lake Peipus, had fallen. General Reinhardt's XLI Panzer Corps had pierced the Stalin Line with its 1st Panzer Division and parts of 6th Panzer Division, and on 4th July, after some fierce tank fighting, taken Ostrov. Continuing its swift advance, the northern Panzer corps of Colonel-General Hoepner's Fourth Panzer Group, with 36th Motorized Infantry Division and parts of 1st Panzer Division, four days later reached the vital turning-point on the way to Leningrad. Hoepner ordered the troops to wheel north-east towards the city. Perhaps Leningrad would fall even before Smolensk. And if it fell Russia's armed might in the Baltic would collapse. Moscow's northern flank would lie exposed. Then the race could start as to who would first drive into the Kremlin—Hoepner, Hoth, or Guderian? Things were looking hopeful. Maybe Hoepner would repeat his 1939 triumph of Warsaw, when the 1st and 4th Panzer Divisions of his XVI Motorized Corps stood west and south of the Polish capital within eight days of the start of operations.

Two hundred miles south of Pskov was Vitebsk, an important railway junction on the upper Western Dvina, the gateway to Smolensk. And Vitebsk fell. The 20th Panzer Division took it by storm on 10th July. Fanatical Komsomol members had set fire to the town. It was blazing. But Hoth's Panzer divisions needed no quarters for the night. They simply drove past the burning town, forward, farther to the east, into the rear of Smolensk.

On Guderian's sector, too, where the spearheads had crossed the Berezina at Bobruysk and Borisov and were now making for the Dnieper, the most important decision of the 1941 campaign was made on 10th July.

"What's your opinion, Liebenstein?" Guderian asked his Chief of Staff every evening when he returned from the forward lines to his headquarters. "Shall we continue our thrust and force the Dnieper with armour alone, or do we have to wait for the infantry divisions to catch up with us?" It was a question that had been discussed for days at Second Panzer Group's headquarters. And every time the same argument developed. Infantry were better suited than tank regiments for forcing river crossings. On the other hand, a fortnight would pass before the infantry arrived. And what use would the Russians make of a fortnight spent idly by the Germans on the Berezina or in front of the Dnieper? Lieutenant-Colonel Bayerlein, the chief of operations, listed through the Intelligence officer's file on enemy movements. The evidence was there: aerial reconnaissance reported strong motorized units moving towards the Dnieper and the formation of a new Soviet concentration to the north-east of Gomel.

The establishment of these new Russian concentrations somewhat damped the optimism of the German High Command as voiced by Colonel-General Halder on 3rd July. Unless the Russians were to be allowed to man the Dnieper line in strength and establish defensive positions speedy action was needed.

In these arguments with his superiors Guderian came out strongly in favour of continuing operations on the central sector, and his staff were unanimously behind him. To-day we know that Guderian's anxieties were justified. According to Yeremenko's memoirs, as well as the most recent Soviet military publications, Timoshenko, acting in accordance with a decision by the State Defence Committee, had reorganized what used to be the Western Front and personally taken command of the newly formed Army Group Western Sector. In the north and south, the "fronts," the defence zones corresponding to the old Military Districts, were reorganized into Army Groups—the north-western sector

Objective Smolensk 83

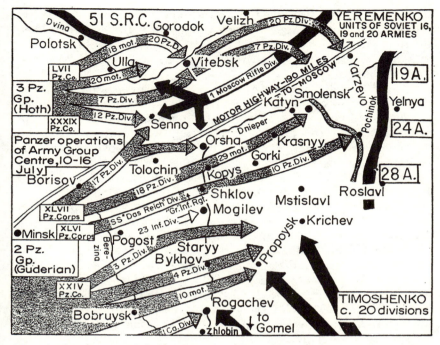

Map 4. The crossing of the Dnieper on 10th/11th July 1941 and the resulting capture of Smolensk were the first decisive operation of the campaign on the Central Front.

under Marshal Voroshilov, and the south-western sector under Marshal Budennyy.

From 10th July onward Timoshenko collected division after division along the Dnieper. On 11th July his Army Group again comprised 31 infantry divisions, seven armoured divisions, and four motorized divisions. To this must be added the remnants of the Fourth Army—those which had escaped from the pocket of Minsk—and parts of the Sixteenth Army, which was being switched from the south to the Central Front. Altogether, 42 combat-ready Soviet divisions were lining up on the upper Dnieper.

The following story used to be told about Guderian and the campaign in France. When the attack was being planned his argument that the success of the armoured formations depended on rapid and ruthless penetration right into the rear of the enemy lines had not been shared by his colleagues. There had been much argument with Colonel-Generals von Rundstedt and Halder. When Guderian had pierced the Maginot Line and wanted to drive through to the Channel coast with his XIX Panzer Corps, in order to cut off the British and

French forces, he was slowed down time and again after wheeling to the west. The headquarters of Army Group A and the Fuehrer's headquarters alike were haunted by the spectre of exposed flanks. That was why they wanted to halt Guderian's rapid advance on 15th May and 17th May 1940.

"You're throwing away our victory," Guderian had pleaded with Colonel-General von Kleist, then his C-in-C. With clever cunning Guderian had time and again managed to get his views accepted, but at Dunkirk he failed. At Dunkirk the victory was really thrown away.

"You're throwing away our victory," Guderian had been shouting down the telephone ever since the beginning of July 1941, whenever he was instructed by Field-Marshal Hans Günther von Kluge, C-in-C Fourth Army, to await the infantry on the Dnieper.

On 9th July Field-Marshal von Kluge personally appeared at Guderian's headquarters in Tolochino. A heated discussion began. "Clever Hans"—"der kluge Hans," as the C-in-C was called in a pun on his name—and "Fast Heinz"—as Guderian was known to his troops—clashed head on. Guderian wanted to cross the Dnieper. Kluge said no. Guderian passionately defended his plan. Kluge remained cool. Thereupon Guderian resorted to a white lie. He maintained that most of his armour was already deployed along the Dnieper bank for an attack across the river—a disposition which could not be kept up indefinitely without risk.

"Moreover, I am convinced of the success of the operation," Guderian implored Kluge. "And if we strike quickly at Moscow I believe that this campaign can be decided before the end of this year."

So much resolution and confidence impressed even the unemotional Kluge. "Your operations invariably hang by a silken thread," he said. But he let Guderian have his way.

The Colonel-General nodded towards his officers. "We're off, gentlemen. We're crossing. First thing to-morrow." To-morrow was 10th July.

Fortune favours the bold. That applied also to Guderian. The development of the action proved him right. His advanced detachments had discovered that the Russians had fortified and were strongly holding the principal Dnieper crossings at Rogachev, Mogilev, and Orsha. Attempts to seize these crossings by surprise had been costly failures. Reconnaissance detachments of the Panzer Corps, however, quickly discovered the soft spots between the enemy's strongpoints on the western bank of the Dnieper. They found these soft spots at Staryy Bykhov, Shklov, and Kopys.

Staryy Bykhov was in the south, in the area of XXIV Panzer Corps; Shklov was in the centre, in the area of XLVI Panzer Corps; and in the area of XLVII Corps in the north there was Kopys. They were miserable dumps, without bridges, and no one had ever heard of them. The

Russians never dreamed that the Germans would attack at these points. But the great secret in war is always to hit the enemy where he expects it least.

In fact, the Dnieper was crossed at all three points without great losses on 10th and 11th July. Above and below Staryy Bykhov the 3rd and 4th Panzer Divisions got across at their first attempt. The 1st Battalion, 3rd Rifle Regiment, as well as the 10th Motorized Infantry Division, crossed the river at Soborovo, secured the bridgehead, and repelled all counter-attacks. At Staryy Bykhov the 2nd Company, Motorcycle Battalion 34, under Captain Rode, forced a crossing and in this way covered the first bridgehead. Engineers Battalion 79 instantly began building an emergency bridge which was ready for use during the night of 10th/11th July.

At Kopys the crossing did not at first succeed. The 29th Motorized Infantry Division had to fight hard to get across the river in the face of enemy air-attacks and artillery-fire. At 0515 hours on 11th July Lieutenant-Colonel Hecker's engineer companies crossed the river in assault boats under cover of self-propelled guns, and ferried the infantry to the other bank. Within 45 minutes four assault battalions had gained the far bank. They underran the enemy fire and dug in.

At Shklov, where 10th Panzer Division crossed, the "Grossdeutschland" Infantry Regiment clashed with the "Stalin Scholars," a crack unit of officer cadets. Lieutenant Hänert's machine-gun company of 1st Battalion, "Grossdeutschland" Infantry Regiment, eventually gained the regiment the elbow-room it needed by driving the Soviets back into the woods. The sappers built their bridge in record time. The heavy weapons were brought across.

As for the strongly fortified towns of Orsha, Mogilev, and Rogachev, Guderian's divisions simply bypassed them and drove farther east. Their objective was Smolensk.

Guderian was pressed for time, for Marshal Timoshenko had already built up a strong concentration of 20 divisions in the south, in the Gomel area. He tried to attack Guderian's units from the flank and thus to save Smolensk. The extremely heavy defensive fighting in which the German units were engaged testified to the seriousness of the situation. However, Timoshenko's plan miscarried. The credit for this must go above all to 1st Cavalry Division, under General Feldt, which hurled itself against Timoshenko's attacks. Together with 10th Motorized Infantry Division and parts of 4th Panzer Division, this cavalry division covered the flank of Second Panzer Group.

This crucial action by the 1st Cavalry Division deserves special mention. The only major German cavalry unit in the Second World War until 1944, Major-General Feldt's cavalry brigades operated along the fringe of the impassable Pripet Marshes, on ground not negotiable by the tanks. The roads were no more than bridle-paths, and shrubs and moorland provided an ideal terrain for enemy ambushes and traps. The 1st Cavalry Division acquitted itself exceedingly well on this ground, protected Guderian's flank, and efficiently maintained contact with the units of Rundstedt's Army Group operating south of the great marshes. It was the successful repulse of all attacks against his flank that enabled Guderian to strike towards Smolensk.

Blow now followed blow. In the evening of 15th July the 7th Panzer Division, which was part of Colonel-General Hoth's Third Panzer Group, drove past Smolensk to the north, with strong Luftwaffe support, and cut both the motor highway and the railway-line from Smolensk to Moscow. The town was thus cut off from supplies and reinforcements, and a new pocket had been formed with 15 Soviet divisions in it.

The Soviet High Command wanted to hold Smolensk at all costs. Smolensk was something rather like Stalingrad—a symbol as well as a vital strategic position. Smolensk was the key to the gates of Moscow, a fortress on the upper Dnieper, one of the most ancient Russian settlements. It was here that, on 16th and 17th August 1812, Napoleon won the victory which enabled him to march to Moscow. It was here that exactly three months later, on 16th and 17th November 1812, the Tsarist General Kutuzov defeated France's Grande Armée. This explains the emotional intensity with which Smolensk was defended. The men of General von Boltenstern's 29th Motorized Infantry Division were to encounter it very soon.

Those were days which the 71st and 15th Regiments, the 29th Artillery Regiment, the engineers, and the Motorcycle Battalion, above all 2nd Company, under Second Lieutenant Henz, who throughout six days hung on to the Dnieper bridge east of the town after it had been taken by a surprise attack, will never forget.

According to General Yeremenko's report the commandant of the Smolensk garrison had been ordered to practise 'total defence.' The streets were barricaded, and concrete pill-boxes had been set up. Every house, every cellar was a centre of resistance. Workers and clerks had been armed and, with units of the State police and militia, had been organized into street-fighting detachments. They had orders to hold their blocks of buildings or die. The military backbone of the city's defence was provided by the rifle regiments of the Soviet XXXIV Rifle Corps.

Nevertheless Smolensk fell. What was more, it fell quickly. The defence was no match for the bold and cunning assault by the Thuringian 71st Infantry Regiment. In the morning of 15th July, at 0700 hours, Colonel Thomas went into action with his regiment. He circumvented the enemy lines by taking a farm track 9 miles southwest of the town. He attacked from the south. At 1100 his 2nd Battalion stormed the heavy Russian batteries on the hills of Konyukhovo.

Objective Smolensk 87

Prisoners reported that the southern exit from the town was also heavily fortified. Thomas therefore wheeled his regiment once more to the right and attacked the town from the south-east. When the defenders spotted the German spearheads at 1700 hours it was too late. By nightfall assault detachments of the 71st Regiment were already in the streets of the southern suburbs.

On the following morning at 0400 the main attack was launched jointly with 15th Infantry Regiment. Heavy artillery, 8·8-cm. AA guns, mortars, self-propelled guns, and flame-throwing tanks cleared the way for the infantry. In the northern part of the town, in the industrial suburbs, police and workers' militia units resisted stubbornly. Every house and every cellar had to be taken separately by pistol, handgrenade, and bayonet. Towards 2000 hours on 16th July the troops reached the northern edge of the town. Smolensk was in German hands.

Thus, on the twenty-fifth day of the campaign, the first strategic objective of Operation Barbarossa had been reached: the first troops of Army Group Centre were in the area Yarzevo–Smolensk–Yelnya–Roslavl. They had covered 440 miles. It was another 220 miles to Moscow.

Only at Mogilev, now far behind the German lines, did fierce fighting continue. This regional centre of the Belorussian Soviet Socialist Republic, a town on the upper Dnieper, with 100,000 inhabitants and a large railway repair-shop, the centre of the West Russian silk industry and the ancient see of the Archbishop of all Catholics in the Russian Empire, was being stubbornly defended by three divisions of the Soviet Thirteenth Army, under Lieutenant-General Gerasimenko.

On 20th July the town west of the river was surrounded by four German divisions, forming VII Corps.

At 1400 hours on the same day Major-General Hellmich's 23rd Infantry Division from Berlin–Brandenburg attacked with two regiments. The 9th Infantry Regiment from Potsdam, successor to the traditions of the old Potsdam Foot Guard Regiments, succeeded in crossing the river, but was presently pinned down in a small bridgehead. The 68th Infantry Regiment was unable to break through the Soviet defences, and 67th Infantry Regiment fared no better the next day.

As the frontal attack had got stuck near the edge of the town, Hellmich attempted to strike at the bridge linking Mogilev with Lupolovo from the south-east—in an upstream direction. He succeeded. In a hard-fought night engagement 9th Infantry Regiment managed to dislodge the skilfully dug-in enemy.

But the German losses were heavy. The 11th Company, 67th Infantry Regiment, under Lieutenant Schrottke, was smashed up. In an orchard it had come under enemy fire from the flank. All its officers were killed. The company lost two-thirds of its combat strength. Meanwhile, on the western side of the Dnieper, Second Lieutenant Brandt, with 10th Company, 67th Infantry Regiment, worked his way right up to the road-bridge under cover of the river-bank. Dodging between Russian vehicles, his men raced over the bridge and established contact with 9th Infantry Regiment pinned down on the eastern bank.

Brandt held the bridge and the bridgehead against furious Soviet attacks, against sudden sharp artillery bombardments, and against the more dangerous snipers, who would pick off any man who so much as put his head out of cover. When Major Hannig stormed into the eastern part of the town with the 1st Battalion, 9th Infantry Regiment, the attack ran into Soviet machine-gun fire. The major fell on the bridge, seriously wounded. He ordered his men to press on. Snipers finished him off.

In the morning of 26th July the Russians, under cover of mist lying over the Dnieper valley, succeeded in blowing up the 200-yard-long wooden bridge into the eastern part of the town and wrecking part of it completely. In this way the Soviet units literally burned their bridges. They were holding out in lost positions. They fought to their last round. Eventually, caught in the stranglehold of 78th Infantry Division, 15th Infantry Division, 23rd Infantry Division, and 7th Infantry Division, the defenders ran out of breath. Some of them attempted to break out to the west in lorries, but they were shot up.

The wooden bridge was quickly repaired, and 23rd Infantry Division crossed to the east. The 15th Infantry Division occupied Mogilev. A strange-smelling brown liquid was flowing down the main streets: the Russians had shot up the huge vats of a big brewery. Streams of beer were running into the Dnieper. It was not to be enjoyed by the conquerors.

The 23rd Infantry Division and 15th Infantry Division took 12,000 prisoners. There were surprisingly few officers among them. The officers had been killed or had fought their way out. The losses of 23rd Infantry Division alone totalled 264 killed, 83 missing, and 1088 wounded. It was a heavy price to pay for a town far behind the front line.

4. Moscow or Kiev?

*Inferno in the Yelnya bend—A visit from the Mauerwald—
Hitler does not want to make for Moscow—Guderian flies
to see Hitler—Dramatic wrangling at Hitler's headquarters—"My generals do not understand wartime
economics."*

NO general, no officer, no rank-and-file trooper on the Eastern Front had any doubts about the further course of operations after Smolensk, or about the next objective. Moscow, of course—Moscow, the heart and brain of the Soviet empire. Anyone looking at a pre-war map of Russia will find that all roads lead to Moscow. The intellectual and political metropolis was at the same time the main traffic junction, the heart of the Red empire. If this heart was stabbed it seemed reasonable to suppose that the vast country would collapse. That was how Field-Marshal von Brauchitsch, Commander-in-Chief of the Army, argued. His opinion was shared by Halder. It was also shared by Guderian, Hoth, Bock, and all the other commanders-in-chief on the Eastern Front. They all agreed with Clausewitz, the father of modern strategy, who had described Napoleon's Moscow campaign, in spite of his defeat in Russia, as logical and correct. The objective in a war is the enemy country, its capital, its seat of political power. However, Clausewitz points out, "the gigantic Russian empire is not a country which can be kept formally conquered—that is to say, occupied. A profound upheaval, extending right into the heart of the state, was needed. Only by dealing a vigorous blow at Moscow itself could Bonaparte hope..." Indeed, only thus could he hope to shake the Russian empire, to precipitate the country into internal disorder, to arouse discord, and to sweep away the regime. The reasons for Napoleon's failure were his inadequate forces, the strategy of deliberate withdrawal successfully practised by the Russians, and the firm, unshakable ties between people and tsar.

The German generals had closely studied their Clausewitz. Was not everything working out in accordance with his precepts? The Russians had not withdrawn into their vast hinterland. They had stood and fought. The German forces had proved superior to them. The Russian people appeared to hate Bolshevism, and in many places in Western Russia the invaders had been hailed as liberators. What could possibly go wrong? Nothing. Well, then, on to Moscow.

But Hitler was reluctant to proclaim Moscow as the strategic objective of the second phase of his campaign. Suddenly he shied away from Stalin's capital. Was he afraid he might suffer Napoleon's fate? Did he distrust the traditional strategic concepts? Or did he fail to understand Moscow and Russia?

Whatever the reasons—he did not want to move against Moscow. And when at Smolensk all the preparations had been made for the thrust at Russia's heart, when the great victory seemed within arm's reach, when all the world was waiting for the order, "Panzers forward! Destination Kremlin!" Hitler suddenly scotched these plans. Flabbergasted, the generals at Army High Command and at Army Group Centre on 22nd August, after five weeks of waiting and five weeks of tug-of-war behind the scenes, read Hitler's order dated 21st August: "The most important objective to be achieved before the onset of winter is not the capture of Moscow but the seizure of the Crimea...."

Towards midnight on 22nd August the telephone rang at Second Panzer Group headquarters in Prudki. A call from Borisov: Guderian was wanted by Army Group headquarters. Field-Marshal von Bock himself was on the line: "Will you please come over here to-morrow morning, Guderian? We're expecting a visit from the Mauerwald," said the Field-Marshal.

Guderian thought quickly. A top-level visit? Had the die been cast? Was the green light for Moscow to be given at last? But Guderian sensed at once that Bock was not in a good mood. He therefore asked briskly, "At what time do you wish me to report to you, Herr Feldmarschall?" "Let's say 10 o'clock," Bock replied, and rang off.

A visit from the Mauerwald. That was the name of the forest in East Prussia in the immediate neighbourhood of the Fuehrer's headquarters, where the Commander-in-Chief of the Army and the Chief of General Staff had their wartime headquarters. Or would Hitler come in person?

Guderian inquired if his chief of staff and his chief of operations were still awake. Two minutes later he was sitting at the map-table in the bus, together with von Liebenstein and Bayerlein. Entered on the big situation map were all the many engagements of the past few weeks—black and red arrows, little flags and numbers, continuous and dotted lines, arcs, and those even stranger shapes, the pockets. It was all neatly drawn, yet it stood for blood and fear and death. But the cost was not entered on the map. There was nothing to show that a great many men had had to die so that this arrow could be drawn across the village of Kruglovka.

Guderian and his staff had been at Prudki, west of Pochinok, for the last four weeks. The German motorized divisions had taken the notorious Desna bend with the little town of Yelnya in the middle of July. Since then they had had only one thought—Moscow. They had reached their jumping-off position—even though they were panting for breath and with tank regiments greatly shrunk and supply columns decimated. But they had achieved their task according to plan. Now

Moscow or Kiev? 91

for a short halt, organization of a new supply base—and off again for the last push of this campaign, into the very heart of the Soviet Union. That was the order they were all waiting for.

On 4th August Guderian and Hoth had had an interview with Hitler —also at Bock's headquarters in Borisov. They had reported to him that the Panzer divisions would be ready to move off again for their attack on Moscow between 15th and 20th August. Guderian had added: "My Fuehrer, we shall take it." But Hitler had shown a strange reserve. He left no doubt about the fact that he had different ideas. He wanted to make for Leningrad first. And perhaps also for the Ukraine. The generals had listened in amazement. They had shaken their heads. They had reacted coldly. Hitler had sensed their opposition and had left the question open. No decision had been taken. He had hesitated ever since. And meanwhile the generals in the field were hoping that he might after all decide to strike at Moscow. In fact, they had made their preparations for the offensive quite deliberately. Since the beginning of August infantry divisions of General Geyer's IX Corps—the 137th Infantry Division and the 263rd Infantry Division—had been in the line. During the night of 18th/19th August they had relieved armoured and motorized units. All was ready for the start. Staying put and defending the line merely meant losses.

"How far is it to Moscow from the most forward lines of 292nd Infantry Division in the Yelnya bend?" Guderian asked. Lieutenant-Colonel Bayerlein did not have to work it out. "One hundred and eighty-five miles to the outskirts of the city," he answered promptly.

One hundred and eighty-five miles. Guderian glanced at his situation map. Like a springboard the Yelnya bend projected from the front line. Right at its tip was the so-called "Graveyard Corner." There, for the past few weeks, the fighting had been more bitter than at any other point on the Eastern Front.

This is borne out by a Corps Order of the Day issued by XLVI Panzer Corps headquarters on 10th August 1941, and read out in all companies:

> After a heavy defensive engagement on the north-eastern front of Yelnya Unterscharführer[1] Förster's section of the 1st Company, SS Motorcycle Battalion "Langemarck" of "Das Reich" Division, whose task it had been to cover the company's left flank, were found as follows: section-leader Unterscharführer Förster, his hand on the pull-ring of his last hand-grenade, shot through the head; his number one, Rottenführer[2] Klaiber, his machine-gun still pressed into his shoulder and one round in the breech, shot through the head; the number two, Sturmmann[3] Oldeboershuis, still kneeling by his motor-cycle, one hand on the handlebar, killed at the moment when

[1] Rank in Waffen SS equivalent to Army corporal.
[2] Rank in Waffen SS equivalent to Army lance-corporal.
[3] Rank in Waffen SS equivalent to Army private.

leaving with his last dispatch; driver Sturmmann Schwenk, dead in his foxhole. As for the enemy, only dead bodies were found, lying at hand-grenade range in a semicircle around the German section's position. An example of what defence means.

That was Yelnya—a desolate wrecked dump on the Desna, 47 miles east of Smolensk. It gave its name to a sector of the front where the battles raged for five weeks. The Soviet resistance at Yelnya was not accidental. Just as it was not accidental that "the high ground of Yelnya" was mentioned alongside Smolensk as the first strategic objective of Army Group Centre in the deployment directives for Operation Barbarossa. What was the reason? As a road junction and a commanding ridge of high ground it represented an important strategic position for whoever wanted to get to Moscow and for whoever defended the city.

The Germans knew it, and so, of course, did the Russians. Ruthlessly, Timoshenko had employed the civilian population on fortification works, in order to develop the Desna sector south of Yelnya into a strong obstacle to armour. Whatever forces Moscow managed to scrape together were directed into the Yelnya area. The Desna position was to become the great new blocking line. German aerial reconnaissance discovered these intentions. It was therefore advisable to strike quickly, before the Russians had strengthened their defences. Lieutenant-General Schaal's 10th Panzer Division and General Hausser's Motorized Waffen SS Division "Das Reich" were assigned the task of taking Yelnya and the area behind it.

That sounded simple enough, but it was anything but simple for Guderian's Panzer divisions, which had by then fought their way across roughly 600 miles—through deserts of dust, over unmetalled roads, and through virgin forest. The artillery's fire-power had also been greatly diminished by the loss of many heavy and medium batteries. Given fresher formations, with stronger armoured and artillery support, the high ground of Yelnya would have been no problem. But in the circumstances it was quite a task.

General Schaal, then commanding the 10th Panzer Division, has described the operation to the author. Beyond the Dnieper, he explained, the Russians no longer stood up and fought openly, but increasingly adopted the tactics which were to be practised later by the large partisan units. General Schaal quoted the following instance:

"Between Gorodishche and Gorki the division's vanguard had driven through a patch of thick forest. The bulk of the division got past the same spot during the night. But the artillery group which followed was suddenly smothered with mortar-fire from both sides and attacked by infantry at close quarters. Fortunately a motor-cycle battalion of the SS Division 'Das Reich' was bivouacking near by. They came to the assistance of the gunners and hacked them free.

"More serious than this kind of skirmish was the wear and tear on armoured fighting vehicles. The shocking roads, the heat, and the dust were more dangerous enemies than the Red Army. The tanks were enveloped in thick clouds of dust. The dust and grit wore out the engines. The filters were continually clogged up with dirt. Oil-consumption became too heavy for supplies to cope with. Engines got overheated and pistons seized up. In this manner the 10th Panzer Division lost the bulk of its heavy Mark IV tanks on the way to Yelnya. They were defeated not by the Russians but by the dust. The men of the maintenance units and engineer officers worked like Trojans. But they were short of spares. And the spares did not arrive because supplies no longer functioned. The distances from the army stores had become too great. Every single ammunition or supply convoy lost about a third of its vehicles en route, either through breakdown or through enemy ambushes. Not only the machines but the men too were overtaxed. It would happen, for instance, that parts of a column on the march failed to move off again after a short rest because its officers and men had dropped off into a comatose sleep."

These conditions applied not only to 10th Panzer Division. It was the same throughout the central sector—on Hoth's part of the front as much as on Guderian's. In a letter to Field-Marshal von Bock, Hoth wrote: "The losses of armoured fighting-vehicles have now reached 60 to 70 per cent. of our nominal strength." Nevertheless, the troops accomplished their task. On 19th July the 10th Panzer Division took Yelnya.

The wide anti-tank ditch which Russian civilians had built around the town in ceaseless round-the-clock work was overcome by the infantry of 69th Rifle Regiment in spite of murderous gunfire. The division suffered heavy losses, but worked its way forward yard by yard. By evening the infantry had pushed through Yelnya and dug in on the far side. Lieutenant-General Rokossovskiy, commanding hurriedly collected reserves, drove his regiments against the German positions. But the line of 10th Panzer Division held. On 20th July the SS Division "Das Reich" took up position on the high ground to the left of them. The troops needed a breather.

The Yelnya bend projected a long way eastward from the German front line. It was its most advanced spearhead. South of it the front ran back as far as Kiev, and north of it there was a kink in the direction of Smolensk and thence a wide semicircle towards Leningrad. A glance at the map made it obvious that the Yelnya bend was a bridgehead, the logical strategic starting-point for an offensive against Moscow. The Soviets understood that too, and therefore determined to smash the Yelnya bend. From the end of July until the beginning of September Army Group Centre was engaged here in its first great defensive battle. Nine German divisions passed through the hell of Yelnya in the course of these weeks—the 10th Panzer Division, the SS Division "Das Reich," the 268th, 292nd, 263rd, 137th, 87th, 15th, and 78th Infantry Divisions, as well as the reinforced "Grossdeutschland" Infantry Regiment.

The Soviet High Command let Timoshenko have whatever reserves were to hand. Parts of four Armies were sent into action on his front. With nine infantry divisions and three armoured formations Timoshenko attacked the Yelnya bend, which was at no time held by more than parts of four German divisions. It was the battle experience, the discipline and, above all, the stolid perseverance of the reduced German battalions and companies that proved decisive in this frightful battle.

The following is an account from the sector of the Motorized Infantry Regiment "Grossdeutschland," generally known as G.D.

First Lieutenant Hänert of 4th (Machine-gun) Company, 1st Battalion, G.D. Regiment, was in his foxhole, looking through his trench telescope. That was in front of the level-crossing at Kruglovka in the Yelnya bend. Russian artillery had been firing ceaselessly for the past three hours. All telephone lines were cut, and no runners or repair parties could leave their foxholes. Now the barrage was being stepped up. But it passed over the battalion's sector.

They are lengthening their range—that means they'll charge in a minute, Lieutenant Hänert thought to himself. And, true enough, there they were in his telescope. He stared in amazement: the Soviet troops were charging in close order, mounted officers in front and behind and on both sides of the uniformed earth-brown mass, like sheepdogs around a flock. Bent double, the Russians were pulling their low two-wheeled carts with their water-cooled heavy machine-guns, the Maksims. Infantry guns and anti-tank guns were also heaved into position on the double, including the dangerous 7·62-cm. field-gun known to the German troops as "Crash-boom" because with its flat trajectory the burst of the shell was heard before the sound of the firing.

That was the moment when the German artillery should have massively intervened. But the guns were firing only sporadically. For the first time since the start of the campaign there was a shortage of ammunition, because supplies had all but broken down. It was the first warning of things to come.

The Russians jumped into the ditch of a small stream and vanished from sight. A moment later they were coming up the bank—in front, the officers, who had now dismounted.

The men of First Lieutenant Rössert's 2nd Company, dug in to the right of 4th Company, looked out of their foxholes. The Russians were still 700 yards away. Now they were at 600 yards. "Why isn't Lieutenant Hänert opening up with his machine-guns?" the men asked Sergeant Stadler. "He's got his reasons," the sergeant grunted.

Hänert had his reasons. He was looking through his telescope. Now he could make out the faces of the Russians. But still he did not give the firing order. The sooner he ordered fire to be opened the sooner the Russians would go to ground and merely creep up under cover. Hänert knew from experience that the Russians must be crushed decisively with the first blow. Their infantry charges were made with a tenacity bordering on insensate obtuseness. Even if ten machine-guns mowed down wave after wave the Russians would come up again. They would cry "Urra!" and be killed.

What was the reason for that? The evidence of captured officers and NCOs supplied the answer. In the Red Army a commander was personally held responsible for the failure of an attack. Consequently he would drive his men time and again against the objective named in his orders. This is not to say that he would be indifferent to the loss of his men, but consideration for the individual is less important in the Soviet Army than in the armies of Western countries. Advanced positions, strongpoints, or encircled units would be sacrificed without much hesitation if such a sacrifice yielded strategic advantages. From his first day as a recruit the Soviet soldier would be told: Action means close combat. For that reason he would seek out close combat. And he was particularly well trained for it. Bayonet practice took up most of a recruit's day. And the Russians were past masters of this gruesome business. They had also been drilled in firing from the hip. And as for handling the spade and the rifle-butt, they were every bit as good as the German assault companies. The Soviet Field Service Manual of 1943 says: "Only an attack launched with savage determination to annihilate the enemy in close combat ensures victory." It was in this spirit that the Russians made their charges.

Lieutenant Hänert, by the railway embankment of Kruglovka, saw them coming. They were still 500 yards away. At last Hänert stood up and shouted, "Continuous bursts!" Like a thunderclap a storm of stuttering broke out. The Russians went down. Past the dead and wounded of the first wave the second wave pushed forward—firing, leaping, using aimed fire with single rounds. And the Russians were excellent marksmen.

The grenadiers of 2nd Company had to push their heads out of their foxholes if they wanted to fire. And they must fire if they did not want to be killed by the Russians. But as soon as a head appeared anywhere the Russian snipers opened up with their excellent automatic rifles with telescopic sights. More and more weapons fell silent in the area of 2nd Company, "Grossdeutschland" Infantry Regiment, by the level-crossing of Kruglovka in the Yelnya bend.

But the last fifty yards defeated the Russians. Night fell. Russian artillery opened up again. The Russian guns killed many of their own men still alive on the open ground, which afforded no cover.

At midnight the pounding ceased. Rössert's and Hänert's men climbed out of their foxholes. There had been two men to each hole when the battle started. But from most of them only one man emerged now. They called for stretchers for the wounded and for the dead by whose sides they had crouched for hours, firing.

The battle was resumed at dawn. It went on for five days. Over hundreds of dead bodies the Russians pushed their way into the positions of 1st Battalion. The machine-gun 20 yards to the right of Sergeant Stadler was silent: the last gunner had got a bullet in his stomach, heaven knew how—probably a ricochet. Sergeant Stadler heard the sharp crack of a pistol: the lance-corporal had preferred this way out to the long and painful death of a stomach wound. Ten minutes later two Russians jumped into the foxhole. Stadler straightened up. He placed three hand-grenades in front of him. He pulled the pin of the first and flung it. Too short. The second hit the lip of the foxhole and showered it with fragments. The third grenade rolled right in. Like fireworks the machine-gun ammunition went up.

During the sixth night, on 27th July, the position by the railway embankment of Kruglovka was abandoned. The 2nd Company withdrew some 800 yards, to the edge of the wood. The Russians followed up. And the same thing began all over again. On 18th August the regiment was relieved by 263rd Infantry Division. The 2nd Battalion, 463rd Infantry Regiment, repulsed 37 Russian attacks in 10 days. On 25th August the reconnaissance detachment of 263rd Infantry Division joined the neighbouring 2nd Battalion, 483rd Infantry Regiment, in an immediate counter-attack against the enemy, who had penetrated into its positions on the fiercely contested "Crash-boom Hill." In this engagement Captain Orschler, commanding the reconnaissance detachment, was killed—the first member of the German Wehrmacht to receive the Gold Cross. On 29th August the companies of 15th Infantry Division dropped into the blood-drenched infantry foxholes. The battle continued. Three Soviet divisions were sacrificed by Timoshenko on the northern sector at Yelnya alone. The Russian doctor in charge of the dressing station at Stamyatka, who was taken prisoner, stated that on the sector of 263rd Division he had tended 4000 wounded in a single week.

On the situation map spread out before Guderian in his headquarters bus at midnight on 22nd August 1941 these human tragedies were not recorded. All the map showed was the triangular pennants, representing the divisional headquarters of the 15th, the 292nd, and the 268th Infantry Divisions, and the black, square pennants of regimental headquarters. In front of the German lines the identified Soviet divisions had been entered. On 22nd August they numbered nine rifle and two armoured divisions.

Yet Guderian, who was continually on the move, and mixed with his men in the fighting line, knew what lay behind the entries made by his staff officers. "Pack up the map; I'll take it along with me to Borisov in the morning," Guderian said. "Good night, gentlemen."

And how did the other Panzer Group on the Central Front fare in the meantime—Colonel-General Hoth's Panzer Group, north-east of the highway?

In General Yeremenko's memoirs we read the blunt statement:

> The recapture of Smolensk proved impossible. The High Command therefore decided at the end of July to order the Twentieth and Sixteenth Armies, which were encircled by Hoth's forces north of Smolensk, to break out of the pocket. The divisions of these Armies had by then been reduced to no more than 2000 men. The whole of Twentieth Army had only 65 tanks and nine aircraft left.

That was the measure of Hoth's triumph. Like Guderian south of the Smolensk–Moscow highway, Hoth had ordered his divisions to keep going. He had reached the Vop, where his now exhausted forces came up against the Stalin Line, which had been fortified in a surprisingly short period of time. With parts of his motorized forces and the infantry divisions which followed behind he put a ring around Yeremenko's 15 divisions which were to have retaken Smolensk.

Yeremenko resisted desperately. He had to fight without supplies and to hang on where he stood. The Soviet High Command was pinning him down with relentless orders. Commanders who retreated had to face courts martial. Soldiers who abandoned their positions were shot. The Soviet High Command was determined to recapture Smolensk at all costs. It was there that the German storm was to be broken. It was to become a dress-rehearsal for Stalingrad.

Moscow's determination was confirmed by the fact that, upon Stalin's personal command, a jealously guarded secret weapon was first employed here, although it was not yet in mass production and could not therefore be expected to play a decisive part. Yeremenko's account is most interesting on this point:

> About mid-July I received a telephone message from headquarters: "It is intended to employ 'yeresa' in the battle against the fascists. A detachment armed with this new weapon will be assigned to you. Test the weapon and let us have your report on it."

'Yeresa' was the name for the first rocket-mortar batteries. Not even Yeremenko had known about them.

> We tested the new weapon near Rudnya [Yeremenko reports]. The rockets streaked through the air with a terrifying whine. They soared up like comets with a red tail and then exploded with a crash like thunder. The effect of the bursts of 320 rockets within a span of 26 seconds in a very limited area exceeded all expectations. The Germans ran away in panic and terror. Admittedly, our own

troops withdrew likewise. For security reasons we had not informed them beforehand about the use of the new weapon.

The victims of this surprise were parts of Hoth's 12th Panzer Division. At first the effect on the troops was really terrifying. The German troops nicknamed the rocket mortar "Stalin's organ-pipes." The Russians called it "Katyusha"—Little Kate. Luckily, Yeremenko had only one unit. Thus the appearance of the howling Katyusha at Rudnya did not turn the tide of the battle, but it was another reminder of the technological capacity of the Soviets. It convinced the optimists in the German High Command of the need for caution—or, to put it differently, for haste.

Shortly before 1000 hours on 23rd August Guderian landed in his Fieseler Storch on the airfield of Borisov and drove over to Army Group headquarters. The commanders-in-chief of the Fourth, Ninth, and Second Armies had also just arrived—Field-Marshal von Kluge, Colonel-General Strauss, and Colonel-General Freiherr von Weichs. The visitor from the Mauerwald was expected at any moment: he was Colonel-General Halder, Chief of the General Staff.

He arrived towards 1100. He looked ill and seemed depressed. The reason was soon obvious to all. Halder announced: "The Fuehrer has decided to conduct neither the operation against Leningrad as previously envisaged by him, nor the offensive against Moscow as proposed by the Army General Staff, but to take possession first of the Ukraine and the Crimea."

Everybody was stunned. Guderian stood stiff as a ramrod. "This can't be true."

Halder regarded him resignedly. "It is true. We spent five weeks wrangling for the drive to Moscow. On 18th August we submitted a plan of attack. Here is the reply." He read from a sheet of paper:

"Fuehrer's Directive, 21.8.1941.

"The Army's proposal for the continuation of operations in the East, submitted to me on 18.8., is not in line with my intentions. I therefore command as follows:

"(1) The most important objective to be achieved before the onset of winter is not the capture of Moscow but the seizure of the Crimea and of the industrial and coal-mining region on the Donets, and the cutting off of Russian oil-supplies from the Caucasus area; in the north it is the isolation of Leningrad and the link-up with the Finns."

The order continued, under item 2, to list the strategic targets for Army Groups South and Centre, and, under item 3, contained the instruction to Army Group Centre to participate in the operations aiming at the destruction of the Russian Fifth Army by making available sufficient forces. Finally, it explained Hitler's plan for the continuation of operations after the battle for the Ukraine. This ran as follows:

(4) The capture of the Crimean peninsula is of paramount importance to ensure our oil-supplies from Rumania. For this reason a rapid crossing of the Dnieper in the direction of the Crimea is to be attempted with all available means, including the employment of fast units, before the enemy has had time to bring up fresh forces.

(5) Only the tight sealing off of Leningrad, the link-up with the Finns and the annihilation of the Russian Fifth Army will provide the prerequisites and the available forces for attacking the enemy's Army Group under Timoshenko with any prospect of success, and of defeating it, in line with the supplementary order to Directive No. 34 of 12.8.

(Signed) ADOLF HITLER.

This then was the decision. It was what the generals had always feared and what they had hoped would never happen. Now it had been uttered.

It has been fashionable to describe Hitler's turning away from Moscow as the key error of the summer campaign. This view cannot be proved wrong, but the author does not believe that Hitler's decision to turn towards Kiev, with the time lost in consequence, was the sole cause of the subsequent disaster before Moscow. Upon objective consideration Hitler's decision seems, in many respects, justified and reasonable. The battles of the summer had shown one thing clearly: the different rate of advance of armour and infantry had inevitably divided the army into two successive parts which not only moved separately but also fought their engagements separately. This represented a serious weakness which the enemy might well exploit as soon as he had realized the German mode of operation. Various attested remarks of Stalin show that he had understood the German method by the end of July 1941. Moreover, the debilitating effect of the geographically vast area and the heavy wastage which resulted mean that no further justification is required. It is also true that owing to the much slower advance of Army Groups North and South the flanks of Army Group Centre remained exposed. The Soviet Fifth Army was a real threat to Bock's extended flank. Something had to be done to protect the flanks. The experience gained in battles of encirclement, moreover, suggested that in future the Russian forces ought not to be crushed in operations involving such great distances, but in closer co-operation between Panzer Groups and infantry. In the light of what has since become known about the strength of Soviet armour and their inexhaustible reserves of manpower, Hitler's caution does not appear unreasonable.

But—and this is an important but—for a strategy of caution it was then too late. On the Central Front Germany was already involved far too deeply in Russian territory. If the idea of a Blitzkrieg against the heart of the Soviet Union was dropped altogether and the enemy given time to recover, then surely the campaign and probably the whole war was lost. Seen in this light, Hitler's decision represented an admission that Yelnya–Smolensk had broken the impetus of the German Blitzkrieg. If the generals accepted that view it meant the basis of Operation Barbarossa had become invalid. It was this view that Halder, the Chief of the General Staff, and the commanders in the field, especially Guderian, were trying to oppose.

"What can we do against this decision?" asked Bock. Halder shook his head. "It is immutable."

"We've got to upset it," Guderian persisted. "If we head for Kiev first we shall inevitably get involved in a winter campaign before we can reach Moscow. What the roads and our supply difficulties will be like then I shudder to think. I doubt that our tanks are up to the strain. My Panzer corps, especially XXIV Corps, have not had a single day's rest since the beginning of the campaign."

Field-Marshal von Bock agreed. There was a heated discussion. Eventually it was decided that Guderian should accompany Halder to the Fuehrer's headquarters, request an interview, and try to change Hitler's mind. Late in the afternoon the aircraft started for Rastenburg, in East Prussia. As Guderian said good-bye to von Bock, the Field-Marshal quoted the words attributed to the officer of the guard at the bishop's palace in Worms on 17th April 1521, to Martin Luther, as he set forth to justify his teaching to the Emperor: "Little monk, little monk, yours is a difficult road."

The Ju-88 droned on above vast harvested cornfields. Guderian was making notes and studying the maps. At dusk they touched down at the airfield of the Fuehrer's headquarters near Lötzen, in East Prussia. They drove across to the "Wolfsschanze," the camp of concrete huts under tall oak-trees where Hitler and the High Command of the Wehrmacht resided. The sentry saluted, raised the barrier, and let the car through. They rolled along an asphalted road. On the left, just inside the compound, was the press office. Scattered on both sides were the low grey huts on whose roofs shrubs had been planted. They passed the *Teehaus*, the canteen. On the left was Keitel's hut. And right at the end of the road, in a small dip, was the "Fuehrer hut"— surrounded by a double fence and guarded by double sentries. A special yellow pass was needed to enter the inner sanctum of Hitler's headquarters.

Hitler's hut was exactly like the others—gloomy, Spartan, with simple oak furniture and a few prints on the walls. Here 'he' sat through the night, bending over maps and reports, over photographs, tables of figures, and memoranda.

Within two hours of his arrival Guderian stood in the map-room of the Fuehrer's hut, making his report on the state of his Panzer Group. The following account is based on information supplied by General

Moscow or Kiev? 101

Bayerlein, to whom Guderian had given a detailed account of his conversation with Hitler for inclusion in the group diary, and on notes left by Guderian himself.

Hitler had not been told what Guderian wanted. Moreover, Field-Marshal von Brauchitsch had specifically forbidden Guderian to broach the subject of Moscow himself. He therefore began by speaking about his Panzer Corps—about engine breakdowns, about the supply situation, about Russian resistance, and about his losses. The picture he painted was not gloomy but realistic. And, as he had hoped, Hitler himself gave him his cue. "Do you consider your troops are still capable of a major effort?" Hitler asked.

Everybody's eyes were on Guderian. He answered, "If the troops are set a great objective, the kind that would inspire every man of them—yes."

Hitler: "You are, of course, thinking of Moscow."

Guderian: "Yes, my Fuehrer. May I have permission to give my reasons?"

Hitler: "By all means, Guderian. Say whatever's on your mind."

The crucial moment had come.

Guderian: "Moscow cannot be compared with Paris or Warsaw, my Fuehrer. Moscow is not only the head and the heart of the Soviet Union. It is also its communications centre, its political brain, an important industrial area, and above all it is the hub of the transport system of the whole Red empire. The fall of Moscow will decide the war."

Hitler listened in silence. Guderian continued: "Stalin knows this. He knows that the fall of Moscow would mean his final defeat. And because he knows this he will employ his entire military strength before Moscow. He is already bringing up everything he has left. We have seen it at Yelnya for weeks. Outside Moscow we shall encounter the core of the Russian military might. If we want to destroy the vital force of the Soviets it is here that we shall encounter it; here is our battlefield, and if we rally all our strength we shall pull it off at first try." Hitler was still silent. Guderian was now in full flight. "Once we have defeated the enemy's main forces before Moscow and in Moscow, and once we have eliminated the Soviet Union's main marshalling yard, the Baltic area and the Ukrainian industrial region will fall to us much more readily than with Moscow still intact in front of our fighting line, able to switch reserves—mainly from Siberia—to the north or to the south." Guderian had warmed to his subject. There was silence in the situation room. Keitel stood leaning against the map-table, Jodl was taking notes. Heusinger was listening intently.

Through the open windows came the cool evening air. Fine mosquito-netting kept out the midges and flies which Hitler detested. Vast swarms of them hovered over the little lakes and ponds outside the compound. The unit of sappers had repeatedly attacked them by spraying petrol over a stagnant pool near the Fuehrer's hut. The smell had hung about the place for days, but the midges had survived.

Guderian strode over to the map. He put his hand on the Yelnya bend. "My Fuehrer, I have kept this bridgehead towards Moscow open till to-day. Deployment plans and operational orders are ready. Routing instructions and transport schedules for the advance to Moscow are already worked out. In many places the troops have even painted the signposts: so-and-so many miles to Moscow. If you give the order the Panzer corps can move off this very night and break through Timoshenko's massive troop concentrations before Yelnya. I need only telephone a code-word to my headquarters. Let us march towards Moscow—we shall take it."

In the long history of the Prussian and German Armies there has never been such a scene between a general and his supreme commander, a scene so packed with exciting drama as this. It was probably the last time that Hitler listened so long and so patiently to a general who disagreed with him. He looked at Guderian. He rose. With a few quick steps he was by the map. He stood next to Jodl, the chief of the operations staff in the High Command of the Wehrmacht. He put his hand on the Ukraine and launched on a lecture to justify his position.

In a sharp voice Hitler began: "My generals have all read Clausewitz, but they understand nothing of wartime economics. Besides, I too have read Clausewitz and I remember his dictum, 'First the enemy's armies in the field must be smashed, then his capital must be occupied.' But that is not the point. We need the grain of the Ukraine. The industrial area of the Donets must work for us, instead of for Stalin. The Russian oil-supplies from the Caucasus must be cut off, so that his military strength withers away. Above all, we must gain control of the Crimea in order to eliminate this dangerous aircraft-carrier operating against the Rumanian oilfields."

Guderian felt the blood rising to his ears. Wartime economics were not strategy. War meant crushing the enemy's military might—not rye, eggs, butter, coal, and oil. That was the approach of a colonialist, not of a Clausewitz.

But Guderian remained silent. After what he had said what else could he, a commander in the field, say to the man who held the supreme political and military power? A decision had been made by the politician, and there was nothing left for the soldiers to do.

At midnight the historic meeting was at an end. When Guderian reported to Halder, who had not been invited by Hitler to be present, the Chief of the General Staff broke down and raved, "Why don't you fling your command in his face?"

Guderian was surprised. "Why don't you?"

"Because there's no point in our doing it," Halder replied. "He'd be glad to get rid of us, but we've got to hold on."

Half an hour later the telephone rang at Second Panzer Group headquarters in Prudki. The chief of operations was on duty and lifted the receiver. Wearily Guderian's voice came over the line: "Bayerlein, the thing we've prepared for is not coming off. The other thing is being done, lower down—you understand?"

"I understand, Herr Generaloberst."

5. Stalin's Great Mistake

Battles of annihilation at Roslavl and Klintsy–Stalin trusts his secret service–Armoured thrust to the south–Yeremenko expects an attack on Moscow.

BAYERLEIN had understood Guderian very well. During the day the first directives had come down from Army Group Centre revealing the new plan: parts of Second Panzer Group were to drive south into the Ukraine.

Immediately after Guderian's telephone call Colonel Freiherr von Liebenstein, chief of staff of Second Panzer Group, summoned the staff officers. He knew Guderian. When he came back from Rastenburg he would expect the new plan to be ready in outline.

There was no one at headquarters who was not deeply depressed by Hitler's decision to turn against the Ukraine instead of against Moscow. Nobody understood it. Every one regarded it as a mistake. The staff officers' trained minds rebelled against the fundamental violation of one of the basic strategic rules in the spirit of Clausewitz—not to be seduced away from one's main objective, always to stick to the basic framework of one's operational plan, and to concentrate all one's forces against the enemy's strong point.

This turning away from Moscow at the very moment when it seemed within reach, barely two hundred miles away and, as far as anyone could predict, almost certain to fall to Guderian's and Hoth's now refreshed armoured forces, was very soon to be seen as a serious error of judgment.

The directives for the new operation were clear. As far as Guderian's two Panzer Corps were concerned, they read: "Drive to the south into the rear of the Soviet Fifth Army, the core of Marshal Budennyy's Army Group South-west Sector, defending the Ukraine beyond the Dnieper to both sides of Kiev." Guderian's first target was the big railway junction of Konotop on the Kiev–Moscow line. The next step would depend on the situation, according to the progress made by Army Group South.

When, on 24th August, Guderian arrived at Shumyachiy, a small village on the Moscow highway where Liebenstein had set up the headquarters of the Panzer Group, he was again full of zest. He greeted Liebenstein, Bayerlein, and Major von Heuduck, his Intelligence officer, who were all patently disappointed, and went with them straight to his headquarters bus.

"I know what you're thinking," he said calmly. "Why didn't he succeed—why did he give in?" He did not wait for an answer. "There was nothing I could do, gentlemen," he continued. "I had to give in. I was out there alone. Neither Field-Marshal von Brauchitsch, the Commander-in-Chief, nor the Chief of the General Staff had accompanied me to the Fuehrer. I was faced by a solid front of the High Command of the Wehrmacht. All those present nodded at every sentence the Fuehrer said, and I had no support for my views. Clearly the Fuehrer had expounded his arguments for his strange decision to them before. I spoke with a silver tongue—but it was in vain. Now we can't go into mourning over our plans. We must tackle our new task with all possible vigour. Our hard-won jumping-off positions for Moscow—at Roslavl, Krichev, and Gomel—will serve us now as a springboard into the Ukraine."

Guderian was right. The operations conducted by his Army Group around Roslavl and Krichev at the beginning of August, resulting in about 54,000 Russian prisoners, now proved a valuable prerequisite also for the new operation. Let us look back at the three weeks which have passed.

On 1st August Guderian had started operations against Roslavl. His plan was a typical battle of encirclement. He operated with two Infantry Corps and one Panzer Corps. The bulk of the infantry divisions attacked the enemy frontally in order to tie him down. The 292nd Infantry Division, acting as IX Corps' striking division, strongly supported by artillery and rocket mortars, pushed to the south in the Russian rear. From the south-western wing 3rd and 4th Panzer Divisions performed a rapid outflanking movement, first to the east, then north across the Roslavl–Moscow road, and closed the ring with 292nd Infantry Division on the Moscow highway. The plan worked. Roslavl became a genuine, if minor, battle of encirclement.

The war diary of Captain Küppers, artillery liaison-officer of 197th Infantry Division, the combat report of VII Army Corps, and the day-to-day reports of engagements of an infantry battalion—all of them extant—provide an impressive picture of the fighting.

H-hour was 0430. Along the entire line of VII Corps the attack was launched without artillery preparation. The spearheads of the infantry regiments worked their way forward—past the communications group of the artillery commander, who had been lying in the front line with Lieutenant-Colonel Marcard since 0300 hours, watching the Russian positions. Everything was quiet on the Russian side. Suddenly the quiet of the morning was broken by the first rifle-shots from the infantrymen who had just moved forward. Triggers were pulled too soon by nervous fingers. They roused the Russian night sentries. At once Soviet machine-guns opened up. Mortars plopped. Major-General Meyer-Rabingen, the commander of 197th Infantry Division, drove in his jeep to the foremost line. Farther down, in the village of Shashki, Major Weichhardt's 3rd Battalion, 332nd Infantry Regiment, had already broken into the Russian positions. It was a case of bayonets, spades, and pistols. Thirty minutes later the white Very lights went up: "We are here!"

"Artillery forward," the advanced observer radioed back. A moment later Captain Bried was on the move. He commanded the 2nd Battalion, 229th Artillery Regiment. His car got as far as the edge of the village. Then there was a flash and a crash—a minefield.

The nearside front wheel of Bried's car sailed through the air. The observer's car, which followed behind, suffered the same fate as it tried to swing off the road. In response to the signal "Sappers forward!" Engineers Battalion 229 cleared the mines. Meanwhile the guns of the 2nd Battalion had moved into position and were supporting the infantry with their fire. The first few prisoners were brought in for interrogation. A short Ukrainian was found to speak German. He looked trustworthy. An interpreter unit supplied him with a denim uniform and a white armlet lettered "German Wehrmacht."

On 2nd August at 0400 hours the infantry went into action again. Their objective was the main road from Smolensk to Roslavl. It was a particularly hard day for 347th Infantry Regiment. Its battalions were stuck in difficult terrain in front of a thick and swampy patch of woodland and were only able to advance inch by inch and at the cost of heavy losses. The Russians again proved their mastery in forest fighting. With sure instinct they moved among the impenetrable undergrowth. Their positions, not on the forest's edge but deep inside, were superbly camouflaged. Their dugouts and foxholes were established with diabolical cunning, providing for a field of fire only to the rear. From in front and from above they were invisible. The German infantrymen passed them unsuspecting, and were picked off from behind.

The Russians were also very good at infiltrating into enemy positions. Moving singly, they communicated with each other in the dense forest by imitating the cries of animals, and after trickling through the German positions they rallied again and reformed as assault units. The headquarters staff of 347th Infantry Regiment fell victim to these Russian tactics.

In the night, at 0200, the shout went up, "Action stations!" There was small-arms fire. The Russians were outside the regimental headquarters. They had surrounded it. With fixed bayonets they broke into the officers' quarters. The regimental adjutant, the orderly officer, and the regimental medical officer were cut down in the doorway of their forest ranger's hut. NCOs and headquarters personnel were killed before they could reach for their pistols or carbines. Lieutenant-Colonel Brehmer, the regimental commander, succeeded in barricading himself behind a woodpile and defending himself throughout two hours with his sub-machine-gun. An artillery unit eventually rescued him.

Meanwhile, 332nd Infantry Regiment had reached the main road from Roslavl to Smolensk. First Lieutenant Wehde blocked the road with his 10th Company and stormed the village of Glinki. The Soviets in Roslavl realized they were in danger of being encircled. They left the town in lorries and tried to run down the positions of 10th Company. They scattered hand-grenades among them by the armful and fired wildly from machine-guns and sub-machine-guns. But 10th Company held out, but only until midday. After that they were unable to stand up to the Soviet attacks. The Russians retook the village.

Now for an immediate counter-attack. Lieutenant Wehde scraped up anyone he could lay his hands on—supply personnel, cobblers, bakers—and dislodged the Russians. But in the afternoon they were back in Glinki. Another immediate counter-attack. House after house was recaptured with flame-throwers and hand-grenades. The place was to change hands many more times.

On Sunday, 3rd August, 197th Infantry Division found itself in difficulties because 347th Infantry Regiment was hanging back a long way. The Soviets tried to break through at the contact point between 347th and 321st Infantry Regiments. The gunners fired from every barrel they had. To make matters worse it started to pour with rain. Roads became quagmires. At 1600 hours Lieutenant Wehde was killed outside Glinki. The 321st Infantry Regiment was fighting desperately. Several groups were encircled and had to defend themselves on all sides.

Things went better on the right wing of VII Corps. Towards 1100 hours 78th Infantry Division had reached the Krichev–Roslavl road with the bulk of its units. Fascinated, the infantrymen watched 4th Panzer Division moving off for its outflanking attack on Roslavl.

On the extreme left wing, meanwhile, in the area of 292nd Infantry Division, the 509th and 507th Regiments were struggling towards the south along soft, muddy roads. In the leading company of 507th Infantry Regiment, the regiment forming the left wing, a man with crimson stripes down the seams of his trousers was marching by the side of the captain—Colonel-General Guderian.

Reports of the difficulties which 292nd Infantry Division had with its advance—difficulties that might affect the overall plan—had induced him to find out for himself by taking the part of an ordinary infantryman. As though this were the most natural thing in the world, Guderian later told his headquarters staff, "In this way I kept them on the move without having to waste words."

"Fast Heinz as an infantryman!" the troops were shouting to each other. They pulled themselves together. When the leading self-propelled gun stopped a few miles from the Moscow highway, the target for the day, Guderian was up on the vehicle in a flash. "What's the trouble?"

"There are tanks along the highway, Herr Generaloberst," the gun-layer reported. Guderian looked through his binoculars. "Fire white Very lights!" The white flare streaked from the pistol. And from the highway in the distance came the reply: also white Very lights. That meant that the 35th Panzer Regiment, of 4th Panzer Division, was already on the Moscow highway. At 1045 hours 23rd Infantry Division penetrated into the northern part of Roslavl.

On 4th August Glinki was lost once more. Stukas attacked the Soviet strongpoint. Russian tank attacks against the left and right flanks of 197th Infantry Division collapsed in the concentrated fire from all available guns. Glinki was taken again. The Russians wavered and withdrew. Hastily they re-formed for desperate break-through attempts along the Moscow highway.

On 5th August it was discovered that a strong Soviet armoured unit had fought its way out of the pocket at Kazaki, in the area of 292nd Infantry Division. The division's regiments were so extended, and, moreover, so involved in heavy defensive fighting, that they were unable to close the gap. The Russians were pouring through—supplies, infantry, artillery units. Guderian at once drove to the gap. Personally he moved a tank company against the Russians streaming through the gap; he organized a combat group from armoured units, self-propelled guns, and artillery; and this group, under General Martinek, the artillery commander of VII Corps, at last closed the gap. The Russians still coming through met their doom.

On 8th August it was all over. Some 38,000 prisoners were counted. Booty included two hundred tanks, numerous guns and vehicles. The Soviet Twenty-eighth Army under Lieutenant-General Kachalov had been smashed. But that was not the main thing. For 25 miles in the direction of Bryansk and towards the south there was no enemy left. A huge gate had been opened towards Moscow. But Guderian wanted to play safe. In order to have truly free flanks for a drive against Stalin's capital he must first eliminate the threat from the deep right flank at Krichev.

General Freiherr Geyr von Schweppenburg, the shrewd and resolute commander of XXIV Panzer Corps, whose divisions had only just closed the trap at Roslavl, ordered his armour to turn about in a bold operation and attack Timoshenko's divisions in the Krichev area by an encircling move. On 14th August this operation too was successfully concluded. Three more Russian divisions were smashed, 16,000 prisoners were taken, and large quantities of guns and equipment of all kinds captured. As with a heavy hammer, Guderian had smashed Timoshenko's bolt on the gate to Moscow.

Guderian's success now whetted the appetite of the High Command of the Wehrmacht. On the very next day it demanded that Timoshenko's strong force in the Gomel area should also be attacked, so as to bring relief to Colonel-General Freiherr von Weichs's Second Army. Guderian was to make one Panzer division available to Second Army. But Guderian's reply was: "If anything, a whole corps must be used. One division alone is not enough for an operation over such a distance." He made sure he got his way.

On 15th August XXIV Panzer Corps moved off again—towards the south—with 3rd and 4th Panzer Divisions in the forefront, followed by 10th Motorized Infantry Division. When the force had successfully broken through the enemy lines the division on the right wing would strike at Gomel. Just the one division—as ordered by the High Command. It was a clever interpretation of the order, and it ensured victory. Guderian made the most of it.

On 16th August the 3rd Panzer Division took the road intersection of Mglin; on the 17th it took the railway junction of Unecha. Thus the railway-line Gomel–Bryansk–Moscow was cut. On 21st August Guderian's two Panzer Corps reached the important jumping-off positions of Starodub and Pochep. All the points were now set for the drive to Moscow. It was on that very day that Hitler called off all plans against the Soviet capital and ordered the advance into the Ukraine.

It was a dramatic turn of events. Its significance was even greater in view of what was happening in the Kremlin.

On 10th August Stalin received a report from his top agent, Alexander Rado, in Switzerland. Rado claimed to have reliable information that the German High Command intended to let Army Group Centre strike at Moscow via Bryansk. The information certainly was reliable: this had been precisely the plan of the German Army High Command.

The effect this report had in Moscow is described in General Yeremenko's memoirs. On 12th August he was instructed by Marshal Timoshenko to come to Moscow at once; he was to take up a new command. Yeremenko writes:

> I arrived in Moscow at night and was at once received at the High Command by Stalin and the Chief of the General Staff of the Soviet Army, Marshal Shaposhnikov. Shaposhnikov briefly outlined

the situation at the fronts. His conclusion, based on reconnaissance and other information [no doubt Rado], was that on the central sector an attack on Moscow was imminent from the Mogilev–Gomel area, via Bryansk.

Following Marshal Shaposhnikov's résumé, I. V. Stalin indicated to me on his map the directions of the main enemy offensives and explained that a new strong defensive front must be built up in the Bryansk area as quickly as possible in order to cover Moscow. At the same time, a new striking force must be created for the defence of the Ukraine.

Stalin then asked Yeremenko where he would like to serve. The argument about this point throws an interesting light on the practices of the Soviet General Staff as well as the manner in which Stalin treated his generals. Here is Yeremenko's account:

> I replied, "I am prepared to go wherever you send me." Stalin regarded me intently and a shadow of impatience flickered across his features. Very curtly he asked, "But actually?" "Wherever the situation is most difficult," I replied quickly.
> "They are both equally difficult and equally complicated—the defence of the Crimea and the line in front of Bryansk," was his reply.
> I said, "Comrade Stalin, send me wherever the enemy will attack with armoured units. I believe I can be most useful there. I know the nature and tactics of German armoured warfare."
> "Very well," Stalin said, satisfied. "You will leave first thing tomorrow morning and start at once on the establishment of the Bryansk front. Yours is the responsible task of covering the strategic sector of Moscow from the south-west. The drive against Bryansk has been assigned to Guderian's armoured group. He will attack with all his might in order to break through to Moscow. You will encounter the motorized units of your old friend with whose methods you are acquainted from the Central Front."

The assurance with which Stalin expounded the plan of Army Group Centre is astonishing if one remembers how badly the Soviet Command had been informed about the German intentions during the first few weeks of the war.

Naturally, the fact that Moscow was within the zone covered by the German offensive plans was obvious even without secret tip-offs. But the German plan of attack might equally well have envisaged a drive down from the north. Indeed, the High Command Directive No. 34 of 10th or 12th August envisaged this very possibility. Guderian, on the other hand, did not wish to strike via Bryansk, but to drive towards Moscow from the Roslavl area along both sides of the Moscow highway. But the plan of operations submitted to Hitler on 18th August by Colonel-General Halder, Chief of the Army General Staff, as the proposal of the High Command, included the Bryansk area and agreed exactly with what Stalin told Yeremenko on 12th August.

Stalin believed in the Bryansk-Moscow operation. He believed in Alexander Rado. He continued to believe in him long after Hitler had overthrown the High Command's plan and ordered Guderian's Panzer Group to turn towards the south.

The stubbornness with which the Soviet Command clung to its idea of Moscow being the objective of the German offensive is reflected in the way in which highly revealing evidence by German prisoners of war and the downright alarming discoveries of Russian aerial reconnaissance were dismissed.

Yeremenko writes:

> Towards the end of August we took some prisoners who stated under interrogation that the German 3rd Panzer Division, having reached Starodub, was to move to the south in order to link up with Kleist's Panzer Group. According to these prisoners, the 4th Panzer Division was to keep farther to the right and move parallel with the 3rd Panzer Division. This information was confirmed by our aerial reconnaissance on 25th August, when a massive motorized enemy column was discovered moving in a southerly direction.

The prisoners' evidence was correct. They must have been well-informed troops who supplied this dangerous information to the enemy. It was quite true that on 25th August Guderian had ordered his 3rd and 4th Panzer Divisions as well as the 10th Motorized Infantry Division to cross the Desna river in the area of Novgorod Severskiy and Korop. The 17th Panzer Division and 29th Motorized Infantry Division provided flank cover against Yeremenko's divisions in the Bryansk area.

But the Soviet General Staff and Yeremenko believed in an offensive against Moscow. They regarded Guderian's drive to the south as a large-scale outflanking movement. Yeremenko notes: "From the enemy's operations I concluded that with his powerful advanced units, supported by strong armoured formations, he was engaged in an active reconnaissance and in a manœuvre designed to strike at the flank of our Bryansk front."

A fatal error. Guderian's Panzer divisions pushing to the south did not intend to wheel round towards Moscow, and the 29th Motorized Infantry Division and 17th Panzer Division, which were fighting against Yeremenko's positions in the dangerous, ambush-riddled forests along the road and railway to Bryansk, were not in fact aiming at Bryansk. They were covering Guderian's drive towards the Desna, a drive which was to close the trap behind the Soviet lines at Kiev. These flank-cover engagements were exceedingly costly. The fierce fighting there is linked with the name of Pochep. There the 167th Infantry Division was involved in heavy defensive fighting. Its 331st Infantry Regiment lost nearly its entire 3rd Company on a single day.

Meanwhile the 3rd Panzer Division, the "Bear" Division from

Berlin, was making rapid progress towards the upper reaches of the Desna, a wide, marshy river-course, where Timoshenko had feverishly built up defences during the past few weeks by pressganging the civilian population. During the day the German troops drove or fought, and at night they slept by the roadside, under their tanks or inside their lorries. Their objective was not Moscow but the towns of the Northern Ukraine.

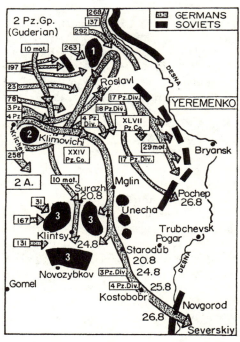

Map 5. Guderian's drive to the South. In a bold operation Panzer and infantry formations of 2nd Panzer Group and Second Army smashed the Soviets at Roslavl (1), Krichev (2), and in the Gomel area (3), forced the Desna river, and thus initiated the pincer operation against Kiev.

But the Soviet High Command was blind. Stalin not only employed his troops in the wrong direction, he did something far worse. He dissolved the Soviet central front with its Twenty-first and Third Armies —the front which formed a barrier across the Northern Ukraine—and placed the divisions under Yeremenko's Army Group for the defence of Moscow. Yeremenko remarks bitterly: "The High Command informed us once more that Guderian's blow was aimed at the right wing of the Bryansk front—in other words, against Moscow. On 24th August Comrade Shaposhnikov notified me that the attack was to be expected within the next day or two."

They waited for it in vain. Yeremenko's account continues:

> However, this assumption was not borne out. The enemy attacked in the south and merely brushed against our right wing. At that time neither the High Command nor the command in the field had any

evidence that the direction of the offensive of the German Army Group Centre had been changed and turned towards the south. This grave error by the General Staff led to an exceedingly difficult situation for us in the south.

Hitler and Stalin seemed to vie with each other in frustrating the work of their military commanders by fatal misjudgments. So far, however, only Stalin's mistakes were becoming obvious.

The date was 25th August. It was a hot day and the men were sweating. The fine dust of the rough roads enveloped the columns in thick clouds, settled on the men's faces, and got under their uniforms on to their skins. It covered the tanks, the armoured infantry carriers, the motor-cycles, and the jeeps with an inch-thick layer of dirt. The dust was frightful—as fine as flour, impossible to keep out.

The 3rd Panzer Division had been moving down the road from Starodub to the south for the past five hours. Its commander, Lieutenant-General Model, was in his jeep at the head of his headquarters group, which included an armoured scout-car, the radio-van, motor-cycle orderlies, and several jeeps for his staff. The infantrymen cursed whenever the column tore past them, making the dust rise in even thicker clouds.

Model, leading in his jeep, pointed to an old windmill on the left of the road. The jeep swung over a little bridge across a stream and drove into a field of stubble. Maps were brought out; a headquarters staff conference was held on the bare ground. The radio-van pushed up its tall aerials. Motor-cycle orderlies roared off and returned. Model's driver went down to the stream with two field-buckets to get some water for washing. Model polished his monocle. Bright and sparkling, it was back in his eye when Lieutenant-Colonel von Lewinski, CO of 6th Panzer Regiment, came to report. A Russian map, scale 1:50,000, was spread out on a case of hand-grenades.

"Where is this windmill?" "Here, sir." Model's pencil-point ran from the hill with the windmill right across on to the adjoining sheet which the orderly officer was holding. The pencil line ended by the little town of Novgorod Severskiy.

"How much farther?"

The Intelligence officer already had his dividers on the map. "Twenty-two miles, Herr General."

The radio-operator brought a signal from the advanced detachment. "Stubborn resistance at Novgorod. Strong enemy bridgehead on the western bank of the Desna to protect the two big bridges."

"The Russians want to hold the Desna line." Model nodded.

Certainly they wanted to. And for a good reason. The Desna valley was an excellent natural obstacle, 600–1000 yards wide. Enormous bridges were needed for crossing the river and its swampy banks. The big road-bridge at Novgorod Severskiy was nearly 800 yards long, and the smaller pedestrian bridge was not much shorter. Both were wooden bridges, and neither of them, according to the division's aerial reconnaissance squadron, had been blown up so far. But they were being defended by strong forces.

"We must get one of those bridges intact, Lewinski," Model said to the Panzer Regiment commander. "Otherwise it'll take us days, or even weeks, to get across this damned river." Lewinski nodded. "We'll do what we can, Herr General." He saluted and left.

"Let's go," Model said to his staff. As the main route of advance was congested with traffic the divisional staff drove along deep sandy forest tracks. Through thick woods their vehicles scrambled thirty miles deep into enemy territory. They might find themselves under fire at any moment. But if one were to consider that possibility one would never make any progress at all.

From ahead came the noise of battle. The armoured spearheads had made contact with the Russians. Motor-cycle troops were exchanging fire with Russian machine-guns. The artillery was moving into position with one heavy battery. Through his field-glasses Model could see the towers of the beautiful churches and monasteries of Novgorod Severskiy on the high ground on the western bank of the river. Beyond those heights was the Desna valley with its two bridges.

Russian artillery opened fire from the town. Well-aimed fire from 15·2-cm. batteries. The artillery was the favourite arm of the Soviets—just as it had been that of the Tsars. "Artillery is the god of war," Stalin was to say in a future Order of the Day. The plop of mortar batteries now mingled with the general noise. A moment later the first mortar-bombs were crashing all around. Model was injured in one hand by a shell-splinter. He had some plaster put on—that was all. But a shell got Colonel Ries, the commander of 75th Artillery Regiment. He died on the way to the dressing-station.

Low-level attack by Russian aircraft. "Anti-aircraft guns into action!"

The enemy's artillery was now finding its range. Time to change position.

The 6th Panzer Regiment and the motor-cycle battalion launched their attack that very evening at dusk. But the tanks were held up by wide anti-tank ditches with tree-trunks rammed in. The infantry regiment which was to have attacked the Russians from the north-west at the same moment had got stuck somewhere on the sandy roads.

Everything stop! The attack was postponed until the following morning.

At 0500 everything flared up again. The artillery used its heavy guns to flatten the anti-tank obstacles. Engineers blasted lanes through them. Forward! The Russians were fighting furiously and relentlessly

114 *Part One: Moscow*

in some places, but in others their resistance was half-hearted and incompetent. The first troops began to surrender—men between thirty-five and forty-five, largely without previous military service and with no more than a few days' training now. Naturally they did not stand up to the full-scale German attack—not even with the commissars behind them. German tanks, self-propelled guns, and motor-cycle infantry drove into the soft spots.

At 0700 hours First Lieutenant Vopel, with a handful of tanks from his 2nd Troop and with armoured infantry-carriers of 1st Company, 394th Rifle Regiment, took up a position north of Novgorod Severskiy. His task was to give support to an engineers assault detachment under Second Lieutenant Störck in their special operation against the big 800-yard-long wooden bridge. First Lieutenant Buchterkirch of 6th Panzer Regiment, who was Model's specialist in operations against bridges, had joined the small combat group with his tanks. Towards 0800 a huge detonation and cloud farther south indicated that the Russians had blown up the smaller bridge.

Everything now depended on Störck and Buchterkirch's operation.

Störck and his men in the armoured infantry-carriers took no notice whatever of what was happening to the right or left of them. They shot their way through Russian columns. They raced across tracks knee-deep in sand. Under cover of the thick dust-clouds they infiltrated among retreating Russian columns of vehicles. They drove through the northern part of the town. They raced down into the river valley to the huge bridge.

"It's still there!" Buchterkirch called out. Driver, radio-operator, and gunner all beamed. "Anti-tank gun by the bridge! Straight at it!" the lieutenant commanded. The Russians fled. Second Lieutenant Störck and his men leaped from their armoured carriers. They raced up on to the bridge. They overcame the Russian guard. There, along the railings, ran the wires of the demolition charges. They tore them out. Over there were the charges themselves. They pushed them into the water. Drums of petrol were dangling from the rafters on both sides. They slashed the ropes. With a splash the drums hit the water. They ran on—Störck always in front. Behind him were Sergeant Heyeres and Sergeant Strucken. Corporal Fuhn and Lance-corporal Beyle were dragging the machine-gun. Now and again they ducked, first on one side then on the other, behind the big water-containers and sand-bins.

Suddenly Störck pulled himself up. The sergeant did not even have to shout a warning—the lieutenant had already seen it himself. In the middle of the bridge lay a heavy Soviet aerial bomb, primed with a time-fuse. Calmly Störck unscrewed the detonator. It was a race against death. Would he make it? He made it. The five of them combined to heave the now harmless bomb out of the way.

They ran on. Only now did they realize what 800 yards meant. There did not seem to be an end to the bridge. At last they reached the far side and fired the prearranged flare signal for the armoured spearhead. Bridge clear.

Buchterkirch in his tank had meanwhile driven cautiously down the bank and moved under the bridge. Vopel with the rest of the tanks provided cover from the top of the bank. That was just as well. For the moment the Russians realized that the Germans were in possession of the bridge they sent in demolition squads—large parties of 30 or 40 men, carrying drums of petrol, explosive charges, and Molotov cocktails. They ran under the bridge and climbed into the beams.

Coolly Buchterkirch opened up at them with his machine-gun from the other side. Several drums of petrol exploded. But wherever the flames threatened to spread to the bridge squads of engineers were on the spot instantly, putting them out. Furiously, Soviet artillery tried to smash the bridge and its captors. It did not succeed. Störck's men crawled under the planking of the bridge and removed a set of charges —high explosives in green rubber bags. A near-by shell-burst would have been enough to touch them off.

Half an hour later tanks, motor-cycle units, and self-propelled guns were moving across the bridge. The much-feared Desna position, the gateway to the Ukraine, had been blasted open. A handful of men and a few resolute officers had decided the first act of the campaign against the Ukraine. Russia's grain areas lay wide open ahead of Guderian's tanks. Under a brilliant sunny late-summer sky they rolled southward.

Second Lieutenant Störck was just getting a medical orderly to stick some plaster on the back of his injured left hand when General Model's armoured command-vehicle came over the bridge.

The Second Lieutenant made his report. Model was delighted. "This bridge is as good as a whole division, Störck." At the same moment the Russian gunners again started shelling the bridge. But their gun-laying was bad, and the shells fell in the water. The General drove down the bank. Tanks of 1st Battalion, 6th Panzer Regiment, followed by 2nd Company, 394th Rifle Regiment, were moving into the bridgehead. The noise of battle in front grew louder—the plop of mortars and the rattle of machine-guns, interspersed by the sharp bark of the 5-cm. tank cannon of Lieutenant Vopel's 2nd Company. The Russians rallied what forces they could and, supported by tanks and artillery, threw them against the still small German bridgehead. They tried to eliminate it and recapture the bridge of Novgorod Severskiy— or at least destroy it.

But Model knew what the bridge meant. He did not need Guderian's reminder over the telephone: "Hold it at all costs!" The bridge was their chance of getting rapidly behind Budennyy's Army Group Southwest by striking from the north. If Kleist's Panzer Group, operating farther south, under Rundstedt's Army Group South, pushed across the lower Dnieper and wheeled north, a most enormous pocket would be formed, one beyond the wildest dreams of any strategist.

6. The Battle of Kiev

Rundstedt involved in heavy fighting on the southern wing—Kleist's tank victory at Uman—Marshal Budennyy tries to slip through the noose—Stalin's orders: Not a step back!—Guderian and Kleist close the trap: 665,000 prisoners.

BUT where was Colonel-General von Kleist? What was the situation on Field-Marshal von Rundstedt's front? Where were the tanks and vehicles with the white 'K,' the mailed fist of Army Group South? What had been happening on the Southern Front while the great battles of annihilation were fought on the Central Front at Bialystok, Minsk, Smolensk, Roslavl, and Gomel?

What Smolensk meant for Army Group Centre, Kiev was to Army Group South. The Ukrainian capital on the right bank of the lower Dnieper, there about 700 yards wide, was to be captured following the annihilation of the Soviet forces west of the river—in exactly the same way as Smolensk had been taken following the battle of encirclement in the Bialystok–Minsk area.

For Rundstedt's Army Group South, however, the plan did not go so smoothly as in the centre. There were several nasty surprises. Since, for political reasons, no operations were to be mounted initially on the 250 miles of Rumanian frontier in the Carpathians, the entire weight of the offensive had to be borne by the left wing—*i.e.*, the northern wing—of the Army Group. There General von Stülpnagel's Seventeenth Army and Field-Marshal von Reichenau's Sixth Army were to break through the Russian lines along the frontier, to drive deep through the enemy positions towards the south-east, and then—with Kleist's Panzer Group leading—turn towards the south and encircle the Soviets, with Kleist's Panzer Corps acting as pincers. Or, rather, as one jaw of a pair of pincers. For, unlike Army Group Centre, Rundstedt had only one Panzer Group. The second jaw of the pincers, very much shorter, was to be provided by Colonel-General Ritter von

The Battle of Kiev 117

Schobert's Eleventh Army, which was in the south of Rumania. This army was to cross the Prut and Dniester and drive to the east, in the direction of Kleist's armour—in order to close the huge pocket behind Budennyy's 1,000,000-strong Army Group.

It was a good plan, but the enemy facing Rundstedt was no fool. Besides, more important, he was twice as strong. Budennyy was able to oppose Kleist's 600 tanks with 2400 armoured fighting vehicles of his own—including some of the KV monsters. And he had entire brigades of the even more terrifying T-34.

On 22nd June the German divisions had successfully crossed the frontier rivers also in the south, and had pushed through the enemy's fortified positions along the border. But the planned rapid breakthrough on the northern wing did not materialize. To make a single Panzer Group the striking force for the conquest of so large and so well defended an area as the Ukraine had been a planning mistake. The rapid successes on the Central Front had been achieved by revolutionary tactical skill. There the two powerful Panzer Groups, boldly led, had encircled and liquidated the bulk of the Soviet defending forces. In the south and in the north, on the other hand, the absence of a two-pronged attack made it impossible to reach the intended targets. There simply were not enough armoured formations for the kind of vast-scale operation Hitler expected his armies in the east to perform along the entire front. Failure to reach the scheduled targets in the south was not due to any lack of skill on the part of the commanders, nor to any lack of courage, nor, least of all, to a lack of staying power on the part of the troops. It was simply due to the fact that there were too few armoured units—to a numerical shortage of the very service branch intended to carry the whole of Operation Barbarossa.

Not till after eight days of very heavy fighting, on 30th June, did the Soviet lines begin to waver. Rundstedt's northern wing rushed forward. But presently it was halted again by a new position—the hitherto unknown "Stalin Line." Heavy thunderstorms had turned the roads into quagmires. The tanks struggled forward. Bale after bale of straw was collected by the grenadiers from the villages and flung down in the mud. Even the infantry got stuck with their vehicles and made only very slow progress.

In the early light of 7th July Kleist's Panzer Group succeeded in penetrating the Stalin Line on both sides of Zvyagel. The 11th Panzer Division under Major-General Crüwell pierced the line of pillboxes and fortifications in full depth and by a bold stroke took the town of Berdichev at 1900 hours. The Russians withdrew. But they did not retreat everywhere. The 16th Motorized Infantry Division got stuck in the line of pillboxes near Lyuban. There the Russians even counter-attacked with armour. General Hube's 16th Panzer Division likewise encountered stiff resistance at Starokonstantinov. The troops ran out of ammunition. Transport aircraft had to bring up supplies for the tanks. Bombers, Stukas, and fighter bombers of Fourth Air Fleet came to the division's assistance and smashed concentrations of Soviet armour. The combat group Höfer of 16th Panzer Division drove farther to the east and over-ran retreating artillery regiments. The 1st Battalion, 64th Rifle Regiment, experienced its most costly day of close combat at Stara Bayzymy. Within two hours the 1st Company lost three successive company commanders.

At long last, with the help of 21-cm. mortars, the bulk of 16th Panzer Division broke through the Stalin Line at Lyuban on 9th July. General Hube heaved a sigh of relief: only another 125 miles to the Dnieper.

The only pleasant feature of these weeks was the abundance of eggs. Early in July the division had captured an enormous Red Army food store with a million eggs. The quartermasters replenished their stocks. For a long time the only worry of the NCO cooks was inventing new ways of serving eggs.

The officers at Divisional, Corps, and Army Group Headquarters had different worries. Anyone thinking that the crack units entrusted by Stalin with the defence of the Ukraine were on the point of collapse was soon disillusioned. Defeated at one moment, they rallied again at the next. They hung on to their positions. They withdrew, and a moment later they stood and fought again. Fierce fighting flared up around Berdichev. The Russians used all the artillery they could lay hands on. The German artillery was completely smothered by the Soviet fire. Only by a supreme effort did Crüwell succeed in keeping them down with his reinforced 11th Panzer Division. Things were similar throughout the southern sector. The Russians were indomitable. Rundstedt did not succeed in trapping Kirponos.

The fighting had been going on for over twenty days, and no decisive success had so far been achieved. There was some impatience at the Fuehrer's headquarters. Things were moving too slowly for Hitler. Suddenly he got the idea that "small pockets" would be a better plan. He therefore demanded that Kleist's Panzer Group should operate as three separate combat groups with the object of forming smaller pockets. One combat group was to form a tiny pocket near Vinnitsa, in conjunction with Eleventh Army coming up from the south. Another group was to drive towards the south-east in order to cut off any enemy forces intending to withdraw from the Vinnitsa area. A third combat group, finally, was to drive towards Kiev, together with Sixth Army, and to gain a bridgehead on the eastern bank of the Dnieper.

Field-Marshal von Rundstedt emphatically objected to having his one and only Panzer Group split up in such a manner. This, he argued, was an unforgivable sin against the spirit of armoured warfare.

"Operations in driblets won't give results anywhere," he telephoned to Hitler's headquarters. Hitler relented.

On the western bank of the Dnieper Kleist's Panzer Group pushed past Kiev towards the south-east with concentrated forces, and thus created the prerequisite either for describing a smaller arc towards Vinnitsa or a larger one towards Uman.

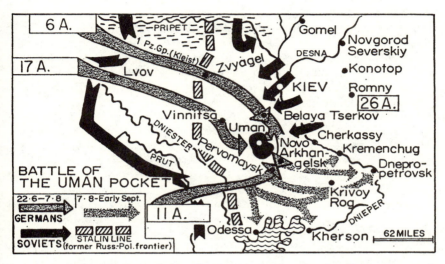

Map 6. The battle of encirclement of Uman was developed with only one prong and in a fluid movement. Kleist's Armoured Corps circumnavigated twenty-five Soviet divisions and forced them against a wall of German infantry, composed of units of Sixth, Seventeenth, and Eleventh Armies. Three Soviet Armies were smashed.

Everything was set for a small and for a large pocket south of Kiev. All the factors had been carefully taken into account, except one—Budennyy. The Marshal with the bushy moustache, who had been appointed Commander-in-Chief Army Group South-west Sector on 10th July, now played his final card. From the impenetrable Pripet Marshes, impassable to armour, he sent the divisions of his Fifth Rifle Army under Major-General Potapov against the northern flank of Reichenau's Sixth Army. These tactics, just as in the area of Army Group Centre, gave rise to severe and critical defensive fighting on Reichenau's left flank. But here, too, all went well in the end.

On 16th July Kleist's tanks reached the key centre of Belaya Tserkov. The first major battle of encirclement offered itself. Rundstedt wanted a large-scale outflanking movement and a big pocket. But Hitler ordered the lesser alternative, and for once he was right. A break in the weather favoured the movements of the Panzer divisions. Kleist struck accurately at the retreating enemy forces. On 1st August he the Eighteenth. The commanders-in-chief of the Sixth and Twelfth Pervomaysk. Then he wheeled to the west and, in conjunction with the infantry divisions of Seventeenth and Eleventh Armies, closed the ring around the Russian forces in the Uman area.

It was not a huge pocket as at Bialystok, Minsk, or Smolensk. Nevertheless three Soviet Armies were smashed—the Sixth, the Twelfth, and the Eighteenth. The commanders-in-chief of the Sixth and Twelfth Armies surrendered. But 'only' 103,000 prisoners were taken as a result of this classical battle of encirclement with reversed front, fought under such exceedingly difficult conditions. Considerable enemy forces succeeded in breaking out, even though the 1st and 4th Mountain Divisions, as well as the 257th Infantry Division from Berlin, repeatedly tried hard to seal the gaps. Major Wiesner's 1st Artillery Battalion, 257th Infantry Division, fired as accurately as if on the practice range and blasted column after column trying to break out of the pocket. The scale of the fighting is reflected in a single figure: the four guns of 9th Battery, 94th Mountain Artillery Regiment, fired 1150 rounds during the four days of the battle of Uman. That was more than the total rounds fired by the battery throughout the campaign in France. The enemy weapons destroyed or captured testified to the fierceness of the fighting: 850 guns, 317 tanks, 242 anti-tank and anti-aircraft guns were abandoned by the Russians on the battlefield.

Yet the significance of the battle of Uman considerably surpasses these numerical results. The strategic implications of the victory won by Army Group South were considerably greater than the number of prisoners suggested.

The road to the east, into the Soviet iron-ore region of Krivoy Rog and to the Black Sea ports of Odessa and Nikolayev, was open. Above all, Kleist's Panzer Corps was now able to push through to the lower Dnieper and to gain the western bank of the Dnieper bend from Cherkassy to Zaporozhye. This move, moreover, offered the chance of a great battle of annihilation around Kiev—a chance which so mesmerized Hitler that he halted Army Group Centre's offensive against Moscow and diverted Guderian's tanks to the south, towards Kiev. These two great armoured prongs were now to embark on a new gigantic battle of encirclement against the Soviet South-west Front with its one million men.

On 29th August Guderian's Fieseler Storch aircraft took off from Novgorod Severskiy and described a bold arc over the Russian front. Above the Russian lines, right on top of Yeremenko's divisions attacking the German bridgehead, he dipped down low, then banked, and across the Desna returned to Unecha, to Panzer Group headquarters. The time was just before 1800.

The Battle of Kiev 121

Guderian had been to see his 3rd and 4th Panzer Divisions which were trying to extend their bridgehead in order to continue their thrust to the south. But the troops were pinned down. He had also been to XLVI Corps, whose 10th Motorized Infantry Division and 17th and 18th Panzer Divisions were busy repelling fierce Soviet attacks from the flank. The situation there was not too rosy either. Too much was being demanded of the men. They were short of tanks and short of sleep.

Next to Guderian sat Lieutenant-Colonel Bayerlein, the situation map spread out on his knees. Thick red arrows and arcs on the map indicated the strong Russian forces in front of the German spearheads and along their flanks. "Yeremenko is going all out to reduce our bridgehead," Guderian was thinking aloud. "If he succeeds in delaying us much longer, and if the Soviet High Command discovers what we are trying to do to Budennyy's Army Group, the whole splendid plan of our High Command could misfire."

Bayerlein confirmed the anxieties of his Commander. "I was on the phone to Second Army yesterday. Freiherr von Weichs seems to be worried about it too. Lieutenant-Colonel Feyerabend, their chief of operations, has had reports from long-range reconnaissance about the Russians beginning to withdraw from the Dnieper front below Kiev. At the same time, work has been observed in progress on positions in the Donets area."

"Well, there you are." Guderian was getting heated. "Budennyy has learnt his lesson at Uman. He's slipping through the noose. Everything now depends on which of us is quicker."

But Guderian and Weichs need not have worried. True, Budennyy had realized the danger threatening his Army Group in the Dnieper bend around Kiev from a German thrust jointly from the north and the south. He had planned his withdrawal and was building up new lines of interception along the Donets. But Stalin would not hear of a withdrawal. On the contrary, he squeezed another twenty-eight major formations into the already packed river-bend. Whatever came off the assembly lines of the famous tank factories in Kharkov was thrown into the Dnieper bend—the modern T-34s, the T-28s, super-heavy self-propelled guns, heavy artillery, and multiple mortars.

"Not a step back. Hold out and, if necessary, die," was Stalin's order. And Budennyy's Corps obeyed. Rundstedt's divisions on the northern wing of his Army Group soon discovered it. The experienced 98th Infantry Division from Franconia and the Sudetenland lost 78 officers and 2300 other ranks within eleven days of fighting for the key point of Korosten. The battle on the Desna between Guderian's and Yeremenko's divisions also went on for eight days. It was a terrible battle— a fierce struggle for every inch of ground. "A bloody boxing match," was Guderian's description. But then came the moment, a lucky accident exploited by a bold operation, when the tide was definitely turned against Budennyy.

In the afternoon of 3rd September the Intelligence officer of XXIV Panzer Corps placed a dirty and charred bundle of papers on the desk of his Corps commander, General Geyr von Schweppenburg. The papers came from the bag of a Soviet courier aircraft that had been shot down. Geyr read the translation, studied the map, and beamed. The papers clearly revealed the weak link between the Soviet Thirteenth and Twenty-first Armies. At once Geyr moved his 3rd Panzer Division against that gap. Guderian was informed by telephone.

The next morning Guderian turned up at Geyr's headquarters. It had taken him four and a half hours by car to cover the 48 miles: such was the condition of the roads after only a short rainfall. But he was cheered by the news awaiting him at Geyr's headquarters. General Model's 3rd Panzer Division had in fact driven into the gap in the Soviet lines. His tanks had torn open the flanks of the two Soviet Armies. As through a burst dam, the rifle regiments and artillery battalions were now spilling through to the south.

Guderian at once drove to Model. "This is our chance, Model." There was no need for him to add anything. Model's units were already racing to the Seym and towards Konotop in a headlong chase. Three days later, on 7th September, the advanced battalion of 3rd Panzer Division under Major Frank succeeded in crossing the Seym and establishing a bridgehead. On 9th September the 4th Panzer Division likewise crossed the river. Stukas, supporting the experienced 35th Panzer Regiment and the 12th and 33rd Rifle Regiments, blasted the way for them through units of the Soviet Fortieth Army which had been freshly launched against the bridgehead. The Russians began to fall back.

Meanwhile Model's 6th Panzer Regiment was still outside Konotop. At the "Wolfsschanze" in East Prussia and in Bock's headquarters in Smolensk Guderian's headlong rush was being followed intently. It was important that Colonel-General von Kleist, down in the south, should be given his starting order at the right moment.

Major Frank had thrust past Konotop.

Army Group at once rang up Guderian: "Final order: drive towards Romny. Main pressure on the right." That meant that the pocket around Budennyy was to be closed in the Romny area. It was there that Guderian's and Kleist's tanks were to meet.

Romny had been the headquarters of King Charles XII of Sweden in December 1708; 93 miles away was Poltava, where Tsar Peter the Great inflicted a crushing defeat on the Swedes in 1709. The battle was the death-blow to Sweden's Nordic empire, and marked the emergence of Russia as a modern great power in history. Was this era now to come to an end again at Romny?

Everything went like clockwork. Guderian's tanks achieved the decisive breakthrough at Konotop. It was pouring with rain. But victory lent the troops new strength. The spearheads of 3rd Panzer Division were racing towards Romny. They were far in the rear of the enemy. But where was Kleist? Where was the second prong of the vast pincers? He had been held back wisely so that the Russians should not realize prematurely the disaster that was about to overtake them.

In the evening of 10th September Kleist's XLVIII Panzer Corps, commanded by General Kempf, reached the western bank of the Dnieper near Kremenchug, where the Soviet Seventeenth Army was holding a small bridgehead. Here, too, the ground and the roads had been turned into quagmires by violent late-summer thunderstorms and torrential rain. Nevertheless a temporary bridge was finished by noon of 11th September. Parts of 16th Panzer Division crossed over. Throughout the night the division from Rhineland–Westphalia drove or marched through complete darkness and heavy rain over to the other bank. On the following morning at 0900 Hube's tanks moved into the attack. Against a stubbornly resisting enemy the division gained 43 miles in 12 hours, over roads knee-deep in mud. It was followed by General Hubicki's 9th (Viennese) Panzer Division.

On 13th September the 16th Panzer Division stormed Lubny. The town was defended by anti-aircraft units and workers' militia, as well as formations of the NKVD, Stalin's secret police. The 3rd Company of the Engineers Battalion, 16th Panzer Division, captured the bridge over the Sula in a surprise coup. Using "Stukas on foot"—howling smoke mortars—they confused and blinded the Russians and in a spirited assault took the suburbs of the town. Behind them came the 2nd Battalion, 64th Rifle Regiment. Savage street fighting developed. The Soviet commander in the field had called the entire Russian civilian population to arms. Firing came from roofs and cellar windows. Behind barricades combat units armed with Molotov cocktails pelted the tanks. The eerie fighting continued throughout the day.

On 14th September, a Sunday, 79th Rifle Regiment joined action. In the afternoon Lubny was in German hands. By the evening the division's reconnaissance detachment was still 60 miles away from the spearheads of 3rd Panzer Division.

The Russians meanwhile had realized their danger. Aerial reconnaissance by the German Second and Fourth Air Fleets reported that enemy columns of all types were on the move from the Dnieper front against Guderian's and Kleist's formations, in the direction of the open gap. That gap had to be closed unless large parts of the Soviet forces were to get away.

Striking from the north, Guderian's divisions had taken Romny and Priluki. Model, with a single regiment, was struggling over muddy

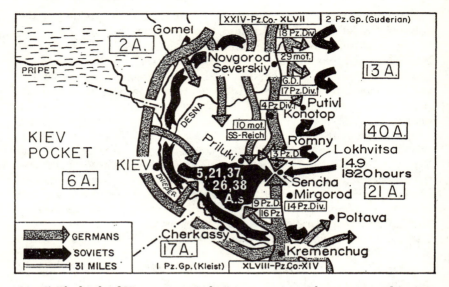

Map 7. The battle of Kiev was a typical pincer operation with two armoured prongs. Guderian struck from the north, while Kleist moved in from the south. While the Russians were involved in heavy defensive fighting in the Dnieper bend, the German fast troops closed the trap behind them.

roads to Lokhvitsa. The rest of the division was still stuck in the mud a long way back. Major Pomtow, the chief of operations of 3rd Panzer Division, was tearing his hair.

There was still a gap of 30 miles between the two Panzer Groups. A 30-mile-wide loophole. Russian reconnaissance aircraft were circling over the gap, directing supply columns through the German lines. Hurriedly assembled groups of tanks were moving ahead to clear a path for them. General Geyr von Schweppenburg at his advanced battle headquarters suddenly found himself under attack by one of these Russian columns trying to break through the German ring. The headquarters turned itself into a strongpoint. An SOS was sent to 2nd Battalion, 6th Panzer Regiment. But they were still 12 miles away. In the nick of time Lieutenant Vopel's 2nd Company succeeded in snatching the general commanding XXIV Panzer Corps from almost certain death. The offensive towards the south continued.

The time was 1200 hours, the scene a muddy road near Lokhvitsa. "First Lieutenant Wartmann to the commander!" The order was passed on through the column. Wartmann, commanding a tank company, waded through the mud to the command tank of Lieutenant-Colonel Munzel, the new OC 6th Panzer Regiment. A quarter of an hour later the tanks started up their engines and the armoured infantry carriers of 3rd Platoon, 1st Company, 394th Regiment, under Sergeant Schröder, moved over to the right to make way for the tanks. The tankmen removed the camouflage from their vehicles: Lieutenant Wartmann was organizing a strong detachment for a reconnaissance towards the south. His orders were: "Drive through the enemy lines and make contact with advanced formations of Kleist's Panzer Group."

At 1300 hours the small combat group passed through the German pickets near Lokhvitsa. Stukas escorted them for a short distance. The sun shone down from a cloudless sky. The undulating country stretched far to the horizon. In front were the dark outlines of a wood. They had to pass through it. Suddenly a hastily retreating Russian column crossed their path—supply vehicles, heavy artillery, engineering battalions, airfield ground crews, cavalry units, administrative services, fuel-supply columns. The vehicles were hauled by tractors and horses. They carried drums of petrol and oil.

"Turret one o'clock. H.E. shells. Fire!" The fuel vehicles burned like torches. The horses broke loose. The Russians scuttled into the forest and behind the thatched farmhouses of a village. There was chaos on the road.

The German detachment moved on. Their job was not to fight the enemy, but to make contact with the forward units of Army Group South. They were still in radio contact with Division. There Major Pomtow was sitting next to the radio operator, intently following the recce unit's report on enemy dispositions, terrain, and bridges. Pomtow read: "Stiffening enemy opposition." Then there was silence. What had happened?

From Wartmann's tank, meanwhile, the situation looked like this. Horse-drawn carts and tractors were standing on the road, abandoned. Machine-gun and anti-tank fire was coming out of sunflower fields. Wartmann halted his tanks. He looked through his binoculars. A windmill on a near-by hill caught his attention. It was behaving rather strangely: one moment its sails would go round one way and then again the other way. Then they would stop altogether. Wartmann let out a soft whistle. Clearly an enemy observer was there, directing operations. "Tanks forward!" A moment later the 5-cm. shells slammed into the windmill. Its sails turned no longer. Forward.

Pomtow's radio operator, earphones clamped on his head, was writing: "1602 hours. Have reached Luka, having crossed Sula river over intact bridges." Pomtow was smiling. That was good news. Wartmann's detachment moved on—through uncanny terrain with deep sunken lanes, swamps, and sparse forest. Whichever way he turned he saw enemy columns.

Wartmann's tanks had covered 30 miles. The day was drawing to its end. Suddenly radio contact was lost. Over to the south the silhouette of a town could be made out against the evening sky. That no doubt was Lubny, the area where 16th Panzer Division was operating. They could hear the noise of battle. Evidently they had got close to the fighting lines of the southern front. But which way was the enemy? Was he in front or were they about to run into his flank?

Cautiously the armoured scout cars accompanying the tanks picked their way across a vast cornfield with the harvested grain piled in stooks. They dodged from one stook to the next. Suddenly an aircraft appeared overhead. "Look—a German reconnaissance plane!" "White Very light!" Wartmann commanded. With a whoosh the flare streaked up from the turret of the tank. White signals always meant: Germans here. A tense moment. Yes, the plane had seen it. He dipped down low. He circled. He circled again. "He's touching down!" And already the machine was rolling to a stop among the stooks in the cornfield—right among the enemy lines. There was much laughter and hand-shaking.

To-day nobody knows who the three resolute airmen were. They informed Lieutenant Wartmann about the situation on the front: less than six miles away were units of Kleist's 16th Panzer Division. A moment later the aircraft took off again. Wartmann's men could see it dip down low beyond the wide ravine, dropping a message.

"Tanks forward!" On they went. Through the ravine and up the far bank.

Infantry in field grey were scrambling up the slope in battle order. "White Very lights!" Wartmann ordered for the second time that day. At once the reply came—also in white. The men shouted with joy and flung up their arms. They were the 2nd Company of the Engineers Battalion, 16th Panzer Division, under First Lieutenant Rinschen. The two officers shook hands among scenes of enthusiasm. Their handshake meant that the trap 130 miles to the east of Kiev had now been closed, even though so far only symbolically.

At Model's headquarters the radio suddenly sprang into life again. "Connection re-established!" the operator shouted. Then he listened. Five minutes later the chief of operations dictated to his map drafts-man the following entry, to be placed next to a tiny blue lake: "14th September 1941, 1820 hours: link-up of First and Second Panzer Groups."

In the orchard outside the headquarters of 2nd Panzer Regiment the tanks and troop-carriers with the white G and the white K were standing next to each other, well camouflaged under trees and hedges. The sky was alive with the flashes of the artillery and the howling of mortar salvos. The curtain was being rung up on the last act of the greatest battle of encirclement in military history.

The very next day the 9th Panzer Division with units of 33rd Panzer Regiment, having moved north on the road east of the Sula river after the capture of Mirgorod, linked up with the most forward parts of

3rd Panzer Division by the bridge of Sencha. Now the ring was properly closed and the trap shut behind fifty enemy divisions.

There was more fierce fighting to come with the encircled armies, as well as with the forces employed by the Soviet High Command from outside with the intention of saving Budennyy. There were some critical situations, especially along Guderian's extended eastern flank. Near Romny on 18th September an attack from the flank launched with four divisions against the German 10th Motorized Infantry Division and a few AA batteries got within 900 yards of Guderian's observation post up on the tower of the town gaol, and was halted only with great difficulty.

At Putivl the cadets of Kharkov stormed singing against the positions of 17th Panzer Division and the "Grossdeutschland" Motorized Infantry Regiment. They were killed to the last man. Near Novgorod Severskiy six Soviet divisions, supported by armoured formations, pounced on the combat-hardened 29th Motorized Infantry Division.

But it was all in vain. The Russian attacks were not aimed at a single focus. They caused some critical situations, but they did not turn the tide. The Russians did not succeed in denting Guderian's 155-miles-deep flank even at a single point.

On 19th September Sixth Army infantry—more particularly, divisions of XXIX Army Corps—took Kiev. By 26th September the great battle was over. Five Soviet Armies had been smashed completely, and two more badly battered. One million men had been killed, wounded, scattered, or taken prisoner. Marshal Budennyy, Stalin's old comrade-in-arms and once a sergeant in the Tsarist Army, had been flown out of the pocket on top-level orders. Stalin did not want this hero of the Revolution to fall into German hands or to be killed. Budennyy's command was taken over again by Colonel-General Kirponos. He was killed in action, together with his Chief of Staff, Lieutenant-General Tupikov, while trying to break out of the ring.

In figures the balance of the battle was as follows: 665,000 prisoners, 3718 guns, 884 armoured fighting vehicles, and a vast quantity of other war material. One single Panzer Corps, General Kempf's XLVIII Corps, which had its three divisions engaged right at the centre of this vast battle of annihilation, alone took 109,097 prisoners—more than the total number of prisoners taken in the battle of Tannenberg in the First World War.

The numerical scale of the battle was unprecedented in history. Five Armies had been destroyed. The reason for this victory was superior direction of operations on the German side, the daring mobility of German units, and the toughness of the troops.

It was a tremendous defeat for Stalin. When Guderian questioned Potapov, the forty-year-old Commander-in-Chief of the Soviet Fifth Army, who had been taken prisoner by Model's Panzerjägers, why he

128   *Part One: Moscow*

had not evacuated the Dnieper bend in good time, the Russian general replied, "Army Group had issued orders for the evacuation. We were, in fact, withdrawing towards the east when an order from the highest quarters—this means Stalin—instructed us to turn back and fight, in accordance with the slogan: 'Stand fast, hold out, and if need be die.'"

Potapov spoke the truth. On 9th September Budennyy had issued orders for preparations to be made for a withdrawal and requested Stalin to agree to his abandoning Kiev and the Dnieper bend. But the dictator had thrown a fit of temper and had issued his famous Stand-fast-and-die order.

Stand fast and die! That order had cost a million men. It had cost the whole of the Ukraine. Now the road to the Crimea and the Donets basin was open. Stalin's mistake and stubbornness had terrible consequences—all but fatal consequences. Yet, in retrospect, it may well be that this mistake resulted in Russia's victory. The rapid progress of the campaign and his belief in the accomplished strategic surprise and in the invincibility of German arms produced in Hitler that spiritual pride which was to lead presently to a string of fateful mistaken decisions.

The first great mistake that sprang from the victory at Kiev was Hitler's conclusion that the Russians in the south were no longer able to build up a serious line of defence. He therefore ordered: "The Donets basin and the Don are to be reached before the onset of winter. The blow at the Soviet Union's industrial heart must be struck swiftly."

Hitler was anxious to gain the Soviet Union's industrial heart as soon as possible in order to make it beat for the German war effort.

But if Stalin's power was indeed reeling after the crushing blows of the summer campaign, then why not strike at its political heart as well? Why not exploit the demoralization in the enemy camp and deliver the coup de grâce by the capture of Moscow? Why not lay low the dizzy, reeling colossus by one last furious assault?

On the final day of the battle of Kiev Hitler therefore ordered the opening of the battle of Moscow. Its code name was "Typhoon." D-Day was 2nd October. The objective was Moscow itself. With bated breath the officers and men of the Eastern Front heard the Order of the Day from Hitler's headquarters being read out to them: "The last great decisive battle of this year will mean the annihilation of the enemy."

7. Code Name "Typhoon"

Caviare for Churchill–The mysterious town of Bryansk–Moscow's first line of defence over-run–Looting in Sadovaya Street–Stopped by the mud–Fighting for Tula and Kalinin–The diary of a Russian lieutenant–Secret conference at Orsha–Marshal Zhukov reveals a Soviet bluff.

MR COLVILLE had scarcely shut his boss's bedroom door behind him when he heard him yell out in rage. He turned back. Mr Churchill was sitting up in bed. Spread out around him were the morning papers. Opened before him lay the *Daily Express*.

Angrily Mr Churchill brought his hand down flat on the paper. "Have a look at this." He pointed to a dispatch from Moscow. Churchill's secretary, too, was speechless when he read the report. Lord Beaverbrook, it said, who had been in Moscow with a mixed British-American delegation since 28th September in order to sign an agreement about military and economic aid to the Soviet Union in its war with Germany, had instructed a man in his entourage to spend a considerable sum of money on caviare—for Mr Churchill.

"That's a dirty trick," Churchill fulminated. Colville knew that no such request had ever come from Mr Churchill.

Britain, after all, had more serious things to worry about in September 1941. In North Africa Rommel had surrounded Tobruk; he had pushed far east to the Halfaya Pass and was threatening to strike at Cairo.

But that was not the worst of it. Hitler's U-boat campaign was making life difficult for Britain. The new German tactics of operating in packs and the employment of larger U-boats had again begun to offset what defensive successes the British had achieved during the summer. The battle in the Atlantic continued to rage with unabated fury. In September alone Dönitz's "grey wolves" had sunk 683,400 tons. Thus the total tonnage sunk since the beginning of the war had risen to 13,700,000—more than half the British merchant fleet. And new building could replace no more than 10 per cent. Britain's supplies were in a critical position. Most Britons considered themselves lucky if they got one egg for their Sunday breakfast. And just then Beaverbrook's mass-circulation paper must announce that the Prime Minister, who daily demanded sweat and tears from his nation, would receive from Moscow a present of caviare by the pound—that very symbol of luxury and good living.

Still in bed, Churchill dictated a furious telegram for transmission

E

by the Foreign Office to his lordship in Moscow. It was handed to Beaverbrook by an Embassy secretary just as he was in conference with Molotov and Harriman.

The Press lord's interview with his Moscow correspondent, summoned to the presence, was noisy but unsuccessful. The correspondent was stubborn. He had got hold of the story, and he maintained it was true. Why shouldn't he report it? Was this not in line with his lordship's principles? Beaverbrook surrendered. Churchill did not get any caviare.

This happened in Moscow on 30th September 1941—the very day when the fate of Stalin's capital seemed already to have been sealed by a thousand action and marching orders. For on that day the entire force of Field-Marshal von Bock's Army Group Centre was set into motion to capture Moscow.

The Muscovites had no suspicion of all this. Since the German Blitzkrieg against the Soviet metropolis had been stopped behind Smolensk in the Yelnya bend and on the Vop about the middle of July, the inhabitants of the city had got used to the idea that the enemy was less than 200 miles away. After a while 200 miles seemed quite a healthy distance. Moscow had been spared. The war had swung to the south. True, something had happened at Kiev, but the Soviet High Command communiqué of 30th September reported laconically: "Our troops are engaged in fierce defensive fighting along the entire front." This was followed by fantastic figures about some 560 German aircraft shot down and destroyed during the preceding six days. It looked as if the Germans were being defeated in the air and were unable to make progress on the ground.

"What does the communiqué say about the situation up at Leningrad?" Ivan Ivanovich asked his father as he returned home on the morning of 30th September from helping to dig an anti-tank ditch far to the north of Moscow. "It doesn't say a thing," said the concierge of No. 5 Kaluga Street. "And what are those liars on the radio saying about the situation in the south, where Grandfather lives?" "They say that we have destroyed many tanks on our South-western Front. And that we have taken up new defensive positions according to plan." "And outside the city here? What's the situation there? Did they say anything on the radio?" "Yes." Ivan's father nodded proudly. "Near Vitebsk our partisans blew up a great many fascists. And they blasted the road. The Hitlerites can't get any further."

Ivan Ivanovich nodded. He went out to the kitchen to look for a piece of bread. His father could hear him grumble. The slice that was left did not seem large enough to his son. "There's some cabbage soup," he called out.

While Ivan Ivanovich Krylenkov was eating his watery cabbage soup on that morning of 30th September in the basement of Moscow's Kaluga Street, some 300 miles away, near Glukhov, in the Northern

Code Name "Typhoon" 131

Ukraine, Second Lieutenant Lohse, commanding 1st Company, 3rd Rifle Regiment, raised his hand in his armoured car: "Forward!" And just as the spearhead of the 3rd Panzer Division moved off towards the east at Glukhov, along with it the 4th Panzer Division, the 10th Motorized Infantry Division, and the whole XXIV Panzer Corps moved into action. To the left was General Lemelsen's XLVIII Panzer Corps with 17th and 18th Panzer Divisions and 29th Motorized Infantry Division. Behind was General Kempf's XLVIII Panzer Corps, another two Infantry Corps with six divisions, and the 1st Cavalry Division for subsequent flank protection. Thus the Second Panzer Group was moving towards the north again in a broad wedge aimed at Moscow. Operation Typhoon had begun—"the last battle of the year for the annihilation of the enemy," as Hitler had put it.

Colonel-General Guderian had been given a three-day lead so that he could play his part in the great offensive at the right moment and at the right spot. It was a bold and carefully calculated plan, designed to outmanœuvre Stalin's strong defensive forces before Moscow. It was perhaps the coolest and most precise battle plan of the whole war, and was now running like clockwork.

This modern battle of Cannae was intended to unroll in two phases. Phase one was to open with a break-through along the Soviet "Western Front" where it was held by the Ninth and Fourth Armies, to the north and south of the Smolensk–Moscow motor highway. Two Panzer groups were to race through the gap—Third Panzer Group forming the northern and Fourth Panzer Group the southern jaw of the pincer movement. These jaws were to close on the highway near Vyazma, thereby surrounding the enemy forces outside the immediate defences of the city. Simultaneously, Guderian's Panzer Corps was to strike towards Orel from the south-west, from the Glukhov area in the Northern Ukraine. After driving deep into the rear of Yeremenko's forces the Corps would wheel towards Bryansk. Three Soviet Armies would thus be encircled. Phase two of the operation then envisaged the pursuit of escaping enemy forces along a broad front by all three Panzer Groups; this to be followed by a drive to Moscow, with either the capture or the encirclement of the city.

It was a considerable force that was moving into battle under Field-Marshal von Bock—three infantry Armies (the Ninth, the Fourth, and the Second), the two Panzer Groups of the Central Front (Guderian's Second and Hoth's Third), to which was now added Hoepner's Fourth Panzer Group, which had been switched down from the Leningrad front and was now in charge of the right jaw of the pincers along the Smolensk–Moscow highway, while its LVI Panzer Corps was stiffening the left wing of Hoth's Panzer Group. In this manner fourteen Panzer divisions, eight motorized divisions, and two motorized brigades, as well as forty-six infantry divisions, had been brought together for the operation. The offensive was supported by two Air Fleets. Strong anti-aircraft units had been assigned to the Armies.

Everything was magnificently planned. Only the weather could not be foreseen. Would it hold? Or would the autumn mud set in before the troops reached Moscow? Moltke had written in 1864: "An operation cannot be based on the weather, but it can be based on the season." But the favourable season on which the operation should have been based was over. Winter was around the corner. Nevertheless Hitler risked the venture. On the morning of 30th September the crash of tank cannon and anti-tank guns ushered in the double battle of Vyazma and Bryansk, the Cannae of the Second World War, the most perfect battle of encirclement in military history.

The infantrymen of 3rd Company, brought right up to the front as reinforcements, were riding on top of the armoured troop carriers of 1st Company, 3rd Rifle Regiment, commanded by Colonel von Manteuffel. Why walk if you could ride?

Second Lieutenant Lohse was in front, in the command car of 1st Company. "Watch out for dogs, Eikmeier," he said to his driver. "Dogs, sir?" the lance-corporal asked in surprise. "Why dogs, Herr Leutnant?" Corporal Ostarek, the machine-gunner, also regarded the lieutenant doubtfully. Lohse shrugged his shoulders. "Three Russian prisoners were brought in at Regiment yesterday, each with a dog. Under interrogation they said they belonged to a special Moscow unit which used dogs with primed demolition charges against tanks." Ostarek giggled. "That's the craziest story I've heard for a long time." Lohse raised his hands apologetically. "I wouldn't have mentioned it if the regimental commander had not personally warned Captain Peschke and myself. Anyway, don't say I didn't tell you."

The vehicles were crossing a vast field. From the left came the stutter of Russian machine-guns: the first Soviet positions were along the edge of a village. The crash of 3·7-cm. anti-tank guns mingled with the rattle of machine-guns. The infantrymen of 3rd Company had jumped down from the vehicles and were now advancing on foot between the armoured troop carriers. Hand-grenades were flung into the peasants' shacks. A wooden fence was steamrollered by a vehicle. They kept going. Near the church there were more Soviet positions among the houses, well camouflaged. They advanced cautiously.

Sergeant Dreger with his machine-gun made the Russians in their dug-out keep their heads down. Suddenly Eikmeier shouted: "A dog!" A Dobermann came loping up. On its back was a curious saddle. Before Ostarek could even swing his machine-gun round Captain Peschke in a vehicle 30 yards away had snatched up his carbine. The dog made another leap and then collapsed.

Just then Corporal Müller shouted: "Watch out, there's another!" A sheepdog, a beautiful animal, was approaching at a careful trot.

Ostarek fired. Too high. The dog pulled in its tail and was about to turn back. Russian voices were heard shouting at it, and the animal once more headed straight for Lohse's vehicle. Everybody fired, but the only one to hit the animal was Corporal Seidinger with his captured Russian rapid-fire rifle, an automatic operated by gas pressure.

"Put out a warning over the radio telephone, Müller, about those dogs," Lohse commanded. And now they heard it in all the vehicles: "Dora 101 to all: Watch out for mine dogs...."

Mine dogs—a term coined on the spur of the moment. A new term for a new and much disputed Soviet weapon. On their backs these dogs carried two linen saddlebags containing high-explosive or anti-tank mines. A wooden rod, about four inches long, acted as a mechanical detonator. The dogs had been trained to run under the tanks. If the rod was bent over or snapped the charge went off.

The 3rd Panzer Division was lucky in its encounter with the four-legged mines of the "Moscow Infantry Company." The Soviet weapon was similarly unsuccessful in the sector of 7th Panzer Division. But two days later General Nehring's 18th Panzer Division was less fortunate. Tanks had over-run Soviet field positions and anti-tank strongpoints on the eastern edge of Karachev. The motorized infantry units broke into the town. The 9th Company, 18th Panzer Regiment, pushed through to the northern edge and then traversed a huge field of maize. A few more anti-tank guns were silenced. There was no more firing.

The tank commanders were leaning in their turrets. The company commander had just given the signal: "Close up on me on the right. Halt. Switch off engines." Hatches were flung open. At that same moment two sheepdogs came racing through the maize. The flat saddles on their backs were plainly visible. "What on earth is that?" the radio operator asked in amazement. "Messenger dogs, I suppose; or maybe medical-corps dogs," suggested the gunner.

The first dog headed straight for the leading tank. It dived under its tracks. A flash of lightning, a deafening crash, fountains of dirt, clouds of smoke, a blinding blaze. Sergeant Vogel was the first to understand. "The dog," he shouted; "the dog!" Already the gunner had whipped out his 8-mm. pistol. He fired at the second dog. He missed. He fired again. Another miss. A machine pistol spluttered from tank No. 914. Now the animal stumbled and its forelegs folded up. When the men reached it the dog was still alive. A pistol bullet put an end to its sufferings.

Soviet writings are silent about this diabolical weapon—the mine dog. But there can be no doubt about its employment, especially as it is mentioned also by the war diaries of other formations, as, for instance, 1st and 7th Panzer Divisions. From the interrogation of dog-handlers captured by 3rd Panzer Division it was learned that the Moscow Light Infantry Company had 108 such dogs. They had been trained with tractors. They had been given their food only underneath tractors with their engines running. If they did not get it from there they had to go hungry. They were also led into action hungry, in the hope that hunger would drive them under the tanks. But instead of food they found death. The Moscow Light Infantry Company was not very successful with its new weapon. Only very few dogs could be trained to stand up to the noise made by a real tank. That, presumably, was the reason why mine dogs were hardly ever used in the later stages of the war, except occasionally by partisan units.

To return to the battle. One might have expected that Guderian's offensive on the Bryansk flank would encounter a well-prepared opponent and therefore stiff resistance. After all, General Yeremenko had begun building up his famous front as early as 12th August, immediately after his conversation with Stalin, when an attack seemed imminent, and he had been reinforcing it ever since.

To this day Marshal Yeremenko maintains in his memoirs that at the end of August Guderian could never have broken through his defensive front, and that his drive to the south, to Kiev, had essentially been a second-best choice. For the fox Guderian the grapes of Moscow had hung too high: that was why he had gone to Kiev instead. Strangely enough, they were now, six weeks later, very much within his reach. Boldly and unconcernedly Guderian reached out for them by his drive to Bryansk, the important rail and road junction.

Even at the time of Guderian's drive into the Ukraine, in August, the town of Bryansk had been a kind of mysterious and sinister threat in his flank. It was known, from the evidence of prisoners, that General Yeremenko and his staff resided there, together with special contingents and crack units. It was known that the town was a key point in the Soviet defence of Moscow. It lay embedded in thick forests, protected by swampy lowlands. From it attacks were launched repeatedly against Guderian's exposed flank. And now that the decisive blow at Moscow was beginning to take shape from the Roslavl–Smolensk area, Bryansk and the Soviet armies in its vicinity again represented a grave threat to Guderian's flank. To remove this threat was as much a prerequisite for the main attack on Moscow as was the liquidation of the strong covering forces in the Vyazma area.

That was the tactical meaning of the double battle of Vyazma and Bryansk.

To every one's surprise Guderian's attack against Yeremenko's defensive front succeeded at the first attempt. The break-through was accomplished in the area of the Soviet Thirteenth Army.

It was fine autumn weather. The roads in the area of Second Panzer Group were still dry. The spearhead of XXIV Panzer Corps, the 4th Panzer Division, raced ahead as if the devil were at their heels. As he was chasing after his advanced detachment—already being led against Dmitrovsk-Orlovskiy by Major von Jungenfeldt—Guderian met the Corps commander and the commander of 4th Panzer Division, Generals Freiherr Geyr von Schweppenburg and Freiherr von Langermann-Erlenkamp. The great question was: should the advance be continued in order to knock out completely the Soviet Thirteenth Army, which was already in confusion, or should the troops be halted and given time to re-form and stock up with fuel? Both generals counselled caution: they had been getting reports that fuel was running short and that the men were tired out.

A little later, near the windmill hill of Sevsk, Guderian met Colonel Eberbach, the commander of the Panzer brigade. "I hear you're forced to halt, Eberbach," said Guderian. "Halt, Herr Generaloberst?" the colonel asked in surprise. He added drily: "We're just going nicely, and it would be a mistake to halt now." "But what about the juice, Eberbach? I'm told you're running out." Eberbach laughed. "We're running on the juice that hasn't been reported to Battalion." Guderian, who knew his men, joined in the laughter. "All right, carry on," he said.

That day the tanks of 4th Panzer Division covered 80 miles, fighting all the way. The Soviet Thirteenth Army was completely dislodged. What Yeremenko had thought impossible happened: the town of Orel, 125 miles behind the Bryansk front, was taken by Eberbach's tanks at noon on 3rd October. The pickets outside the town were taken so much by surprise that they did not fire a single round. The first vehicle the German tanks encountered was a tram full of people. The passengers clearly thought that Soviet troops were moving into the town and waved delightedly.

Things were now going badly for Yeremenko's Bryansk front. The 17th and 18th Panzer Divisions of XLVII Panzer Corps wheeled towards Karachev and cut the Bryansk–Orel road behind Yeremenko's headquarters. On 5th October the 18th Panzer Division took Karachev. The trap was closing. Yeremenko saw the impending disaster. He telephoned the Kremlin and asked for permission to break out. But Shaposhnikov, the Chief of General Staff, put him off. He urged him to wait a little longer.

Yeremenko waited.

But Guderian's armoured spearheads did not.

With an advanced detachment of the reinforced 39th Panzer Regiment Major Gradl struck towards Bryansk from Karachev—that is, from behind, 30 miles past Yeremenko's headquarters. And on 6th October General von Arnim's 17th Panzer Division did what not even the greatest optimist would have considered possible: they took the town of Bryansk and the bridge over the Desna by a swift coup. Bryansk was taken. The town crammed full with troops, heavy artillery, and police units had quite simply fallen. In vain were 100,000

Molotov cocktails lying in the stores. In vain had the strict order been issued: not a single house to be surrendered without opposition. One of the most important railway junctions of European Russia was in German hands. The link-up between Guderian's Second Panzer Group and the Second Army, which was coming up from the west, had been accomplished. Around Karachev and north of it the 18th Panzer Division and, subordinated to it, the "Grossdeutschland" Motorized Infantry Regiment were providing cover. Farther south, to both sides of Dobrik, the 29th Motorized Infantry Division was covering the Corps' flanks. The trap had been closed behind three Soviet Armies—the Third, the Thirteenth, and the Fiftieth. The date was 6th October.

During the following night the first snow fell. For a few hours the vast landscape was shrouded in white. In the morning it thawed again. The roads were turned into bottomless quagmires. The great highways became skid-pans. "General Mud" took over. But he was too late to save Stalin's armies in the Vyazma–Bryansk area. Entire infantry divisions were detailed for road-clearing. They worked like men possessed in order to keep the advance going.

Farther north, along the Smolensk–Moscow highway, the offensive had likewise started successfully. Hoepner's Fourth Panzer Group sluiced three Panzer Corps—XL, XLVI, and LVII Corps—through the Soviet front south of the highway at Roslavl behind the 2nd Panzer Division. They fanned out and with their left wing thrust northward in the direction of the motor highway.

On 6th October the spearhead of 10th Panzer Division was only 11 miles south-east of Vyazma, skirmishing with retreating Soviet units. The battle of Vyazma had reached its peak. During the night there was a succession of Soviet attempts to break out of the ring. At nightfall the whole vast forest area seemed to come to life. Firing came from everywhere. Ammunition was blowing up. Ricks of straw were blazing. Signal flares eerily lit up the scene for a few seconds. The area swarmed with Red Army soldiers who had lost their units. The advance command post of XL Panzer Corps had to fight for their lives. Where was the front line? Who was surrounding whom? When the long night at last drew to its end a Soviet cavalry squadron tried to break through in the grey light of dawn of 7th October. Behind it came a convoy of lorries carrying Red Army women. Machine-gun positions of 2nd Panzer Division foiled the attempted break-out. It was a painful and sickening picture—horses and their riders collapsing and dying under the bursts of machine-gun fire.

In the morning of 7th October the most forward parts of General Fischer's 10th Panzer Division penetrated through the slush into the suburbs of Vyazma and finished off Soviet resistance inside the burning town. Beyond the northern edge the men of the 2nd Battalion,

Code Name "Typhoon" 137

69th Rifle Regiment, crawled into the abandoned Russian fox-holes. The spearheads of General Stumme's XL Panzer Corps, followed by 2nd Panzer Division and 258th Infantry Division, had thus reached the objective of the first phase of the operation.

South of them followed XLVI Panzer Corps under General von Vietinghoff, with 11th and 15th Panzer Divisions as well as 252nd Infantry Division. Behind them, in turn, were LVII Panzer Corps under General Kuntzen with 20th Panzer Division, the "Reich" Motorized SS Infantry Division, and the 3rd Motorized Infantry Division.

Hoth's two Panzer Corps—LVI and XLI Corps—and VI Infantry Corps, having broken through on high ground west of Kholm, encountered very stiff resistance north of the Moscow highway from several well-dug-in infantry divisions as well as Russian armoured brigades. Because of the extremely unfavourable terrain Colonel-General Hoth united the tanks of LVI Panzer Corps—most of them Mark IIIs—into the "Panzer Brigade Koll," which after fierce fighting pierced the Soviet positions on the Vop along a firm causeway made from branches and planks thrown into the mud. Following behind, XLI Panzer Corps provided cover for the northern flank by attacking Sychevka with 1st Panzer Division and 36th Motorized Infantry Division.

Meanwhile the 6th and 7th Panzer Divisions reached the undamaged Dnieper bridges at Kholm, and likewise wheeled round towards Vyazma. In the evening of 6th October the battle-hardened 7th Panzer Division, Rommel's old striking force from the French campaign, was on the Moscow highway behind the enemy's rear, facing west for the third time in fifteen weeks. On 7th October Hoth's tanks linked up with Hoepner's in Vyazma. The pocket around six Soviet Armies with 55 divisions was closed.

Simultaneously with the breakthrough towards Vyazma, von Manteuffel's combat group had reached the Moscow highway by a surprise advance and cut it. Field-Marshal von Brauchitsch, Commander-in-Chief of the Army, thereupon sent a signal to the division: "I express my special commendation to the splendid 7th Panzer Division which by its swift advance to Vyazma has, for the third time in this campaign, made a major contribution to the encirclement of the enemy."

At Bryansk also, Guderian's two Corps had meanwhile trapped Yeremenko's three Armies of 26 divisions in a northern and a southern pocket. Now followed difficult days for the infantry—opposing fierce Russian attempts to break out of the pockets, splitting up the pockets, reducing individual strongpoints, and dealing with prisoners as, towards the end, the Soviets surrendered in entire regiments. The fighting continued until 17th October. Naturally, parts of the trapped forces succeeded in breaking out, especially from the southern pocket

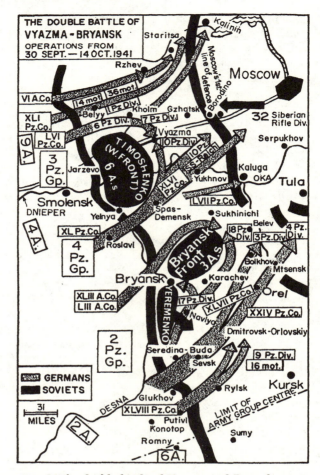

Map 8. The double battle of Vyazma and Bryansk was a perfect pincer operation. The jaws of the pincers were formed by the fast formations of three Panzer Groups. Infantry divisions of three Armies co-operated. The forces defending the Soviet capital were encircled and smashed: the road to Moscow was clear.

at Bryansk. Among those who succeeded were General Yeremenko and his staff. Yeremenko himself was seriously wounded and had to be flown out by aircraft.

The great battle was over. The first act of Operation Typhoon had been played out. Some 663,000 prisoners had been taken, and 1242 tanks and 5412 guns destroyed or captured.

Only three weeks after the battle of Kiev, when half a dozen Soviet

Armies of Budennyy's Army Group had been crushed in the south and more than 665,000 Soviet troops taken prisoner, another vast fighting force of nine Armies with 70 to 80 divisions and brigades had now been annihilated on the Central Front.

These were the Armies which were to have protected Moscow. In endless, pitiful columns they were now trudging into captivity over the muddy roads. Moscow had lost its shield and its sword. A wide gap had been torn in its defences. The German Army Group Centre had gained freedom of manœuvre for the bulk of its armoured and motorized formations against Stalin's capital. The second phase of Operation Typhoon could now begin—the pursuit of the enemy right into the city. Tank rally in Red Square!

On they drove. Or, rather, they did not drive—they struggled through the mud. Entire companies were pulling bogged-down lorries out of the mud of the roads. The motor-cyclists made wooden skids for their machines from boards and planks and pulled them along behind them.

Major Vogt, commanding the support units of 18th Panzer Division, was in despair. How did the Russians manage with these muddy roads year after year? He hit on the answer. He got hold of the small tough horses he had seen the local peasants use, as well as their light farmcarts, and used them for sending his divisions' supplies forward, a few hundredweight on each cart. It worked. The motorized convoys were stuck in the mud, but the small peasant carts got through. The prize of Moscow spurred the men to a supreme effort.

Kaluga, 100 miles south-west of Moscow, fell on 13th October. On 14th October Eckinger's advanced detachment of 1st Panzer Division took Kalinin, 93 miles north-west of Moscow, cut the Leningrad–Moscow railway, and captured the only Volga bridge that was to fall into German hands intact during the Second World War. A small bridgehead on the eastern bank, held by 1st Panzer Division and Motorized Training Brigade 900, covered the bridge. Thus the cornerstones of the 190-mile-long first line of defences covering Moscow had been brought down. The centrepiece of this line, however, the barrier across the motor highway some 60 miles outside Moscow, was between Borodino and Mozhaysk. There, at Borodino, 62 miles from Moscow, the "Reich" Motorized SS. Infantry Division was in position on 14th October. It was historic ground. Here, in 1812, Napoleon was brought to the brink of defeat. Here, in 1941, Stalin intended to bring Hitler to a halt. To do this he had hurriedly brought up the best forces he had—a crack unit from Siberia, the 32nd Siberian Rifle Division from Vladivostok, with three infantry regiments and two armoured brigades newly equipped with T-34s and KV-2s. Stalin began to denude his Far Eastern frontier ruthlessly. He could afford to do so. He knew that Japan would not attack. Japan, after all, was planning to strike at America in the Pacific. Stalin had reliable information from his spy Dr Sorge, the adviser to the German Ambassador in Tokyo. Dr Sorge was worth more than a whole army to Stalin.

At Borodino the regiments of the "Reich" SS Infantry Division and the "Hauenschild Brigade" of 10th Panzer Division with the 7th Panzer Regiment, as well as a battalion of 90th Motorized Artillery Regiment and the motor-cycle battalion of 10th Division, had their first encounter with the Siberians—tall, burly fellows in long greatcoats, with fur caps on their heads and high fur boots. They were most generously equipped with anti-aircraft and anti-tank guns, and even more so with the dangerous 7·62-cm. multi-purpose gun nicknamed by the German troopers the "Crash Boom." They fought impassively. There was never any panic. They stood fast and held on. They killed and let themselves be killed. It was an appalling battle.

The Russians employed their multiple mortars, the "Katyushas," known to the German forces as "Stalin's organ-pipes," which invariably caused havoc by their high-fragmentation effect. At Borodino also the heavy T-34 tanks were used for the first time in massed formations. Since 8·8-cm. anti-aircraft guns were not always available, the infantry had to tackle the T-34s with high-explosive charges. More than once the outcome of the battle hung in the balance. The casualties suffered by the "Reich" Motorized SS Infantry Division were so alarmingly high that its 3rd Infantry Regiment had to be disbanded and the survivors divided up between the "Deutschland" and "Der Fuehrer" Regiments. The entire army artillery available on the sector of the Panzer Group was concentrated under the command of the artillery commander 128th Division, Colonel Weidling, with instructions to blast a hole through the Soviet defences for the Waffen SS grenadiers, who charged with death-defying courage. First, the flame-thrower batteries with their remote-control electric firing devices had to be taken. Then came the minefields. Then the barbed wire. Then the pillboxes. Experienced assault parties under-ran the defensive fire of massed anti-aircraft, anti-tank, and mortar batteries and repulsed immediate counter-attacks by Russian tanks in close combat. Hell was let loose. Overhead roared the Soviet low-level bombers. German fighters of VIII Air Corps tore in and out of the billowing clouds of smoke.

The dressing stations were kept busy. Lieutenant-General Hausser, the commander of the "Reich" SS Infantry Division, was seriously wounded. Row upon row of injured lay on the ground—the tank-men in their black uniforms, the grenadiers in torn field tunics, the men of the Waffen SS in their blotchy camouflage smocks. Dead, gravely wounded, burnt, or beaten to death. Anger had made the troops see red—on both sides. No quarter had been given.

Code Name "Typhoon"

At last a breach was torn through the strong positions held by the Siberians. The two infantry regiments of the "Reich" SS Division, the "Deutschland" and "Der Fuehrer" Regiments, charged through. There was no time to fire their guns. Spades and rifle-butts were the weapons used. The Siberian batteries were taken from behind. Their crews, behind the breastworks of anti-aircraft, anti-tank, and machine-guns resisted stubbornly and were cut down in hand-to-hand combat. The infantry regiments of 10th Panzer Division were engaged in the same kind of fighting. They fought on the battlefields where Napoleon had stood 130 years before them; they stormed the stubbornly defended historic scarp of Semenovskoye. The Siberians resisted in vain.

The 32nd Siberian Rifle Division died on the hills of Borodino. The great bolt of Moscow's first line of defence on the Moscow highway had been blasted open. The 10th Panzer Division and the "Reich" Division advanced across snow-covered fields towards the Moskva. There the last resistance of the Russian combat groups was broken. On 19th October 1941 Mozhaysk fell. Mozhaysk—right outside the gates of Moscow! A mere 60 miles of motor highway. And that highway led straight from Mozhaysk into the Soviet capital.

"Mozhaysk has fallen!" The news spread through the streets of Moscow. "Mozhaysk has fallen. The Germanski are coming."

Clouds of smoke were rising from the Kremlin chimneys, just as though the outside temperature was 30 degrees below zero Centigrade. They were burning the secret papers which they could not evacuate.

The Muscovites were flabbergasted. Only a fortnight previously they had been full of confidence in victory in view of America's promises of help. On 2nd October Churchill's representative, Lord Beaverbrook, and Roosevelt's representative, Mr Harriman, had gone to the Kremlin to sign the protocol on Anglo-American arms deliveries. Although the United States of America was still neutral and a non-combatant, it was announced that the three Great Powers were determined to co-operate towards victory over the German arch-enemy of all nations. For the first ten months of the agreement, starting with 1st October, the following supplies were promised and also delivered: 3000 aircraft—2000 more than the total number of operational machines available to the Luftwaffe on the Eastern Front on 30th September—4000 tanks—three times as many as all three German Panzer Groups had at their disposal on 30th September—and 30,000 motor vehicles.

But would these deliveries come in time? Was Hitler not again winning his race against the Western Powers, just as he had won it in the Kremlin before in 1939?

On 10th October a dinner was given for the foreign diplomats and journalists at Moscow's Hotel National. On the menu were bliny, caviare, vegetable soup, roast beef, creamed potatoes, steamed carrots, chocolate pudding, and mocha. Toasts were drunk to Stalin and to the defence of Moscow. And to victory. That was the day when Timoshenko was relieved of his command and replaced by a man known then to only very few people—Army General G. K. Zhukov. He was made Commander-in-Chief Western Front; Lieutenant-General V. D. Sokolovskiy became his Chief of Staff, and N. A. Bulganin, as member of the Military Council, became the political head of the front.

Five days later, at 1250 hours on 15th October, Foreign Minister Molotov received the US Ambassador, Steinhardt, and informed him that the whole Government, with the exception of Stalin, were leaving Moscow, and that the Diplomatic Corps was being evacuated to Kuybyshev, 525 miles east of Moscow. Each person was allowed only as much luggage as he or she could carry.

When the news spread through the city, and in particular when it became known that Lenin's coffin had been removed from the Mausoleum in Red Square, panic broke out. "The Germans are coming!"

Those who lived on the Mozhaysk Road in Moscow pricked up their ears at any noise that sounded like tanks. Had they got here already? Anything was considered possible in Moscow at that time.

Cities have nerves too. And if the strain on them becomes too much they give way. On 19th October 1941 Moscow's nerves were strung to breaking-point. Alarmist rumours were flying about the city. The Government had fled. The Diplomatic Corps had left Moscow. Lenin's coffin, the glass coffin with the father of the Revolution, had been removed to an unknown destination. And the postscript to all these stories and rumours was: "The Germans are already outside the city." And in a whisper they added, "Their tanks might be here any minute now." The possibility of this happening had the most astonishing effect on the population. The people suddenly lost their fear of Stalin's secret police, the militia, and the security detachments. There was a rumble of angry voices in the bread queues outside the bakers' shops: "We've had enough of the war—put an end to it!"

Presently the first shop was stormed in Sadovaya Street. A lorry loaded with tinned food was ransacked, overturned, and set on fire. Rebellion was lurking in the dank, cold streets, cowering in ill-heated flats, sharing the table of starving people. Stalin's power was tottering. His portrait was being removed from the walls; the first Party cards were being burnt. Handbills, crude hurriedly printed sheets, suddenly appeared in people's letter-boxes in the morning: "Death to the Communists!" they said. They also contained anti-Semitic slogans. Horrified, the recipients stared at the seditious text. Moscow, Mother

Moscow, was reeling. The heart of the Soviet Union was missing a beat. Yet the skies had not fallen.

A. M. Samsonov, the official Soviet chronicler, describes the situation in his book *The Great Battle of Moscow*. He says:

> A mood of alarm spread in the city. The evacuation of industrial undertakings, Ministries, authorities, and institutions was speeded up. There were also, at that time, sporadic cases of confusion among the public. There were people who spread panic, who left their place of work and hastened to get out of the city. There were also traitors who exploited the situation in order to steal socialist property and who tried to undermine the power of the Soviet State.

The dictator in the Kremlin struck with a mailed fist. On 20th October he declared a state of emergency in Moscow. The capital was declared a zone of military operations. The law of the fighting front now governed its life.

Samsonov writes: "The decree laid it down that all enemies of public order were to be handed over at once to courts martial, and that all provocateurs, spies, and other enemies calling for rebellion were to be shot out of hand." And so they were. The capital had become the front line. Its inhabitants were virtually incorporated in the Army. As early as 11th July People's Defence Divisions totalling 100,000 men had been recruited from among the city's population by decree of the Defence Committee and posted along the city's western outskirts. In the subsequent winter operations the German divisions encountered this People's Army at all crucial points of the Central Front. Frequently these men fought fanatically—by Lake Seliger, at Rzhev, outside Dorogobuzh, and at Maloyaroslavets. Starting on 1st October, the lists of inhabitants were combed through once again. Another 100,000 Muscovites were called to the colours. They passed through a 110-hour training course—*i.e.*, twenty days—and were then sent to the front.

Between 13th and 17th October the Moscow City Soviet finally raised a further twenty-five independent Workers' Battalions—men who at the same time worked at their jobs and served in the forces. They numbered 11,700 men, the equivalent of a division. They were employed mainly on the eastern bank of the Moska-Volga canal. At the same time the 1st and 2nd Moscow Rifle Divisions were set up from reservists with experience of active service, and twenty-five Local Defence Battalions, numbering 18,000 men, formed to maintain order in the city. It was a truly total mobilization of a metropolis.

Every man and every woman was integrated into the military machine. Some 40,000 boys and girls under seventeen were mobilized to dig earthworks on Moscow's second line of defence and organized under military command. Together with 500,000 women and old men, they worked three shifts, day and night, under atrocious conditions, constructing 60 miles of anti-tank ditches, 177 miles of wire obstacles, and 5010 miles of infantry trenches.

By the end of October, however, neither the fanaticism of the Party nor courts martial and executions were able to check the progressive disintegration of the city. The flats of evacuees were looted or taken over by deserters. Wounded men, juveniles who had run away from labour detachments, and young children roamed the streets. Security units had to comb underground tunnels, railway stations, and bomb sites continually. Moscow seemed finished. These harsh but unquestioned facts have been described by Mendel Mann, a Jewish village schoolmaster from Poland who had escaped to Russia. His book *At the Gates of Moscow*, first published in Israel and since translated into nearly all languages of the Western world, is called a novel, but its setting is based on the author's own experiences.

In this book we find the following scene, which is typical of the situation in Moscow at the end of October 1941:

> Two wounded soldiers came tumbling out of a small side-street. One of them, tall and angular, had an arm in plaster, the other, short and plump, was moving very adroitly on his crutches. He had a knee injury. They had reached the middle of the all but empty main street and shouted, "German tanks are in Kaluga Street and in Psochnaya! They are in the city already! They are here! Save yourselves, Russians!" A patrol of six armed men, three of them militia and the other three NKVD, stopped by a gateway and then slowly retreated across Sadovaya Street. They did not exchange a word, but regarded each other silently.... Suddenly the shops were being locked up. Iron shutters were pulled down with a clatter. The doors of houses opened and curious spectators collected in the doorways.
>
> The two wounded servicemen stopped at a corner. The gaunt one pointed to something with his good arm and shouted:
>
> "There they are, the Germans!"
>
> The patrol disappeared in a dark doorway. A short while later the six men came out again, bareheaded and unarmed. They had removed the militia flashes from their army greatcoats.
>
> "The rats are leaving the sinking ship!" a woman screamed.
>
> "Let them scram! They'll be caught!"
>
> Slowly the crowd formed into a procession. At its head marched the two wounded, followed by a few women, and then the crowd.
>
> Some boys of fourteen or fifteen—boys who worked in the factories—came out from the side-streets. Jeering, they joined the grown-ups. Suddenly a man unfurled a white cloth and waved it above his head like a flag. At its centre was a black swastika.
>
> The crowd fell back and stood rooted to the ground.
>
> "Death to the Communists!" cried the man with the flag. "Down with the Jews!"
>
> Silence hung underneath the grey sky of Moscow. Like a sheet of rigid fear the sky hung over the people.
>
> "The war is over!"

"Thanks be unto you, Holy Virgin, Mother of God!"

The sub-machine-gun of a security patrol put an end to the eerie scene. And the Germans did not come. Why not? After all, they had been seen crossing the motor highway and the Mozhaysk road on the approaches to Moscow—roughly half an hour's drive from the city. Where were they?

Where, indeed, were they?

Lieutenant-Colonel Wagner had spread out his map on a case of hand-grenades. The officers of the Engineers Battalion, 19th Panzer Division, were standing around their commander. "Here"—Wagner indicated a spot on the map—"here is Maloyaroslavets, 12 miles ahead. That's where our tanks have to get by to-morrow. And here, Podolsk, 21 miles from Moscow, is the division's target for next week."

Wagner looked up from the map: "That's why we must break through this damned pillbox position in front of us and open up the road. The tanks can't drive over the sodden fields, and the infantry who have pushed ahead south of the road are in need of supplies."

The date was 16th October. The scene was outside Ilyinskoye, the kingpin of the first line of defence before Moscow. The positions were held by the cadets of the Podolsk Military College. The 19th Panzer Division from Lower Saxony had got stuck in front of these Soviet pillboxes, manned by officer cadets, young fanatical Communists. The Stukas had been unable to smash the pillboxes. The gunners had been no more successful. It was therefore up to Wagner's sappers now.

An assault party with two flame-throwers and high-explosive charges cautiously filtered into the flat, swampy terrain in front of the Russian lines. Bomb and shell craters provided useful cover. The German artillery put down a heavy barrage immediately in front of the Russian pillboxes. Under its cover the sappers crept right up to the concrete blocks.

The shellburst was uncomfortably close in front of them. Sergeant Tripp, leading a section of the Engineers Battalion, 19th Panzer Division, was flattening himself against the edge of a shellhole. He raised his Very pistol. One white flare went up—the arranged signal. It meant: Have reached objective. Abruptly the artillery-fire ceased.

"Now!" The flame-throwers hurled their searing jets of burning oil against the two pillboxes in the middle and on the right. The fire roared through the embrasures. Black smoke blotted out everything. The Russians had no hope of firing small-arms or throwing hand-grenades. The pillbox on the left was kept down by machine-pistol fire on the embrasures, while Lance-Corporal Vogel climbed up on its top. From above he shoved his charge through the embrasure and leapt back. There was a loud crash, a sheet of flame, and black smoke.

The second obstacle was reduced in much the same way. But then, from the concrete passage linking the pillboxes, came sudden machine-gun fire. The flame-thrower party on the right was mown down. Tripp raced across to the communication trench from the left and opened up with his machine pistol. The Russians raised their hands. Only the commissar continued to throw hand-grenade after hand-grenade until he was mown down.

They fired another Very light—a white one. The infantry farther back cheered: "They've done it." The barrier of Ilyinskoye was breached.

The 27th Panzer Regiment under Lieutenant-Colonel Thomale, together with 2nd Battalion, 19th Artillery Regiment, and a battery of 8·8-cm. anti-aircraft guns, now moved off, and along the cleared road advanced towards Maloyaroslavets. In front was the 1st Company under First Lieutenant von Werthern. The companies of 74th Rifle Regiment were moving along either side of the highway.

It was 60 miles to Moscow.

The Protva river was crossed without difficulty. They kept moving. They were aiming at Verabyi on the Istya river.

The bridge was intact. The crossing was furiously defended by a Soviet anti-tank gun. "All weapons, fire—and get across that bridge," von Werthern radioed to his unit commanders. Second Lieutenant Range commanded the lead tank. Driver Kurt Wiegmann had heard the order in his earphones and needed no instruction. He engaged gear and moved off.

They had just cleared the bridge when a Soviet anti-tank gun, emplaced to the left of the steep bank, caught them. There was a crash, and the tank filled with smoke. "We're getting out!" Second Lieutenant Range ordered. They all managed to clamber out of the tank and leap into the ditch. They saw the second tank receive a direct hit and burst into flames. Only the commander got out. But already the third was crossing the bridge, swivelling its turret to 10 o'clock and firing. A direct hit on the Soviet anti-tank gun. In the teeth of Russian artillery fire from the edge of the forest a tractor with an 8·8-cm. anti-aircraft gun raced over the bridge. It went into position and at once opened fire at the Russian batteries. All right so far.

Werthern's 1st Company established a bridgehead in the face of furious Russian opposition. The Russian troops were officer cadets who fought with unbelievable bravery, time and again attacking the German tanks at close quarters.

Lieutenant-Colonel Thomale ferried over the bridge whatever parts of his 27th Panzer Regiment he could lay hands on. He was now 25 miles in front of his division, and the Istya bridgehead had to be held until the bulk of the troops came up. Thomale's combat group managed to do it. By nightfall the Russian position, built hastily during the past days but nevertheless held by strong anti-tank and artillery forces, was smashed.

The commander of 19th Panzer Division, Lieutenant-General von Knobelsdorff, drove up to the spearhead. "We mustn't give the Russians time to dig in again," he said. "Keep going. The new objective is the Nara."

The Nara river marked the second, and presumably the last, line of defences outside Moscow.

Rain was falling. It was cold. The roads were getting muddier and muddier. The tanks were churning to a standstill. With increasing frequency the shout went up: "Russian tanks!" The T-34s struck down swiftly from the hills on their broad tracks. They were ideally constructed for mud and snow. Their toll of victims was heavy. Often it was only the 8·8-cm. anti-aircraft guns which saved the situation at the last moment. Nevertheless, the motor-cycle units and tanks of 19th Panzer Division reached the Nara. They crossed it north of the highway after the sapper battalion had built a pontoon bridge overnight in record time, though under the costly fire of Soviet mortar batteries. Would it be possible to widen the breach into a dam burst?

In a surprise coup the tanks took the high ground east of the Nara. "It's working!" the men were calling to one another. The 59th Rifle Regiment, 20th Panzer Division, temporarily subordinated to 19th Panzer Division, was switched across the river. Everything now depended on the motor highway being reached and the strong barrier position between Gorki and Nikolskoye being smashed. The road to the Kremlin would then be open.

In spite of the soft roads the 98th Infantry Division had come up by forced marches. At Detchino it had fought its way through cunningly devised field positions and pillbox lines arranged in deep echelon and manned by Mongolians and Siberians. These men took no prisoners because they had been told that the Germans would first cut off their ears and then shoot them. For five days the furious fighting raged. The battalions suffered heavy casualties. The 282nd, 289th, and 290th Infantry Regiments were greatly reduced in number; most of the battalion and company commanders had been killed or wounded. The sapper battalion lost 100 men. But Moscow, the great objective, spurred the men on. True, the horses were finished. And so, for that matter, were the gunners and infantrymen. To the severity of the fighting were now added the hardships of rain, cold, and lice. So far no winter clothing had arrived for the troops. But the knowledge that they were now fighting the decisive battle kept them going. They were giving the last, the very last, ounce of their strength.

On 23rd October the 290th Infantry Regiment crossed the Nara at Tarutino, south of the motor highway. The division instantly turned towards the north in order to give support to the 19th Panzer Division in clearing the Moscow highway.

The 1st and 2nd Battalions, 289th Infantry Regiment, under Lieutenant-Colonel von Bose and Captain Ströhlein respectively, stormed the thickly wooded hills outside Gorki. The Russians made an immediate counter-attack and dislodged the 289th Regiment again. On the following day the struggle continued. Every inch of ground had to be gained in bitter hand-to-hand fighting. In the end only 200 yards remained to the motor highway.

First Lieutenant Emmert, Acting O.C. 1st Battalion, 282nd Infantry Regiment, personally led the charge of his 1st Company. Its commander, Second Lieutenant Bauer, was killed at once. Men were dropping right and left. By a supreme effort the men reached the houses of Gorki and flung themselves down. The Russians fell back. True, the German troops were only in the southern part of the town, but at least they had got behind Moscow's last line of defence. From Gorki it was only 40 miles to Moscow.

"Forty miles—that's as far as from Nuremberg to Bamberg," remarked Second Lieutenant Frey, a troop commander in the Panzerjäger Battalion, 198th Division. But he himself managed only another three. His grave is just outside Gorki, at Kusolevo.

The offensive against Moscow was essentially a battle for the roads. Throughout the summer they had been vital arteries for supplies. But now, during the period of winter mud, when no farm track, let alone rough ground, was negotiable, not only the movement of supplies but in fact all operations of tanks and infantry depended on road conditions. This was a serious handicap for the attacker, but a fortunate circumstance for the defenders. A road junction covered by pillboxes and field positions could only very rarely be bypassed. It had to be taken by frontal assault. Thus the road junctions became the battlefields of the drive towards Moscow.

Gorki on the Nara was one such junction, and so was Naro-Fominsk on the Smolensk–Kaluga–Moscow railway. Krimskoye, between the Moscow motor highway and the famous postal road, was another.

Other keypoints still were Zvenigorod, Istra, Dmitrov, Tula, and Kalinin, forming a large semicircle around Stalin's capital. These localities represented the keypoints of Soviet opposition in Moscow's second line of defence: behind them with its numerous lines of communication lurked the Red capital like a spider in its web.

More than sixty German divisions were involved in the costly fighting for Moscow. That meant sixty times an average of 5000 to 10,000 men. Every single division could deserve individual mention. But we can attempt to trace the fate of only a few—their fate along a terrible, murderous road full of human and military drama. They came so near their aim that it seemed within arm's reach. They saw the towers of the Kremlin; they stood at the bus-stops in the outer suburbs. One unit got within five miles of Moscow, and its tanks stood within nine to 18 miles of the Kremlin at the beginning of December 1941.

On they marched, the infantrymen of 78th Infantry Division, along a road pockmarked with craters and water-holes, from Vyazma towards Moscow. It was raining. Presently, for a change, snow fell. Their stomachs were rumbling. Their field kitchens were stuck somewhere behind in the mud. Their uniforms were sodden and stiff with dirt. This was no longer the sweeping advance of the hot summer days. How long ago was that? It seemed a lifetime ago. They had marched through the summer and through the autumn. And now they were marching through the mud and slush into the winter.

While the 78th Division was moving along the right-hand side of the highway in a long, unending column, the companies of 87th Infantry Division were trudging along to the left of it. The middle was kept free for traffic in the opposite direction.

South of the motor highway, between Yukhnov and Gzhatsk, the 197th Infantry Division was struggling eastward along a bad road. On 19th October, a Sunday with rain and snow, its regiments clocked up their 930th mile of foot-slogging. Nine hundred and thirty miles!

Captain Küppers, the commander of 1st Battalion, 229th Artillery Regiment, was impatient with the rate of progress. The road he was on was so rutted and deep in mud that his artillery vehicles were hardly able to make any headway at all in the deep morass. With the permission of Lieutenant-Colonel Ruederer, the leader of the column, he turned off along the transversal road from Yukhnov to Gzhatsk, with the intention of reaching the motor highway. There, he thought, progress would be easier and faster.

The artillerymen reached the highway. But they had not expected the picture they saw. Mud-hole after mud-hole, and pitted with deep craters, the highway was jammed tight with vehicles. There was no hope here for his horse-drawn batteries. Along the highway sector from Gzhatsk to Mozhaysk alone between 2000 and 3000 vehicles were stuck.

After what they had seen the artillerymen of 197th Infantry Division hurriedly turned back again. Back into the mud. Their speed, which had averaged 28 miles a day in the summer, had dropped to sometimes less than one and never more than three miles per day. At night, worn out with fighting, battered, filthy, lice-ridden, hungry, and weary to death, they crowded around the stoves of the miserable peasant shacks in the small villages. The wretched horses outside were pressing against one another, nibbling the ancient mossy straw on the low roofs. Inside the troops were drying their uniforms. And if any of them asked, "Any idea where we are?" he would get the plain soldier's answer: "Right up the arsehole of Europe!" The following morning they would trudge on again, on and on, in the track of the motorized divisions. On and on, towards Moscow.

By the second half of October Moscow's first line of defence had been pierced everywhere on a broad front between Kaluga and Kalinin. The German divisions were advancing against Moscow's second and last line of defence along three main roads—from Maloyaroslavets to Moscow, from Naro-Fominsk to Moscow, and from Mozhaysk to Moscow. This second line of defence ran, reading from south to north, from the town of Tula to Serpukhov, thence along the Nara via Naro-Fominsk to the Nara Lakes on the motor highway, then along the Moskva valley via Zvenigorod, Istra, the Istra reservoir, and Klin to the "Moscow Sea," south-east of Kalinin.

This line of defence was not in fact a line, but a system of positions organized in considerable depth. Towards the west, moreover, all road junctions and railway stations, even well outside the defences proper, were strongly fortified. Towards the rear—*i.e.*, in the direction of Moscow—the anti-tank ditches and field positions went right back to the edge of the capital. Thence they continued all the way to Red Square in the form of barricades, road-blocks, tank-traps, and buried armoured fighting vehicles.

By the end of October Moscow's doom seemed sealed. In the north, in the area of what used to be Hoth's Third Panzer Group and was now the Third Panzer Army commanded by General Reinhardt, the Thuringian-Hessian 1st Panzer Division had succeeded in crossing the Volga in an eastward direction at Kalinin. The combat group Heydebrand, with Training Brigade No. 900 attached to it, thrust along the Torzhok road as far as Mednoye and blocked the roads and railway to the north. A few days later, however, the combat units had to be taken back again to the outskirts of Kalinin, following fierce fighting with fresh reinforcements of Siberian armoured formations. By violent counter-attacks the Russians were trying to regain this important cornerstone of Moscow's defences on the Upper Volga. Their efforts were in vain. Parts of 6th Panzer Division, of 14th and 36th Motorized Infantry Divisions, and later also of 129th Infantry Division, jointly succeeded in holding the vital bridgehead. It was consolidated by XLI Panzer Corps, now under the command of General Model.

The main pressure of the German offensive, however, was along both sides of the Moscow motor highway. Fighting was heaviest there in the area of General Stumme's XL Panzer Corps. Its 10th Panzer Division had taken Shelkovka, an important and strongly fortified road junction, and had advanced across the Moskva river into the area north of Ruza. The task of the Corps was to strike at Moscow from the north-west with the "Reich" SS Infantry Division and 10th Panzer Division. The 10th Panzer Division was determined to be the first in Red Square.

It was halted 49 miles outside Moscow—not by the Russians but by the mud. General Fischer's division had to be supplied along a nine-mile causeway of timbers placed on the mud. Along both sides of this wooden road the vehicles, guns, and tanks lay motionless, stuck fast. Infantrymen, sappers, Panzerjägers, and motor-cycle troops were entrenched in the villages and forests. The tanks had no fuel left. The guns received only a dozen shells each day. Meanwhile the Russians were launching ceaseless attacks with their T-34s, which remained manœuvrable on the soft ground. The 10th Panzer Division lay bogged down, slowly bleeding to death. Those who survived remember and curse the villages of Prokovskoye and Skirminovo to this day.

The men sat in the peasant houses, in despair, praying for the ground to freeze so they could move again. But the frost was slow in coming that year. Meanwhile the division was bleeding to death. When Major-General Fischer reported his effective strength to the Corps commander, General Stumme exclaimed, horrified, "Good God, this is no more than a reinforced reconnaissance patrol."

Thirty miles south of XL Panzer Corps the 78th Infantry Division had likewise driven a wedge 20 miles deep into the enemy's positions, driving from Ruza along the Zvenigorod–Moscow road, and had thus come up close to the main fortifications of Moscow's second line of defence. In difficult forest fighting and against strong road-blocks the 195th and 215th Regiments, in particular, mastered superhuman difficulties. They succeeded in gaining the fortified ground west of Lokotnya, within 40 miles of Moscow.

But then the mud took over and brought the attack to a standstill on that sector too. They had to wait for the frost.

South of the motor highway, in the area of Kluge's Fourth Army, the offensive at first went very well. The 7th and 292nd Infantry Divisions gained the Kryukovo area, just outside Moscow's second line of defence. They launched an attack on the defences—and got bogged down in the mud. The attack on the main positions along the Nara had to be called off.

Had nature conspired against the German forces? Would nothing be successful any more? Oh, yes, some actions were. The 258th Infantry Division and the 3rd Motorized Division were luckier. The 258th succeeded, by means of a daring stroke of Major Lübke's 2nd Battalion, 479th Infantry Regiment, in taking Naro-Fominsk on the Roslavl–Moscow main road on 22nd October. A penetration had thus been made into Moscow's second line of defence, 43 miles from the city itself.

South of Naro-Fominsk on 22nd October the 3rd Motorized Division thrust across the Nara with 29th Motorized Infantry Regiment and gained a seven-mile-wide bridgehead. "Things are moving again!" the men called out to one another. Yes, they were moving again. The

8th Motorized Infantry Regiment, the sister regiment of the 29th, not only repulsed all Russian counter-attacks, but itself mounted an immediate counter-attack and annihilated a strong Russian combat force. They took 1700 prisoners, including 52 officers. They were members of battalions raised in Moscow, or workers' militia, or Ukrainians. Many of them shouted, "Voyna kaputt"—the war is lost—and later they denounced their political commissars who had torn their insignia of rank off their shoulders.

Another 20 miles farther south the 98th Infantry Division likewise succeeded in leaping across the main obstacle of Moscow's second line of defence—the strongly reinforced Nara river. On its eastern bank the division swung north in order to clear the big road bridge of Gorki on the highway to Podolsk and Moscow, in co-operation with the 19th Panzer Division.

The 19th Panzer Division from Lower Saxony had crossed the river north of Gorki—as already reported—and its 27th Panzer Regiment successfully repulsed all Soviet counter-attacks. With the capture of Naro-Fominsk and the crossing of the Nara above and below Gorki the last rampart to the south-west of Moscow was breached in three places. The dam built with the sweat, blood, and tears of half a million women, old men, and children, the dam that was to have stopped the German flood, was riddled.

Would the dam burst? The Muscovites feared that it would.

But as they were waiting for the German tanks to arrive, the tanks which now had no other obstacles to face except the ragged and half-starved local defence levies, the weather came to their rescue in this sector too. Rain turned the ground into mud. The mud became impenetrable. Field-Marshal von Bock had to concede victory to the morass. He ordered his forces to halt and wait for the ground to freeze hard so that their vehicles could move again. If only they had had 5000 track-laying vehicles with tracks as wide as the T-34s, then Moscow would have been lost.

But where was Guderian, that successful leader of the attack of Army Group Centre? Where were the spearheads of his combat-hardened divisions?

His Panzer Group had likewise been promoted to the rank of a Panzer Army, the Second Panzer Army, and reinforced to $12\frac{1}{2}$ divisions. It formed the southern wing of Army Group Centre, and its task was to drive towards Tula and seal off Moscow from the south. The High Command had again based its plans on Guderian's aptitude for lightning-like operations and had envisaged the early strangulation of the Red capital from the south-west. To begin with all went well.

On 30th September XXIV Panzer Corps moved off towards the north-east with 3rd and 4th Panzer Divisions in the van. The town of

Sevsk was reached on the following day. That day the spearheads of the attack covered no less than 80 miles. On 3rd October Orel fell suddenly to a surprise coup by 4th Panzer Division. By 5th October the bridgehead over the Oka north of Orel had been extended.

Meanwhile the 3rd Panzer Division had left the main road behind in order to drive towards the north. After a night march through a hurricane-force blizzard the division crossed the Tson river. They marched on and on, to the north. Bolkhov fell—800 prisoners were taken. By mid-October parts of 3rd and 4th Panzer Division with the "Grossdeutschland" Infantry Regiment were ready to strike across the Suzha north-west of Mtsensk. The river was crossed on 23rd October, and the defeated Russian forces vigorously pursued. Chern was taken —only 56 miles from Tula. But then the mud took command here too.

The road to Tula was not equal to the heavy vehicles. Its surface broke up. Deep pot-holes, filling with water and mud, soon turned the road into a quagmire. Supplies got stuck. Fuel failed to arrive. The advance was slowing down. That in turn gave the Russian rearguards time to destroy the bridges along the road and to lay minefields on both sides of it. Miles of firm causeways had to be built here too by placing logs and timbers on the ground, so as to enable supplies to reach the spearheads of the attack.

But Guderian refused to be defeated by nature and made a characteristic decision: he united all armour of XXIV Corps, parts of the 75th Artillery Regiment and the 3rd Rifle Regiment, as well as the "Grossdeutschland" Infantry Regiment, into a fast vanguard formation under the energetic Colonel Eberbach and instructed them to disregard everything else but go ahead and take Tula.

Eberbach's force drove, scrambled, slithered, and fought its way through the mud and the Russians. Wherever resistance was encountered, wherever the Russians tried to block their advance, Stukas first swooped down, screaming, upon the enemy's positions, followed presently by Eberbach assaulting with tanks and grenadiers. Mtsensk was taken. Chern fell. On 29th October the spearhead was within three miles of Tula, the industrial centre with 300,000 inhabitants.

The Soviets had strongly fortified this southern cornerstone of Moscow's last line of defence with numerous anti-tank and anti-aircraft guns. Their reason was obvious: once Guderian had pushed past Tula Moscow would lie to the west of him, and Stalin's capital would find itself in a stranglehold. The old silver-mining town of Tula, though 100 miles distant from the capital, was therefore in a sense a suburb of Moscow. The Russians were well aware of it. Guderian was aware of it. And Eberbach was aware of it. Tula must fall. Tula was half of Moscow. Tula was a symbol. It even had its own Kremlin.

The 2nd Company of the "Grossdeutschland" Infantry Regiment had 60 men left. Sixty out of 150. But Second Lieutenant von Oppen was set on getting into the town. "Forward, men!" He pushed his steel helmet to the back of his head. "Forward!"

The 2nd Company "Grossdeutschland" was thus the vanguard of Guderian's entire Panzer Army. It was an inspiring thought. Things seemed to go well for the company. Tula was lying before them in the haze of an October evening. The dust-clouds of demolitions hung over the town. The men made their way through an enemy-held sunken lane, with pistol and hand-grenade, one leap at a time, man by man. The Russians caught their hand-grenades and flung them back.

"Delay throwing and let them burst in the air!" the sergeant yelled. It worked. They got as far as an industrial housing estate on the southern edge of the town. The Russians were falling back. But Eberbach was unwilling to take risks. "Everybody halt," he ordered over the radio. He then went round the positions in person and pacified the grumbling company: "We'll bag the town to-morrow morning." To-morrow morning. At 0530 hours.

Punctually the next morning Colonel Eberbach was again in the forward positions. He made a personal reconnaissance. He ducked from one house to another through the small industrial estate and spoke to the men of 2nd and 3rd Companies.

"Over there, behind that timber-stack, are the Russian outposts," Second Lieutenant von Oppen reported. "And that red-brick building, probably a barracks, is crammed full with anti-tank guns, mortars, and snipers."

Eberbach nodded. Colonel Hoernlein, the commander of the "Grossdeutschland" Regiment, also arrived on the scene. He glanced at his wrist-watch. "0530," was all he said. Bombast and formalities were unnecessary; indeed, they would have been ridiculous among the men who were now flattening themselves against walls and doorposts—with several days growth of beard, their uniforms and boots caked with dirt, their pockets bulging with hand-grenades, their steel helmets pushed right back, a cigarette shielded in the hollow of the hand, so the Russians should not spot the glowing point.

The second lieutenant stubbed out his cigarette, pulled his trusted 8-mm. pistol from its holster, and cocked it. "We're off!" Words of command issued in a low voice. Somebody clearing his throat. Then the clank of a gas-mask tin. They were moving. In line abreast the 2nd Company made its way through the gardens of the housing estate. A platoon of the 4th (Machine-gun) Company linked up with them on the right.

Von Oppen glanced across to them. It seemed as though he was looking for his friend, First Lieutenant Hänert. But he was not—for, after all, he had been present when they buried him. That was on 17th October, by a little forest stream near Karachev.

Lieutenant Hänert, commanding the 4th Company, had been the

first man in the "Grossdeutschland" Regiment to be decorated with the Knights Cross of the Iron Cross. He was twenty-seven when he was killed in action during the night of 14th October, shot in the abdomen by a hidden Soviet sniper in a tree. Hänert had been a typical product of the Berlin Guards Regiment school. In the Yelnya bend he had held a position against ceaseless attacks by two Soviet divisions with no more than his machine-gun company, one infantry company, and parts of other units of the "Grossdeutschland" Regiment. Under a continuous artillery barrage he gave his orders calmly and coolly, although he had been wounded three times in his arm and legs.

When the news of his death spread through the battalion during the night of 14th October the phenomenon occurred which old soldiers call "going zombie." Suddenly the Russian bursts of fire had lost their terror. The thought that this war was so cruelly, so indiscriminately killing men like Hänert, or his comrades First Lieutenant Daijes, Second Lieutenant Lemp, Second Lieutenant Baumann, Second Lieutenant Ehrmann, and Sergeants Schneider and Jonasson, and so many other splendid fellows, had turned them fatalistic. The men fought fiercely and bitterly. The Soviet attack was repulsed, and the threatened flank of the "Grossdeutschland" Infantry Regiment was covered again.

Meanwhile Second Lieutenant von Oppen with his leading group had got close to the timber-stack. From the left, where the road ran, came the noise of tank engines. Advanced artillery observers were moving forward alongside the machine-gun platoon. Over to the right, arranged in echelons, the long lines of the 3rd Battalion came into view in the grey light of dawn.

Then the first Russian Maksim machine-gun opened up. The men took cover. Suddenly the flood-gates of war were opened: artillery, mortars, 'crash-boom' guns, rifle-fire. Every yard became a trial of courage. Small groups of men collected behind every house.

Wait for it! The first man made his dash. Then the next. And then the rest. They had gained the cover of the next house. Right in front were the daredevils and the experienced old soldiers. They worked their way from one corner to the next. Finally, they reached the last houses of the estate. In front of them were some 200 yards of flat ground. Then came a wide anti-tank ditch. And some 300 yards beyond that was the large new red-brick building.

One by one they scurried over the open ground. Those who made it let themselves drop into the anti-tank ditch. From the brick building came continuous fire. If only they could get at that building. But the tanks could not clear the ditch. The advanced artillery observer had his telephone-wire severed by shell-fire and was unable therefore to direct his batteries to shell the building.

The remainder of 2nd Company was pinned down in the anti-tank ditch. The 3rd Company was farther to the left, on the far side of the road, in front of the brickworks. The moment a head was raised Russian snipers opened up with their semi-automatic rifles. More and more men were killed. More and more were crying out: "Stretcher, stretcher!" At last the artillerymen, though suffering from a severe shortage of ammunition, managed to place a few howitzer salvos among the brickworks. The 3rd Company stormed and gained possession of it. But at once the men came under murderous machine-gun and mortar fire from the first tenement blocks on the outskirts of the town. They were forced to take cover.

The 3rd Battalion was likewise unable to make any headway. "If only we could get at that shed we could plaster this damned brick building from the flank," Sergeant Wichmann thought aloud. The three men operating his heavy machine-gun nodded.

"Let's be off then," said Wichmann. He leapt up, and scuttled across the empty ground in front of the shed. Thirty yards. Fifty yards. The Russians opened fire. The machine-gun crew were panting behind the sergeant. Only a few more steps—scarcely half a dozen. Wichmann crumpled up, severely wounded by a bullet in the abdomen. He died later on the way to the field hospital. But the men with the machine-gun made it. They assembled the gun and sprayed the windows of the red-brick building.

The 2nd Company managed to gain 50 yards. But then it was pinned down again. When the sun set on 30th October it was apparent that the attack against Tula had got stuck. The assault on Moscow from the south had lost its impetus. There were not enough armoured forces, not enough artillery, not enough grenadier battalions.

The other formations of XXIV Panzer Corps had likewise been unable to make progress. Eberbach's tanks were halted on the road in front of heavy Russian anti-tank barriers. The armoured infantry carriers of 3rd Panzer Division, the 1st (Infantry Carrier) Company, 3rd Rifle Regiment, and Major Frank's Panzerjägers were fighting it out with brand-new T-34s. The duel continued until late at night.

Thus, on 29th October 1941, Colonel Eberbach with the armoured spearhead of XXIV Panzer Corps got only to within three miles of Tula. The attempt to take this important town in a swift coup failed in the face of strong anti-tank and anti-aircraft defences, and the cost to the attackers was heavy. On 30th October a more carefully prepared attack with combat groups of 3rd and 4th Panzer Division and the "Grossdeutschland" Infantry Regiment likewise failed to achieve any worthwhile success. True, the 3rd Panzer Division under Major-General Breith succeeded, after heavy and costly fighting, in gaining a little ground. But the troops were utterly exhausted by the end of the day, and, because of the shocking road conditions, it had become extraordinarily difficult to supply them. An attempt was made to drop ammunition and petrol-drums from aircraft flying only 15 to 30 feet above the ground, but it did not help. Most of the drums burst on hitting the hard ground. Attacks by the Luftwaffe failed in the face of the Soviet anti-aircraft ring around Tula. On 31st October the 3rd Panzer Division at Tula had only 40 tanks left—40 out of an original 150. Thus the attack by Breith's 3rd Panzer Division once more ground to a halt on the southern edge of Tula.

The Russians were defending Tula with the utmost ferocity. They employed all available formations and service branches in order to halt Guderian's advance. For the first time major units of multiple rocket mortars, "Stalin's organ-pipes," were employed.

The down-at-heel German formations simply could not go on. They were down-at-heel and starved beyond belief. The spearhead of XLIII Infantry Corps under General Heinrici—as the general himself reported to Colonel-General Guderian—had received no bread for the past eight days. The gunners of XXIV Panzer Corps had to ration their salvos because hardly any shells were coming up along the mud-bound roads. The troops were cold and hungry, out of fuel and almost out of ammunition. Tula was saved not by the strength of the Russian defences but by the breakdown in German supplies.

General J. F. C. Fuller, one of the most authoritative of Anglo-Saxon war historians, confirms this in his book about the Second World War. He says there: "In all probability it was not so much the resistance of the Russians—strong though it was—or the effect of the weather on the Luftwaffe that saved Moscow, as the fact that the vehicles of the German front were bogged down in the mud."

Things were not much better for the infantry of Second Panzer Army. Thus a war diary of 112th Infantry Division reports:

> On 22nd October 1941 the advance began, and with it the period of the greatest difficulties of movement ever experienced by 112th Infantry Division. Even though the division had a good deal of experience in poor road conditions, what was now demanded of it vastly exceeded anything known in the past. The completely sodden forest paths, the areas of swampy marsh, and the sticky clay on open ground simply defy description. On 26th October 1941, when the division's vanguard reached the Oka sector near Utkino, the picture was as follows: all motorized vehicles were hopelessly bogged down. Those which were not actually stuck in the swamp or on soft roads were unable to move for lack of fuel. The infantry regiments had spread out into unendingly long columns: the heavy vehicles were unable to keep up and had to be manhandled along. It was even worse for the artillery, which continually had to leave guns behind. Any normal supply of foodstuffs, fodder for the horses, and fuel was out of the question. It was therefore decided to unite all motorized vehicles of the division, the Panzerjäger Battalion, all fourteen companies, the heavy squadron of the Reconnaissance Battalion 121, and the signal units of Communications Battalion 112,

under the command of Major Wildhagen, who gradually collected them in Nizina, subsequently transferred them to Orel, and did not rejoin the bulk of the division until the beginning of December. From 26th to 30th October 1941 a halt was called west of the Oka to enable the units to rally again and to allow, at the same time, for the building of a bridge over the Oka at Ignatyevo. Supply difficulties were overcome as the troops gradually learned to live off the land. Oats for the horses were found in sufficient quantity along the road of advance, although naturally the units farther behind had greater difficulty in meeting their requirements than those farther forward. The field kitchens, in addition to meat, potatoes, and occasionally cabbage, made use of the lentils cultivated locally. The greatest problem was bread. The local Russian bread was too heavy and produced digestive upsets. For that reason some battalions set up so-called baking details which moved ahead with the advanced formations, requisitioned what flour they could, and baked their own bread. Gradually this improved in quality. During the further advance east of the Oka the roads were slightly better as the first frosts set in, but on the other hand the terrain was intersected by deep so-called rain ravines which the tired horses had the greatest difficulty in negotiating.

On 5th November 1941 the division at last reached the Plavsk–Tula highway, the divisional staff officers on horseback. The advance had been a colossal feat in view of the exceedingly difficult road and weather conditions, and this was specially mentioned in the commendation addressed to Second Panzer Army by Colonel-General Guderian.

The Panzer Army's motorized and armoured units had been almost entirely left behind on the soft roads, so that the advance was maintained exclusively by the infantry divisions. Only the onset of frosty weather enabled the motorized units once more to resume their movement.

This account is typical of conditions among all infantry divisions on the Central Front towards the end of October 1941.

Shortly before midnight on 31st October the medical orderlies collected the wounded and killed outside the first houses on the edge of Tula. The platoon commanders dodged behind the corners of walls, into cellars, behind heaps of rubble—wherever parties of riflemen or machine-gunners were in position. They were organizing the picket lines. "Hold on!" was the order. "Hold on until the offensive is resumed!" No one suspected that it would be three weeks until then.

On the northernmost point of Moscow's line of defence, at Kalinin, in the bridgehead which XLI Panzer Corps had established over the Volga, the divisions and combat groups of General Reinhardt's Third Panzer Army were likewise getting into difficulties. On 18th October 1941 Lieutenant-General Maslennikov over and over again drove the Siberian battalions of his Twenty-ninth Army, reinforced by numerous artillery units, mortar batteries, and tanks, against the most forward parts of the reinforced German 1st Panzer Division which was pushing over the Volga towards the north, along the road to Torzhok. On 19th October Heydebrand's armoured combat group—the reinforced 1st Rifle Brigade—was compelled to give up the railway bridge over the Volga at Mednoye after its partial destruction, and Maslennikov presently tried to recapture also the important railway and road junction of Kalinin itself.

The Red commissars had established 'security companies' behind the attacking formations and were threatening to open fire at them if they retreated.

On the north-western edge of the town fighting was also very fierce. Time and again the Russians made penetrations across the Volga, either with a view to recapturing the railway bridge or to cutting the XLI Panzer Corps' supply lines to Kalinin, the roads from Staritsa and Latoshino to the Volga bridgehead. More than once the situation was saved only by a hurried switch-round of the last reserves. It was a savage trial of strength. On several occasions it was General Freiherr von Richthofen's VIII Flying Corps that saved the situation by massive Stuka attacks against Russian tank concentrations and mortar batteries.

The 129th Infantry Division and the 36th Motorized Infantry Division, the latter reinforced by a motorized training brigade, defended the northern and south-eastern parts of the town. Between them the 1st Panzer Division held the Volga sector with the two bridges in the north-western district. Its 73rd Panzer Artillery Regiment, whose gunners were from Weimar, Erfurt, and Hamburg-Wandsbek, was on the southern bank of the Volga, providing artillery support for the bitterly fighting combat units and, together with several Army Artillery battalions subordinated to it, keeping down the Russian batteries on the northern edge of the town.

On the Upper Volga General Model's divisions maintained their position, but they too had become too weak to continue the offensive in a northerly direction in order, as had been planned, to meet the divisions of Army Group North who were advancing over the Valday Hills. The troops were exhausted from fighting, the battalions of 1st and 11th Panzer Regiments as well as the Special Purpose Panzer Battalion, 101st Division, were greatly reduced in numbers, and infantrymen and grenadiers were discovering that the heavy weapons they had lost could no longer be replaced. In this way the mud remained victorious at Kalinin too. The offensive of Army Group Centre gradually petered out. The formations of Third Panzer Army were likewise ordered to halt until the infantry of Ninth Army caught up with them again.

"Hold on until the frost comes!" They held on. The military cemetery behind the church by the southern ramp of the road bridge over the Volga was becoming increasingly full of wooden crosses. There, on 20th October, the first man in 1st Panzer Division to be awarded the Oak Leaves to the Knights Cross of the Iron Cross, Major Dr Joseph Eckinger, was laid to his last rest. A native of Styria, he had commanded a battalion of 113th Rifle Regiment in the bold action on 14th October which resulted in the capture of both Volga bridges intact.

That then was the picture at Tula and Kalinin at the beginning of November 1941. It was the same along the entire 600-mile front of Army Group Centre.

Things were no better with the Armies engaged in the frontal advance towards Moscow—Fourth Panzer Group and Fourth Army. A typical description of the actions in that area during the last ten days of October is provided by the diary of an infantry division operating there.

On 25th October 195th Infantry Regiment, 78th Infantry Division, at Ruza was ordered to prepare for the capture of Zvenigorod, a strongpoint in Moscow's second line of defence. As the 2nd Battalion emerged from the woods around Vorontsovo it came under heavy fire from the high ground on both sides of Panovo. After quick redisposition the battalion attacked, over-ran three guns, captured a quadruple machine-gun and three multiple mortars, and took Panovo at nightfall. During the night the battalion pushed through the deep forest towards Krivosheino. On 27th October the entire regiment moved off from Krivosheino via Apalchino towards Lokotnya. Presently it came up against a strongly fortified line of enemy pillboxes, no doubt intended to cover the Ruza–Zvenigorod–Moscow road. The Russians resisted stubbornly. Fierce fighting ensued. Tanks appeared on the scene. Nevertheless the German troops succeeded in taking Apalchino and Kolyubakino in the evening.

During the night of 27th/28th October fierce fighting took place for the two villages after the enemy had mounted a counter-attack from the south, with tanks and infantry. All the battalions of the regiment, as well as the assault guns assigned to it, were obliged to join in. In view of the situation on the southern flank—especially at VII Corps immediately on the right—any further advance had to be called off. Since, however, possession of Lokotnya with its dominating high ground was essential as a jumping-off point for the resumed attack, the troops were ordered to take the village. This attempt led to bitter fighting for enemy positions in the woods west of Lokotnya on 29th October. It was not possible to take the village. All further attacks were therefore called off. The division reorganized itself for defence along a line from Osakovo via Kolyubakino to Apalchino. The enemy had proved too strong in the area of IX Corps too. As elsewhere, the end of the muddy season had to be awaited.

The divisions were thus bogged down in the mud and slush on and along the roads. Their lines of supply were not only tremendously long, but they were also barely negotiable. The fast-moving German divisions, accustomed to Blitzkrieg operations, had become clumsy and slow, almost as clumsy as the Napoleonic armies in 1812. The first thing they did was try to solve their problems by switching supplies to local types of vehicles. Next they reorganized their debilitated units into smaller but more vigorous formations. Thus the tanks of XLI Panzer Corps were regrouped into 'action units,' instead of the former two or three battalions with eight to twelve companies to each regiment, and the remnants of eight companies of infantry were reorganized into the three companies of a divisional carrier-borne rifle battalion. Reconnaissance battalions and motor-cycle battalions were amalgamated to make new battalions, and the armoured scout car troops were united in a single company directly under Division. In this way the troops in the field attempted to overcome their difficulties by improvisation, inventiveness, and sheer guts. Everybody was hoping that the High Command would meet the changed situation at the front with new measures. But the Fuehrer's headquarters were far, far away—many hundred miles behind the front, at Rastenburg, in East Prussia.

The Soviet High Command, on the other hand, made full use of the fact that it was waging its war on Moscow's doorstep. It enjoyed the advantage of what are called interior lines. From his seat of government Stalin was able, by suburban trains or even on foot, to dispatch new formations from the eastern part of his empire and tanks straight off the assembly-line to wherever he wished, to switch them rapidly from one point of the front to another, and thus to concentrate them, time and again, at the keypoints or on critical sectors of the battle. As a result, no sooner had a German combat group anywhere broken through the Soviet lines than it found itself faced by numerically superior Soviet formations and strong armoured tactical reserves. Yet the fighting morale of most of the Soviet formations was by no means good. With the exception of Far Eastern and Siberian Guards Divisions, and a few cavalry divisions, the Russian troops in the fighting line before Moscow were nothing like the unflinching heroes portrayed by Soviet military historians.

The following passage is from the diary of a Soviet second lieutenant whose name shall remain unpublished for the sake of his parents or children. He was killed in the Tula area on 12th November. On 31st October he made the following entry:

> During the night of 30th/31st we crossed the Orel–Tula highway in the area of Gorbachevo–Plavsk and reached the village of Fedorovka. Cases of desertion assumed unbelievable proportions before we crossed that road. The deputy commander, Lieutenant

Alaportsev, and others grabbed some officers' horses, including my own, and rode back to the spirit factory. A fine lot of officers! I am still sick with influenza, terribly weak, with fits of giddiness and aching temples. In our battalion 80 per cent. have deserted, including some seemingly reliable people in No. 3 platoon. They go into the villages, throw away their weapons, their equipment, and their uniforms, and put on rags. In the villages the collective farms are forcibly liquidated, and horses, harness, and carts are shared out. Grain is driven away from the stores and seed stock divided among the people. There is much talk that the war is lost anyway and that very soon there will be no collective farming.

That then was the picture. But it was like a boxing match when both opponents have no strength left in their fists. The exhausted and poorly supplied German units in the front line no longer had the strength to deal a knock-out blow to the reeling Soviet colossus. "If only the frost would come!" they moaned. "If only the roads were usable again!" If only...

The frost came during the night of 6th/7th November. All along the front of Army Group Centre winter suddenly set in. It was a gentle, welcome frost which made the ground hard again and usable by vehicles. The troops along the roads heaved a sigh of relief. True, they had no winter clothes, and many of them were still wearing their summer uniforms. But at least it was the end of that dreadful mud.

They dragged their guns free from the frozen ground. Here and there the result was broken wheels and axles. But what did it matter? Supplies were getting through again—troops' comforts, cigarettes, mail, spirits, and spares. Tanks were rolled out of mobile repair shops. Ammunition was delivered again in the line. Slowly the war machine began turning again. And with it the hope was revived that Moscow might yet be taken.

Needless to say, if that was to be done, the final push had to be started at once. The Army High Command called for urgent action. The Commander-in-Chief Army Group Centre, Field-Marshal von Bock, was equally anxious to get a decision on the resumption of operations. But the armies were so burnt up that they needed time for recovery. The first few days therefore were busy times for the supply troops. On lorries, on sledges, and on farm carts they ferried to the fighting line the *matériel* needed for the resumption of operations. In that first fine flush of doing everything possible for the fighting front a few strange things happened as well—things which caused a great deal of anger among the fighting forces. Some supply authority in France, for instance, had conceived the no doubt praiseworthy idea of giving the Eastern Front a special treat and at the same time boosting the French wine business. As a result, two goods trains full of French red wine in bottles were dispatched from Paris. Wine trains instead of the desperately needed ammunition trains! Heaven knows who authorized these deliveries. Anyway, when they arrived in Yukhnov in the area of Fourth Army, the temperature was 25 degrees Centigrade below freezing. All that the unloading squads found in the wagons was large chunks of red ice intermingled with glass splinters. "Frozen Glühwein instead of winter clothing," the men cursed. General Blumentritt, then Chief of Fourth Army General Staff, has stated that he had never seen the troops so angry as in the face of this truly deplorable faux pas.

On 12th November the thermometer stood at 15 degrees below zero Centigrade. On 13th November it dropped to 20 degrees. It was a lively day for the airfield of Orsha. Halder's machine from Rastenburg and the planes of the Army Group staffs and Army C-in-Cs arrived one after another: Colonel-General Halder, Chief of the Army General Staff, had summoned the Chiefs of Staff of the three Army Groups and of all Armies on the Eastern Front to a secret conference.

The subject of the conference was: What was to be done? Should the divisions dig in, take up winter quarters, and wait for the spring? Or should the offensive—mainly against Moscow—be continued in spite of the winter?

The conference of Orsha is of particular significance in the history of war. It probably provides the answer to a question argued with much passion to this day: Who was ultimately responsible for the resumption of the ill-fated winter offensive?

Was it Hitler? Was it the General Staff? Or—and this is the most recent and most sensational theory—was it all a trick of Stalin, who, by means of false reports planted on the German secret service, lured Hitler into resuming his offensive and thus into a trap? It is an interesting theory, and its source cannot lightly be dismissed.

In his book *Soviet Marshals Explain* Kyrill Kalinov, a Soviet General Staff officer who emigrated to the West from Berlin in 1949 and who worked in the Soviet High Command during the war, quotes an interesting statement by Zhukov—admittedly without a precise source reference. According to Kalinov, Marshal Zhukov made the following claim in 1949, apparently in a lecture:

"The Germans estimated the total of Soviet forces annihilated by them at the fantastic figure of 330 divisions. They did not therefore believe that we had any fresh reserves at our disposal, and consequently expected that all they would encounter was contingents of workers' militia hurriedly raised in Moscow. That was the decisive reason why Hitler took the gamble of mounting his final offensive against our capital.

"In this connexion I can now disclose an important detail which has hitherto been kept secret. The report about the allegedly destroyed

330 divisions was launched by us deliberately to find its way to Germany through the Military Attaché of a neutral country whom we knew to be in touch with Germany's military intelligence service. Our aim was to support Hitler against his General Staff. The generals, as we know, recommended that the German troops, as in 1914, should dig in wherever they stood and set up winter quarters.

"It was, however, to our advantage that the Germans should not give up their intentions with regard to Moscow, but should press forward into the flat wooded country where we should be able to inflict on them a final defeat.

"I was emphatically supported by Comrade Stalin, who was even prepared to risk the surrender of our capital. For four days, therefore, we only employed divisions of the militia in the fighting line immediately outside the capital. The Germans were to gain the impression that these formations were all we had left to put up against their experienced and usually victorious divisions."

In view of the position of the author, this theory cannot be dismissed out of hand. The possibility is too disturbing and too important. It deserves careful examination. The decision to resume the offensive against Moscow was made in Orsha on 13th November. About the proceedings of the Orsha conference a number of reliable reports exist, including one by Major-General Blumentritt, then Chief of Staff of Kluge's Fourth Army, himself a participant in the conversations.

According to him, Halder reviewed the general situation on the 1250-mile-front from Lake Ladoga to the Sea of Azov. His report culminated in the question: Should the offensive be maintained or should defensive positions be taken up? General of Infantry von Sodenstern, Field-Marshal von Rundstedt's representative, speaking on behalf of Army Group South, demanded the cessation of the offensive and going over to the defensive. Rundstedt, after all, was on the Don, outside Rostov, some 220 miles farther east than the front line of Army Group Centre before Moscow.

Lieutenant-General Brennecke, Chief of Staff of Field-Marshal Ritter von Leeb, had no difficulty in arguing that Army Group North had been so weakened by having its entire armoured forces detached from it that all offensive operations were out of the question. In fact, it had long gone over to the defensive.

Army Group Centre did not share this view. It pleaded for the continuation of the offensive against Moscow. Major-General von Greiffenberg supported his Field-Marshal's view that the capture of Moscow was necessary both militarily and psychologically. There was, of course, the danger that they might not pull it off, but this would be no worse than lying on open ground in the snow and cold only 30 miles from the tempting objective.

Bock's arguments were in line with the views held by the High

Command. In the Fuehrer's headquarters it was believed that the Russians were at the end of their tether and that one last effort would be enough to defeat them completely. This optimism was not shared by Bock and his staff—neither by Greiffenberg nor by Lieutenant-Colonel von Tresckow, the Chief of Operations; they knew the condition of the troops and realized that only a short span of time was left before the onset of the severe winter weather. But Bock nevertheless regarded the offensive as the preferable alternative to spending a desolate winter in the field, a winter which might give Stalin plenty of time to get his second wind.

Halder was pleased with the attitude of Army Group Centre, as indeed was Field-Marshal von Brauchitsch, the Commander-in-Chief Army. They both favoured a resumption of the offensive, since they regarded this as the only chance to conclude the campaign victoriously.

Halder already had the operation orders in his pocket, and now he announced them. The objectives were mapped out ambitiously. Guderian's Second Panzer Army was to take the traffic junction of Tula and its well-equipped airfield, and then drive south-east of Moscow through Kolomna to the ancient city of Nizhniy Novgorod on the Volga, now called Gorkiy—250 miles beyond Moscow.

In the north Ninth Army was to move east across the Volga–Moskva Canal with Third Panzer Army, and then wheel towards Moscow as the left prong of a pincer movement.

In the centre a frontal attack was to be made by Fourth Army on the right and 4th Armoured Group on the left.

The date for the start of the offensive was not yet laid down. Field-Marshal von Bock was in favour of starting at once, but the supply situation demanded that it be delayed for a few more days.

This account shows that the German High Command, though it may have had some cause to doubt the usefulness of this last great offensive operation of 1941, did not resume the offensive against Moscow solely at Hitler's pressure—as Zhukov claims. Field-Marshal von Bock, whatever his reasons, was a determined champion of the new offensive. Moscow had been his objective at all times and at every phase of the campaign. On this point he found himself in full agreement with the Army High Command, which time and again declared Moscow to be the most important objective. The anxiety to reach Moscow before the end of the year was entirely understandable.

For one thing the general strategic situation demanded it. Was Army Group Centre to dig in along a front line a thousand miles long? With only a single infantry division as a reserve behind the fighting line—and otherwise a vast empty hinterland controlled by partisans? Was the initiative to be left to the Russians to launch continuous local attacks? Were the German troops to watch Stalin use Moscow as an ideal marshalling yard for bringing up fresh forces from all parts of his great empire and employing them against the thin and frozen German lines? That, surely, would have been the wrong solution.

But there was yet another important consideration. Field-Marshal von Brauchitsch, the Commander-in-Chief Army, his Chief of the General Staff, and more particularly Field-Marshal von Bock and Colonel-General Guderian, had been urging Hitler ever since the battle of Smolensk to give them the green light for the attack on Moscow. They had resisted his plan of first pressing ahead with the battle of Leningrad in order to clear the flank for the offensive against Moscow. They had opposed his operation against Kiev, and had ceaselessly implored, persuaded, and warned him that Moscow itself must be the principal military objective.

Hitler, on the other hand, had from the outset opposed the views of his General Staff. He did not believe that the capture of Moscow was all-important. He maintained that the course of operations must show whether Moscow could be taken. "Russia will be defeated when we possess Leningrad and the Gulf of Finland in the north, and when we have the grain of the Ukraine and the industrial area of the Donets in the south," was his argument. Strangely enough, and contrary to his usual custom, he had in the end allowed himself to be swayed away from his favourite target of Leningrad.

At any rate, Moscow was not his pet objective. It was and continued to be the pet objective of his General Staff. Now he had given in to his generals. Were Brauchitsch, Halder, von Bock, and Guderian now to go to him and say, "Sorry, we cannot make it: because of the unfavourable terrain and winter weather we shall have to dig in 30 or 20 miles from our target."

No: they wanted the offensive to be continued. They wanted to take Moscow. And they believed that they were able to do so, whether or not 330 Russian divisions had been destroyed.

Zhukov is mistaken in thinking that Hitler ordered the resumption of the winter offensive against Moscow in defiance of the wishes of his High Command. Consequently the sensational theory that he had supported Hitler against a war-weary High Command and by planting on him faked figures of prisoners had lured Army Group Centre to its doom—as Prince Kutusov had done to Napoleon—must fall to the ground.

8. Final Spurt towards Moscow

*"The days of waiting are over"–Cavalry charge at Musino
–On the Volga Canal–Within five miles of Moscow–Panic
in the Kremlin–Stalin telephones the front–40 degrees
below zero Centigrade–Battle for the motor highway–
Men, horses, and tanks in ice and snow–Everything stop.*

D-DAY for the "Autumn offensive 1941" was 19th November. The troops did whatever they could to prepare themselves for this last difficult battle. The determination to make one more all-out effort is reflected in the Order of the Day of Fourth Panzer Group announcing the launching of the offensive. It is typical of a great many others.

> To all commanding officers in Fourth Panzer Group.
> The days of waiting are over. We can attack again. The last Russian defences before Moscow remain to be smashed. We must stop the heart of Bolshevik resistance in Europe in order to complete our campaign for this year.
> This Panzer Group has the great fortune to be able to deal the decisive blow. For that reason every ounce of strength, every ounce of fighting spirit, and every ounce of determination to annihilate the enemy must be summoned.

One of the key-points of the battle of Moscow was situated in the area of Fourth Panzer Group, between Shelkovka and Dorokhovo. It was there that the old postal road—the historic road taken by Napoleon—the modern motor highway, and the Smolensk–Moscow railway intersected with the great north–south route from Kalinin to Tula. Whoever held Shelkovka and Dorokhovo and the high ground outside controlled this vital communications centre.

The 10th Panzer Division had taken Shelkovka at the end of October. But the Russians were still established on the high ground. Just as the 7th Infantry Division from Munich relieved it—this division also included the first volunteers of the "French Legion," known as 638th Infantry Regiment—the first Soviet counter-attack burst right into the middle of these movements and opened a string of exceedingly ferocious engagements.

Stalin had brought up his 82nd Motorized Rifle Division from Outer Mongolia for the recapture of Shelkovka. The attack of this Mongolian crack unit was effectively supported by two armoured brigades, also fresh troops brought into the fighting line, and by multiple mortars and army artillery. The German 8·8-cm. anti-aircraft guns, now used against ground targets, could not be everywhere at the same time. The men from Munich were simply helpless against the solid swarms of T-34s, and thus the 7th Infantry Division had to give up the crossroads after suffering heavy casualties. The fact that the Soviets once more controlled the Shelkovka–Dorokhovo area was to have far-reaching consequences.

All the troops of XL Panzer Corps in the Ruza area found their only supply road cut. The 10th Panzer Division, engaged in costly fighting on the causeway between Pokrovskoye and Skirminovo, was left without ammunition, without fuel, and without food; it was also unable to send its wounded out of the fighting line. Units of the "Reich" SS Division, urgently needed for supporting 10th Panzer Division, were held up idly at Mozhaysk, unable to reach their destination.

The way in which this dangerous situation was cleared up is described by Captain Kandutsch, Intelligence officer at XL Panzer Corps headquarters, whose original report is extant:

"The same evening I was ordered by Colonel von Kurowski, the Chief of Staff, to reconnoitre towards the crossroads at 0400 the following morning and to report as quickly as possible whether the Motorcycle Battalion, 'Reich' SS Division, could be moved up. At 0400 hours I set out from our headquarters at Ruza, accompanied by Corporals Schütze and Michelsen on a motor-cycle with sidecar. Since no armoured scout car was available I had to make my reconnaissance in a staff car. As far as the Moskva bridge at Staraya Russa everything was quiet; the road to Makeykha was under sporadic harassing fire from enemy artillery, and Makeykha itself was the target of repeated sudden artillery bombardments. At 0515 I picked up a maintenance party NCO of Communication Battalion 440 in order to have a telephone-line laid in the direction of the crossroads. At 0540 communication was restored with Captain Gruscha, commanding Mortar Battalion 637, about two miles south of Makeykha. I found the mortar crews hard-pressed, dug in around their battery, ready to defend it against enemy attacks. After making a telephonic report to my Chief of Staff I proceeded at 0600 hours to the headquarters of the newly brought up Infantry Battalion, 267th Infantry Division, about a mile north of the crossroads, and had the telephone-line laid to it. At that moment the German counter-attack for the recapture of the crossroads was in full swing. The noise of battle was increasing all the time. The battle area was under heavy gunfire. The road itself was also being continually raked by Russian machine-guns. By having the telephone-line extended as and where the infantry gained ground I was able at 0730 hours to report to the Chief of Staff that the crossroads had been cleared of the enemy, and at 0800 I reported the arrival of the first parties from the Motorcycle Battalion, 'Reich' SS Division, who had got across the road intersection with comparatively light casualties."

Final Spurt towards Moscow 169

At the beginning of November General Fahrmbacher's VII Corps went into action with the Bavarian 7th, the Middle Rhine–Saar 197th, and the Lower Saxonian 267th Divisions with the intention of dislodging the Russians at long last from their high ground, and of making the crossroads usable for the impending offensive. The attack was supported by 2nd Battalion, 31st Panzer Regiment, of the Silesian 5th Panzer Division.

Advancing rapidly, the tanks broke into the positions of the Mongolian brigade. But the sons of the steppes did not yield: they attacked the tanks with Molotov cocktails. The infantry regiments in the wake of the tanks had to take position after position at bayonet point. Wherever they achieved a penetration they were instantly showered with rocket salvos. Losses were heavy on both sides.

However, after two days' fighting the Russians were definitely thrown back on this sector. Wheeled traffic again flowed freely over the crossroads of Shelkovka. The supply route on the right wing of Fourth Panzer Group was open once more.

Between 15th and 19th November the divisions of Army Group Centre mounted their final assault on Moscow, one by one in carefully timed succession. The officers, all the way down to the smallest unit, knew what was at stake. Colonel-General Guderian writes in his memoirs that he explained to his Corps commanders that no more time must be lost. He implored them to do everything in their power to make sure the objective was reached. Colonel-General Hoepner likewise endeavoured to rouse his troops to a supreme last effort in his Order of the Day of 17th November addressed to his unit commanders:

> Arouse your troops into a state of awareness. Revive their spirit. Show them the objective that will mean for them the glorious conclusion of a hard campaign and the prospect of well-earned rest. Lead them with vigour and confidence in victory! May the Lord of Hosts grant you success!

This Order of the Day is reproduced here not because of its bombast and the kind of magniloquence that is customary in a war: the significance of the document lies on an entirely different plane. It reveals that so outstanding a military leader as Hoepner, a man of great personal courage who was later to die on the gallows as one of the active conspirators against Hitler, was still convinced on 17th November 1941 that Moscow could be captured.

On 16th November Hoepner's V Infantry Corps mounted its attack against the town of Klin, north-west of Moscow on the road to Kalinin. On its left the LVI Panzer Corps of Third Panzer Army was scheduled to move forward.

Dawn was breaking near Musino, south-west of Klin—the dawn of

17th November. It was a grey and hazy morning. Towards 0900 the sun appeared through the fog as a large red disc. The observer post of a heavy battery was on a hill. About two miles farther ahead the edge of a broad belt of forest could just be made out. Everything else was flat fields under a light cover of snow. It was cold. Everybody was waiting for the order to attack.

1000 hours. Field-glasses went up. Horsemen appeared on the edge of the wood. At a gallop they disappeared behind a hill.

"Russian tanks!" a shout went up. Three T-34s were approaching over the frozen ground. From the edge of the village the anti-tank guns opened up. It was odd that the tanks were not accompanied by infantry. Why would that be? While the artillery observers were still busy puzzling out the mystery another shout went up: "Look out—cavalry to the right of the forest." And there they were—cavalry. Horsemen approaching at a trot. In front their reconnaissance units, then pickets of forty or fifty horsemen. Now the number had grown to one or two hundred. A moment later they burst out of the forest on a broad front—squadron next to squadron. They formed up into one gigantic line abreast. Another line formed up behind them. It was like a wild dream. The officers' sabres shot up into the air. Bright steel flashing in the morning sun. Thus they approached at a gallop.

"Cavalry charge in regiment strength. Spearhead of attack at 2500 yards!" The artillery spotter's voice sounded a little choked as he passed the information back over the telephone. He was lying in a hole in the ground, on a sheet of tent canvas. His trench telescope had been painted white with a paste of chalk tablets immediately after the first fall of snow. Now it did not show up against the snow blanket which, still clean and white, covered the fields and hills of Musino. Still clean and white. But already the squadrons were charging from the wood. They churned up the snow and the earth: the horses stirrup touching stirrup, the riders low on the horses' necks, their drawn sabres over their shoulders.

The machine-gun crew by the artillery observation post had their gun ready for action on the parapet. The gunner pulled off his mittens and put them down by the bolt. The gun commander's eyes were glued to his field-glasses. "2000 yards," they heard the artillery spotter shout down his telephone. He followed up with firing instructions for his battery.

Barely a second passed. And across the snowy fields of Musino swept a nightmarish vision such as could not be invented by even the most fertile imagination. The 3rd Battery, 107th Artillery Regiment, 106th Infantry Division, had opened fire at close range. With a crash the shells left their barrels and burst right among the charging squadron. The HE shells of the anti-tank guns in the village, which had only just been attacked by T-34s, landed amid the most forward Russian group. Horses fell. Riders sailed through the air. Flashes of lightning. Black smoke. Fountains of dirt and fire.

The Soviet regiment continued its charge. Their discipline was terrific. They even pivoted about their right wing and headed towards the village. But now salvo after salvo of the heavy guns burst amid the squadrons. The batteries were firing shrapnel which exploded 25 feet above the ground. The effect of the splinters was appalling. Riders were torn to pieces in their saddles; the horses were felled.

But the terrible spectacle was not yet over. From out of the forest came a second regiment to resume the charge. Its officers and men must have watched the tragedy of their sister regiment. Nevertheless they now rode to their own doom.

The encircled German batteries smashed the second wave even more quickly. Only a small group of thirty horsemen on very fast small Cossack animals penetrated through the wall of death. Thirty out of a thousand. They charged towards the high ground where the artillery observer was stationed. They finished up under the bursts of the covering machine-gun.

Two thousand horses and their riders—both regiments of 44th Mongolian Cavalry Division—lay in the bloodstained snow, torn to pieces, trampled to death, wounded. A handful of horses were loose in the fields, trotting towards the village or into the wood. Slightly wounded horsemen were trying to get under cover, limping or reeling drunkenly. That was the moment when Major-General Dehner gave the order for an immediate counter-attack.

Out of the village and from behind the high ground came the lines of infantrymen of 240th Infantry Regiment. In sections and platoons they moved over the snowy ground towards the wood.

Not a shot was fired. Sick with horror, the infantrymen traversed the graveyard of the 44th Mongolian Cavalry Division—the battlefield of one of the last great cavalry charges of the Second World War. When they reoccupied the village of Spas Bludi the grenadiers found that their comrades of 240th Infantry Regiment, taken prisoner there after being wounded, had been done to death.

The Russian attack had been senseless from a military point of view. Two regiments had been sacrificed without harming a hair on the opponent's head. There was not a single man wounded on the German side. But the attack showed with what ruthless determination the Soviet Command intended to deny the German attackers the roads into the capital, and how stubbornly it was going to fight for Moscow.

Another illustration is found in the diary of the young Soviet lieutenant mentioned earlier, the commander of a mortar platoon on Moscow's southern front. Under 17th November we read as follows:

The battalion received the categorical order to take the fascist

172 Part One: Moscow

position on the high ground outside the village of Teploye. However, we were unable to make a single step forward because the fire of the Germans was too strong. Kryvolapov reported to Regiment that without artillery support we could not make any progress. The reply was: You will have taken that position in twenty minutes, or else the officers will face a court martial. The order was repeated six times. We attacked six times. The commander was killed. Tarorov, the adjutant, and Ivashchenkov, the Party Secretary, are also dead. The battalion has only twenty guns left.

That was how Stalin made his troops fight. He employed everything he had for the defence of his capital. Whatever human or material reserves were left in his empire he mobilized for the defence of Moscow.

Stalin knew what Moscow stood for and what its loss would mean. He confessed as much to Harry Hopkins, Roosevelt's representative, when he said to him, "If Moscow falls the Red Army will have to give up the whole of Russia west of the Volga." Nothing can illustrate his desperate mood more clearly than his request to Roosevelt, reported by Hopkins: "He, Stalin, would welcome it if American troops appeared on some sector of the Russian front, and, what is more, under the unrestricted command of the US Army."

Isaac Deutscher, Stalin's biographer, very rightly points out: "This is one of the most revealing remarks of Stalin that have been recorded by the chroniclers of the Second World War." Indeed, it shows as nothing else how desperately Stalin saw his own position.

Roosevelt did not send any troops to the Soviet front, and Stalin had to make do with what he could scrape together within his empire. Not all the units were willing to go into action. Many of the regiments had passed through the searing fire of the summer battles. Entire divisions could only be made to fight by the threat that, in the event of their withdrawal, they would be mown down by reliable security formations.

The Mongolian and Siberian divisions, on the other hand, switched by Stalin to the west from the Far East of his country, were vigorous and full of fighting spirit. It was largely due to them that Moscow in the end was saved. And, of course, also to the fact that Stalin could calmly denude his 5600-mile sea frontier from the Bering Straits to Vladivostok and his 1900-mile land frontier from Vladivostok to Outer Mongolia, without having to fear that Japan's Kwantung Army would cross the USSR's eastern frontier and help its German allies by stabbing the Russians in the back. He was able to do that because he knew from his master spy Dr Sorge that the Japanese, the allies of Germany, were preparing instead to attack the Americans in Pearl Harbor in order to capture for themselves the islands of the Pacific. It was this decision that saved the Soviet Union. Japan was to reap a poor reward from Stalin for this service.

Final Spurt towards Moscow 173

The appearance of Siberian crack divisions before Moscow was of decisive importance, even though Marshal Zhukov disputes this fact in order not to have to share his glory with the Siberian tactical reserves. According to Kyrill Kalinov, Zhukov declared: "Reinforcement by Siberian troops was exceedingly useful to us. But the Siberians did not amount to more than 5 per cent. of the troops engaged in the battle. It would be ludicrous to describe their part as decisive."

Soviet military history refutes the Marshal. In Samsonov's book *The Great Battle of Moscow* we read: "During the muddy period the High Command concentrated strong strategic reserves in the Moscow area; they had been brought up from the deep hinterland, from Siberia and Central Asia. New operational units were formed."

These reserves were so considerable that, according to Samsonov, the Russian defending forces at Moscow, at the resumption of the offensive on the Central Front in November, were for the first time numerically superior to the Germans. Samsonov gives the proportion of infantry divisions as 1 to 1·2 in favour of the Soviets. If one remembers that the German infantry divisions had lost 30 to 50 per cent. of their combat strength after their ceaseless marching and heavy fighting, and that the armoured divisions were mere shadows of their former selves, operating with barely one-third of their normal strength, one begins to understand what happened at Moscow between 18th November and 5th December, and what it was that the Russian war historians call "the miracle of Moscow."

The cavalry charge at Musino was the bloody overture to the thrust to be made by V Corps on the left wing of 4th Panzer Group against Moscow's vital artery in the north-west—the Kalinin–Klin–Moscow road. General of Infantry Ruoff was to open the way to the capital between that road and the Moskva–Volga Canal.

In the mild winter weather of the first few days of the offensive Lieutenant-General Veiel's 2nd Panzer Division struck swiftly and confidently across the Lama river. Russian resistance was broken. The division bypassed Klin in the south, while LVI Panzer Corps of the Third Panzer Army was moving against that town from the north-west. The first meagre consignments of winter clothing arrived at the front— one greatcoat to each gun crew. One greatcoat! That was on 19th November. On that day the weather broke. The thermometer dropped to more than 20 degrees below zero Centigrade. Snow fell. Freezing fog formed even in daytime. The severe Russian winter had arrived, earlier than in many previous years, but by no means as exceptionally early as is often claimed.

On 23rd November Lieutenant-Colonel Decker's combat group, moving ahead of the spearheads of V Corps with parts of the reinforced 3rd Panzer Regiment, penetrated into Solnechnogorsk from the west. The 2nd Rifle Brigade under Colonel Rodt attacked the town from the north-west with 304th Rifle Regiment. The strong Russian defences were overcome and more than two dozen enemy tanks destroyed. The bridges over the canal were secured intact. Things were moving again. As a result, General Veiel's Viennese 2nd Panzer Division stood 37 miles from Moscow on an excellent road.

On 25th November Colonel Rodt took Peshki, south-east of Solnechnogorsk, another six miles nearer Moscow. Standing on a hill, the colonel saw through his binoculars three tanks approaching. "What type of tank are those?" he asked his orderly officer. "No idea, Herr Oberst," was the reply.

The first shots were fired. The spearhead of 1st Battalion, 3rd Panzer Regiment, appeared from behind undulating ground and opened up at the surprised enemy tanks with its 7·5-cm. guns. Two of the tanks were hit; the third withdrew. When Colonel Rodt inspected the wrecks he was much surprised—British Mark III tanks, which could be effectively opposed even with the German 3·7-cm. anti-tank gun. Russian translations of the original English lettering and instructions were chalked up on the sides of the tank. They were the first items of British aid for Stalin to appear in the fighting line.

The infantry divisions of V Corps were likewise driving along both sides of the great road, southward towards Moscow and south-eastward towards the Moskva–Volga Canal. They were the 106th, 35th, and 23rd Infantry Divisions. The canal was the last natural obstacle to Moscow's being outflanked in the north. If it was overcome the northern attacking force—Fourth Panzer Group and Third Panzer Army—would have the worst behind them. The Potsdam 23rd Infantry Division headed for the canal via Iksha with 9th Infantry Regiment. The division's other infantry regiment, 67th Infantry Regiment, and the Reconnaissance Battalion 23 were likewise fighting their way to the canal north-east of Krasnaya Polyana. Farther south the reinforced 2nd Rifle Brigade, moving past Krasnaya Polyana, gained Katyushki on 1st December. This village changed hands several times. Patrols of 2nd Company, Panzer Engineers Battalion 38, were advancing in the direction of the railway station of Lobnya. It looked as though the Blitzkrieg was in full swing again.

At first the Russians were confused. And, as always in such a situation, a great many opportunities presented themselves. One of these is illustrated by the following episode. Motor-cycle patrols of Panzer Engineers Battalion 62—originally operating under 2nd Panzer Division, but moved forward by Hoepner himself on 30th November beyond the most forward units of 2nd Panzer Division, to strike at the railway station of Lobnya and the area south of it—roared forward on their machines and, without encountering any opposition, got as far as Khimki, the small river port of Moscow, five miles from the outskirts of the city. They spread alarm and panic among the population and

Final Spurt towards Moscow   175

raced back again. It was these motor-cyclists and Corps sappers who got closest to Stalin's lair. But units of 106th Infantry Division, attacking on the right of 2nd Panzer Division, got almost as close to the Kremlin when a combat group of 240th Infantry Regiment, reinforced by a combat detachment of 52nd Anti-aircraft Regiment, reached Lunevo. Russian sources relate these events with an air of horror to this day—the same horror that swept the Kremlin more than twenty years ago when the news came: "The Germans are at Khimki!"

In the General Staff citadel inside the Kremlin there had in fact been grave dismay ever since 27th November. Stalin was pacing up and down along the great map table, scowling. There was disastrous news from the front: "Enemy forces of the German Third Panzer Army have crossed the Moskva–Volga Canal at Yakhroma, 43 miles north of Moscow, and have established a bridgehead on its eastern bank. There is the danger of a break-through to Moscow from the north." Since there were no further defences beyond the canal, the words "danger of a break-through from the north" were tantamount to an admission that, unless major enemy forces were prevented from crossing over to the eastern bank, Moscow would be lost.

What had happened?

The battle-hardened LVI Panzer Corps under General Schaal—at the beginning of the campaign Manstein's striking force—had been operating to the left of V Corps with 6th and 7th Panzer Divisions as well as 14th Motorized Infantry Division. On 24th November it had taken Klin, and shortly afterwards Rogachevo; it had pressed forward through the burst seam between the Thirtieth and Sixteenth Soviet Armies as far as the Moskva–Volga Canal, and had immediately established a bridgehead on the far bank. In a bold stroke Colonel Hasso von Manteuffel seized the canal bridge at Yakhroma with the reinforced 6th Rifle Regiment and units of 25th Panzer Regiment, stormed across the waterway, and dug in for all-round defence of the bridgehead. A Soviet armoured train which appeared on the scene was immediately attacked by a tank company of 25th Panzer Regiment under Lieutenant Ohrloff, an officer decorated with the Knights Cross of the Iron Cross, and quickly destroyed. The Russians, in utter confusion, were taken prisoners, and Moscow's big electric power station was occupied undamaged. Manteuffel had thus gained possession of the most easterly point of the Moscow front, and, in addition to setting up a bridgehead for Third Panzer Army on the eastern bank of the canal, also seized the Kremlin's light-switch.

From his fortified room in the Kremlin Stalin continually telephoned to Zhukov, Voroshilov, and Lieutenant-General Kuznetsov, the C-in-C of the First Striking Army.

These telephone calls were Stalin's way of influencing the strategic and even the tactical decision of his military leaders—a practice which has been much criticized by Khrushchev and his friends as the reason for many of the Soviet defeats during the first year of the war. On the other hand, it cannot be denied that Stalin's authority secured many a decision which would probably otherwise not have been taken.

This was certainly true of 27th November. Stalin ordered that two brigades should at once be employed against Manteuffel's bridgehead, regardless of all other considerations. That bridgehead was to be liquidated at all costs.

Hans Leibel well remembers that day over twenty years ago, near Yakhroma. The weather favoured the Russians. On that afternoon of 27th November, within the short span of two hours, the thermometer dropped to 40 degrees below zero Centigrade. Against this Arctic cold the men of Manteuffel's combat group had only their simple balaclava helmets, their short cloth coats, and their much too tight jackboots. In this kind of outfit it was impossible to fight at 40 degrees of frost—even against a weak enemy.

Their unpreparedness for the Russian winter had to be paid for dearly. Not only were there no fur jackets and no felt boots—what was even worse, the German High Command did not know, or failed to apply, certain perfectly simple and easily practicable rules of winter warfare. If any proof were needed that this war against Russia had not been carefully prepared over a long period—at least not by the German General Staff—then it is provided by the evidence of total ignorance of the simplest facts of winter warfare. Thus when, after the first snowfalls, the Finns saw that the German troops were still wearing their jackboots with steel nails, they shook their heads in amazement: "Your nailed boots are ideal conductors of the cold—you might just as well walk about in your stockinged feet!"

In a lecture to the Moscow Officers' Club towards the end of the war, Marshal Zhukov stated that his respect for the German General Staff had first been shaken when he saw the German prisoners taken during the winter battle. "Officers and men all had closely fitting footwear. And, of course, they had frost-bitten feet. The Germans had overlooked the fact that ever since the eighteenth century the soldiers of the Russian Army had been issued with boots one size too large, so that they could pack them with straw in the winter, or more recently with newspapers, and thus avoid frostbite."

The Russians certainly avoided frostbite. Among the German frontline troops, on the other hand, the incidence of frost-bitten feet was as much as 40 per cent. in many divisions during the winter of 1941–42.

But the frost struck not only at the troops' feet. The oil froze in the machines. Carbines, machine pistols, and machine-guns packed up. Tank engines would not start. In these circumstances it is hardly

Final Spurt towards Moscow 177

surprising that Manteuffel's combat group was unable to hold the Yakhroma bridgehead, in spite of the defenders' stubborn resistance, when two Soviet brigades, the 28th and 50th Brigades of the Soviet First Striking Army, wearing winter greatcoats and felt boots, attacked them. The Russians' sub-machine-guns peeped out of fur cases, and the locks of their machine-guns were lubricated with winter oil. There were no stoppages or jammed bolts on the Russian side. The Russians were able to lie in the snow, if necessary for hours, to creep up to the German outposts at a suitable moment and silence them. Their infantry was supported by T-34s, whereas all that the 25th Panzer Regiment, 7th Panzer Division, had left were some 48-ton Skoda Mark III tanks with 3·7-cm. cannons and a few Mark IVs with 7·5-cm. cannons.

Thus, on 29th November, Manteuffel had to relinquish his bridgehead. He took up covering positions on the western bank of the canal. To the south-west the 6th Panzer Division covered the right wing of LVI Panzer Corps. The Corps' left wing was covered by 14th Infantry Division and 36th Motorized Infantry Division. The chance of a lightning blow at Moscow from the north had been lost.

Twenty miles south of Yakhroma, on the other hand, the situation took a dramatic turn. South of Rogachevo the XLI Panzer Corps, which had been brought up from Kalinin, was attacking the canal crossings north of Lobnya on the right wing of Third Panzer Army on 1st December. First of all, units of the Potsdam 23rd Infantry Division, surrounded south of Fedorovka, had to be relieved. Farther south, to the north-west of Lobnya, General Veiel's 2nd Panzer Division was threatening Moscow from the north-west. One of its combat groups, under Lieutenant-Colonel Decker, picked its way through blizzard and icy cold as far as Ozeretskoye along the mined road from Rogachevo to Moscow. The village was taken. "All aboard for the Kremlin—on the Red Square route," the outposts wisecracked to each other. They were standing in the bus-stop shelters on the suburban route to Moscow, beating their arms round their bodies and stamping their feet to keep warm. "Where's that damned bus?" they joked. "Late as usual."

As Second Lieutenant Strauss of the 1st Company Panzerjäger Battalion, 38th Division, passed the bus-stop in his car, driving down the road to Gorki, his driver turned to him with a giggle: "Why don't we take the bus, Herr Leutnant? Only a forty-five-minute journey to Comrade Stalin's home."

The sergeant had a somewhat optimistic idea about Soviet buses. The distance to Red Square was, after all, 24 miles.

However, the combat group of the reinforced 2nd Rifle Brigade under Colonel Rodt got much nearer to their objective. On 30th November the brigade's rifle battalions and sappers had taken Krasnaya Polyana against stubborn resistance by Siberian cavalry fighting dismounted, and Moscow workers' militia; they had taken Pushki, and, on the following day, Katyushki. Now Major Reichmann's 2nd Battalion, 304th Rifle Regiment, got as far as Gorki. That was a mere 19 miles to the Kremlin or 12 miles to the outskirts of Moscow. An assault party of Panzer Engineers Battalion 38 actually penetrated as far as the railway station of Lobnya and blew it up in order to prevent its use by Soviet tactical reserves. That was 10 miles from the outskirts of the city and 17 from the Kremlin.

Moscow's heart seemed to stop for a moment when the news reached the city. It was the day when *Pravda* carried two exciting and typical reports on its front page—one about the shooting of marauders in the city's main streets and the other about sentences of death for foodstuff speculators.

Moscow had become the fighting line. Through the town rumbled the new T-34s from the factories on the city's eastern outskirts; lorry-loads of workers' militia and Komsomol members rattled to the railway stations—tactical reserves to be thrown against Katyushki and Gorki. Siberian battalions drove to the front line in taxis and in the requisitioned private cars of Party and State officials. Ammunition was carried to the danger spots in requisitioned vans and buses. A workers' battalion from a tractor plant on the eastern outskirts of Moscow could be in action in the west or the north-west within an hour. It was the use of what strategists call the inner lines that enabled Stalin to halt the German spearheads at Katyushki and Gorki by employing sufficient tactical reserves at the crucial spots.

On the road leading from Staritsa via Volokolamsk to Moscow lies the little town of Istra. This town had been chosen as the key-point of Moscow's second line of defence. It was held by Siberian infantry regiments.

The XL and XLVI Panzer Corps of Fourth Panzer Group had to fight hard for every village and every patch of wood. Inch by inch the advanced formations and combat groups of 5th and 10th Panzer Divisions and the "Reich" SS Motorized Infantry Division struggled forward—over windswept fields and through forests deep in snow. On 23rd November they succeeded in reaching the Istra river and the Istra reservoir. The reservoir was 11 miles long and, on an average, a mile and a half wide. It fed the Istra river, which was about 100 feet wide and flowed into the Moskva. The ground on the eastern bank of the Istra was high and thickly wooded. The Russians were well established there in favourable positions, with a wide view over the snow-covered fields of the western bank. Anyone wishing to attack them had to cross the river or the reservoir.

Nevertheless 11th and 5th Panzer Divisions succeeded on 24th and 25th November in crossing the river and the reservoir and forming bridgeheads. Motorcycle Battalion 61 of 11th Panzer Division, led by

Major von Usedom, made a daring rush over the ice of the Istra. The Russians opened up at them with artillery. The air was filled with splinters of steel and ice. But the motor-cyclists fought their way across to the far bank and gained a precarious foothold on the frozen ground. The reservoir itself was crossed near Lopatovo, at its narrowest point. There were some anxious minutes as the men headed for the dam of the reservoir. It must have been wired for demolition. What would happen if the dam suddenly burst and gigantic masses of water were released?

But the assault units of 11th Panzer Division were lucky. Their surprise came off. There was no time for the Russians to press the button. Lieutenant Breitschuh's sappers removed 1100 mines and two tons of high explosive from the reservoir dam.

Farther to the south the crossing of the important Istra river was likewise successfully accomplished. Lieutenant-Colonel von der Chevallerie seized the bridge of Busharovo with the reinforced 86th Rifle Regiment, 10th Panzer Division. The operation was carried out under cover of a thick blizzard. Chevallerie's group was the remnants of the once proud 10th Panzer Division. Now its 7th Panzer Regiment had no more than twenty-eight tanks left, and the 69th and 86th Rifle Regiments had shrunk to four weak rifle battalions of 120 men each. Boehringer's artillery battalion was down to one single tractor and ten guns. Nevertheless the remnants of 10th Panzer Division fought with spirit.

The enemy put up furious resistance and brought up whatever he could lay his hands on, according to a diary account of one of the men in the action. The self-sacrificing way in which the Russians fought was admirable, but for the time being unavailing, since the attacking units of Army Group Centre continued to nibble their way towards Moscow in spite of all difficulties.

On 26th November, a cold hazy day 20 degrees below zero Centigrade, the combat group of 10th Panzer Division attacked the town of Istra from the north. It was a costly engagement. In the forest fighting which ensued the attackers suffered heavily from the shrapnel of Soviet multiple mortars, but they succeeded in pushing the Soviets—Manchurian units from Khabarovsk—out of the woods and, with a last supreme effort, reaching the northern edge of Istra.

Meanwhile the battalions of the "Reich" SS Infantry Division had come up. The SS Motorcycle Battalion Klingenberg first of all had to burst through a fortified line in the forest immediately west of Istra on the Volokolamsk–Moscow road, held by units of the famous 78th Siberian Rifle Division. The men of that division had a reputation for the fact that they neither took prisoners nor allowed themselves to be taken. In hand-to-hand fighting, with hand-grenades and spades, pillbox after pillbox had to be taken. Klingenberg's motor-cyclists fought with spectacular gallantry, and many of the young men of the Waffen SS paid with their lives. When Captain Kandutsch reported on the engagement to his C-in-C, General Stumme, there were tears in his eyes. Many of the eighteen-to-twenty-year-olds who lay dead on the battlefield were barefoot inside their boots. Yet the temperature was 15 degrees below freezing.

Just outside Istra, in a loop of the river, was the fortress of the town, dominating its western approaches. The "Reich" SS Division succeeded in taking the citadel by surprise. The "Deutschland" and "Der Fuehrer" SS Infantry Regiments, supported by the "Reich" SS Artillery Regiment, had broken in from the south and infiltrated into the barricaded streets. Hitler's and Stalin's guards gave each other no quarter. The Siberians were forced to withdraw. Istra, the keypoint of Moscow's last line of defence, was taken.

On 27th November Polevo fell. That day the Soviet air force began its ceaseless attacks on Istra. The Russians were determined not to yield undamaged that vital transport centre before Moscow. The German staffs—this was learned from monitored commands by radio—were not to find accommodation. The church towers with their onion domes were reduced to rubble. House after house was shattered by the Red Air Force. Two thousand bombs were dropped on the little town. No roof was left intact for the German staffs.

On the morning of 28th November the Waffen SS took Vysokovo and continued its advance towards Moscow. By then the assault units were within a 20-mile radius of the Kremlin.

The thermometer stood at 32 degrees below zero Centigrade. The men had to spend the nights in the open. They put on everything they had—but it was not enough. They had no sheepskin jackets, no fur caps, no felt boots, no fur gloves. Their toes froze off. Their fingers in the thin woollen mittens turned white and stiff.

But in spite of all the hardships there were moments of comfort and ease. During the black, eerie, tense nights at the turn of November–December 1941, when the whole land seemed rigid in the grip of the ringing frost, when the Junkers planes overhead droned towards Moscow and the skyline was lit up by Soviet anti-aircraft fire, the troops would switch on the German forces programme from Belgrade and listen to Lale Andersen's dark voice singing *Lili Marlene*. It seems hardly credible, but whoever was in that campaign against Moscow that winter and got away alive will always remember that sentimental, nostalgic song that brought tears of homesickness to the men's eyes.

On 2nd December the spearheads of the "Reich" SS Infantry Division were outside Lenino. Second Lieutenant Weber, the orderly officer of the OC Army Artillery 128, Colonel Weidling, wrote in a letter to his mother in Hamburg:

These Russians seem to have an inexhaustible supply of men. Here they unload fresh troops from Siberia every day; they bring up fresh guns and lay mines all over the place. On the 30th we made our last attack—a hill known to us as Pear Hill, and a village called Lenino. With artillery and mortar support we managed to take all of the hill and half of the village. But at night we had to give it all up again in order to defend ourselves more effectively against the continuous Russian counter-attacks. We only needed another eight miles to get the capital within gun range—but we just could not make it.

Fourth Panzer Group just could not make it any longer. Its offensive formations advanced only a few more miles. The situation of 10th Panzer Division was typical. The combat group of the combat-hardened 69th Rifle Regiment had reached the village of Lenino on 1st December, supported by the division's last tanks. But they could only seize the western fringe of the village from the Russians. In its eastern part, separated from the Germans by a small stream, the enemy was firmly established, as if concreted into the ground. For four days they lay opposite each other. Russian artillery ceaselessly shelled the German positions. The handful of men of 69th Regiment became fewer and fewer—and they did not gain an inch of ground. This was only 21 miles from the Kremlin, 14 miles from the north-western outskirts of Moscow, and 11 miles from its northern river port.

But other divisions were still worming their way forward through ice and snow towards the capital. South of Istra, to both sides of the Ruza–Zvenigorod road and along the Moskva river, the IX Corps under General Geyer tried its luck with 252nd, 87th, and 78th Infantry Divisions. Their first objective was the Zvenigorod–Istra road and the town of Zvenigorod itself, an arsenal and ammunition dump for the western sector of Moscow's defences.

The town was situated amid a virgin forest area deep in snow. Inside it, in countless well-camouflaged dugouts and concrete pillboxes, were the regiments of the Soviet Fifth Army. The first obstacle to be taken was Lokotnya. There the 78th Infantry Division from Württemberg got stuck in the mud towards the end of October. Now it intended to thrust past the enemy barrier.

In a bold outflanking movement Colonel Merker led his reinforced 215th Regiment on a "tiptoe advance" secretly along small paths, in single file, through snowbound virgin forest and across exposed clearings into the rear of the Russian positions, over-ran them, and on 20th November captured Lokotnya.

By 24th November the infantry regiments, reinforced by sappers, fought their way right up to Aleksandrovskoye, a veritable fortress, and by noon of 2nd December to the eastern end of Yershovo. By then the division had spent its strength. It did not succeed in taking Zvenigorod.

Between the 78th Infantry Division's left-hand neighbour, the 87th Infantry Division (IX Infantry Corps), and the "Reich" SS Infantry Division (XL Panzer Corps) the 252nd Infantry Division drove forward and penetrated into the Soviet defences. Heavy fighting ensued in the pathless forests, and the regiments found themselves in great difficulties. The 461st Infantry Regiment was cut off and had to rely on its own resources for the next two days. Stukas battered the Russians until their resistance was broken. The 7th Infantry Regiment reached Prokovskoye. On 1st December a combat group of 2nd Battalion pushed the fighting line a few miles beyond Prokovskoye against repeated enemy attacks. Beyond that they were unable to advance. The snow, the cold, exhaustion, and Russian opposition forced a halt.

The Russians proved themselves masters of rapidly improvised defence, especially in the wintry forests and swamps. Four months earlier the forces they employed outside Moscow would very probably have been smashed by the German divisions. But against the overextended, down-at-heel, and half-frozen German spearheads, lacking armour and heavy weapons, the Russians were strong enough. The old adage was proved true once more: it is the last battalion that matters. The best illustration was provided by the fighting for the motor highway.

The motor highway from Smolensk was the shortest, the fastest, and the best route into Moscow. Where it threaded its way between the Nara Lakes, together with the old postal road, east of the Shelkovka–Dorokhovo crossroads, the Russians had dug in and blocked this most important artery of the German offensive.

In vain did 4th Panzer Group, together with General Fahrmbacher's VII Corps, try to break through the barrier running from the Nara Lakes via the motor highway and the postal road to the Moskva bend. The 267th Infantry Division from Hanover, fighting north of the Moskva, was stuck in deep snow and bitter cold. The experienced 197th Infantry Division, known as the "Highway Clearing Division," and the Bavarian 7th Infantry Division tried unsuccessfully, together with the gallant French Legion, to break the enemy's stubborn resistance along the Nara Lakes–motor highway–postal road–Lake Poletskoye–Moskva bend line by bypassing it on the left. But the neck of land at Kubinka remained barred.

In order to gain the modern motor highway to Moscow after all, at a point south-east of Naro-Fominsk, Field-Marshal von Kluge on 1st December mounted a bold operation with XX Infantry Corps of his Fourth Army, at its junction with Fourth Panzer Group.

It very nearly came off. Colonel P. A. Zhilin, the official Soviet military writer, reports in his book *The Most Important Operations of the Great Fatherland War*:

Final Spurt towards Moscow 183

At the beginning of December the enemy made his last attempt at breaking through to the capital from the West. For this purpose the tanks and the motorized and infantry divisions of his Fourth Army were concentrated in the Naro–Fominsk area. The enemy succeeded in penetrating deep into our defensive positions.

That was exactly what happened. Kluge intended to gain the motor highway behind the Nara Lakes by means of a sweeping encircling movement, and then to cover its flank. Towards 0500 hours on 1st December the XX Corps under General Materna mounted its attack against the motor highway east of Naro-Fominsk with 3rd Motorized Infantry Division, 103rd, 258th, and the reinforced 292nd Infantry Divisions—the main effort being with 258th Infantry Division, which already held the bridge over the Nara at Tashirovo. In sub-zero weather the extensive field fortifications south-east and north of the town were pierced. The 292nd Infantry Division, reinforced by units of 27th Panzer Regiment, 19th Armoured Division, wheeled to the north. Colonel Hahne gained Akulovo with his headquarters troops and 2nd Battalion, 507th Infantry Regiment; this village was only four miles from the motor highway and 35 miles from Moscow.

On the right wing of XX Corps the 183rd Infantry Division fought its way right up to the motor highway west of Shalamovo with two battalions of 330th Infantry Regiment on 2nd December, and dug in for all-round defence. On the morning of 3rd December 330th Infantry Regiment, without being pressed by the enemy, was ordered to withdraw to its starting positions on the Nara, south of Naro-Fominsk.

The 3rd Motorized Infantry Division and 258th Infantry Division launched an outflanking attack against Naro-Fominsk. The temperature was 34 degrees below zero Centigrade, and there was an icy wind which made the troops' bones ache. The first instances occurred of men throwing themselves down in the snow, crying, "I can't go on." The battalions shrank more and more—through frost injuries rather than enemy action. Some battalions were down to eighty men.

In the Brandenburg 3rd Motorized Infantry Division the 1st Battalion, 29th Infantry Regiment, lost all its company commanders during the first few days of fighting. The 5th Company, which started this final offensive with seventy men, had only twenty-eight men left by the first evening. The company commander was wounded, the two sergeants had been killed, and of the other nine NCOs four had been killed and three wounded. Nevertheless the 29th Infantry Regiment took Naro-Fominsk and drove another three miles to the east along the highway. But then the attack ground to a standstill at 38 degrees below zero Centigrade.

The only progress made towards the east was on the division's left, in the area of 258th Infantry Division. There a mobile combat group under the command of Anti-aircraft Battalion 611, operating on the division's left wing, punched its way through to the north-east, via Barkhatovo and Kutmevo, to Podazinskiy. Indeed, the "advanced detachment Bracht," with Motorized Reconnaissance Battalion 53, 1st Company, Panzerjäger Battalion 258, two platoons of 1st Company Anti-aircraft Battalion 611, and a few self-propelled guns, succeeded in getting as far as Yushkovo, to the left of the highway. From there it was only 27 miles to the Kremlin.

On the other side of the road was the village of Burzevo. This miserable place with its thirty thatched houses on the far side of a snow-covered drill square was the target of the spearheads of 258th Infantry Division.

In the late afternoon of 2nd December the 3rd Battalion, 478th Infantry Regiment, likewise penetrated into the village of Burzevo along the Naro-Fominsk to Moscow road. Units of 2nd Battalion had been holding their ground desperately for several hours against enemy attacks. The twenty-five or thirty straw-covered houses of the little village exercised a mesmeric attraction on the troops. The smoke rising straight into the icy sky from their chimneys promised hot stoves. There was nothing the men longed for more than a little warmth. They had spent the previous night in the old concrete pillboxes of a tank training ground west of the village, and had been caught there by a sudden drop of temperature to 35 degrees below zero Centigrade.

The collective farmers had been using those pillboxes as chicken-houses. The chickens had gone, but the fleas stayed behind. It was an appalling night. The only way to escape from the fleas was to cower behind the chunks of concrete. And there the frost was lurking. Before the men realized it their fingers had turned white and their toes were frozen into insensitivity inside their boots. In the morning thirty men reported at the medical post, some of them with serious frostbite. But there was no point even in taking the boots off—the skin would merely be left behind frozen to the insoles together with the rags they had wrapped their feet in. There were no medical supplies for the treatment of frostbite. Nor was there any transport to take the casualties to the main dressing station. Thus the frost-bitten men remained with their units and longed for the warm houses of Burzevo.

The battalion had launched its attack at dawn, without artillery preparation. They were supported by three self-propelled guns and one 8·8-cm. anti-aircraft gun. The Russians in their positions outside and in Burzevo were clearly also suffering from the cold. They were equally badly supplied with winter clothing as the German troops, and seemed unwilling to engage in any major fighting. The Russian wounded and those who surrendered were patently under the influence of vodka. They maintained that behind them there were no further defences this side of Moscow, except for a few anti-aircraft positions.

Final Spurt towards Moscow 185

At two points only did the Russians try to set fire to the village. The terrible meaning of Stalin's scorched-earth order was first made apparent.

Major Staedtke reduced sentries and pickets to the bare minimum and allowed the rest of his men to go into the houses with their warm stoves. There they sat, crouched, or lay, crowded together like sardines with the Russian civilian population. They piled bricks into the stoves. And every hour, as a few men went out to relieve the sentries, they would take a brick with them—but not to warm their feet or hands. The heat had to be saved for something more important. The hot bricks were wrapped in rags and placed on the locks of the machine-guns to prevent the oil from freezing. If a Russian suddenly emerged behind a snow hummock, where he might have lain for hours, the sentries could not afford a jammed gun. Thus they carted their hot bricks and stones outside every hour to keep their weapons warm. Those who had been relieved and came inside felt as though they were entering paradise.

But paradise was short-lived—six hours in all. The OC 258th Infantry Division withdrew the reinforced 478th Infantry Regiment to Yushkovo; the 3rd Battalion covered the movement as the rearguard. At 2200 hours the Russians made another attack with T-34s. They knew what they wanted. Systematically they fired at the straw roofs to set the houses on fire. Then they broke into the village. Fighting continued in the light of the burning farmhouses. The 8·8-cm. gun finished off two Soviet tanks, but then received a direct hit itself. Self-propelled guns and T-34s chased each other among the blazing houses. The infantry lay in the gardens, behind baking ovens, and in storage cellars. Second Lieutenant Bossert, with an assault detachment of 9th Company, tackled the T-34s with old Russian anti-tank mines.

Half a dozen of the blotchy monsters were lying motionless in the village street, smouldering. But two of the three German self-propelled guns were also out of action. One of them stood in flames just outside the garden where Dr Sievers, of the Medical Corps, had organized his regimental dressing station in a potato store cellar. Pingel, his medical NCO, was ceaselessly injecting morphia or SEE—a combination of Scopolamin, Eukodal, and Ephetonin—in order to relieve the pain of the wounded. He carried his equipment in his trouser pocket because otherwise the ampoules would freeze up. Of course, it was not sterile—but what was the point of asepsis in those conditions? The main thing was to help the wounded lying on the ground in such weather.

When the day dawned the 23rd Battalion was still hanging on to the ruins of Yushkovo. Six T-34s lay in the village, gutted or shot up. The Russian infantry did not come again. The attack had been repulsed. But it was also clear that there could be no question of a further advance towards Moscow. The men were finished. Seventy seriously wounded were lying in the icy potato cellars. The order came through to abandon Yushkovo and to withdraw again behind the Nara. It was the hour when the whole of Fourth Army suspended its offensive and recalled its spearheads to the starting-lines.

Dr Sievers ordered the wounded to be loaded on the horse-drawn carts which had arrived in the line at night with ammunition and food supplies. But there was not enough room for them. The shattered vehicles were likewise loaded with wounded and hitched like sledges to the tractor of the 8·8-cm. gun. The most serious cases were placed on the self-propelled guns. The dead had to be left behind, unburied. It was almost a Napoleonic retreat.

The columns had no sooner left the village than the Russians began shelling them. Hits were scored among the columns. The horses drawing two carts of wounded fell. The carts overturned. The wounded cried out desperately for help. Suddenly the silhouettes of Soviet tanks appeared on the edge of the wood in front.

"Russian tanks!" There was panic. Escape was the only thought. For the first time Dr Sievers drew his pistol. "Pingel, Bockholt, over here!" The three men—the doctor and the two medical NCOs—positioned themselves across the road, their pistols drawn. The gesture was enough. Abruptly, reason once more prevailed. The wounded were loaded on the carts again. Twelve men harnessed themselves to each of the carts. Pingel led one of them and Bockholt the other.

At a trot they made for the patch of wood where the last self-propelled gun had gone into position and where the horse-drawn columns were waiting for them. On 4th December they were back behind the Nara river.

On 5th December the assault formations of Third Panzer Army and Fourth Panzer Group on the left wing of Army Group Centre were engaged in heavy offensive fighting along a wide arc north and north-west of Moscow. On the Moskva–Volga Canal, just over 40 miles north of the Kremlin, the 7th Panzer Division held its barrier position west of Yakhroma. About 25 miles farther south the combat group Westhoven of 1st Panzer Division, operating in conjunction with units of 23rd Infantry Division, was attacking via Belyy Rast towards the south-east and east in the direction of the canal crossing north of Lobnya. The Motorcycle Battalion, reinforced by tanks and artillery, took Kusayevo, a little over a mile west of the canal and about 20 miles north of the Kremlin, in the late afternoon. At Gorki, Katyushki, and Krasnaya Polyana—at their most easterly point still about 10 miles from Moscow—the troops of the Viennese 2nd Panzer Division were engaged in bitter fighting. The same kind of heavy defensive fighting was also going on in the neighbouring sectors, with XLVI Panzer

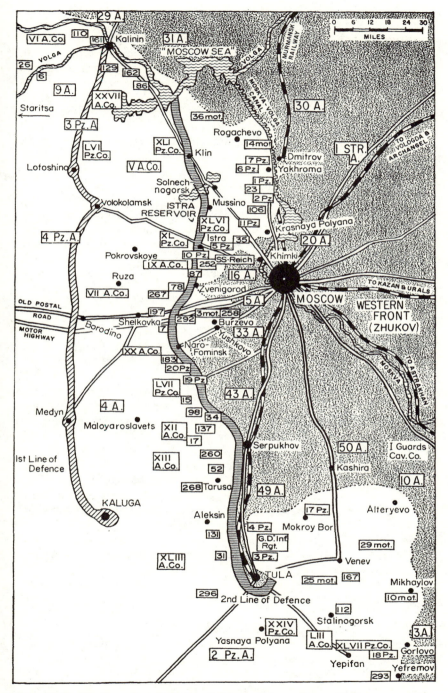

Map 9. On 5th December 1941 the divisions of Army Group Centre were at the gates of Moscow. The city's two lines of defence were breached. Forward units had reached Khimki, 5 miles from the outskirts of the city.

Corps and XL Panzer Corps, as well as IX and VII Infantry Corps of Fourth Panzer Group.

At Katyushki—one of the most south-easterly advanced strongpoints of 2nd Panzer Division—units of the 2nd Rifle Brigade, the reinforced 1st Battalion, 304th Rifle Regiment, under Major Buck, were engaged in fierce fighting. Katyushki was so near Moscow that through his trench telescope in the loft of the farmhouse by the churchyard Major Buck was able to watch the life in the streets of the city. It all seemed within arm's reach. But their reach was too short. Their strength was insufficient.

On 4th December a few more winter coats arrived and a few pairs of long, thick woollen stockings. Over the radio, simultaneously, came the announcement: "Attention, frost warning. Temperatures will drop to 35 degrees below zero Centigrade." By no means yet had all the men of 1st Battalion been issued with overcoats. There were also many days when they scarcely got a mouthful of hot food. But even that was not the worst. The worst was the shortage of weapons and ammunition. The Panzerjägers had only two 5-cm. anti-tank guns left per troop, and the artillery regiment was down to one-third of its guns. With this kind of equipment they were expected to capture Moscow in 30 to 40 degrees below zero Centigrade.

What the men in the open went through during those days, with their machine-guns or anti-tank guns, cowering in their snow-holes, borders on the fantastic. They cried with cold. And they cried with fury and helplessness: they cried because they were within a stone's throw of their objective and yet unable to reach it. During the night of 5th/6th December the most advanced divisions received orders to suspend offensive operations. The 2nd Panzer Division was then 10 miles north-west of Moscow.

At much the same time, during the night of 5th/6th December, Colonel-General Guderian likewise decided to break off the attack on Tula on the southern flank of Army Group Centre, and to take back his far-advanced units to a general defensive line from the Upper Don via Shat to Upa. It was the first time in the war that Guderian had to retreat. It was an ill omen.

To begin with, the new offensive had gone well on his sector. The Second Panzer Army had gone into action with twelve divisions and the reinforced "Grossdeutschland" Infantry Regiment. But the strength of 12½ divisions was only a figure on paper: in terms of fighting strength it amounted to no more than four.

On 18th November Second Lieutenant Störck of the 3rd Panzer Division again pulled off a masterly stroke with the Engineers platoon of Headquarters Company, 394th Rifle Regiment. South-east of Tula he seized the railway bridge over the Upa in a surprise coup. This time Störck, the bridge specialist, had thought up a very special trick.

Final Spurt towards Moscow

The main fighting line ran four miles from the bridge. Four miles of flat, frozen ground with no cover for creeping up to the bridge and taking it by surprise. However, Störck realized that the Russians, just as the Germans, clung to the villages at night because of the cold, and he therefore believed that it might be possible in the dark to filter through a thin enemy picket-line.

The plan was put into effect. An assault party of altogether nineteen men with three machine-guns, relying only on their compass, moved softly forward through the black night and the Russian lines. When dawn broke they were 500 yards from the bridge. Then came the second part of the plan.

Störck, Sergeant Strucken, and Lance-corporal Beyle took off their battle equipment and dressed themselves up as German prisoners. Pistols and hand-grenades were stowed away in their overcoat pockets. Two Ukrainians, Vassil and Yakov, who had been with the sapper platoon for the past two months, shouldered their rifles. In their Russian greatcoats and forage caps they looked absolutely genuine. Talking Russian in loud voices, they led their three 'prisoners' towards the bridge while Sergeant Heyeres and his men were waiting under cover.

The first Soviet bridge guard of four men were asleep in two foxholes. The action took only a few seconds, and there was no sound.

Now the five were heading for the 80-yard-long bridge. Their footsteps rang on the hard ground. Vassil and Yakov, talking at the top of their voices, acted their part magnificently. They had nearly got to the bridge when a shadow detached itself from it. A sentry was coming towards them. "Just the man we want," Vassil said loudly. "We are from the next sector, but perhaps you can take these fascists off our hands."

It was all over before the Russian suspected anything. But the second sentry at the end of the bridge was watching intently. And as they came nearer he challenged them, became suspicious, jumped down the bank under cover, and raised the alarm. Too late.

Störck fired two white Very lights. Sergeant Heyeres was already on top of the bridge with his machine-gun and fired for all he was worth. Beyle and Strucken flung their hand-grenades against the dugouts of the bridge guard. The Soviets came staggering out, still dazed with sleep, and raised their hands: 87 prisoners, five machine-guns, two heavy anti-tank guns, three mortars, and an intact assault bridge were the bag made by the handful of German troops. Their cunning and courage had gained the equivalent of a victorious battle.

On 24th November Guderian's 3rd and 4th Panzer Divisions and the "Grossdeutschland" Regiment had encircled Tula from the southeast against stubborn resistance by Siberian rifle divisions. The advanced detachments of 17th Panzer Division were approaching the town of Kashira. Just then Lieutenant-General I. V. Boldin threw his Soviet Fiftieth Army into battle against Guderian's weakened forces. Pressure against the thin and extended German front line became dangerous, since Guderian's adage "We tankmen are in the fortunate position of always having exposed flanks" applied to Blitzkrieg but not to positional warfare.

In a letter to his wife Guderian wrote with bitterness and pessimism:

> The icy cold, the wretched accommodation, the insufficient clothing, the heavy losses of men and *matériel*, and the meagre supplies of fuel are making military operations a torture, and I am getting increasingly depressed by the enormous weight of responsibility which, in spite of all fine words, no-one can take off my shoulders.

Nevertheless the 167th Infantry Division and the 29th Motorized Infantry Division on 26th November surrounded a Siberian combat group in the Danskoy area, beyond the Upper Don. Some 4000 prisoners were taken, but the bulk of the Siberian 239th Rifle Division succeeded in breaking out.

The encircling forces—in the north the 33rd Rifle Regiment, 4th Panzer Division, in the south and west units of LIII Corps with 112th and 167th Infantry Divisions, in the east units of 29th Motorized Infantry Division—were simply too weak numerically. Magnificently equipped, with padded white camouflage overalls and even their weapons whitewashed, the Siberians again and again attacked the weak encircling forces in night raids, wiped out whoever opposed them, and fought their way out towards the east between the 2nd Battalion, 71st Motorized Infantry Regiment, and the 1st Battalion, 15th Motorized Infantry Regiment. The German formations were no longer strong enough to prevent a breakout. The battalions of 15th and 71st Infantry Regiments suffered exceedingly heavy losses. Thus, in spite of all efforts, it proved impossible to capture the surrounded town of Tula—"little Moscow," as it was called—or to drive beyond Kashira, let alone to reach the distant objective of Nizhniy Novgorod, now Gorkiy. True enough, on 27th November the 131st Infantry Division in an easterly push succeeded in taking Aleksin. Likewise, the 3rd and 4th Panzer Divisions succeeded on 2nd December in advancing as far as the Tula–Moscow railway-line and blowing it up. Indeed, on 3rd December the 4th Panzer Division succeeded in reaching the Tula–Serpukhov road at Kostrova. The XLIII Corps thereupon tried once more to link up with the 4th Panzer Division to the north of Tula, and to throw the enemy back to the north. On 3rd December the most forward units of the Corps—the 82nd Infantry Regiment, 31st Infantry Division—got within nine miles of the 4th Panzer Division, but they could not realize their aim. On 6th December the offensive had to be suspended on this sector too. The troops and their vehicles stuck fast in a truly Arctic frost of 30 degrees, and in places 45 degrees, below zero Centigrade.

Desperate, Guderian sat over his maps and reports at his headquarters nine miles south of Tula, in a small manor house famous throughout the world—Yasnaya Polyana, Tolstoy's estate. In the grounds outside, overgrown with ivy and now deep in snow, was the grave of the great writer. Guderian had allowed the Tolstoy family to keep the rooms in the big house and had moved with his staff into the museum; even there two rooms were set aside for the exhibits and sealed up.

There, in Tolstoy's country house during the night of 5th/6th December, Guderian decided to recall the advanced units of his Panzer Army and go over to the defensive. He was forced to admit: "The attack against Moscow has failed. We have suffered a defeat."

9. Why couldn't Moscow be taken?

Cold weather and Siberian troops–The miracle of Moscow was no miracle–A chapter from the history of German–Soviet collaboration after the First World War– The unknown army–Tukhachevskiy's alliance with the Reichswehr–Himmler's grand intrigue–Stalin beheaded the Red Army.

IN April 1945, when the Russian troops were in Oranienburg, Potsdam, Hennigsdorf, and Grossbeeren, the doom of Berlin was sealed. But in 1941 the Germans were just as close to the gates of Moscow and were yet defeated.

Why? What were the reasons for this defeat which was of such crucial importance for the further course of the war? For whatever victories were yet to come, the divisions of Army Group Centre never recovered from the blows they suffered before Moscow. They were never again brought up to full strength; they never recovered their full effectiveness as a fighting force. At Moscow the strength of the German Army was broken: it froze to death, it bled to death, it spent itself. At Moscow also Germany's faith in the invincibility of the Wehrmacht was shaken for the first time.

What were the causes of this defeat? Was it "General Winter," with his 30, 40, or 50 degrees below zero, that defeated the German Army in the east?

Was it the Siberian crack divisions with their splendid winter equipment and the cavalry from Turkestan? Undoubtedly the exceptionally cold weather played a disastrous part, with its record thermometer reading of minus 52 degrees Centigrade—a temperature for which no German soldier was prepared and no weapon fitted. And undoubtedly the vigorous Siberian divisions played a decisive part.

But cold weather and Siberian troops were only the more obvious reasons for the German defeat. The "miracle of Moscow," as the Soviets call the turn of the tide outside their capital, was due to a simple fact which was anything but a miracle—a fact that can be summed up in very few words. There were too few soldiers, too few weapons, too little foresight on the part of the German High Command, in particular an almost total lack of anti-freeze substances and the most basic winter clothing. The lack of anti-freeze lubricants for the weapons was particularly serious. Would the rifle fire or wouldn't it? Would the machine-gun work or would it jam when the Russians attacked? Those were questions which racked the troops' nerves to the limit. Improvised expedients were all very well while the troops were on the defensive, but to launch an attack or even an immediate counter-attack with weapons functioning so unreliably was out of the question.

Adolf Hitler and the key figures of his General Staff had underrated their opponent, in particular his resources of manpower and the performance and morale of his troops. They had believed that even their greatly debilitated armies would be strong enough to deal him his *coup de grâce*. That was the fundamental error.

Liddell Hart, the most important military writer in the west, in *The Soviet Army*, attributes the salvation of the Soviet Union above all to the toughness of the Russian soldier, to his capacity to endure hardships and ceaseless fighting under conditions which would have finished off any Western army. Liddell Hart then adds that an even greater advantage for the Russians was the primitive nature of the Russian roads. Most of them were no more than sandy country lanes. Whenever it rained they turned into quagmires. This circumstance contributed more to the repulse of the German invasion than any sacrifice by the Red Army. If the Soviet Union had had a road system such as the Western countries, Russia would have been over-run as quickly as France. All that Hitler had failed to consider; like most Western military men he had been ignorant of these facts. The final resistance at Moscow could have been overcome only by a fresh, well-equipped, adequately supplied force of about the strength of that which mounted the offensive on 22nd June. But what was that force like now? Five months of ceaseless fighting had reduced the regiments of the front-line divisions to a third of their nominal strength, and often less. The frost did the rest. Before Moscow casualties from frost-bitten limbs were higher on average than casualties through enemy action.

We still have the original schedule of the losses suffered by XL Panzer Corps. Between 9th October and 5th December the "Reich" Division and the 10th Panzer Division, including Corps troops, lost 7582 officers, NCOs, and men. That was about 40 per cent. of their nominal combat strength.

Total casualties on the Eastern Front as of 5th December 1941 were 750,000, or 23 per cent. of the average total strength of 3,500,000 troops. Nearly one man in every four was killed, wounded, or missing.

The Russians had suffered considerably greater losses, but they also had the greater resources. Army Group Centre did not receive a single fresh division in December 1941. The Soviet High Command, on the other hand, switched to the Moscow front thirty fresh rifle divisions, thirty-three brigades, six armoured divisions, and three cavalry divisions.

The question "Why did not the German forces reach Moscow?" will, of course, be answered differently by the strategist, the commander in the field, and the airman. The economist, no doubt, will have a different answer again.

General Blumentritt, for instance, the Chief of the General Staff of Fourth Army, and subsequently Quartermaster-in-Chief of the Army General Staff, sees the reason for the disaster in Hitler's strategic planning error in failing to tackle Moscow and Leningrad as the priority objectives in good time—*i.e.*, immediately after Smolensk. That is the view of the strategist.

Anyone remembering the wartime enemy air raids on German towns will ask: What about the Luftwaffe? He will note with surprise that the German Luftwaffe did not succeed in interfering with the passage of Soviet troops to the front through the Moscow transport network, nor in preventing the arrival of the Siberian divisions, nor generally in paralysing Moscow itself as an area immediately behind the lines. Nothing of that kind happened. The last German air raid on Moscow was made during the night of 24th/25th October with eight machines. After that only nuisance raids were made in December. Thus during the decisive phase of the operation the nerve centre of Russia's defence, the mainspring of Russian resistance, remained unharassed from the air. Why?

Every German airman who was at Moscow knows the answer. The Russians had established tremendously strong anti-aircraft defences around the city. The forests were thick with AA batteries. Moreover, the German Luftwaffe in the east had been decimated in ceaseless operations, just as much as the ground forces, and had to yield the air to the Soviet Air Force, which, before Moscow, was numerically twice as strong. Besides, the Soviet Air Force had numerous well-equipped airfields near the front, with heated hangars, enabling any unit to take off swiftly and repeatedly regardless of the weather. The German machines, by way of contrast, were based on primitive air strips, a long way behind the fighting line, which permitted operations only in favourable weather. Thus Moscow was virtually spared from the air.

Marshal Zhukov, it is true, does not regard the German weakness in the air as decisive. In a lecture to Soviet officers he said: "The Germans were defeated at Moscow because they had not ensured sufficient locomotives of suitable gauge to move supplies and reserves to the front line in large quantities, regardless of mud and snow, in what is the Soviet Union's best and most comprehensive railway network, that of the Moscow area."

Certainly there is some truth in that. But the decisive fact was that Stalin won the race for fit manpower—for both the fighting forces and the armaments industry.

The struggle for manpower had become the most serious problem of the war. The irreparable losses of the German side, and the resulting shortage of combatant troops, decided the battle of Moscow. The subject has hitherto not received the attention it deserves, but some interesting facts are revealed in the more recently published papers and letters of Field-Marshal Keitel, the former Chief of the High Command of the Wehrmacht.

Keitel wrote:

> I had to force upon Speer, the new Minister for Armaments and Ammunition, a programme enabling me to call up again for active service 250,000 servicemen exempted for armament production. The struggle for manpower began at that moment and has never ceased since.

The German Wehrmacht—*i.e.*, Keitel—lost that struggle. The number of men who remained exempted from active service without good reason has been estimated at half a million. Keitel writes:

> What would these men have meant to the armies in the East? The calculation is simple. With 150 divisions of 3000 men each, they would have meant a reinforcement of their combat strength by half their nominal establishment. But instead the shrunken units were replenished with grooms and farriers and suchlike, and these in turn replaced by willing Russian prisoners of war.

Keitel quotes two figures which illustrate the problem:

> The monthly losses of the land forces alone, in normal conditions and excluding major battles, averaged 150,000 to 160,000 men. Of these only 90,000 to 100,000 could be replaced. Thus the army in the field was reduced in numbers by 60,000 to 70,000 men each month. It was a piece of simple arithmetic to work out when the German front would be exhausted.

And how do the Russians see the miracle of Moscow? Their answer

Why couldn't Moscow be taken? 195

in all military reviews is simple: We won because we were bound to win. We were better, we were stronger, because Bolshevism is better and stronger than all other systems. This is how Samsonov formulates it: "The Soviet people and its army... wore down the attacking Army Group Centre in heavy fighting and halted its advance along the approaches to the capital."

How then do they explain the victorious German advance right up to the very gates of Moscow? How do they explain the fact that even Stalin's Government expected to lose the capital? This has remained to this day the weak point of the Soviet theory of the invincibility of the army of workers and peasants—an army in which even Stalin himself placed no excessive hopes at certain times. Nikita Khrushchev has tried to remove this discrepancy by putting forward an explanation for the Russian defeats during the first six months of the war which had long been secretly advanced by the Soviet Officer Corps, but which had previously lacked official authority. Khrushchev announced it at the 22nd Party Congress in Moscow in October 1951. He declared: Only because Josef Stalin had robbed the Red Army's Officer Corps of its best men by his insane purges in 1937-38, only because his executions and incarcerations of allegedly anti-Party commanders almost completely denuded the troops of their leaders and disorganized them, did the Germans succeed in 1941 in getting to the gates of Moscow.

It is a spectacular theory. After the grave charge of having presented to Hitler the advantage of surprise by his gullibility, Stalin is now also blamed for the military defeat. How convincing is the historical evidence for this theory?

It is quite true that in his purges during 1937 and 1938 Stalin, on reliable evidence, liquidated 20,000 to 35,000 active officers of the Red Army. Khrushchev's theory therefore makes sense. For if a man kills off his marshals, generals, and officers he must not be surprised if his army loses its military efficiency. Removing a General Staff officer is like felling a tree: it takes eight to ten years on an average to train a major in the General Staff who could organize a division's supplies or direct its operations. But Stalin had at least half of all his General Staff officers executed or imprisoned.

But why did the Red dictator kill off nearly half his Red Army Officer Corps? Why did he get his NKVD henchmen to liquidate 90 per cent. of all generals and 80 per cent. of all colonels by a bullet in the back of their necks? Why did three of his five marshals, 13 of his 15 Army commanders, 57 of his 85 Corps commanders, 110 of his 195 divisional commanders, 220 of his 406 brigade commanders, as well as all the commandants of his Military Districts have to die by the bullets of his green-uniformed NKVD execution squads?

The sensational answer provided by Khrushchev at the 22nd Party Congress was: The tens of thousands of officers liquidated on charges of high treason and hostility to the Party were all innocent; not one of them was an enemy of the Party, not one of them attempted to overthrow the regime, not one of them was a spy in German pay, as Stalin maintained. No—it was Hitler who had staged it all. Through his secret service he had planted fake evidence on Stalin—evidence about a conspiracy headed by Marshal Tukhachevskiy and other prominent military leaders. Evidence, moreover, of Tukhachevskiy's and his friends' collaboration with the German Wehrmacht. Khrushchev concluded literally: "With deep sorrow mention has been made here of the many famous Party and State functionaries who lost their lives innocently. But prominent Army leaders also fell victim to persecution, such as Tukhachevskiy, Yakir, Uborevich, Kork, Yegorov, Eydemann, and others. They were men who had served our Army well—especially Tukhachevskiy, Yakir, and Uborevich. They were famous Army leaders. Later victims of persecution were Blyukher and other well-known Army leaders. The foreign Press once published a rather interesting report, to the effect that Hitler, while preparing to attack our country, got his secret service to plant on us a faked document showing Comrades Yakir, Tukhachevskiy, and others to be agents of the German General Staff. This allegedly secret 'document' fell into the hands of President Beneš of Czechoslovakia, and he, evidently with the best intentions, passed it on to Stalin. Yakir, Tukhachevskiy, and other comrades were arrested and subsequently liquidated. Many outstanding commanders and political workers of the Red Army were murdered."

Thus far Khrushchev. Although as Premier and Party leader of the Soviet Union all archives and records are available to him, he submitted no evidence to support his theory, but referred to foreign Press reports. No doubt he had good reason, to avoid giving away too many secrets. Certainly, in spite of its fantastic implication, his claim is not new.

The sensational story has been cropping up here and there for over a decade. President Beneš of Czechoslovakia, who died in 1948, and Sir Winston Churchill have both supplied evidence in connection with it in their memoirs, as have also two of the leading officials of Himmler's Secret Security Service, Dr Wilhelm Höttl—alias Walter Hagen—and Walter Schellenberg. These pieces of evidence, together with responsible reports by German and Czech diplomats dating back to 1936 and 1937, add up to a sinister Machiavellian play acted out in our own century. The play, perhaps, is not quite as simple as Khrushchev now presents it, or as Beneš, Churchill, and Himmler's lieutenants make out.

Certainly, these dark threads deserve following up. After all, the

Why couldn't Moscow be taken? 197

Tukhachevskiy affair was the most important scandal in modern history and the one with the most fateful consequences. Many actors were involved and many settings, right from the first years of the Soviet Union and the secret collaboration between Reichswehr and Red Army in the years 1923 to 1933. Himmler and Heydrich appeared only in its final act, but for the sake of better understanding we shall take this last act first. It began about the middle of December 1936.

Paris, 16th December 1936: The former White Russian General Skoblin, who was working both for Stalin's secret service and for Himmler, passed on two pieces of information to a representative of the German intelligence service. Item No. 1: The Soviet Army Command is planning a coup against Stalin. The leader of the conspiracy is Marshal Tukhachevskiy, the Deputy War Minister. Item No. 2: Tukhachevskiy and his closest followers are in touch with leading generals of the German High Command and the German intelligence service.

It was a sensational story. After all, the man named as the leader of an imminent revolt against Stalin was his Deputy Minister of War, his former Chief of the General Staff, the ablest and most outstanding military figure of the Soviet Union. The Marshal, then forty-three, represented the increasingly powerful Red Army. He was a man of aristocratic origin, a former guards officer. He had been trained as a general staff officer at the exclusive Tsarist Alexander Academy. Released from German captivity, he had gone over to Lenin's troops. In 1920 he had defeated General Denikin, the key leader of the White Russian counter-revolution. Ever since he had been the celebrated civil war general, the saviour of the Red revolution.

Heydrich,[1] a cold man but with a fine sense for grand intrigue, instantly realized the possibilities of the information from Paris. If Skoblin's news was correct the Soviet Union might become a military dictatorship. The enormous empire might be ruled by a supremely able organizer and strategist, a Red Bonaparte, a Russian Napoleon. Would that be to the advantage of Hitler's Germany?

Heydrich's answer was: No. It may be assumed that he made sure of Hitler's agreement with his view. Certainly there is no doubt that he discussed the matter with Hitler at once; there can likewise be no doubt that Hitler did not want a strong Russia.

In the circumstances, what could be more natural than to allow the information from Paris to reach Stalin and thus to deliver Tukhachevskiy, the best military mind in Russia, together with his followers, to the executioners?

But Jahnke, a member of Heydrich's staff, was against it. Skoblin, he argued, was in contact with the Soviet secret service, and it was therefore not impossible that the Kremlin may have planted the whole

[1] Formerly Deputy Chief of the Gestapo and 'protector' of Czechoslovakia.

198 *Part One: Moscow*

story on the Tsarist general in Paris. For what purpose? Maybe to make Hitler suspicious of his own generals. Or maybe in order to lure Hitler's secret service into a trap and to manœuvre the German leaders into mistaken decisions. Who could tell?

But Heydrich put Jahnke under house arrest and began to implement his plan. Tukhachevskiy was to be handed over to the executioner. With this end in view Heydrich performed a few secret service moves which testified to his natural gift for intrigue.

With a cold smile he lectured to his friend, SS Standartenführer Hermann Behrens: "Even if Stalin merely wanted to bluff the German leaders with Skoblin's information—I will supply the old man in the Kremlin with enough evidence to show that his own lies were the purest truth."

He ordered a secret squad of expert burglars to break into the secret archives of the Wehrmacht High Command and to steal the Tukhachevskiy file. This contained the papers of the so-called Special Detachment R, a camouflage organization of the Reichswehr which existed from 1923 to 1933 under the official designation of GEFU—Gesellschaft zur Förderung gewerblicher Unternehmungen, meaning Association for the Promotion of Commercial Undertakings. It came under the Armaments Department, and its task was to manufacture in the Soviet Union all the weapons and war materials which the Treaty of Versailles forbade the German Wehrmacht to possess. The file contained the records of many conversations between German officers and representatives of the Soviet military authorities, including, of course, Tukhachevskiy, who was Chief of the Red Army General Staff from 1925 to 1931. Heydrich had the GEFU file altered, he extended the correspondence by shrewd additions, he added some new letters and notes, so that in the end a perfect file was available, with authenticated documents and seals, a file that would have got any general in any country before a court martial on charges of high treason.

In the cellars of the Prinz Albrecht Strasse, Heydrich inspected the work of his specialists with approval. The first step had been accomplished. Now came the second: how could the file be played into Stalin's hands?

To fake a document and to make it look convincing is not particularly difficult for the experts of any secret service. But to get such a document to the proper address, without arousing suspicion, is a problem indeed. It must have been more than ordinarily difficult when the addressee was Josef Stalin. But Heydrich solved the problem.

During the course of 1936 the German Foreign Office had been in touch with the Czechoslovak Minister in Berlin and had from time to time tentatively aired the question of Czechoslovakia's attitude in the event of a German–French war.

This was Heydrich's point of attack. At the end of January 1937—

Why couldn't Moscow be taken? 199

President Beneš records in his memoirs—the Czechoslovak Minister in Berlin, Mastný, sent a cable to Prague, with every sign of surprise, to the effect that his interlocutor in the Foreign Office was suddenly showing a lack of interest in the subject. From certain hints it must be concluded that the Germans were in touch with an anti-Stalin group within the Red Army. Berlin was evidently expecting a change of regime in Moscow, a change which would shift the balance in Europe in favour of Nazi Germany. President Beneš was seriously alarmed at the prospect of losing his Soviet backing against Germany. Czechoslovakia, with its explosive minorities problem, with its restive Sudeten Germans, owed its existence largely to the antagonism between Germany and the Soviet Union. A reconciliation, perhaps even an alliance, between a Russian military dictatorship and German fascism would spell serious danger. Beneš's Republic was a product of the Treaty of Versailles: to liquidate the consequences of that treaty was Hitler's avowed aim. With Russia on his side he would not find it difficult.

What could be more natural than President Beneš instantly summoning the Soviet Ambassador in Prague, Aleksandrovskiy, and passing on to him Mastný's report. Conspiracy of the generals against Stalin. Hitler involved. Wehrmacht generals involved.

The Ambassador listened carefully, hurried back to his Embassy, picked up his suitcase, and immediately flew off to Moscow. Heydrich's mail was being delivered to its addressee.

But Heydrich was a careful man. He did not confine himself to his Prague postman, but acted on the sound principle that a thing worth doing is worth doing well. He therefore supported his Prague action by one in Paris.

At a diplomatic reception in Paris, two or three days after the conversation between Beneš and Aleksandrovskiy, Edouard Daladier, several times French Premier, but just then, for a change, Minister of War, genially linked his arm under that of the Soviet Ambassador, Vladimir Potemkin, and led him to a niche by a window. After a quick glance to make sure there were no unwelcome eavesdroppers, Daladier anxiously told Potemkin that France was worried. There was news about a possible change of course in Moscow. There was talk about arrangements between the Nazi Wehrmacht and the Red Army. Could His Excellency put his mind at rest? Potemkin was poker-faced. He escaped from the situation with non-committal phrases. Ten minutes later he left the reception, drove back to his Embassy, and sent an urgent coded signal to Moscow containing Daladier's information.

How Heydrich managed to play the information into Daladier's hands it is impossible to establish with certainty to-day. Probably there was a contact through a Deuxième Bureau man at the French Embassy in Moscow.

After these preparatory moves Heydrich staged his second act. He sent his special confidential representative, Standartenführer Behrens, to Prague, where he made contact with a personal representative of the Czechoslovak President and drew his attention to the existence of documentary evidence against Tukhachevskiy. Beneš was informed, and immediately passed on his information to Stalin. Shortly afterwards Beneš's contact suggested to Heydrich's representative that he should get in touch with a member of the Soviet Embassy in Berlin, by name of Israilovich. Israilovich was the NKVD representative at the Russian Embassy in Berlin.

Heydrich's man met him and let him see two genuine letters from the faked-up file. Israilovich, in the accepted manner, feigned indifference. He asked the price. Behrens shrugged his shoulders. Israilovich promised to meet him again in a week, together with an authorized person.

The arrangement was kept. The authorized person was a representative of Yezhov, the chief of the Soviet secret service. His first question too was the price. Heydrich, in order to prevent his business partners from getting suspicious, had fixed the price at the fantastic sum of 3,000,000 gold roubles. "But you are authorized to let yourself be beaten down," he had instructed his man.

But there was no question of beating down. Yezhov's representative merely nodded as Behrens, in the most matter-of-fact way, named the sum, the highest ever paid for a file in the history of secret service activities.

No plan of military operations, no treason, and no traitor in history ever achieved such a high price. The deal was clinched within a day. Yezhov's man left for Moscow with Heydrich's file. That was about the middle of May 1937.

Three weeks later, on 11th June 1937, the world was stunned by the news, put out by the official Soviet Tass Agency, that Marshal Tukhachevskiy and seven leading generals had been sentenced to death by shooting by the Soviet Supreme Court under the chairmanship of the President of the Military Tribunal Ulrich. The sentence had been carried out at once.

"The defendants were accused," it was explained in the report, "of having violated their duty as soldiers, of having broken their military oath of allegiance, and of having committed treason against the Soviet Union in the interests of a foreign country." An official communiqué added the following details:

> In the course of investigations it was established that the defendants, together with the Deputy Defence Commissar Gamarnik, who recently committed suicide, had organized an anti-State movement and had been in contact with the military circles of a foreign country pursuing an anti-Soviet policy. In favour of that country the defendants conducted military espionage. Their activity was

aimed at ensuring the defeat of the Red Army in the event of the country being attacked. The ultimate aim of the accused was the restoration of big land ownership and capitalism. All the accused made confessions.

The sensation was complete when Tass put out an Army order by Voroshilov to be promulgated to the forces in all Military Districts. This demanded that suspect persons should be denounced.

The order said:

> The ultimate objective of the traitors was the annihilation of the Soviet regime at any cost and by all means. They strove for the overthrow of the workers' and peasants' Government and had made preparations for murdering the leaders of the Party and the Government. They expected help from the fascist circles of a foreign country and, in return, would have been prepared to hand over the Soviet Ukraine. The principal organizers were in direct contact with the General Staffs of the fascist countries.

Tukhachevskiy's execution and Voroshilov's Order of the Day released an avalanche against which there was no protection. Every disgruntled soldier, every injured subordinate, now settled his accounts by denouncing as suspect any superior he disliked. In that orgy of political purging there was no acquittal. And every man who was condemned dragged down with him his followers, his friends, and his acquaintances—all to their doom. In hundreds at first, but presently in their thousands, and eventually in tens of thousands, the officers made the terrible journey into the NKVD cellars, for a bullet in the back of their necks, or for deportation to the prison camps of Siberia. Within a year the Officers Corps of the Red Army had been reduced by 50 per cent. and its higher ranks almost completely liquidated.

These facts would seem to prove conclusively that, by means of the cunning intrigue of SS Obergruppenführer Reinhard Heydrich, Hitler destroyed the entire command apparatus of the Red Army three years before the attack on the Soviet Union—in other words, that he prepared his victories in the NKVD cellars and the tiled execution rooms of the Lubyanka prison. Can this monstrous thesis really stand up to a thorough examination? Did 30,000 to 40,000 officers of the workers and peasants' army really die through a political swindle by the underworld of the secret service?

Appearances strongly support this conclusion—but appearances are superficial. Heydrich was not the author of the spectacle; he was himself only an assistant. His faked-up file was not the cause of the trial and conviction of Tukhachevskiy and his friends, but only Stalin's alibi. The roots of the tragedy which wiped out the flower of the Soviet Officers Corps were far deeper. They sprang from a genuine ruthless struggle for power between two mighty rivals. It was the savage end of the only power which could have overthrown Stalin. It marked the fateful victory of Georgian despotism over the Russian Bonaparte Tukhachevskiy, who—even if his hand was not yet stretched out to grasp supreme power—was already standing by to take over from the lunatic dictator and, supported by the strength of the Army, put an end to the Stalinist mismanagement. The slaughter of the Officers Corps was the result of a dramatic process, not merely of a low trick.

As such it takes its place in history as the tragic culmination of German–Soviet relations after the First World War, and as one of the factors in that most appalling tragedy of modern history—Operation Barbarossa. It began long before Hitler with playing at war, and it ended with war in earnest. A proper understanding of the whole tragedy of the German–Soviet war requires acquaintance with that earlier chapter.

In April 1925 a strange incident took place in the free port of Stettin. A customs officer newly posted to Stettin—he is still alive, so we shall call him Ludwig, although that is not his real name—was making his nightly routine inspection when he came upon a few men trying to remove a large crate from Shed 1. When challenged the men abandoned the crate and melted into the shadows. Ludwig raised the alarm. A fellow customs officer, who turned up surprisingly suddenly, seemed anxious to dismiss the whole incident. Ludwig became suspicious and ran his torch over the packing-case. "Machinery spares" was stencilled on it in large black letters, first in German and below in Russian. A label glued to the case gave the addressee as GEFU, Berlin, Germany. The sender was GEFU, Lipetsk, USSR. When Ludwig wanted to investigate the mysterious case his colleague suddenly asked him, "Were you in the Army, colleague Ludwig?"

Ludwig was astonished. "Of course." The other man nodded. "And did you see active service?"

"Want me to show you my Iron Cross?" Ludwig retorted angrily. "Or would you like to see my Free Corps service book?"

The other man smiled and tried to pacify him. "No, no, colleague Ludwig; but I think I may tell you now what this case contains. A tin coffin with a body. An Air Force officer of the Reichswehr."

Ludwig stepped back in terror. "What are you saying? A dead man? An Air Force officer? But it says on the case 'machinery spares.' And it's from Russia."

"That's right," the other man nodded. Then they talked for half an hour outside Shed 1 in the free port of Stettin. After that Ludwig was satisfied, saluted, and left. His colleague whistled lightly. Out of the shadows of the shed appeared four men.

"Everything's all right," the customs officer said softly. "A new chap, didn't know the ropes yet. But we've got to get a move on now, gentlemen; it's getting late." They put the case on a trolley and wheeled it to a pier. Alongside a small boat was made fast. Cautiously they lowered the freight into it. Then they jumped in themselves. They saluted and quietly rowed away, towards the bank of the Oder.

If the customs officer Ludwig had been a man of the political left rather than the right this incident would probably have triggered off a political scandal that would have reverberated round the world. The episode in the port of Stettin, with a dead body in a packing-case from Lipetsk, in Russia, declared as machinery spares, would have torn the curtain of silence which hid one of the most astonishing chapters of the Weimar Republic—the chapter of secret collaboration between the German Reichswehr and the Red Army. This collaboration formed the background to the Tukhachevskiy trial. It marked a dramatic period in the German–Soviet alliance, an alliance whose champions and representatives were murdered by Stalin, but are to-day being rehabilitated by Khrushchev.

Germany was the great loser of the First World War. But Russia too, Germany's former adversary, was not on the side of the victorious Powers. She stood aloof, isolated from the rest of the world, as did Germany—for the October Revolution and the establishment of a Communist Soviet State had given rise to a coalition of capitalist countries aiming at the overthrow of the Bolsheviks. They tried to achieve this aim by military intervention. When that failed an attempt was made to put economic pressure on the Soviets and coerce them into recognizing the obligations undertaken by the Tsarist empire. But Lenin's Government resisted, and the Soviet Republic refused to pay the Tsarist empire's debts to the Western 'capitalist' democracies.

Germany similarly resisted reparation payments, and in particular opposed the suggestion of Western statesmen that she should also pay the Tsar's old debts to the Western Powers. Out of this common opposition to the victorious Western Powers sprang the alliance of the defeated and have-nots. Logically enough, it began in the economic field. Its first fruit was the Treaty of Rapallo—a quickly arranged agreement signed between German and Soviet negotiators in the little Italian riviera resort of Rapallo on Easter Day 1922. Rapallo swept away the legacy of the war between the Soviet Union and Germany. The two Powers waived compensation claims in respect of war costs and war damage. It was decided to resume diplomatic relations, to regard one another as equal partners, and to apply in matters of trade the principle of most favoured country treatment. There were no secret military clauses in the Treaty of Rapallo, although this suggestion is occasionally heard to this day. This misconception springs from the fact that the agreement on common economic interests soon gave rise to further agreements. It was a logical development.

Rapallo had put an end to both Germany's and the Soviet Union's diplomatic and economic isolation. Why should not an attempt be made to develop the spirit and the letter of this Treaty into a kind of blockade runner against the military impositions and prohibitions with which Versailles had hamstrung the German Reichswehr? The Reichswehr, for instance, was forbidden to possess any tanks or anti-tank guns, any heavy motorized guns, any aircraft, and any means of chemical warfare. Under such restrictions it was impossible to build up a modern army. In particular, the strict ban on all armour was cutting Germany off from the military developments which were bound to follow from the introduction of armoured fighting vehicles in the First World War, developments which were generally regarded as being of decisive importance. Indeed, for that very reason the victorious Powers had insisted, in Article 171 of the Treaty of Versailles, that Germany must neither manufacture fighting vehicles nor "import armoured cars, tanks, or any similar constructions suitable for military purposes." In the circumstances, what could Germany do? Unless these prohibitions could be circumvented every single mark paid out on the Reichswehr was a pointless waste of money.

It was Karl Radek, the brilliant intellectual in Lenin's Old Guard, who brought about the first contacts between the Soviets and Colonel-General von Seeckt of the Reichswehr Directorate, and thereby helped Germany to cast off the shackles of Versailles.

Radek, a convinced Bolshevik, a genuine people's tribune, one of the founders of the Communist Party of Germany, and an associate of Lenin during the latter's Swiss exile, was an ardent champion of the idea that "the common enemy, the victors of Versailles," must be defeated by an alliance between the Soviet Union and Germany. Radek did not consider it necessary that, for the purpose of such an alliance, Germany had to be Communist. In fact, he regarded the German nationalists as a transitional stage on the way to Bolshevism. Thus, when Albert Leo Schlageter, a second lieutenant in one of the irregular German Free Corps, an underground fighter against the French occupation of the Ruhr, was sentenced to death and shot by the French in May 1923, for an act of sabotage, Radek paid tribute to him before the Communist International on 20th June 1923 in a sensational speech entitled "Leo Schlageter, Traveller into Nothingness."

Karl Radek assisted at the birth of the military alliance between the Red Army and the Reichswehr. He was also to become its gravedigger.

The Soviets were interested in letting their young armed forces profit from the experience of German officers and in rebuilding their utterly derelict armament industries with German help. The Reichswehr, on the other hand, needed weapons whose manufacture was prohibited in Germany; it also needed training grounds where men might learn to use these forbidden weapons. On this basis a number of secret agreements were concluded between the Reichswehr and the Red General Staff. On the German side these activities were entrusted to "Special Group R"—R standing for Russia—a top-secret department of the German Army Directorate. Its executive organ was an economic front organization, the firm named GEFU, the Association for the Promotion of Commercial Undertakings.

This camouflage firm had an office in Berlin and another in Moscow. It was financed from the secret funds of the Reichswehr. It concluded contracts with Soviet authorities, maintained daughter companies in the most diverse parts of Russia, and set up German–Russian production units for the purpose of secret rearmament, with a production programme confined not only to aerial bombs, tanks, aircraft, and means of chemical warfare, but including even submarines—in short, everything that Germany, under the Treaty of Versailles, was forbidden to manufacture or use.

Geoffrey Bailey, the American expert on this backstage work of the Red Army, says in his book *Conspirators*:

> By 1924 the firm of Junkers was building several hundred all-metal aircraft a year in the Moscow suburb of Fili. Very soon more than 300,000 shells a year were coming from the reconstructed and modernized Tsarist arsenals in Leningrad, Tula, and Zlatoust. Poison gas was manufactured by the firm of Bersol in Trotsk (now Krasnogvardeysk), and U-boats and armoured ships were being built and launched in the yards of Leningrad and Nikolayev. In 1926 more than 150,000,000 marks, nearly one-third of the Reichswehr's annual budget, went on purchases of armaments and ammunition in the USSR.

The directing body which controlled these German activities in the Soviet Union was a secret organization under the code name ZMO, short for "Zentrale Moskau," or Moscow Central Office. ZMO was the German Army Directorate's 'Foreign Office' in Russia. Its representatives, von der Lieth-Thomsen and Professor Oskar Ritter von Niedermayer, known as Neumann, conducted all negotiations with the top officials of the Red Army and the Soviet Government. ZMO was ever present. ZMO, in fact, was a kind of shadow Government of the Weimar Republic functioning in Russia. Yet its representatives carefully kept out of the limelight.

Naturally, the manufacture of the forbidden war material was only one side of this collaboration. Since the importation of such weapons into Germany was likewise forbidden and, in the circumstances, would have been impossible to keep secret, it was equally important to make arrangements for the establishment of training centres outside Germany for the use of these weapons. The Soviet Union thus became the Reichswehr's training-ground.

Between 1922 and 1930 the following facilities were set up or extended for German use: a German Air Force centre at Vivupal near Lipetsk, 250 miles south-east of Moscow; a school of gas warfare in Saratov on the lower Volga, in operation since 1927; and a school of armoured fighting vehicles with training-grounds at Kazan on the middle Volga, in use since 1930.

As a collateral, Soviet officers who were being groomed for the Red Army staffs—former NCOs of the Tsarist Army, meritorious civil war fighters, and decorated political commissars—sat side by side with German General Staff aspirants in the classrooms of the German military academies and listened to lessons on Moltke's, Clausewitz's, and Ludendorff's art of warfare.

The spacious military airfield near the spa of Lipetsk was situated on high ground overlooking the town. Since 1924 it had been developed into an entirely modern air base. Officially the 4th Squadron of a Soviet Group was stationed there—but the language in the 4th Squadron was German. Only the liaison officer and the airfield guard were Russians. And, of course, the few ancient Soviet reconnaissance machines with their conspicuously large Soviet markings outside the hangars were Russian. Otherwise everything was German. Lipetsk was listed in the Reichswehr budget with 2,000,000 marks annually. The first hundred fighters used for the training of German pilots were bought from the Fokker works in Holland. Between 200 and 300 German airmen were stationed at Lipetsk. Here the first German fighter-bombers were tested. In realistic manœuvres, under simulated wartime conditions, the "Lipetsk fighters" practised the technique of low-level bombing and thereby laid the foundations for the much-feared German Stukas of a later date.

The first types of light bombers and fighter aircraft, which were fully developed for serial production when the build-up of the German Luftwaffe started in 1933, had all been developed and tested in Lipetsk. The first 120 magnificently trained fighter pilots, the core of the German fighter force, came from Lipetsk; the same was true of the first hundred officer observers. Without Lipetsk Hitler would have needed another ten years to build up a modern air force. Lipetsk was the kind of adventure that could scarcely be imagined nowadays. While the mistrustful eyes of the Western allies and the pacifist-minded German Left wing were searching Germany for the slightest indication of any prohibited rearmament, far away, in the Arcadia of the German Communists and Left-wing Marxists, the Lipetsk fighter squadrons were roaring over the Don, dropping dummy bombs on practice targets, testing new bomb-sights, screaming at low level over Soviet villages in central Russia, right up to the edge of Moscow itself, and co-operating as artillery spotters with Soviet ground forces in full-scale manœuvres on the army training-grounds of Voronezh. The military achievement of Lipetsk was equalled by the organizational one. Everything, down to the last nail, had to be supplied from Germany. The Russians supplied the earth and the stones—nothing more.

The necessary materials and supplies were shipped to Leningrad from the free port of Stettin. Particularly secret or dangerous equipment, or goods which could not easily be camouflaged, could not be loaded in Stettin. They were put on board small sailing-ships, manned by officers, and sailed across the Baltic in secret. Naturally, now and then an entire shipment would be lost. Traffic in the opposite direction included such items as the coffins containing the bodies of airmen crashed at Lipetsk: they were packed in cases, declared as machinery spares, and shipped to Stettin. Customs officers in the confidence of the Reichswehr helped to smuggle them out from the port.

All officers leaving for Russia were first discharged and officially crossed off the army lists. They were, of course, promised reinstatement upon their return, but there was no legal claim of any kind. Certainly it would have been impossible to enforce such a claim in the courts, especially if the camouflage system should have broken down in the meantime. That was the personal risk each officer underwent by going to Russia for training.

What Lipetsk meant for the Air Force, Kazan meant for the tankmen. There, on the middle Volga, the foundations were laid for Guderian's, Hoepner's, Hoth's, and Kleist's armoured divisions. This fact was one of the main reasons why, until Hitler's rise to power, no Russian and no German military leader ever considered the possibility of war between Germany and the Soviet Union, let alone worked out a plan for this contingency. The Reichswehr, with its founder and guiding spirit, Colonel-General von Seeckt, was anxious to liquidate the results of the Treaty of Versailles in alliance with Russia. It wanted to wipe out the consequences of the defeat in the West and to restore the old Western frontier of Germany. Even more so it wanted to re-establish the ancient frontier in the east by smashing Poland.

In the summer of 1922, when the newly appointed German Ambassador to Moscow, Count Brockdorff-Rantzau, opposed a unilateral pro-Russian policy on the part of Germany and warned against military collaboration with the Red Army, von Seeckt replied to him in a memorandum dated 11th September:

> The existence of Poland is intolerable; it is incompatible with Germany's vital needs. Poland must disappear, and it will disappear through its own internal weakness and through Russia—with our help. Poland is even more intolerable to Russia than it is to us; no Russian Government will ever reconcile itself to the existence of Poland. With Poland, one of the strongest pillars of the peace of Versailles will have fallen—the hegemony of France.

And what about the Soviets? What did the alliance with the

Prussian generals mean to them? To them it meant the strengthening, development, and modernization of the Red Army for "the last battle" —an end for which they were prepared to do anything. They were, moreover, interested in preventing at all costs an alliance between Germany and the Western Powers, since both Lenin and Stalin regarded a renewed Western intervention with German troops as a mortal danger. And finally, the aim of the German Right wing, the destruction of Poland, was Moscow's aim too. Thus the Reichswehr's anti-Western attitude suited Lenin's political concept, and later that of Stalin. Above all, it suited the man who, on the Soviet side, was the military partner of the German Army Directorate, the man who was increasingly becoming the personification of the Red Army—Marshal Tukhachevskiy.

Who was that man Tukhachevskiy? A hero and military genius, as was claimed for a whole decade until 1936? A traitor, a spy for the German Reichswehr, a "mangy dog," as Stalin called him after he had ordered him to be shot? Or a patriotic anti-Stalinist, the first and most fateful victim of the wicked old man, as Khrushchev maintains today? Which of these pictures is true?

On 5th December 1941, when Colonel-General Guderian from his snowed-in headquarters at the Tolstoy estate of Yasnaya Polyana instructed his Second Tank Army to suspend its attack on Moscow, the 45th Infantry Division, which was Second Army's linking division on its right wing, was furiously fighting for the possession of the town of Yelets. It was an unimportant little town, but it stood on the intersection of the great road from Moscow via Tula to the Don region and the east–west railway-line from Orel via Lipetsk to Stalingrad. Lipetsk, the former secret training base of the Reichswehr, where the young men of the German Luftwaffe had been taught their craft prior to 1933, was 40 miles away.

The combat-hardened regiments of 45th Infantry Division, mentioned earlier in this account during the fighting for Brest-Litovsk, penetrated into Yelets in fierce street battles in severe cold, and dislodged the Russians. The division was thus 15 miles from the upper Don—1300 miles from its starting-point. Thirteen hundred miles of marching and fighting, covered within five months and two weeks.

Two days before the attack on Yelets the monitoring unit of 135th Infantry Regiment succeeded in hooking itself into a Russian telephone-line and listening in to the conversations between the Soviet commanders in the field. There were frequent references in these conversations to a formation on the western edge of the town—"the Khabarovsk lot." At 135th Infantry Regiment's headquarters this designation was at first regarded as a code name until it was discovered from the evidence of a few prisoners that this was in fact part of a once top-secret military formation, long since dissolved. The officers of this unit were nicknamed "the Khabarovsk lot": they were the so-called Special Corps of the Far Eastern Army, the keystone of Marshal Tukhachevskiy's long-forgotten military policy.

The history of this Corps holds the key to the Tukhachevskiy mystery. It began in the summer of 1932. Germany then had 6,000,000 unemployed. The Soviet Union was in the grip of the greatest famine in modern history. Stalin's compulsory collectivization of agriculture, the expropriation and mass deportation of the wealthier peasants, had led to the complete collapse of agricultural production. Millions of Soviet citizens were dying of starvation. The domestic disaster was made worse by an international crisis.

In Asia, in 1931, the Japanese had leapt from their poor overpopulated islands on to the Chinese mainland in order to conquer a market for their manufactures and raw materials for their industry. In 1932 they occupied Manchuria with its fertile soil and rich ore deposits and made this country on the frontier of Eastern Siberia a Japanese satellite—the Manchurian Empire. In this way Tokyo demonstrated to the world that it was determined, if need be by force of arms, to set up a greater East Asian economic bloc.

This was a serious threat to the Soviet Union's Far Eastern interests. A Russo-Japanese conflict along the Far Eastern border became a real possibility. This happened at the very moment when Stalin's empire was faced with starvation.

At that moment in Moscow the First Deputy War Commissar, General Gamarnik, conceived an idea and put it into effect with General Tukhachevskiy's help. He set up the Far Eastern "Special Corps," also known as the Collective Farm Corps, whose officers very soon began to call themselves the "Khabarovsk lot," after the town of Khabarovsk on the Manchurian frontier.

Gamarnik and Tukhachevskiy's idea was both simple and ingenious: the members of this Corps were soldiers and at the same time peasants—peasants in uniform, as it were. In the event of war with Japan they were to make the Far Eastern Army independent of food and fodder supplies along the single-track trans-Siberian Railway. It was the only possible solution of the vital problem of supplies. Marshal Blyukher, the autocratic Commander-in-Chief of the Far Eastern Army, had prohibited the expropriation of the rich peasants and the collectivization of agriculture in Siberia because he was worried about the morale of his recruits, 90 per cent. of whom came from peasant homes. Thus the only way to ensure a reliable agricultural supply basis for the Far Eastern Army was the one chosen by Gamarnik—military settlements which the soldiers would join with their families after completion of their normal active service. They formed large farming communities, but at the same time retained their military organization and were kept under arms and ready for action. Many farm labourers and sons of peasants from Central Russia volunteered for the Special Corps. Here they received their own house, a large plot of land for their own private use, complete with one cow, chickens, exemption from taxes for ten years, and other privileges.

By 1936 the "Collective Farm Corps" numbered 60,000 men on the active list and 50,000 reservists settled on the army farms. That was a battleworthy force of altogether ten divisions, with its own structure, virtually independent of the Red Army chain of command, and a long way removed from the heart of the regime in Moscow—an ideal tool for a general with political ambitions. And Gamarnik certainly was that. But even more so his friend Tukhachevskiy. Tukhachevskiy was likewise Deputy War Commissar, and, ever since the famine and the liquidation of the peasants, had been a resolute opponent of Stalin, the leader of a clique of generals actively opposed to the dictator. A man waiting for the moment when the despot could be overthrown. The "Collective Farm Corps" was ideally suited to his plans and played a decisive part. In the event of an armed clash with pro-Stalin forces of the Army and Party, the remote East Siberian Special Corps would hold a kind of rebel fortress and, if necessary, a safe area of retreat.

In the light of all these facts Marshal Tukhachevskiy acquires features differing from the picture painted either by Stalinist propaganda or by superficial Western biographers. Anyone viewing this man merely as a "fallen angel," as a Tsarist Guards officer who had embraced Bolshevism although the blood of French counts and Italian dukes coursed through his veins, closes his eyes to a proper understanding of this fascinating and in a way outstanding personality in Soviet history.

He was a worthy opponent of Stalin. He alone would have been able to overthrow the tyrant, to replace him, and to turn the course of Soviet and world history into a different direction. Tukhachevskiy's whole life showed him to be an exceptional man. Born in 1893, he was taken prisoner in August 1915 as a second lieutenant in the battle of Warsaw—the very city where, almost exactly five years later, he was to suffer once more a military defeat. He was taken to No. 9 POW camp near Ingolstadt. In 1917 he escaped and made his way to St Petersburg. When he arrived there the city on the Neva was no longer the capital of Russia. The Tsar had been deposed. The war was over. Lenin's Bolsheviks were in power, and fighting against the counter-revolution of the White generals.

Tukhachevskiy, the Guards officer, the kinsman of half a dozen West European noble families, did not join the Whites but the Reds. Why? It has been said that it was pure accident. Others have attributed this surprising decision to the political inexperience of a young man. Others yet have ascribed it to pure opportunism. None of these explanations is true. Tukhachevskiy went Red from conviction and ambition.

The revolution against the bourgeois world, just because of its ruthless challenge of the existing order, was in line with his own impetuous rejection of Western tradition, Christianity, and the European spirit. Tukhachevskiy's dreams were concerned with the East, not the West. He had seen the West in his POW camp. The West to him was the Tsar and his corrupt and decadent regime. The West and Tsarism, for whose restitution the Whites were fighting, were not Tukhachevskiy's party. To him the future of new ideas and a new power was in the East.

Moreover, on the side of the Reds there were great opportunities for the military ambitions of a young officer to whom the Army meant everything. Trotskiy, the creator of the Army of the Red Revolution, needed professional soldiers, leaders, and staff officers for his wild hordes. Tukhachevskiy therefore joined the Communist Party and became a general staff officer. In May 1918, at the age of twenty-five, he was Commander-in-Chief of the First Army. He threw the Czechoslovak legions back over the Volga. In 1919 he led the Fifth Army in the Urals. The Reds then controlled only one-sixth of the Russian empire. Things looked bad for Lenin. But Tukhachevskiy defeated Admiral Kolchak's White divisions which had got as far as Kazan and chased them over the Urals. In 1920 he drove General Denikin's White southern army into the Black Sea.

Just then the young Soviet Union was facing its greatest military threat. The Poles, taking advantage of Russia's weakness, burst into the Ukraine, occupied Kiev, and presently controlled the grain areas of the starving young Soviet State. Once again Tukhachevskiy was the saviour. He outmanœuvred the Poles by a brilliant operation. They were forced to withdraw. Tukhachevskiy pressed after them and marched on Warsaw. He advanced towards the West. Would Warsaw be the first stage in the victorious advance of the Red revolution into Europe?

Marshal Pilsudski writes in his memoirs that the fate of Poland had then seemed to him gloomy and hopeless. But not till twenty-four years later did the Red Army in fact get to Warsaw and into Europe. Then, in the summer of 1920, the Poles and Europe generally were saved from Lenin's banner by "the miracle on the Vistula." But this miracle was no accomplishment of Europe; it was the result of Josef Stalin's stupidity and disobedience.

Tukhachevskiy was within artillery range of Warsaw. The Revolutionary War Council in Moscow, the supreme authority of the Red Army, invested him with the Supreme Command of all armed forces on the Western front, including the South-western Army, whose cavalry units were commanded by Yegorov and Budennyy. The political commissar of the South-western Army was Josef Stalin. Tukhachevskiy gave the correct order to the South-western Army—to wheel towards the north, towards Lublin, so as to cover the flank of his striking army aiming at Warsaw.

But Josef Stalin had different ideas. He wanted to capture Lvov. He talked the two commanders, Budennyy and Voroshilov, into ignoring Tukhachevskiy's order and into marching not against Lublin but against Lvov. This they did. The French General Weygand, who was the adviser of Pilsudski, the Polish Commander-in-Chief, realized his chance. Through the gap the Poles struck at Tukhachevskiy's left flank and rolled up the wing of his army. Panic broke out. The troops fled. Poland was saved.

It is not difficult to guess what sentiments Tukhachevskiy nurtured for Stalin since those days. If nevertheless he was promoted, under the dictator's rule, to the rank of marshal and to the post of Chief of the General Staff and then Deputy War Minister, this is evidence of his self-control and his military abilities which Stalin could not do without.

The creation of a modern Red Army, above all its motorization and the introduction of armour, was the work of Tukhachevskiy. His avowed model was the Chief of the German Reichswehr, Colonel-General von Seeckt. Seeckt the Prussian and Tukhachevskiy the revolutionary general—were they not like fire and water? Naturally, the two were divided by a whole world, but there was also a lot that they had in common. Stalin's system of spies within the Army, a system which was like a cancer on the morale of the Officers Corps, and the dictator's economic experiments, with their collectivization and slaughter of the peasants, had turned Tukhachevskiy into a bitter enemy of Stalinism. But the decisive motive for his political opposition, presumably, was Stalin's foreign policy. Tukhachevskiy became increasingly convinced that an alliance between Germany and the Soviet Union was an inescapable commandment of history, so that the struggle could be waged against the "decadent West."

Tukhachevskiy knew, of course, that this aim could only be reached against Stalin and his narrow-minded bureaucracy. He therefore had to be armed for the event of a clash. His private army was the Khabarovsk Corps.

Since 1935 Tukhachevskiy had maintained a kind of revolutionary committee in Khabarovsk, the centre of Eastern Siberia. Its members included senior administrative officials and Army commanders, but also some young Party functionaries in high posts, such as the Party leader in the Northern Caucasus, Boris Sheboldayev. This composition is important. It proves that Tukhachevskiy was not out to create an anti-Communist movement, but to mobilize the progressive and patriotic wing of the Bolsheviks against Stalin's tyranny.

In the spring of 1936 Tukhachevskiy went to London as the leader of the Soviet delegation attending the funeral of King George V. Both his outward and homeward journeys led him through Berlin. He used the opportunity for talks with leading German generals. He wanted to make sure that Germany would not use any possible revolutionary unrest in the Soviet Union as a pretext for marching against the East. What mattered to him most was his idea of a German–Russian alliance after the overthrow of Stalin. What evidence is there for this?

Geoffrey Bailey in his above-mentioned book quotes an attested remark by Tukhachevskiy, made at about that time to the Rumanian Foreign Minister Titulescu. Tukhachevskiy said: "You are wrong to tie the fate of your country to countries which are old and finished, such as France and Britain. We ought to turn towards the new Germany. For some time at least Germany will assume the leading position on the continent of Europe."

That was in the spring of 1936. The date is important. For nine months later Skoblin, the OGPU agent in Paris, arranged for information about the Red generals' imminent coup against Stalin to fall into the hands of the contacts of SS Gruppenführer Heydrich. Hitler believed that here was his chance to deliver the Red Napoleon to his executioners and deprive the Soviet Army of its head. But in reality Heydrich was merely doing Stalin's work for him. The dictator had long decided to act against Tukhachevskiy.

Here is the proof. In January 1937 Prosecutor General Vyshinskiy, the Soviet Grand Inquisitor, opened the political purge trial of the old anti-Stalin Bolshevik Guard in the great hall of the former Nobles' Club in Moscow.

The main figure in the dock was Karl Radek, the man who, between 1919 and 1921, had arranged the collaboration between the Reichswehr and the Red Army. He was now to be the man to bring it to its end. At the morning session of 24th January he suddenly introduced Tukhachevskiy's name in reply to a question fired at him by Vyshinskiy. The name came up quite casually. Vyshinskiy probed a little further. And Radek said, "Naturally Tukhachevskiy had no idea of the criminal part I was playing."

An icy silence descended over the court room. And in this silence Radek muttered the name of one of Tukhachevskiy's confidants, General Putna. "Putna was my fellow conspirator," Radek confessed. But Putna was the foreign affairs expert of the Tukhachevskiy group, and had made numerous contacts as Military Attaché in Berlin, London, and Tokyo. What was more, Putna was already under arrest at the time of the hearing. He had been arrested towards the end of 1936.

Thus the moves against Tukhachevskiy had been taking place quietly since the end of 1936. Naturally, the Marshal and his friends realized their danger. Supposing Putna talked? It did not bear thinking of. Swift action was necessary.

In March 1937 the race between Tukhachevskiy and Stalin's secret agents was becoming increasingly dramatic. Like the rumble of an approaching storm was Stalin's remark at a meeting of the Central Committee, at which Tukhachevskiy himself was present: "There are spies and enemies of the State in the ranks of the Red Army."

Why did the Marshal not act then? Why was he still hesitating? The answer is simple enough. The moves of General Staff officers and Army commanders, whose headquarters were often thousands of miles apart, were difficult to co-ordinate, especially as their strict surveillance by the secret police forced them to act with the utmost caution. The coup against Stalin was fixed for the 1st of May 1937, mainly because the May Day Parades would make it possible to move substantial troop contingents to Moscow without arousing suspicion.

However, chance or Stalin's cunning brought about a postponement. It was announced from the Kremlin that Marshal Tukhachevskiy would lead the Soviet delegation to London to attend the coronation of King George VI on 12th May 1937. Tukhachevskiy was to be reassured. And he was reassured. He postponed the coup by three weeks. That was his fatal mistake. He did not go to London, and the coup did not take place. About 25th April he was seen at the Spring Ball at the Moscow Officers Club. On 28th April he attended a reception at the US Embassy. That was his last reliably attested public appearance. Everything that happened afterwards is known only from rumours and unverifiable second- and third-hand reports.

The last official announcement about the Marshal was the Tass report of 11th June 1937, announcing that Tukhachevskiy and seven other generals had been arrested, sentenced, and shot. General Gamarnik was reported to have committed suicide. In fact, he was beaten to death during interrogation.

A large number of stories circulated about the trial and the execution. Nearest the truth, probably, is the version to the effect that a hearing took place with Vyshinskiy as prosecutor. Marshals Blyukher and Budennyy, as well as other senior generals, were members of the tribunal. No witnesses were called. Vyshinskiy needed no witnesses: his surprise move was the submission of the faked-up Reichswehr file supplied by Heydrich. To Stalin and the Party these papers were evidence of the espionage conducted by Tukhachevskiy and his friends. The documents, moreover, made it impossible for the senior generals and marshals to do anything for the conspirators. The first breach was torn in the solid front of the generals. They sat in judgment over their comrades, and in the eyes of the rest became culpable themselves. Each evil deed begot another. Before long Tukhachevskiy's judges were in the dock facing new judges, and the new executioners presently faced newer executioners still. Thus it went on.

There is no evidence to show whether Tukhachevskiy and his seven fellow-accused were present at the main trial, or, indeed, whether they were still alive then. A reliable witness, the NKVD official Shpigelglass, quotes the Deputy OGPU Chief at the time, Frinovskiy, for the remark: "The entire Soviet regime hung by a thread. It was impossible to proceed as in normal times—to have the trial first and the execution afterwards. In this case we had to shoot first and sentence afterwards."

And how was Tukhachevskiy done to death—the man who had done more to save Lenin's revolution than Stalin and all his henchmen together? That too is not known for certain. Most probably he was shot from behind with an eight-round automatic pistol in the tiled cellars of the Lubyanka prison and flung into a mass grave with his comrades.

Day after day and week after week the mass graves grew. Stalin decimated the corps of General Staff officers, he executed the experienced commanders, and, above all, he wrecked the military discipline which Tukhachevskiy had so laboriously built up by now enthroning the political commissars and consolidating the Party's control over the Army.

The settling of accounts came two years later, in the winter of 1939–40. Three months after Hitler's attack on Poland Stalin mounted a "punitive expedition" against his small neighbour Finland. The Soviets had demanded the cession of the Hangö Peninsula in the southwestern part of the Gulf of Finland "for the protection of Leningrad and Kronshtadt." When the Finnish Government refused Moscow replied with the complaint that Finnish artillery had shelled the Soviet frontier village of Mainila.

The Finns guessed Stalin's intention. They offered to conduct a joint inquiry. Stalin's answer was a full-scale attack on land, on the sea, and in the air. The notorious Finnish–Russian winter war had begun. However, it soon took a different course from the one envisaged by Stalin and his military advisers. Stalin had pictured a Blitzkrieg on the model of his ally Adolf Hitler. What ensued instead was a savage and costly campaign with shameful Soviet defeats which amazed the world and which were to have a disastrous effect on world history.

To this day one still encounters the theory that Stalin deliberately waged his war against Finland with weak and poorly equipped troops in order to deceive Germany. But that is a fairy-tale.

Russia attacked with its Seventh, Eighth, Ninth, and Fourteenth Armies. A force of 150,000 to 200,000 troops faced the 700,000 Soviet troops. Nevertheless the Soviets were defeated. The Red Army displayed poor tactics, worse strategy, and an appalling fighting morale. These were the consequences of the purges.

The Finns made a tactical virtue of the necessity of having to face the Soviet Armies with numerically greatly inferior forces. They introduced the tactics of the Motti, or pocket, the precursor of the great German battles of encirclement. Fast Finnish ski troops severed the Soviet divisions' lines of communication, forced them into the forests, and at night pounced on their scattered columns. As a rule they struck silently, with the Puuko, the Finnish dagger. The Soviets lost division after division.

Naturally the Finns could not halt the Red colossus single-handed in the long run. When Marshal Timoshenko mounted his large-scale offensive on 11th February 1940 he deployed thirteen divisions in deep echelon against 12 miles of Finnish defences. Some 140,000 men along a 12-mile front, or seven men to each yard. All this was supported by armour, artillery, and mortars.

In this way Stalin eventually achieved victory and seized the bases he wanted. But he dared not impose a Communist regime on Finland. A Russian general declared: "We were glad to get out of the affair. We had captured just about enough land to bury our dead in."

Stalin learnt his lesson from the Finnish disaster and tried as quickly as possible to remove the shortcomings which had shown up. Hitler, on the other hand, was confirmed by the Red Army's disastrous defeat in his belief that an attack on the Soviet Union would be a military walkover, and that, without any great risks, he could gain control of the Soviet sources of raw materials in order to see the war through against the Western Powers. In this sense the disastrous attack against the Soviet Union on 22nd June 1941 was a belated result of Stalin's murder of Tukhachevskiy.

Stalin's crime against this military genius brought the Soviet Union to the brink of disaster; the memory of his legacy, a return to his principles and to the virtues of military leadership ultimately saved Russia and Bolshevism. An inkling of this truth was felt in the German fighting lines on the last day of the offensive against Moscow.

In the forest of Takhirovo, a wooded area in the Nara bridgehead before Moscow, heavily reinforced with concrete strongpoints, the 2nd Battalion, 508th Infantry Regiment, took an interesting prisoner early in December—the commander of the Soviet 222nd Infantry Division. A party of sappers had brought him out, wounded, from his dug-out, the only survivor.

Captain Rotter, the OC 2nd Battalion, interrogated the colonel. At first the Russian was dejected and apathetic, but gradually he thawed out. This was the fifth war he had been mobilized for, he explained. Did he think Russia could still win the war, Rotter asked him. "No," was the answer. All his calls for reinforcements had produced the same reply: We have nothing left; you've got to hold out to the last man. Behind his division, the Soviet colonel explained, there were only a few

Siberian units left this side of Moscow, apart from workers' battalions. But surely, Captain Rotter objected, very stubborn resistance was being offered everywhere? The colonel nodded. During the past few weeks, he said, many new officers had joined the troops—middle-aged men for the greater part, and all of them from Siberian penal camps. They were men who had been arrested during the great Tukhachevskiy purge, but who had survived in prisons and camps. "Active service at the front is their chance of rehabilitation. And if a man has a penal camp behind him death holds no terror for him," the colonel said. And softly, as though still fearing the ears of Stalin's OGPU, even in captivity, he added: "Besides, they want to prove that they were no traitors, but patriots worthy of Tukhachevskiy."

When the records of this interrogation reached Army headquarters someone on Kluge's staff remarked, "The late Tukhachevskiy is in command before Moscow." A bon mot, but profoundly true.

PART TWO: *Leningrad*

1. Chase through the Baltic Countries

Ostrov and Pskov–Artillery against KV-1 and KV-2 monsters–Hoepner is held back by the High Command–The swamp of Chudovo–Manstein's Corps cut off–The road to Leningrad is clear–Unsuccessful bathing party at Lake Samro.

AN old Finnish proverb says: "Happy the man who does not have to eat his words of the previous day." Many Germans in Finland during the summer of 1941 had this proverb quoted to them. The "previous day" was Germany's attitude during the Finno–Russian winter war and the equivocal declarations of German politicians and diplomats in connection with the Russian aggression. Hitler had in effect observed a friendly neutrality towards the Soviets. Yet on 22nd June 1941 Hitler's proclamation, blared out from all public loudspeakers and announced in huge banner headlines in all newspapers, and read out to the troops along the front from the Arctic to the Black Sea, contained the phrase: "German troops are standing side by side in alliance with Finnish divisions, guarding Finland."

When the author of the present book interviewed Marshal Mannerheim at his secret headquarters in the idyllic little forest town of St Michel, the Marshal criticized this particular phrase in the Fuehrer's proclamation. He said, "The Reich Chancellor's formula did not take full account of the situation in international law, quite apart from the fact that it anticipated later developments." Mannerheim pointed out that, at a Press conference at the Berlin Foreign Office on 24th June, it was publicly stated that Finland was not yet formally at war with Russia. Mannerheim, however, hastened to add, "This was of no importance for the further development of the situation, for I am certain that Stalin would have attacked us in any case in order to cover his flank—Leningrad and the Baltic—no matter how much we tried to remain neutral." He paused for a moment and then added, "Only by going over to the Soviet camp could we have escaped the attack. And that would have meant the same as being defeated."

In support of his view Mannerheim then quoted a remark made by

Stalin to the Finnish Minister in Moscow shortly after the winter war: "I can well believe that you would like to remain neutral," Stalin had said to him, "but a country situated as your little country is cannot remain neutral. The interests of the Great Powers forbid it." Marshal Mannerheim made one more interesting remark: "I realized ever since January [1941] that the Soviet leaders envisaged the possibility of an open rupture with Germany, that they expected an armed clash, and that they were merely playing for time to postpone its outbreak."

All this the Marshal said very gravely, almost impassively. He spoke softly, with resignation in his voice—a *grand seigneur* calmly facing the inevitable and prepared to see the consequences through to the end.

Mannerheim missed no opportunity to point out that Finland was not an ally of Germany, but, as he put it, "a fellow traveller in a war which Finland is waging for its own active defence." He said so to various officials of the German Foreign Office and the Wehrmacht, and he said so also to the clever German Minister in Finland, Herr von Blücher.

"We don't want to conquer anything," he would repeat time and again, "not even Leningrad." There was no doubt that this gentleman, who spoke Russian with less of an accent than he spoke Finnish, whose style of life had been moulded at the Cadet Academy of the Grand Duchy of Finland, in the Corps of Pages at the Tsar's Court, and as a Guards officer in St Petersburg, did not have his heart in the German war against Russia. He was fighting on Hitler's side for reasons of political expediency, against a common enemy.

With a secret smile Mannerheim would relate a story which made the rounds of Helsinki just before the outbreak of the war and caused much amusement. At a tea-party in the drawing-room of a well-known Finnish lady in the autumn of 1940 a British Legation Counsellor complained that Finland had permitted German troops to travel through the country on their way to Northern Norway. His hostess retorted, "We are in a difficult situation. The Russians have extorted from us the right of passage to their strongpoint at Hangö. On what grounds could we deny the Germans the right of passage to their bases in Northern Norway?" "That's quite correct," the Englishman replied. "But most Finns are welcoming the Germans with open arms!" The old lady laughed and replied, "I'm afraid I do that myself. For the more Germans we have in our country the more peacefully I sleep at night."

That in fact was the situation. Since the winter war, which had given Stalin only half a victory, the Finns naturally feared Moscow's revenge. That was why they were greatly relieved when in November 1940 they learnt that Hitler had firmly refused his agreement to Molotov's request in Berlin for a renewed Soviet operation against Finland.

At a private luncheon the Finnish Foreign Minister, Witting, observed, "When Minister von Blücher reported to me in cautious terms the outcome of Molotov's visit to Berlin, and it became clear that in contrast to his earlier attitude Adolf Hitler now opposed the Russian intentions, a great load was taken off all our minds!"

It is important to realize this background in order to understand the subsequent decisions of Germany's military 'fellow traveller' in the far north. The Finns were splendid, brave, and uncomplicated people of unequalled patriotism. One need only remember the almost legendary General Pajari who won the Finnish Knights Cross in the winter war by operating an ancient captured Soviet anti-tank gun singlehanded against an enemy tank attack. The sights and the firing mechanism were out of order. Pajari aimed the gun by sighting along the barrel and fired it by hitting the bolt with an axe. In this manner he destroyed three of the four Soviet tanks. Or when his headquarters was under enemy bombardment and his staff advised a change of position he would put his hand behind his ear, pretend to be listening, and say, "I can't hear anything. You must be mistaken."

Men like this were the secret of the almost unbelievable resistance put up by the Finns in the winter war. In the end they had to yield to an enormous superiority and to agree to a harsh peace treaty with severe losses of territory and towns. Not one of the Western Great Powers had come to their aid; even their Swedish brothers had left them in the lurch. It is not surprising that to them 22nd June 1941 represented a chance, under the powerful shield of the German Wehrmacht, to recapture from the Russians their lost territories, above all the ancient town of Viipuri, and to restore the former Finnish–Russian frontier. The German High Command, admittedly, had rather more extensive hopes of Mannerheim.

When Army Group North under Field-Marshal Ritter von Leeb mounted its offensive on 22nd June between Zuvalki and Klaipeda it had before its eyes a clear operative objective—Leningrad.

In the deployment directive for Operation Barbarossa it was laid down that, following the annihilation of enemy forces in Belorussia by Army Group Centre, strong units of mobile troops were to wheel to the north, where, in conjunction with Army Group North, they were to annihilate the enemy's forces in the Baltic countries and, this task completed, take Leningrad. The attack on Moscow was not scheduled until after the capture of Leningrad.

It is important to remember the sequence of events in this military time-table. Failure to keep to it was one of the reasons for the disaster at Moscow in the winter.

Leningrad was the jewel of European Russia. Pushkin says in a poem: "Novgorod the father, Kiev the mother, Moscow the heart, and

St Petersburg the head of the Russian empire." Since St Petersburg had ceased to be St Petersburg, or even Petrograd, and had become Leningrad, the city on the Neva estuary, built on more than a hundred islands in the low-lying marshes, was no longer the head but it still was the conscience of the Red empire. It bore the name of the father of the revolution, and it was there that the revolution had started. From its munitions plants, its shipyards, its tank assembly lines, its footwear factories, and its textile mills, from its merchant ships and its naval units, had come the revolutionary vanguard of the Bolsheviks. There Lenin had begun his struggle.

If, moreover, the strategic role of Leningrad is considered, as a fortress in the Gulf of Finland, as the naval base of the Baltic Fleet, it becomes clear that this city was an important military, economic, and political objective. To capture it would have been an inestimable victory for Hitler; to lose it would have been a terrible blow to the Bolshevik regime.

Lieutenant Knaak did not live to see his operation succeed at the road bridge of Daugavpils. He had been mown down by a machine-gun on the right-hand ramp of the bridge, and there, though dead, he watched over his thirty-odd men standing up to furious Russian counter-attacks. If ever a Knights Cross was deserved as a reward for an operation which decided the outcome of a battle, then it was the one posthumously awarded to the assault party leader. The swift seizure of the Daugava crossings was of vital importance to the battle for the approaches to Leningrad.

By its thrust over the Daugava Hoepner's Panzer Group provided flank cover for Colonel-General von Küchler's Eighteenth Army operating along the Baltic coast and enabled it to advance across the Baltic countries. Colonel Lasch, commanding 43rd Infantry Regiment, led an advanced formation of mobile units of I Army Corps—cyclists, anti-tank gunners, AA gunners, sappers, and assault guns—straight on through a disintegrating enemy for some 60 miles via Bauska to Riga in order to bar the river crossings there too to the retreating Soviet divisions. Admittedly, there were heavy losses and the Russians succeeded in blowing up the bridges, but the objective was nevertheless achieved: the Soviet columns fleeing from Courland were unable to get across the Daugava and met their doom before Riga.

While Küchler's Eighteenth Army was penetrating into the Latvian-Estonian area, Hoepner's Fourth Panzer Group drove across the old Russian-Estonian frontier south of Lake Peipus. The frontier had been fortified as the so-called Stalin Line—a full-scale line of defences with pillboxes and heavy field fortifications. Colonel-General Kuznetsov hurriedly tried to get some reinforcements to the key-points of the line, in particular to the railway junction of Ostrov. German aerial reconnaissance spotted the move. It was vital that Hoepner should get to Ostrov before the Soviets. Thus began General Reinhardt's great tank race to Ostrov.

Just as 8th Panzer Division had formed the spearhead of Manstein's race to the Daugava, the spearheads of Reinhardt's XLI Panzer Corps were represented by 1st Panzer Division. And this division, under Lieutenant-General Kirchner, won the race from the Daugava bridgehead at Jekabpils across the southern part of Estonia to Ostrov. On 4th July Major-General Krüger's 1st Rifle Brigade penetrated into the town from the south with 113th Rifle Regiment reinforced by units of 1st Panzer Regiment. While the 1st Motorcycle Battalion was coming up from the south-west, Major Eckinger with his Armoured Infantry Carrier Battalion, supported by 7th Battery, 73rd Artillery Regiment, pushed through to the north. The road bridges across the Velikaya river were taken.

Russian reinforcements, including heavy armour, spotted and reported by aerial reconnaissance, arrived exactly twenty-four hours too late to save Ostrov. They now launched their super-heavy KV-1 and KV-2 tanks against the northern part of Ostrov, but were repulsed.

When the combat group Krüger, the vanguard of 1st Panzer Division, launched its attack against Pskov towards 1400 hours on 5th July it came under a heavy attack by massed Soviet tanks. The motorized anti-tank guns of 1st Company, Panzerjäger Battalion 37, with their 3·7-cm. guns, were simply crushed by the heavy Soviet armoured vehicles. Riflemen and Panzerjägers alike again found themselves helpless, as they had been at Raseiniai and Saukotas, against these huge crawling fortresses. They fell back. The Russians rolled past the German tanks—towards Ostrov. Was there nothing to stop them?

That was the great hour of Major Söth, commanding the 3rd Battalion, 73rd Artillery Regiment, formerly the 2nd Battalion, 56th Artillery Regiment, from Hamburg-Wandsbek. He got one of his heavy field howitzers of 9th Battery into position on the road. Its gun-layer, Corporal Georgi, allowed the first KV-2 to get within the correct range. Georgi had loaded a concrete-piercing shell, such as was used against heavy pillboxes. "Fire!" As if hit by a giant fist the KV-2 was flung sideways and remained motionless. Reload—aim—fire! Another twelve Russian tanks were shot up by the gallant corporal and his crew. Other guns similarly intervened in the struggle against the Russian tanks. The action not only halted the enemy attack, but also restored to the infantrymen their self-confidence. Presently they tackled the enemy tanks with demolition charges, supported by the gunners of 3rd Battalion. Shortly afterwards Major-General Krüger was able to report: The advance continues.

Two days later, on 7th July, 1st Panzer Regiment, heading the combat group Westhoven and forming the vanguard of 1st Panzer

Division and, immediately behind it, 6th Panzer Division, launched an attack against the remainder of the Soviet armoured formations before Pskov. Farther back on their left, in echelon, came the 36th Motorized Infantry Division, the third mobile division of XLI Panzer Corps. Captain von Falckenberg, commanding the lead company, now reinforced by armoured infantry carriers of 1st Battalion, 1st Rifle Regiment, was leaning in the turret of tank No. 700 at a crossroads north of the village of Letovo, his binoculars at his eyes.

Through his glasses he watched Second Lieutenant Fromme, whose first troop formed the vanguard of 2nd Battalion, 1st Panzer Regiment, open fire with his tank No. 711 at an approaching Soviet tank. He scored a direct hit. Smoke issued from the enemy tank, but it continued to move. It made straight for Fromme's tank and rammed it. Three Russians leapt out. Fromme too jumped down from his tank, pistol in hand. The Russians raised their hands. At that moment two other Soviet tanks came rumbling up across the field. The three prisoners, taking fresh heart, ran behind their tank. Fromme tried to fire, but his pistol jammed. One of the Russians charged him. Quick as lightning, Fromme reached behind him and snatched up the axe clipped to the caterpillar track guard. Brandishing it, he went for the Russians. They fled. Fromme scrambled back into his tank.

Captain von Falckenberg let himself drop down into his tank, pulling the hatch shut behind him. "Forward!" he called to his driver. "To the crossroads!" Second Lieutenant Köhler of 2nd Troop had likewise watched Fromme's axe duel and roared forward to support him. He moved into position on the right of Fromme's troop and at once joined the action. His four Mark III tanks were just in time to take the next lot of Soviet tanks in the flank. At nightfall eighteen tanks lay disabled in front of Falckenberg's sector. The Soviet counter-attack south-east of Lake Peipus had had its back broken. The road to Pskov was clear.

In an eastward sweep by the combat group Westhoven the reinforced 1st Rifle Regiment drove on as far as the airfield of Pskov, which had evidently been abandoned in a hurry by a senior Soviet Air Force headquarters. The maps in the situation room revealed some interesting information about the enemy's intentions. Major-General Krüger captured a bridge in the Tserjoha sector intact, with a surprise coup by 1st Rifle Brigade. The town of Pskov, by then in flames, was taken by 36th Motorized Infantry Division in a frontal attack on 9th July.

Twenty miles to the south-east the 6th Panzer Division had also broken through the Stalin Line. Twenty heavy pillboxes had been cracked by the sappers and strong enemy armour thrown back. Hoepner's Panzer Group had thus reached its first great objective. The Russian barrier south of Lake Peipus had been pierced, the Russians' southern exit from the Baltic area had been blocked, and the jumping-off position for an attack on Leningrad had been gained.

Chase through the Baltic Countries 225

The swift blow against the city was to be struck in a northerly direction across the narrow neck of land between Lakes Ilmen and Peipus. The aim was still to 'take' Leningrad—i.e., to capture it. The operation was to be supported by the Finnish Army driving from the north across the Karelian Isthmus and simultaneously attacking east of Lake Ladoga; in this way the city with its 3,000,000 inhabitants was to be sealed off from the north and east against the arrival of relief or attempted breakouts from within.

Under the terms of their general orders the 4th Panzer Group intended to make Reinhardt's Panzer Corps drive towards Leningrad along the Pskov–Luga–Leningrad road, and to send Manstein's Panzer Corps along the second road to Leningrad, that from Opochka via Novgorod. Those two great roads were the only ones leading through the extensive marshy area which shielded Leningrad towards the south and south-west.

On 10th July 1941 the Panzer Group mounted its attack along the whole front. The LVI Panzer Corps, which had pierced the Stalin Line at Sebezh on 6th July with the motorized "Death's Head" SS Infantry Division, and after hard fighting had taken Opochka on the Velikaya river, was now to make an outflanking movement to the east and, advancing via Porkhov and Novgorod, cut the big lateral road from Leningrad to Moscow at Chudovo. The 8th Panzer Division and the 3rd Motorized Infantry Division were employed in the front line. Their task was to advance across very difficult wooded ground.

The XLI Panzer Corps, with 1st and 6th Panzer Divisions in front and 36th Motorized Infantry Division behind, moved off along the main road via Luga. To begin with enemy resistance was confined to rearguard actions. The enemy was giving ground. Had the Russians really given up in the north? Nothing of the kind. Voroshilov was not prepared to abandon Leningrad or the Gulf of Finland. On the very next day the advance of Reinhardt's Panzer Corps was slowed down. Its divisions had got into difficult swampy forest terrain which offered the enemy excellent opportunities for defence.

When General Reinhardt tried to move his tanks and armoured infantry carrier battalions, in particular the combat groups Krüger and Westhoven of 1st Panzer Division, off the Pskov–Luga road for an outflanking action, with a view to cracking the Russian roadblocks from the rear, he was to discover that to the right and left of the road the ground was swampy and virtually impassable for armour.

The 6th Panzer Division too had to be brought back from its wretched secondary roads to the Corps' main road of advance behind the 1st Panzer Division because its vehicles were continually getting stuck. No large-scale operations were possible. The tanks lost their advantage of mobility and speed. On 12th July the Corps' offensive ground to a standstill along the Zapolye–Plyusa line.

H

Enemy resistance was even stronger in front of Manstein's Corps—*i.e.*, on the right wing—where in accordance with High Command orders the main weight of the attack was to be concentrated. It was found that the Russians had built up a new fortified zone covering Leningrad and Shimsk on the western shore of Lake Ilmen, and along the Luga river as far as Yamberg on the Narva. The town of Luga, as a bridgehead on the Daugavpils–Leningrad highway, was the keystone of the position and had been strongly fortified.

Ground and aerial reconnaissance of 4th Panzer Group, on the other hand, discovered that the left wing, on the lower Luga, was held by weak enemy forces only. Clearly, because of the bad roads there, the Russians did not expect an attack. The only other enemy force of any size was on the eastern shore of Lake Peipus, near Gdov.

Colonel-General Hoepner was faced with a difficult decision: was he to stick to his orders and keep the main weight of his attack on the right, in the direction of Novgorod, and allow Reinhardt's Panzer Corps to batter their heads against the strong defences at Luga, or should he make a bold left turn towards the lower Luga, strike at the enemy where he was weak, and in this way promote an attack on Leningrad from the west, parallel to the Narva–Kingisepp–Krasnogvardeysk railway?

Hoepner decided on the latter alternative. He switched the 1st and 6th Panzer Divisions to the north under cover of the combat group Westhoven, which was fighting east and north of Zapolye, and replaced them with infantry divisions along the main road to Luga. The two Panzer divisions, followed by the 36th Motorized Infantry Division, then moved off to the north, on 13th July, over difficult roadless terrain.

In a forced march of 90 to 110 miles the three motorized divisions struggled painfully forward, in some places dangerously extended and in others crowded together on a single boggy road, struggling hard to keep up with their vanguards. The small bridges collapsed. The road became a swamp. Sappers had to build wooden causeways. Reconnaissance detachments and covering groups of motor-cyclists, Panzerjägers, and forward batteries scrambled through the mud along the flanks in order to take up covering positions in the most exposed places or to ward off repeated enemy attacks mounted from out of the vast marshes. But the risky manœuvre succeeded. The spearhead of 6th Panzer Division—the advanced detachment of 4th Rifle Regiment reinforced by armour and artillery under the command of Colonel Raus—took Porechye on 14th July. The two bridges fell undamaged into the hands of a special detachment of the "Brandenburg" Regiment —so much was the enemy taken by surprise.

On the same day 1st Panzer Division reached the Luga river at Zabsk with the reinforced Armoured Infantry Carrier Battalion of

113th Rifle Regiment under Major Eckinger, and by 2200 hours had established a bridgehead on the eastern bank against enemy opposition. The ford was extended, and during that same night the bridgehead was enlarged and the bulk of the 113th Rifle Regiment was brought up. In this way 1st Panzer Division succeeded in holding

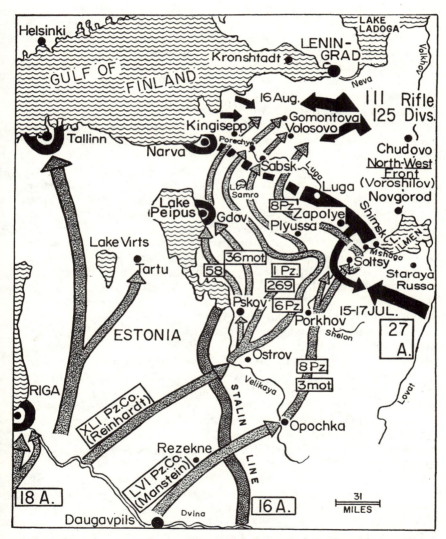

Map 10. The operations of Army Group North from the end of June to the middle of August 1941. The Stalin Line had been pierced. Hoepner's Panzer Group was striking towards Leningrad across the lower Luga.

Zabsk with the combat group Krüger against fierce enemy counter-attacks throughout 15th July. The bridge, however, had been destroyed. But on the following day the bridgehead was further consolidated. The enemy grouping on the western flank of the 4th Panzer Group, on Lake Peipus near Gdov, was smashed by the 36th Motorized Infantry Division and the 58th Infantry Division.

The obstacle of the lower Luga was overcome. A springboard for the final assault had been established 70 miles from Leningrad. In two extensive bridgeheads the riflemen and armour of Reinhardt's Corps were standing by for the assault. The Soviets had been taken completely by surprise by this operation. At first they had no forces of any importance opposite the new German front line. With hurriedly collected formations, including officer cadets from Leningrad, they tried in vain to clear up the bridgeheads. However, the German troops succeeded not only in repelling all attacks, in savage fighting, but, indeed, in extending their jumping-off positions and in improving their supply roads. Thus they awaited the order to resume their attack. Leningrad lay before them, unprotected, only two days' march away.

But now the same tragedy occurred on the northern front, before Leningrad, that we witnessed at Army Group Centre after the swift capture of Smolensk. The German High Command held back Hoepner's tanks in the Luga bridgeheads for three weeks. Three long weeks. Why? Why was not the focus of the offensive formed on this sector? Why was advantage not taken of the chance that offered itself? Once again the High Command bureaucracy frustrated a swift and very probably successful blow against a major objective.

Hitler and the Wehrmacht High Command had made up their minds about having the main weight of the operation on the right—in other words, Leningrad was to be taken by a wide outflanking attack from the south-east. The flank cover for the operation presumably was to be provided by Sixteenth Army coming up from the west; for the time being the gap between it and 4th Panzer Group was covered by the two divisions of LVI Panzer Corps alone.

In this way the Russian divisions streaming back from the Baltic countries were to be caught in a huge arc whose flank would be ideally protected by the marshy river Volkhov. It was a good plan. But it contained one important mistake: because of the wooded and swampy ground on the right of the offensive the tanks could not be used to full advantage. After all, that was why Hoepner had switched his XLI Corps to the left. Some strong infantry divisions, artillery, and air force units remained on the right wing, but manœuvrable armoured forces were lacking because the 8th Panzer Division and the 3rd Motorized Infantry Division were tied down between 15th and 19th July in bitter fighting with strong formations of three to five Soviet Corps.

The newly created focus of attack on the left, on the lower Luga,

Chase through the Baltic Countries 229

on the other hand, had armour, bridgeheads, jumping-off positions, and no enemy in front of it—but it lacked infantry divisions to cover an extended armoured thrust towards Leningrad. Hoepner tried everything to get Manstein's Corps to the north to make up for the infantry he himself lacked and which it would take too long to bring up from the rear. But Army Group would not or could not stand up to the Fuehrer's headquarters. There it was held that Reinhardt's forces were too weak to make the attack on Leningrad by themselves. Further reinforcements were therefore sent to the right wing of the offensive, to Lake Ilmen, where fighting continued under great difficulties.

Why, Colonel-General Reinhardt rightly asks to-day, should it have been impossible to switch Manstein's Corps to his wing? Would it not have been more correct to transfer the main weight of the offensive to the left, to block the narrow passage at Narva as quickly as possible, and then to wheel east and strike the enemy, who was still holding out along the middle Luga, in the rear with strong forces?

When Guderian found himself in a similar situation on the Dnieper before Smolensk Field-Marshals von Kluge and von Bock let him have his own way. Probably, if Kluge or Bock had been in Leeb's place they would have allowed Reinhardt to move off too. But Leeb was no von Bock. Admittedly, he toyed with the idea of giving Hoepner the green light, and he tried to get the High Command directive "main weight on the right" rescinded—but, in fact, he neither did the former nor achieved the latter. Thus a fatal tug-of-war ensued and continued for weeks on end—weeks which the Russians used to scrape together what forces they could and concentrate them opposite Reinhardt's bridgeheads on the Luga. A workers' division appeared on the front. Two further divisions, the 111th and units of the 125th Rifle Divisions, were brought up by rail. The trains moved unconcernedly within sight of the German troops and unloaded the reinforcements along the track. Finally several armoured formations appeared on the scene with heavy KV-1s and KV-2s.

Some of these brand-new super tanks were still manned by their civilian test crews from the factories. Among the infantry in their wake was an entire works brigade of women—students of Leningrad University. Women were also found dead or wounded inside shot-up tanks.

The increasing enemy opposition around the bridgehead perimeters was reflected also in the air. There were no German bomber or fighter formations to oppose the Soviet air attacks; the German machines were in the Lake Ilmen area, in accordance with the "main weight on the right" directive. Only the Trautloft fighter group occasionally intervened in the fighting on the Luga, with one or two flights of ME-109s, from their forward airstrip west of Plyusa.

This Russian superiority in the air gave rise to a bitter humour among Reinhardt's formations, which found expression in little messages in verse sent to division HQ and thence to Corps, asking for air support. But all the higher commands could do was radio back more rhymed couplets.

There was no doubt that the Soviets had gained time to reinforce what used to be the weakest points in the Leningrad defences. The chance of capturing the city in one fell swoop from the north-west had been lost. Colonel-General Reinhardt later remarked, "That the offensive could not be continued immediately was obvious. The road system would first have had to be improved to ensure supplies and the movement of reinforcements. That would have taken several days." Several days, certainly, but not three weeks. Bitterly Reinhardt continued: "Time and again our Corps urged a speedy resumption of the attack and asked that some units at least of Manstein's Corps should be switched over to us, especially as they were bogged down where they stood. But all was in vain."

General Reinhardt's diary shows the following entry under 30th July, when he had been waiting for the resumption of the attack for a whole fortnight: "More delays. It's terrible. The chance that we opened up has been missed for good, and things are getting more difficult all the time."

Events were to prove Reinhardt right. While XLI Corps, favoured by good fortune, had crossed the lower Luga, but was pinned down by orders from above, a crisis was brewing up in the eastern sector of the Panzer Group, at Manstein's LVI Corps. Manstein's orders were to capture Novgorod and then to tackle the important traffic junction of Chudovo in order to cut the road and railway from Leningrad to Moscow.

The 8th Panzer Division had pushed forward beyond Soltsy to form a bridgehead over the Mshaga. The 3rd Motorized Infantry Division had moved up on its left, covering the flank of 8th Panzer Division and fighting its way forward to the north-east and north. Enemy opposition, however, was getting stronger and stronger, and the marshy ground here too was getting less and less negotiable. Moreover, the shunting away of XLI Corps from Luga had released Soviet forces in that area, with the result that Manstein's Corps, which had run well ahead of the general line, although consisting only of 8th Panzer Division and 3rd Motorized Infantry Division, without any reserves and without flank cover, suddenly found itself under attack by numerous divisions of the Soviet Eleventh Army. Voroshilov hurled himself with all available forces against the dangerous German armoured spearhead which was aimed at Novgorod, his command post, and at Chudovo, a vital traffic junction. The Soviet 146th Rifle Division succeeded in making a penetration between the two German divisions and in cutting off their supply route. Manstein instantly made the correct counter-move: he withdrew 8th Panzer Division and prepared for all-round defence.

Three critical days followed. Voroshilov needed a success and tried at all costs to annihilate the surrounded German divisions. He employed half a dozen rifle divisions, two armoured divisions, and strong artillery and air force units. But the steadfastness of the German formations and Manstein's superior generalship prevented a catastrophe. The fierceness of the fighting is attested by the operations report of 3rd Motorized Infantry Division, which had to repel seventeen enemy attacks in a single day. Even the artillery was fighting in the foremost line.

The 1st Battery, 3rd Artillery Regiment, under First Lieutenant von Tippelskirch was able to survive a massed enemy attack. In a forest clearing, two miles behind the foremost infantry line near Gorodishche, the battery was in position. Impassable swamp lay to the right and left of the road. Was it impassable also for the Soviets?

To protect themselves against surprise attack from the swamp the artillerymen had put out sentries and pickets on rapidly made wooden paths, and that was what saved them. For Voroshilov got some locals to guide a newly equipped battalion of his 3rd Armoured Division through the swamp with a view to cutting off the spearheads of the German division. On 15th July the battalion encountered the German pickets. The pickets raised the alarm. The Russians evidently thought they were dealing with an infantry unit and attacked overhastily, without identifying the position of the heavy battery. With shouts of "Urra" the Soviets charged. Machine-guns out in the swamp gave them covering fire. The artillerymen leapt to their guns. The crew of No. 2 gun was mown down by machine-gun fire as they sprang from their dug-outs. The battery officer, Second Lieutenant Hederich, worked his way over to the gun with his troop leaders and manned it himself. The Russians had got within 300 yards. "Fire!"

At point-blank range the 10-cm. shells slammed into the charging ranks. The battery machine-gun raked the attackers. The first wave collapsed on the edge of the clearing. But now the Russians got heavy machine-guns into position. The gun shields were riddled. Mortar shells put the German battery's machine-gun out of action. A dozen Soviet troops got within ten yards of Hederich's gun, leapt to their feet, and charged. Hederich and his men resisted with spades, pistols, and bayonets. Four Russians were killed. Three or four disappeared into the scrub. Lieutenant Hederich and the entire gun crew were wounded. The fighting continued for two hours. Nearly all the ammunition was spent. Most of the officers and NCOs had been killed or wounded, and tractor-drivers and other general service personnel were roped in for combat duty. A mere 120 men were fighting against an entire battalion. At the last minute the battery commander arrived on the scene with a motor-cycle platoon of 8th Infantry Regiment and launched an outflanking attack from the right. This confused the Russians. They withdrew, taking with them some of their wounded, but leaving behind their heavy equipment and fifty dead.

After 4th Panzer Group command had again placed the "Death's Head" SS Division at General von Manstein's disposal the LVI Panzer Corps succeeded in overcoming its critical situation by 18th July and in clearing the Corps' supply route.

The danger had passed by 18th July, but Manstein took the opportunity to urge Army Group, and through General Paulus the High Command, to bring the two Corps of the Panzer Group together again at long last and use them jointly as the strong-point of the coming offensive. It was not a question of Manstein pleading his own case, but of recommending that the bridgeheads established by Reinhardt's Panzer Corps should be made the starting-point for the assault on Leningrad.

But Manstein did not succeed either. Army Group and the High Command insisted on having the main weight of the attack on the right. All they were prepared to do was to detach Manstein's Corps from the Mshaga front and to employ it instead on the middle Luga opposite the important town of Luga. In the impending general offensive it would be Manstein's task to gain the main road at Luga, to destroy the enemy and then drive towards Leningrad.

It was an incomprehensible plan. For weeks the strength of the enemy's fortifications in the Luga area had been well known. And although the ground had proved to be almost entirely unsuitable for armour, it nevertheless remained a mystery why the LVI Panzer Corps employed as the southern striking force was assigned merely the 3rd Motorized Infantry Division, the 269th Infantry Division, and the newly brought up SS Police Division, while the "Death's Head" SS Division was kept back at Lake Ilmen and the 8th Panzer Division was sent to hunt partisans in the rearward areas.

The attack began on 8th August. At 0900 hours, in pouring rain, Reinhardt's divisions moved off from the Luga bridgeheads, but because of the bad weather they had no air support. The two Panzer divisions and the 36th Motorized Infantry Division were to occupy the open ground south of the Leningrad–Kingisepp–Narva railway-line by a swift thrust. The 8th Panzer Division and the bulk of 36th Motorized Infantry Division were then to be brought forward, and the entire force was to wheel eastward beyond the railway-line and strike towards Leningrad. It was a good plan.

But where three weeks earlier there had been only weak Soviet field pickets, there were now the reinforced 125th and 111th Soviet Rifle Divisions in solidly built field fortifications constructed by tens of thousands of civilians—women, children, and members of the Party's youth organizations—in ceaseless round-the-clock work.

Facing the Porechye bridgehead was a Soviet combat unit with extremely strong artillery; according to interrogated prisoners this force had likewise planned to attack the bridgehead on 8th August. However, 6th Panzer Division got their blow in first. In this way they prevented what might have been a disastrous setback to the German offensive. Things were bad enough as they were. After the first day of fighting Corps seriously considered whether, in view of the casualties suffered, the offensive could be maintained. It was maintained only because of the optimistic appraisal of the situation by 1st Panzer Division. Lieutenant-Colonel Went von Wietersheim, commanding a combat unit, in particular was most reluctant to give up his hard-won ground. The optimism of Lieutenant-Colonel von Wietersheim and Lieutenant-Colonel Wenck, the Chief of Operations of 1st Panzer Division, proved justified. On the following morning the regiments made good progress, broke through the enemy line, brought some relief to 6th Panzer Division in its difficult attack from its bridgehead towards Opolye, and pierced the 30-mile-deep belt of forest south of the Leningrad railway, the last natural obstacle before the metropolis on the Baltic Sea.

The fighting continued. On 14th August all the divisions had gained the favourable open ground beyond the swampy forests. The enemy had been defeated. Only minor formations were now encountered. The battlefield was dotted with dozens of brand-new super-heavy Soviet tanks.

The road to Leningrad was once more clear. Only on the left flank was there still some threat from enemy forces withdrawing from Estonia in the direction of Leningrad. That was why Reinhardt did not advance right up to the edge of the city, although, as far as frontal opposition was concerned, he could have done so.

What then was needed? "We've got to have some forces to cover our flank," Hoepner requested, implored, and threatened. "Two divisions—even one division at a pinch—would be enough," he pleaded with Field-Marshal Ritter von Leeb. Hoepner was in a very similar situation to Guderian five weeks earlier, when he extorted from Kluge permission to continue his thrust from the Berezina over the Dnieper to Smolensk: "You're throwing away our victory if you don't let me go ahead," Guderian had implored Kluge. "You are throwing away our victory," was what Hoepner might have said to Field-Marshal Leeb.

On 15th August Leeb arrived in person at Hoepner's headquarters. After a heated discussion the Field-Marshal agreed to detach the experienced and combat-hardened 3rd Motorized Infantry Division from Manstein's Panzer Corps and to place it under Reinhardt's command.

This division could well be spared at Luga. Although, as planned, Manstein had also mounted his offensive on 10th August, with the object of capturing Luga, the inevitable happened: he was halted in front of the strong Russian defensive lines. The 3rd Motorized Infantry Division, scheduled to cover the Corps' flank at a later stage, had thus not gone into action at all by then. It was now decided to move Manstein's headquarters also to the north, into Reinhardt's zone of operations.

Leeb's decision triggered off a mood of victory at Hoepner's headquarters. "Leningrad can't escape us now," the officers were saying to each other. There was much relief also at Manstein's headquarters: the period of piecemeal moves seemed to be over at last, and the Panzer Group would now go into action again as a massive force.

On 15th August Manstein handed over his command at Luga to General Lindemann's L Corps. He then climbed into his command car with his officers and drove off—to Lake Samro, where Hoepner also had his headquarters. The road was frightful, full of potholes and deep sand, so that the 125-mile journey took them eight hours. Covered with dust, Manstein and his staff arrived late in the evening.

"On with your swimming-trunks, gentlemen, and into the lake!" he ordered. But at that moment a runner came racing up from the communications van. "A call from Panzer Group, Herr General!"

Manstein frowned. The runner apologized. "It's very urgent, Herr General; the Commander-in-Chief is on the line in person." Quickly Manstein strode over to the field telephone.

2. Break-through on the Luga Front

Critical situation at Staraya Russa–The battle of Novgorod–A Karelian supplies Russian maps–German 21st Infantry Division against Soviet 21st Armoured Division–Through the forests near Luga–On the Oredezh–The Luga pocket–On top of the Duderhof Hills–Radio signal from Second Lieutenant Darius: I can see St Petersburg and the sea.

THE sun was setting behind Lake Samro in a blood-red western sky. General von Manstein arrived at the communications van. The radio operator held out the telephone receiver to him. "The Herr Generaloberst is on the line, sir."

"Manstein," the General said.

"Hoepner here," the voice came over the line. "I have bad news, Manstein. Our attack on Leningrad is off. A serious crisis has developed for Sixteenth Army on Lake Ilmen, in the Staraya Russa area. You'll have to act as fire brigade. You will halt your 3rd Motorized Infantry Division at once and make it turn about. Move off again to the south. The 'Death's Head' SS Division is being switched over to you additionally from XXVIII Corps from the Luga front. As for yourself, you will drive over with your headquarters to Sixteenth Army headquarters at Dno first thing to-morrow morning. Any further instructions you will get there from Colonel-General Busch."

Manstein was not too pleased. Hoepner sensed the disappointment of his Corps commander. "Field-Marshal Leeb wouldn't stop our advance on Leningrad unless the situation was pretty serious," Hoepner said. "Anyway, best of luck, Manstein—hope you'll soon get back north again!"

It was to prove a vain hope.

When Manstein informed his staff of the new order there were long faces. Was it conceivable? A moment ago they had all been talking about the inevitable fall of Leningrad. And now this! "Everything into reverse," groaned Major Kleinschmidt, the Quartermaster, and started reorganizing the Corps' transport and supplies from scratch.

On the following evening, 16th August, Manstein arrived in Dno, at Sixteenth Army headquarters. This time the 160-mile journey took him thirteen hours.

The situation he found there was, as he himself put it in blunt Army language, "shitty."

A fortnight earlier, at the beginning of August, X Corps, with its three divisions—the 126th, 30th, and 290th Infantry Divisions—had started its attack against the important transport centre of Staraya Russa, south of Lake Ilmen.

The experienced 30th Infantry Division from Holstein had broken into the strong defences nine miles outside the town, but in spite of desperate efforts the 6th and 26th Infantry Regiments were unable to pierce the deeply echeloned system of defences. The regiments of the 290th Infantry Division from Lower Saxony likewise got stuck in front of and inside the wide anti-tank ditch which formed the backbone of the Russian defences.

Young workers from Leningrad who had never seen action before, together with experienced units of the Soviet Eleventh Army, offered stubborn resistance at close quarters. Every foot of ground had to be fought for with rifle-butt, spade, pistol, and flame-thrower. Buried Soviet tanks, enfilading machine-guns, and very heavy shelling eventually brought the German attack to a halt.

A nasty surprise also was the wooden mines encountered here for the first time. Electrical mine detectors did not react to them. In some places the German sappers had to clear as many as 1500 of these dangerous contraptions.

The 126th Infantry Division from Rhineland-Westphalia, operating in the north of the attacking front, along the road from Shimsk to Staraya Russa, was luckier than the 30th and 290th Divisions. After three days of fierce fighting its regiments penetrated the Soviet defences with infantry combat groups made more mobile by the inclusion of Panzerjägers, artillery, sappers, and cyclists. An immediate Russian counter-attack with tanks was repulsed in the sector of 426th Infantry Regiment by Second Lieutenant Fahrenberg's 12th heavy machine-gun company, whose men tackled the enemy armour with demolition charges.

When, after the deep penetration made by 126th Infantry Division, the 30th Infantry Division mounted an attack from the flank the Russians withdrew from their last positions before the town.

At the head of 3rd Battalion, 426th Infantry Regiment, Major Bunzel charged into the western part of Staraya Russa towards noon on 6th August. The penetration was made so unexpectedly that the chief of operations of the Soviet Eleventh Army was wounded and captured.

Following a heavy air attack on the strongly fortified eastern part of the town, beyond the Polstiy river, where every house had been turned into a fortress, the regiment succeeded in penetrating as far as the eastern outskirts. The Russians were still resisting, making immediate counter-attacks and engaging the Germans in savage hand-to-hand fighting in the blazing streets.

During the next four days of continuous fighting against furiously resisting Soviet forces the Lovat river was reached on a broad front. Thus the right flank of Army Group North seemed adequately covered for the attack on Leningrad.

But Marshal Voroshilov, the C-in-C of the Soviet North-west sector, had realized the significance of the German operation. Using all available forces, including units of his newly brought up Thirty-fourth Army, he launched an attack on 12th August against the funnel between Lake Ilmen and Lake Seliger, where the town of Demyansk was situated. This funnel, which positively invited attack by the Russians, had been formed by the diverging directions of the operations of Army Group North and Army Group Centre—the one towards Leningrad and the other towards Moscow. With numerically vastly superior forces—eight rifle divisions, one cavalry Corps, and one armoured Corps—the Soviet Thirty-fourth Army launched an outflanking attack against the three divisions of the German X Corps and threatened to push them back into Lake Ilmen.

Voroshilov, moreover, intended, after the elimination of X Corps, to drive on to the west, block the neck of land between Lakes Ilmen and Peipus, and thus cut off the German armies operating against Leningrad from their rearward communications. It was a highly critical situation that Manstein had been sent to cope with. But cope with it he did.

While General Hansen with his X Corps was holding out in heavy defensive fighting, facing southward, with Lake Ilmen at his back, Manstein led his two fast divisions, unnoticed by the enemy, into the exposed flank and rear of the Soviet Thirty-fourth Army.

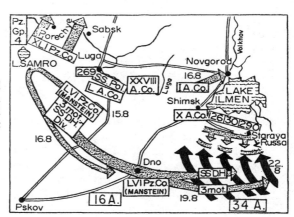

Map 11. 15th–23rd August 1941: Manstein saved X Corps and smashed the Soviet Thirty-fourth Army.

Like a thunderstorm the 3rd Motorized Infantry Division and the "Death's Head" SS Division struck at the Russians on 19th August. They rolled up the Army's flank and shattered its rearward communications. Among the most advanced units of LVI Panzer Corps the reconnaissance battalion of the "Death's Head" Division, which had raced a long way ahead of the bulk of the division, arrived in the most critical sector and with its motor-cyclists dislodged the enemy. They pressed on at once and forced the Soviet spearheads back across the Lovat. The commander of the bold reconnaissance battalion, Sturmbannführer[1] Bestmann, who was subsequently killed in action, was the first member of the "Death's Head" SS Division to win the Knights Cross.

At that moment, just as the Soviet Command was paralysed by shock and surprise, the regiments of X Corps launched their attack. This completed the disaster to Voroshilov's Thirty-fourth Army. It was smashed.

The vast booty of 246 guns included also the first intact multiple mortar, the dreaded "Stalin's organ-pipes," as well as a brand-new 8·8-cm. anti-aircraft battery of German manufacture, dated 1941. Where had it come from? Once before, in Daugavpils, a considerable

[1] Rank in Waffen SS equivalent to Army major.

amount of military equipment of German manufacture had been found in a Soviet Army depot. How these German weapons came into Soviet possession has never been established. The German troops had their own ideas.

The success of Sixteenth Army meant that the threat to the right flank of Army Group North was averted for the time being. But there could be no question of Manstein's Panzer Corps returning to Leningrad to rejoin Hoepner's offensive forces, for Voroshilov did not give up trying. He brought up three more Soviet Armies in order to reach his operational objective—blocking off the neck of land between Lakes Peipus and Ilmen. It was another alarming illustration of Russian resources. The bulk of one Army had just been annihilated, yet units of three new Armies, reinforced to full strength, were being employed at the focal point of the defensive fighting between Luga and Lake Ilmen.

And what had happened meanwhile outside the much contested town of Novgorod "the Golden," situated on the northern shore of Lake Ilmen, exactly opposite Staraya Russa?

There, at the original focal point of the German offensive against Leningrad, at the southern cornerstone of Leningrad's defences, the German Command had been trying for weeks to pierce the Soviet lines in order to reach Chudovo, a railway junction on the Leningrad–Moscow line. At Chudovo the Murmansk railway, coming down from the Arctic Sea, ran into the so-called October Railway. Along this lifeline came the supplies and aid shipped by the Western allies to Murmansk, the supplies of British and, even more, of American tanks, lorries, foodstuffs, ammunition, and aircraft for the entire Soviet front from the Baltic to the Black Sea.

During the night of 9th August, a clear, starry summer night, the divisions of I Corps from East Prussia silently moved into their jumping-off positions for the offensive across the wide, marshy Mshaga river. The cornerstone of Leningrad's defences was to be overturned at last.

The main weight of the attack was borne by General Sponheimer's 21st Infantry Division, which, reinforced by 424th Infantry Regiment, 126th Infantry Division, was to advance along the strongly fortified main road towards Novgorod. The ground was tricky even for infantry. Swamps, thick undergrowth, and numerous streams and river-courses made movement difficult. The Russians, moreover, had developed the whole area into a fortress: there were pillboxes, minefields, machine-gun nests, and mortar positions blocking what few roads and paths led through the swampy ground.

In the grey light of dawn formations of VIII Air Corps had set out from their bases and had been dropping their bombs since 0400 hours on the enemy positions on the far bank of the Mshaga. Stukas made screaming low-level attacks, skimming across the river at barely 150 feet, dropping their bombs on dug-outs, gun positions, and machine-gun posts.

The military machine was working with great precision. No sooner had the last bomb been dropped than 200 guns of all calibres opened up. It was a classic preparation for an attack.

At 0430 hours exactly the company commanders of 2nd and 3rd Battalions, 3rd Infantry Regiment, as well as 1st Battalion, 45th Infantry Regiment, leapt out from their hide-outs. The men dragged inflated dinghies to the river-bank and, under cover of the artillery umbrella, ferried themselves across. Together with the infantry the sappers also crossed the Mshaga, and on the far bank cleared lanes through the minefields for the assault detachments following hard on their heels.

To start with everything went surprisingly smoothly. The enemy seemed to have been utterly shattered by the preliminary aerial and artillery bombardment. His heavy weapons and artillery were silent.

Ducking low, the assault detachments ran along the white tapes with which the sappers had marked out the cleared lanes through the minefields. The bridgehead was secured. The first heavy weapons were ferried across the river. Then the barges were linked to form a bridge. By twelve noon it was ready. The division moved into the bridgehead.

The 24th Infantry Regiment was now also brought forward. Slowly the enemy recovered from his shock. Resistance was getting stiffer. In the late afternoon the 24th Infantry Regiment took the village of Mshaga. By nightfall the Soviet defences had been pierced to a depth of five miles. The following day Shimsk, at first to be bypassed, fell to the Germans.

On 12th August the Ushnitsa river was forced by a frontal attack. The infantrymen were weighed down by their weapons and ammunition-boxes. Everything had to be carried. The Russians were resisting stubbornly. Along the railway embankment especially they contested every inch of ground.

The Soviet soldiers continued to fire until they were killed in their foxholes or blown up by hand-grenades. In the face of such opposition how was progress possible? Furious battles were waged for every inch of ground.

The regimental headquarters of 45th Infantry Regiment was in a roadside ditch before Volinov. The mood was despondent. Reports of casualties were shattering. Colonel Chill, the regimental commander, used the field telephone, which had been laid right up to that point, to speak to division. "The Stukas must go in once more," he implored his superiors.

Just then a runner jumped down into the ditch—Lance-corporal Willumeit. Somewhat out of breath, he saluted the regimental commander. "Message from 2nd Battalion, sir: Lieutenant-Colonel

Matussik sends this captured enemy map. It was taken from a Soviet major killed in action. Evidently he was ADC to a senior commander."

Colonel Chill cast one glance at the map and looked up in amazement. "My friend, for that you shall have my last cigar but one," he said to the runner, pulling out his cigar-case.

Willumeit beamed, accepted the cigar, and said, "I shall take the liberty of swapping it, Herr Oberst—I don't smoke." Everyone joined in the laughter.

The map was a precious find. It showed the Soviet Forty-eighth Army's entire position along the Verenda, until then unknown, complete with all strong-points, dummy positions, gun emplacements, and machine-gun posts.

It was largely due to this captured map that on the following day these positions were pierced in a bold action. That is how fate—or, if you prefer it, blind luck—takes a hand in battle. That was what Frederick the Great, the King of Prussia, meant when he said, "Generals must not only be brave, they must also have *la fortune*."

General Sponheimer could not complain of any lack of *la fortune* before Novgorod. In addition to the captured map, Fortune—again in the form of the 45th Infantry Regiment—sent him a priceless prisoner. He had been found with a column of Soviet supply lorries by a bicycle reconnaissance detachment. He was a sapper officer from the staff of the Soviet 128th Rifle Division—a man from Karelia, Finnish by birth, and with no love for the Bolsheviks.

"Nix Bolshevik," he kept assuring the German second lieutenant. Shortly afterwards, when an interpreter had been fetched, an amazing sequence of events began. "I know all the fortifications," said the Karelian. "The papers are hidden in the forest," he added slyly.

"You trying to pull our leg?" the second lieutenant asked.

The Karelian raised three fingers. "I swear by my mother!"

The lieutenant threatened him with his pistol. "Don't try anything funny—an ambush or something of that kind! Or you'd better start praying."

The interpreter translated. The Karelian nodded. "Let's go then," the lieutenant decided. He himself led his platoon into the near-by forest, cautiously, covering the Karelian all the time. The Karelian did not have to search long. In a thick clump of shrubs, underneath a large boulder, was his sailcloth bag—a big parcel. It contained all the fortification maps of Novgorod as well as the plans of the minefields.

The lieutenant took the packet, complete with the Karelian, straight to the divisional Intelligence officer. The Intelligence officer grabbed it and raced across to the chief of operations, Major von der Chevallerie. The major was almost beside himself with delight. The maps clearly showed the entire defences outside Novgorod, including the defences of the city itself and the fortifications on the small island in the Volkhov between the two main parts of the city.

After that it was not difficult to pierce the Russian positions at the crucial points and to get to the edge of the city itself without too many casualties.

On the morning of 15th August the 3rd Infantry Regiment saw the famous "Novgorod the Golden" spread out in front of them in the morning sun. Novgorod—one of the most ancient Russian settlements, founded by Rurik the Conqueror as his residence in the ninth century, administered in the Middle Ages in accordance with Lübeck city law, depopulated several times by black death and cholera, always rising anew from its ashes. Novgorod, known as "the Golden" because of its important and profitable fur and salt trade with the Hanseatic cities of Germany. Because of its wealth the city was twice sacked completely within a century, by Ivan III and Ivan the Terrible, and its citizens deported or slaughtered. Forty-seven magnificent churches with fine old frescoes surrounded the Kremlin of Novgorod which commanded the bridges over the Volkhov. A proud city, never conquered. Throughout its thousand-year history Novgorod had never, until 1941, been occupied by a foreign enemy, apart from a very brief episode in the Nordic War at the beginning of the seventeenth century. But now Russia's golden city was about to suffer that humiliation.

On 15th August 1941 the 21st Infantry Division from East Prussia intercepted a signal from Moscow to the Soviet Forty-eighth Army. It ran: "Novgorod is to be defended to the last man." As chance would have it, it was the Soviet 21st Armoured Division which was to defend Novgorod to the last man, against the attack of the German 21st Infantry Division.

At 1730 hours on 15th August VIII Air Corps began a heavy air raid on the Russian positions along the city's battlements, and kept it up for twenty minutes. Novgorod stood in flames. The three infantry regiments of 21st Infantry Division lined up for the assault. From the edge of the ancient moat came the stutter of machine-guns, the crash of guns, and the plop of mortars.

To be held to the last man! "To the last man," repeated the commissars. With their pistols drawn they stood at their posts until death relieved them of their duty.

At first light on 16th August the German assault companies were inside the blazing city. At 0700 the 1st Battalion, 424th Infantry Regiment, of 126th Infantry Division—for this attack under the command of 21st Infantry Division—hoisted the swastika over Novgorod's Kremlin.

But there was no time for victory celebrations. The objective was Chudovo and the October Railway.

"Keep going," Major von Glasow, commander of the reconnaissance detachment and now leading the hurriedly formed vanguard of 21st Infantry Division, urged his men. The men of the bicycle companies of 24th and 45th Regiments pedalled for all they were worth. The cavalry squadrons moved off at a trot, followed by the motorized platoon of Panzerjägers and by heavy motorized batteries of 2nd Battalion, 37th Artillery Regiment. There were no tanks at all, and only a few self-propelled guns of Assault Gun Battery 666. The brunt of the fighting was borne by 37th Artillery Regiment, as well as the heavy artillery battalions, Mortar Battalion 9, and Army AA Battalion 272, all of them grouped under Artillery Commander 123.

In that way the companies of 45th Infantry Regiment made their assault. On 20th August, towards noon, Sergeant Fege with his platoon rushed the road bridge leading over the Kerest stream towards Chudovo from the south-east and seized it by a surprise coup. Second Lieutenant Kahle occupied the railway bridge over the Kerest before the Soviet bridge guard was able to touch off the demolition charge.

Meanwhile the 24th Regiment took the bridge which carried the October Railway. They captured it intact. And that was not all. That day seemed an unending string of lucky incidents. Lieutenant-Colonel Matussik with his 2nd Battalion, 45th Infantry Regiment, with great presence of mind seized the chance to drive on towards the east. There lay the huge railway bridge over the Volkhov, the line to Moscow.

In a captured lorry Matussik drove right up to the bridge. There was no guard. On and across! The battalion raced over to the other side of the river. It was shortly to become a fateful river for Army Group North.

Carl von Clausewitz, the great preceptor of the Prussian General Staff, never ceased to impress upon his disciples that a well-prepared strategic plan should be departed from only in quite exceptional circumstances. But should such a departure really become necessary, then it must be made without hesitation, radically and resolutely.

At Luga, where an insuperable Soviet defensive force had been blocking the vital main road from Daugavpils to Leningrad ever since mid-July, the German High Command followed Clausewitz's advice neither in its former nor in its latter injunction.

The original plans of the High Command envisaged the main drive towards Leningrad to be carried out along both sides of this road, which, being the only paved highway in the area, was then to serve as a supply-line. Presently, however, Colonel-General Hoepner detached Reinhardt's Corps, as already related. And later still the bulk of von Manstein's LVI Panzer Corps had to be turned round and switched to the east, to Staraya Russa. Since then the battle for the

Break-through on the Luga Front 243

town of Luga had been waged only by XXVIII Army Corps with the SS Police Division and 269th Infantry Division.

A frontal attack by these two divisions against the heavily fortified Luga bridgehead, defended by five Soviet divisions, yielded no success to begin with, in spite of hard fighting and heavy losses. Fighting in the forests and swampy river valley was tricky and costly. The SS Police Division alone lost over 2000 killed and wounded. Even though, strategically speaking, the Luga position had been outmanoeuvred by the fall of Novgorod and Chudovo, the Russians nevertheless clung to their strategically worthless position.

The German Command, on the other hand, urgently needed the highway, chiefly in order to improve supplies for the northern sector. Sixteenth Army was therefore to attempt to take the strongly fortified town of Luga by a tactical outflanking move. The job was assigned to XXVIII Corps under General Wiktorin. On 13th August the Corps mounted its attack across the Luga east of the town with 122nd Infantry Division, which had meanwhile been brought up to the line.

The following incident is reported in an account of the division's attack. Private Lothar Mallach, a reserve officer aspirant of 1st Company, 410th Infantry Regiment, ran across a forest clearing with the men of his No. 1 Platoon. They came under fire from all sides. The Russians sat in well-camouflaged foxholes and opened fire only after the German infantrymen had passed them. The Russian foxholes were virtually invisible until the men were within a yard of them. They advanced with the sickening knowledge that they might be picked off from behind at any moment.

"Look out!" shouted Sergeant Pawendenat. He flung himself behind a tree-trunk and opened up with his captured Soviet machine pistol. Less than ten feet from him a Russian had fired from a foxhole.

Sergeant Tödt, leading the 1st Company because the company commander, First Lieutenant Krämer, had taken over the Battalion, was waiting behind a woodpile, directing the fire of his machine-guns at the Russian foxholes. From the far right-hand corner of the clearing came the intermittent muzzle-flashes of an automatic Russian rifle.

"Where the hell is that bastard?" Tödt grunted. He was fuming with anger. Behind him Corporal Schmidt was holding Lance-corporal Braun, the machine-gunner of 2nd Section, trying to comfort him. The lance-corporal was writhing in agony: he had been shot through his thigh and abdomen by the invisible Russian sniper to the right of the clearing.

There was another flash from the same spot. Then three more. But this time Lance-corporal Hans Müller, the gun's No. 2, who had taken over the machine-gun, had been watching intently. He opened up with his machine-gun. At the very spot where the flashes had come from the moss was torn to shreds, branches splintered, and a Russian steel helmet spun through the air. There were no more bullets from that quarter.

Sergeant Tödt ordered his company to rally. The men waited another minute. Lance-corporal Braun, the machine-gunner of 2nd Section, died in Schmidt's arms. They wrapped him in a tarpaulin. Three men gave a hand. They must move on now. They would bury him in the evening.

Panting heavily, the troops dragged their ammunition-boxes with them. Under cover of a German heavy field howitzer battery they worked their way forward into the ruins of an old Schnapps distillery.

"Look out—Russian tanks!" a shout went up. "Anti-tank gun forward!"

The 3·7-cm. gun was brought up at the double, hauled by its crew, and manœuvred into position. Already the Russian tanks were on top of them. They were light armoured fighting vehicles—infantry support tanks of the T-26 and T-28 types. One of them started shelling the anti-tank gun. Its crew rolled under cover. The company scattered. The first tanks rumbled past.

At that moment Second Lieutenant Knaak, the Battalion Adjutant, raced forward through the undergrowth. He grabbed the carriage of the anti-tank gun and jerked it round. Aim! Fire! After the third round a T-26 was in flames.

His action was like a signal. The men of the company emerged from behind trees everywhere, clutching demolition charges and flinging them in front of the tracks of the Russian tanks. The machine-guns gave them cover. A second T-26 was immobilized. Up on top of it— open the turret hatch—shove in a hand-grenade. Crash! The third tank was in flames. Three more turned back. The Russian infantrymen fell back with them.

Firing the machine-gun from his hip, Corporal Schmidt with Sergeant Pawendenat charged across the road, after the retreating Russians. In this way the companies of 410th, 411th, and 409th Infantry Regiments forced their way across the Luga.

The villages of Chepino and Volok, the notorious railway embankment, the wrecked distillery, the swampy patches of woodland, and the old wooden hunting lodge of the Tsars right in the middle of the forest, which was reduced to ashes by heavy shellfire—all these were scenes of exceedingly heavy fighting for General Macholz and his 122nd Infantry Division.

During the next seven days the battalions fought their way forward to the last natural obstacle of their offensive—the Oredezh river, up to 500 yards wide in some places, between marshy banks. Once that river was crossed it would be possible to drive through to the great Leningrad highway far behind Luga, cut the highway, and take the strongpoint of Luga from the north. That then was the plan.

The first wave of the attack was to be provided by 1st Battalion,

409th Infantry Regiment. The idea was if possible to get across the river unnoticed, to take the village of Panikovo in a surprise move, and to roll up the Russian defences covering the highway.

In the garden of a fisherman's cottage Captain Reuter, the battalion commander, was sitting with his company commanders, discussing the operation. The ground was favourable. The German river-bank was higher than the northern bank held by the Russians. As a result, there was a good view of the ground across the river: a freshly dug anti-tank ditch ran from one edge of the wood to the other, in front of the village, but there was no indication of what happened inside the wood. Nor, of course, what lay behind it.

The German bank dropped down to the river fairly steeply. But there were shacks, gardens, sheds, and shrubs providing sufficient cover to approach the river unnoticed.

Nothing moved on the far bank. It was noon. It was a scorching day, and the air shimmered with the heat. Shortly before 1400 hours the sappers with their assault boats had reached their starting positions down by the river. Not a shot had yet been fired. A last glance at the watch. Another minute to go.

At 1400 exactly came a short blast on a whistle. The first groups leapt to their feet. Together with the sappers they pushed the boats into the water. With a whine the motors sprang into life. Like arrows the assault boats streaked across the river.

The machine-gunners of 1st and 2nd Companies, 409th Infantry Regiment, lay tense on the bank, their fingers on the triggers. The moment the first shot was fired at the boats from the far bank they would open up for all they were worth in order to keep the Russians down. But there was no shot.

Ten seconds had passed. The boats with the first four groups were moving across the river at speed. Thirty seconds. The next groups leapt into their boats and moved off. The assault sappers were standing by the tillers of their outboard motors, stripped to the waist. The rest of the men were crouching low, with only their steel helmets showing above the gunwales. Fifty seconds had passed. The first boat had another 30 yards to cover before reaching the bank.

In the crossing sector of 1st Company the first shot rang out. Everybody held his breath: surely hell would now be let loose and the boats be shot to pieces. But nothing happened. Desultory fire from a few carbines brought two rapid bursts from a German machine-gun. After that everything was quiet again. The Russian pickets vanished. But no doubt they would raise the alarm.

Strangely enough, nothing happened during the next half-hour. The battalion had crossed the river. Quickly patrols were formed. They reconnoitred as far as the edge of the wood and returned. "No enemy contact."

Were the Russians asleep? Let's go!

At 1515 hours the battalion began its drive through the forest of Panikovo.

There was sporadic harassing fire by light enemy guns. The interval between firing and shell-bursts was very brief. The officers pricked up their ears. These could be tanks. They could only hope for the best. They were indeed tanks.

Some 80 yards in front of the company, on the left wing, the whine of engines suddenly came from a nursery plantation of fir-trees. Bushes were flung aside. Crashing out over snapping young fir-trunks, three, four, five, six Russian tanks, light T-26s, struck at the deep flank of the German units, firing continuously. The worst thing that could happen to infantry. So that was why the Russians had lain silent. They had laid a trap—a deadly trap for the whole battalion.

The men of 2nd Company flung themselves under cover. Accompanying Russian infantry came bursting out of the wood with shouts of "Urra." Hand-grenades exploded. Fiery lines of tracer zoomed to and fro.

Zigzagging among the trees, the tanks tried to wipe out the German infantrymen who were hiding behind tree-trunks and in the thick undergrowth. It was like a hunt with beaters. Wherever a tank appeared the German troops dived or rolled behind trees and bushes. "Damn," they cursed.

They had every reason for cursing: the battalion did not have a single anti-tank gun with it. They had shunned the difficulties of manhandling the guns through swamp and forest. Now they had to pay for it. The T-26s were able to drive around unmolested.

To add to their misfortunes, both the battalion's transmitter and that of the artillery spotter attached to it were put out of action. There was nothing left for Captain Reuter but to order: "Form hedgehog and hold out!"

The Russian infantry attacked under cover of their tanks. Hand-to-hand fighting developed. But fortunately the Russians were weak and it was possible to hold them off. Only the tanks were driving around at will in the battle area.

If some competent Soviet commander had quickly supported his half-dozen tanks with major rifle formations the doom of Captain Reuter's 1st Battalion would have been sealed. But that Russian commander somewhere did not see his chance. And some German runner from the headquarters of Lieutenant Neitzel's 3rd Company somehow managed to get through to the battalions which had crossed the river farther east and report what was happening in the wood.

Thus, towards 1900 hours, as the German resistance was weakening, a metallic clank was heard through the forest. Again, and a third time. With a flash of flame a Soviet tank was flung aside. Another crash.

Break-through on the Luga Front 247

The old soldiers raised their heads out of cover. "Listen—7·5s! German tanks!"

And already the grey monsters were pushing their way through the undergrowth—self-propelled guns. The Russian tanks disappeared. As if to make up for past omissions, the remnants of the company rallied quickly and hurriedly followed the self-propelled guns, out of the forest, against the Russian positions which now lay clearly before them.

The following noon Panikovo fell. The road was open into the rear of the Soviet barrier around Luga. The SS Police Division and the 269th Infantry Division, which had worked their way close to Luga in frontal attack, went into action once more. They advanced right and left of the town for an enveloping attack.

The reinforced 2nd Rifle Regiment of the SS Police Division, which had been brought forward into the Luga bridgehead behind 122nd Infantry Division, was able to make a northward penetration and push ahead as far as the edge of Luga.

On the right wing the attack of 96th Infantry Division likewise went well. On 11th August the men from Lower Saxony crossed the Mshaga sector, wheeled towards the north, and then pierced the Soviet positions in their deep left flank. In the course of their further advance a forward unit forced the Oredezh river at Pechkova and cut yet another rearward supply line of the Soviet units still holding out at Luga. The Chief of Staff of the Soviet Army at Luga was taken prisoner, wounded, by 96th Infantry Division.

The situation now turned critical for the five divisions of the Soviet XLI Corps. In their rear the battalions of 9th and 122nd Infantry Divisions were reaching out for the only road leading through the swamp. On their right and left they were in danger of being outflanked. The Russian commander therefore gave his units the only correct order—to try to fight their way through to Leningrad in small formations.

For that, however, it was too late. The retreating Soviet forces were pushed into the swamps east of the highway and subsequently annihilated in the so-called Luga pockets through the co-operation with 8th Panzer Division and 96th Infantry Division. The spoils of battle were 21,000 prisoners, 316 tanks, and 600 guns. Even more important was the fact that the only hard main road to Leningrad was now clear for the infantry of L and XXVIII Corps, as well as for supply traffic.

"On 3rd September the highway was taken over with a deep sigh of relief from all operational and supply headquarters of the Army Group," recalls General Châles de Beaulieu, the Chief of Staff of Hoepner's Panzer Group. One can understand the sigh of relief. A vital lifeline had at last been secured for the final attack against Leningrad.

248 Part Two: Leningrad

But what had happened meanwhile in the area of Reinhardt's XLI Panzer Corps? What was the position of the spearheads of 4th Panzer Group, poised as they were for the final attack on Leningrad from the west, with hardly any appreciable enemy forces between them and the great objective of the campaign? This question contains in itself the real tragedy of the battle of Leningrad, a tragedy of errors with fateful consequences for the entire course of the war.

After General von Manstein's LVI Panzer Corps had been detached from 4th Panzer Group in mid-August, because of the crisis near Staraya Russa, Colonel-General Hoepner found himself compelled again to put the brakes on his successfully developing attack against Leningrad. The flanks were getting too extended. In particular, the northern flank of 4th Panzer Group had to be protected against the enemy divisions streaming back from Estonia via Narva and Kingisepp. To begin with, the 1st Infantry Division from East Prussia was used for covering the Group's wide-open left flank, while 58th Infantry Division, following behind it, wheeled north and advanced towards the Kingisepp–Narva railway-line. Before long, however, General Reinhardt had to employ nearly all his motorized formations on flank cover.

The reinforced 6th Rifle Brigade under General Raus, and subsequently Lieutenant-General Ottenbacher's 36th Motorized Infantry Division, had to cover the left flank. The 8th Panzer Division, following on the other wing of the Corps, was gradually turned towards the south-east, and eventually wheeled right to the south for the final attack on Luga. Thus all that was left for the attack on Leningrad proper, from the west, was the reinforced 1st Panzer Division and the combat group Koll (the reinforced 11th Panzer Regiment, 6th Panzer Division). To try to take a city of several million inhabitants with such slight forces would have been foolhardy—especially as the striking force of 1st Panzer Division on 16th August, apart from two weakened armoured infantry carrier battalions, was down to 18 Mark II tanks, 20 Mark IIIs, and 6 Mark IVs. In these conditions the most exemplary offensive morale was no use. Nor, for that matter, was the employment of short-range squadrons of VIII Air Corps. Naturally, Colonel-General Hoepner took advantage of the fact that no effective Russian divisions of the line were left between him and the city, and cautiously advanced by about six miles each day. In this manner by 21st August the vanguards of 4th Panzer Group reached the area north-west and south-west of Krasnogvardeysk—25 miles from Leningrad.

In this situation there was only one decision for Army Group North— a decision which Hoepner had been urging upon Field-Marshal Ritter von Leeb ever since 15th August: Colonel-General Küchler's Eighteenth Army must at last be switched from Estonia to the Luga front in order, at the very least, to take over the Panzer Group's northern flank cover and thus to free its mobile formations for the final attack on Leningrad.

The C-in-C Army Group North could not in the long run turn a deaf ear to this justified request. But instead of assigning to Eighteenth Army a clear and unambiguous objective, Field-Marshal Ritter von Leeb gave it a dual task on 17th August: on Estonia's Baltic coast it was to destroy the Soviet Eighth Army, then withdrawing from Estonia via Narva—in other words, eliminate the threat to the flank of Reinhardt's Panzer divisions before Krasnogvardeysk; at the same time Küchler was ordered to capture the coastal fortifications along the southern edge of the Gulf of Finland, where Soviet covering forces had been digging in. This proved to be a downright disastrous double order. While giving Eighteenth Army the chance of scoring spectacular successes, these victories would cost a great deal of precious time and, measured by the final objective of the campaign, would be unnecessary. The Russian strongpoints to both sides of Narva could have been equally well cut off by covering forces and starved out. There was no need to waste time and fighting men by engaging them in battle and tying down strong forces on a secondary front at the very moment when the Army Group's striking forces before Leningrad were desperately in need of every battalion they could get.

Eighteenth Army needed a full eleven days to move from Narva to Opolye, a distance of 25 miles as the crow flies. In a study of the battle of Leningrad the Chief of Staff of 4th Panzer Group observes correctly: "And that at a time when every single man was needed outside Leningrad!"

If formations of Eighteenth Army had been made available to 4th Panzer Group in good time and on a sufficient scale Colonel-General Hoepner would have had a chance of taking Leningrad with his mobile forces by a coup as early as the second half of August. That Hoepner, an old cavalry man and one of the most experienced tank commanders in the Wehrmacht, had it in him to pull off such an operation is proved by the great successes of his XVI Panzer Corps in the Polish and French campaigns, as well as by the successful drive of his armour through very difficult country right up to the gates of Leningrad. Why then was this chance missed?

General Châles de Beaulieu believes—and the present author agrees —that Field-Marshal Ritter von Leeb was anxious to let the Commander-in-Chief of Eighteenth Army, who was a personal friend of his, take a prominent share with his infantry divisions in the victory over Leningrad—a psychologically understandable consideration, but one that was to have disastrous consequences. Each day that Stalin gained on the northern sector he used for reinforcing Leningrad's defences with reserves hurriedly scraped together from his vast hinterland, and for reforming in the Oranienbaum area the troops which he had pulled out from the Baltic countries beyond the Luga, thus maintaining his threat to the German northern flank. Every day the German striking formations were held up north-west of Krasnogvardeysk meant that Stalin was getting stronger outside Leningrad. Every day the stubborn defence of Luga continued to tie down German armoured formations reduced the advantage which Hoepner had gained when his fast formations had crossed the Daugava, burst through the Stalin Line, and broken out of their bridgeheads on the Luga. The chances of taking the second biggest city of the Soviet Union, in terms of morale the most important Soviet city, the great metropolis on the Baltic, by a surprise move were steadily fading away.

At last, at the beginning of September, the final attack on the "White City" on the Neva was decided upon—the moment Hoepner's divisions and the forward regiments of the Infantry Corps of Eighteenth Army had so long awaited! Leningrad was the great objective of the campaign in the north. It was an objective every soldier could understand, an objective which fired every man's fighting spirit.

The signal for the attack was given on 8th and 9th September 1941. The brunt of the attack was to be borne by General Reinhardt's XLI Panzer Corps.

The ground had been very thoroughly reconnoitred, especially from the air. There was no doubt that Zhdanov, Leningrad's Political Defence Commissar and regarded as Stalin's Crown Prince, who shared with Marshal Voroshilov the supreme military command of the Leningrad front, had made good use of the time given him by the continuous postponements of the German attack.

About mid-August the morale of the Soviet troops and the civilian population had been at a dangerously low point following the lightning-like German victories. No one then believed that the city could be defended. Even Zhdanov appears to have toyed with the idea of evacuating it. The delays in the German attack subsequently provided the respite needed by the propaganda machine for stiffening Soviet resistance.

General Zakhvarov was appointed Commandant of the city. For the defence of the city centre he raised five brigades of 10,000 men each. From Leningrad's 300,000 industrial workers some twenty divisions of Red Militia were formed. These factory legionaries continued to be armament workers, but at the same time they were soldiers—workmen in uniform, available for military action at a moment's notice.

In ceaseless day and night toil troops and civilians, including children, were made to build an extensive system of defences around the city. Its main features were two rings of fortifications—the outer and the inner defences.

The outer or first line of defence ran in a semicircle, roughly 25 miles from the city centre, from Peterhof[1] via Krasnogvardeysk to the Neva river. The inner or second line of defence was a semicircle of

[1] Now Petrodvorets.

fortifications in considerable depth, barely 15 miles from the city centre, with the Duderhof[1] Hills as their keypoint. The industrial suburb of Kolpino and the ancient Tsarskoye Selo were its cornerstones.

Aerial reconnaissance had identified a vast number of field fortifications, and behind them enormous anti-tank ditches. Hundreds of pillboxes with permanently emplaced guns supplemented the systems of trenches. This was real assault-troop country, the proper terrain for the infantry. Armour could do no more than drive through the breached defences as a second wave, providing fire cover for the advancing infantry.

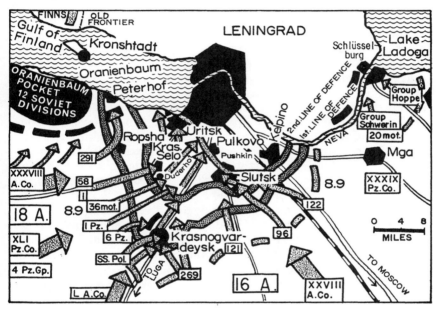

Map 12. The battle of Leningrad between 8th and 17th September 1941.

The main thrust of Hoepner's Panzer Group against the centre of Leningrad's defences in the area of the Duderhof Hills was to be made by Reinhardt's XLI Panzer Corps. The 36th Motorized Infantry Division formed its spearhead. Behind it 1st Panzer Division was standing ready to follow up the first strike. On the right the regiments of 6th Panzer Division were standing ready for assault. Along the highway from Luga the old Luga divisions—the SS Police Division and the 269th Infantry Division—were to attack towards Krasnogvardeysk under L Army Corps. On the left wing the East Prussian 1st, the 58th, and the 291st Infantry Divisions were employed, as the leading divisions of Eighteenth Army. On the right wing, on the Izhora river, the

[1] Now Mozhayskiy.

121st, the 96th, and the 122nd Infantry Divisions were standing ready under the command of XXVIII Army Corps as the striking force of Sixteenth Army. On the extreme eastern wing, along the southern edge of Lake Ladoga, the reinforced 20th Motorized Infantry Division, together with the combat groups Harry Hoppe and Count Schwerin, as part of XXXIX Panzer Corps, had the task of clearing up the bridgeheads of Annenskoye and Lobanov. Their eventual aim was the capture of the town and area of Schlüsselburg.[1]

It was on the Duderhof Hills that the Tsars of Russia used to watch the Guards regiments of St Petersburg hold their manœuvres outside the city. The Guards and the Tsars had long passed away, but their experience was alive in the Red Army: every dip in the ground, every patch of woodland, every rivulet, every approach route, and all distances were known with the greatest accuracy. The artillery had the exact range of all the principal points in the terrain. In the infantry dugouts, in the concrete pillboxes, and in the anti-tank ditches all round the Duderhof Hills Zhdanov, Leningrad's Red Tsar, had deployed his Guards—active crack regiments, fanatical young Communists, and the best battalions of Leningrad's Workers' Militia.

Step by step the assault companies of the German 118th Infantry Regiment, 36th Motorized Infantry Division, had to fight their way forward. The entire Corps artillery as well as 73rd Artillery Regiment, 1st Panzer Division, were pounding the Soviet positions, but the Russian pillboxes were magnificently camouflaged and very solidly built.

"We need Stukas," radioed the division's 1st Battalion from where it was pinned down. Lieutenant-General Ottenbacher rang XLI Panzer Corps. The 4th Panzer Group sent an urgent signal to First Air Fleet through its liaison officer. Half an hour later the squadrons of JU-87s of Richthofen's VIII Air Corps came roaring over the sector of 118th Infantry Regiment, banked steeply, plummeted down almost vertically, with an unnerving whine skimmed quite close above the ground, and dropped their bombs on the Soviet pillboxes, machine-gun posts, and infantry gun emplacements. Flashes of fire shot skyward. Smoke and dust followed, forming a dense curtain in front of the still intact enemy strongpoints.

That was the right moment. "Forward!" shouted the platoon commanders. The grenadiers leapt to their feet and charged. Machine-guns clattered. Hand-grenades exploded. The flame-throwers of the sappers sent searing tongues of burning oil through the firing-slits of the pillboxes. Strongpoint after strongpoint fell. Trench after trench was rolled up. The men leapt into the trenches. A burst of machine-gun fire along the trench to the right, and another to the left. "Ruki verkh!" ("Hands up!") As a rule, however, the Russians continued to fire until

[1] Shlisselburg, now Petrokrepost.

they were hit themselves. In this fashion the 118th Infantry Regiment broke into Leningrad's first line of defence and took Aropakosi. Only when darkness fell did the fighting abate.

On the morning of 10th September the infantry and sappers of the assault battalions had the towering Duderhof Hills in front of them— the bulwark of Leningrad's last belt of defences. This was the key of the second ring round the city. Heavily armed reinforced-concrete pillboxes, casemates with naval guns, mutually supporting machine-gun posts, and a deeply echeloned system of trenches with underground connecting passages covered the approaches of the two all-commanding hills—Hill 143 and, east of it, "Bald Hill," marked on the maps as Hill 167.

Progress was again only yard by yard. Indeed, a dangerous crisis developed for 6th Panzer Division, which was attacking on the right of 36th Division. Alongside 6th Panzer Division, the SS Police Division had been held up in front of a heavily fortified blocking position. But 6th Panzer Division, under Major-General Landgraf, had driven on. The Russians grasped the situation and struck at its flank. Within a few hours the gallant division lost four commanding officers. At close quarters the Westphalians and Rhinelanders struggled desperately to hold the positions they had gained.

From this situation developed the great opportunity of 1st Panzer Division. General Reinhardt turned the 6th Panzer Division towards the east, against the flanking Soviets, and moved 1st Panzer Division into the gap thus created on the right of 36th Motorized Infantry Division.

Lieutenant-General Ottenbacher, with his headquarters staff, was meanwhile close behind the headquarters of 118th Infantry Regiment. His assault battalions were pinned down by heavy fire from the Russians. Ottenbacher once more concentrated his divisional artillery and 73rd Artillery Regiment for a sudden heavy bombardment of the northern ridge of the Duderhof Hills.

At 2045 hours the last shell-bursts died away. The company commanders leapt out of their foxholes. Platoon and section leaders waved their men on. They charged right into the smoking inferno from which rifle and machine-gun fire were still coming. The grenadiers panted, flung themselves down, fired, got to their feet again, and stumbled on. A machine-gunner heeled over and did not rise again. "Franz," his No. 1 called. "Franz!" There was no reply. In a couple of steps he was by his side and flung himself down next to him. "Franz!"

But the second machine-gunner of 4th Company, 118th Infantry Regiment, was beyond the noise of battle. His hands were still clutching the handles of the boxes with the ammunition belts. The box with the spare barrels had slipped over his steel helmet as he fell.

Twenty minutes later No. 1 Platoon of 4th Company leapt into the sector of trench along the northern ridge of the Duderhof Hills. The penetration was immediately widened and extended. A keystone of Leningrad's defences had been prised open with Hill 143.

The 11th September dawned—a brilliant late-summer day. It was to be a great day for 1st Panzer Division. Colonel Westhoven, commanding 1st Rifle Regiment and an experienced leader of combat groups, led his force against Bald Hill. The main thrust was made by Major Eckinger with his lorried infantry in armoured personnel carriers, the 1st Battalion, 113th Rifle Regiment. It was reinforced by 6th Company, 1st Panzer Regiment, and by one platoon of Panzer Engineers Battalion 37, and supported by 2nd Battalion, 73rd Artillery Regiment.

Major Eckinger enjoyed the reputation of having a good nose. He could smell an opportunity, scent the most favourable spot, and, moreover, had that gift of lightning-like reaction and adaptable leadership that won battles.

Plan and execution of the coup against Hill 167 were a case in point. While 1st Rifle Regiment provided flank cover to the east, the reinforced 113th Rifle Regiment drove along the road to Duderhof and threw back the Russian defenders to the anti-tank ditch of the second line. Eckinger's foremost carrier-borne infantry drove right in among the withdrawing Russians. Sergeant Fritsch with his Panzer sapper platoon burst into the great anti-tank ditch, dislodged the Soviet picket covering the crossing, leapt over it, prevented them from blowing it up, and kept it open for the German units. With the aid of trench ladders they negotiated the steep faces of the ditch to the right and left. They put down beams and planks, and provided crossings for the bulk of the armour and armoured infantry vehicles which followed hard on their heels. The companies of Eckinger's battalion were riding into the line on top of the tanks and armoured troop carriers.

It was a thrilling spectacle. Above the battalion's spearhead, as it raced forward, roared the Stukas of VIII Air Corps. They banked and accurately dropped their bombs 200 to 300 yards in front of the battalion's leading tanks, right on top of the Russian strongpoints, dugouts, ditches, tank-traps, and anti-tank guns.

Luftwaffe liaison officers were in the tanks and armoured infantry carriers of the spearhead and also with the commander of the armoured infantry carrier battalion. A Luftwaffe signals officer, sitting behind the turret of Second Lieutenant Stove's tank No. 611, maintained radio contact with the Stukas. A large Armed Forces pennant on the tank's stern clearly identified him as the "master bomber." In the thick of enemy fire the Luftwaffe lieutenant directed the Stuka pilots through his throat microphone.

The attack unrolled with clockwork precision. The village of Duderhof, into which the enemy had again penetrated behind the vanguards of 36th Motorized Infantry Division, was taken once more. Eckinger turned his battalion to the south, then again to the east, and with inverted front charged Bald Hill.

The hill, sparsely covered with low trees, was a fortress belching fire. But the Soviets were jumpy, taken by surprise, and made unsure by Eckinger's ingenious and unpredictable method of attack.

An entire Panzer company and the leading company of armoured troop carriers succeeded in getting into the dead angle of the westward-pointing Russian naval batteries without receiving a single hit. Guns to the right and left of the road were silenced with a few shells from a half-troop of tanks of 8th Company, 1st Panzer Regiment, under Second Lieutenant Koch. Under cover of fire from these tanks the sappers fought their way right up to the massive naval-gun emplacements. Hand-grenades were bursting all round. Flame-throwers shot their tongues into the batteries. The crews were overwhelmed in hand-to-hand fighting.

At 1130 the headquarters staff of 1st Panzer Division overheard a signal sent by Second Lieutenant Darius, commanding 6th Panzer Company, to his battalion commander. Its wording produced a sigh of relief from the division's chief of operations, Lieutenant-Colonel Wenck, who had followed the armoured infantry carrier battalion in Major-General Krüger's signals tank, but it also made them chuckle at the romantic soul of a young tank commander in the middle of a battle. Darius radioed: "I can see St Petersburg and the sea." Wenck understood. Darius was on Hill 167, the top of Bald Hill, and Leningrad was lying at his feet, within reach. The citadel of the last defensive position, on the very "generals' hill" of the Tsars, had fallen.

3. In the Suburbs of Leningrad

"All change—end of the line!"–In the gardens of Slutsk–Harry Hoppe takes Schlüsselburg–Order from the Fuehrer's Headquarters: Leningrad must not be taken–Hitler's great mistake.

FROM the top of Bald Hill Darius had a unique panoramic view of the battle for Leningrad. Through the captured Soviet trench telescopes the busy traffic in the city's streets could be clearly made out. The Neva glistened in the sunlight. The factory chimney-stacks were smoking, for Leningrad was still working feverishly.

In the north, on the extreme left wing, German formations were seen advancing towards Peterhof and Oranienbaum. These were the 291st Infantry Division, the "Elk Division," under Lieutenant-General Herzog, which, together with the East Prussian 1st Infantry Division, had broken through a heavily fortified line of strongpoints at Ropsha. On 11th September the battalions of 505th Infantry Regiment alone had to knock out 155 concrete pillboxes, some of them with built-in guns. The division was then turned to the north, towards Peterhof, in order to cover the left flank against the twelve Russian divisions caught in the Oranienbaum pocket.

On 20th September the 1st Infantry Division reached the coast at Strelnya.

The view from Bald Hill extended as far as Kronshtadt. One could see the port and the powerful Soviet battleship *Marat*, which was shelling land targets with its heavy guns. The hits of the 30·5-cm. shells sent up fountains of earth as high as houses, especially in the sector of 58th Infantry Division, which was making a hell-for-leather drive for the coast, in order to close the Leningrad trap in the direction of Oranienbaum.

The regiments of 58th Infantry Division had broken through the fortified line at Krasnoye Selo. The battalions of 209th Infantry Regiment fought their way through the town and dislodged the Soviets. They continued to advance—always to the north, towards the rooftops of Leningrad's suburb of Uritsk.

The time was 2000 hours on 15th September. First Lieutenant Sierts, commanding 2nd Company, 209th Infantry Regiment, Second Lieutenant Lembke, and Sergeant Pape had worked their way forward with the spearheads of 1st Battalion as far as the big coastal road from Uritsk to Peterhof, and were now lying in the roadside ditch. Within a few feet of them ran the rails of the tramway leading to Leningrad. Civilians on bicycles and with hand-carts were coming from Peterhof. Evidently they had no idea that the enemy was so near. And then, almost unbelievably, there came a tram, crowded with civilians travelling into the city.

"Up!" Sierts ordered. Pape and his men leapt up on to the road.

The driver clanged his bell: Out of the way there—make room for the Leningrad tram. But suddenly he realized that these men with steel helmets on their heads and machine pistols under their arms were no mere traffic obstacle. He slammed on his brakes. The wheels screeched. The passengers were thrown all of a heap.

Pape stepped up to the platform and, chuckling, called out in German, "All change, please—end of the line!" And then he called across to Lembke: "Shall we get on, Herr Leutnant? It's a unique opportunity—we've even got a driver."

"We'll keep the driver till to-morrow morning," Lembke replied. "To-morrow morning we might need him."

Everybody was understandably optimistic. The distance to the centre of Leningrad was only six miles. Sierts, Lembke, Pape, and the men of Colonel Kreipe's 209th Infantry Regiment were practically in the city. And Leningrad was already cut off in the west.

By swivelling the trench telescope on top of Bald Hill over to the other side, to the east, one could make out the Chudovo–Leningrad main road and the deep-cut valley of the Izhora river along which Leningrad's first line of defences ran. The thirteen-foot-high northern bank of the river had been cut off steeply by the Russians and made almost unscalable. This was the sector of Lieutenant-General Schede's 96th Infantry Division.

The Izhora had to be forced. To tackle this heavily fortified obstacle Lieutenant-General Schede on 12th September employed the combat groups Arntzen and Hirthe of the 284th Infantry Regiment under Lieutenant-Colonel von Chappuis. Artillery and Richthofen's indefatigable Stukas again did the preliminary work and enveloped the river-bank in thick clouds of smoke. Under cover of this screen Hirthe's companies crossed the river, which was about 28 yards wide.

"Ladders forward!" came a shout. Instantly the special assault detachments appeared with their assault ladders, of which Engineers Battalion 196 had manufactured hundreds. As in a medieval attack on a fortress, the ladders, each of them 15 to 20 feet long, were propped against the steep bank. Under covering fire from the machine-guns the assault detachments of 2nd Battalion, 284th Infantry Regiment, clambered up on to the high northern bank. Once up, Major Arntzen's grenadiers and the sappers attached to them charged the Soviet machine-gun posts and infantry foxholes on the steep bank with handgrenades, flame-throwers, and those lobbed bombs which were nicknamed "Stukas on foot."

The combat group under von Chappuis likewise got across the river in this manner. Presently, however, under surprise attack by heavy Soviet tanks, they had to fall back to a Soviet anti-tank ditch, since the German 3·7-cm. anti-tank guns were useless against the Kolpino-made T-34s and KVs. Only a last-minute intervention by Stukas saved the situation and prevented the grenadiers from being crushed one by one by the heavy enemy tanks.

Throughout 13th and 14th September heavy fighting continued against attacking Soviet armoured formations. Only the 8·8-cm. anti-aircraft guns and a heavy 10-cm. gun which had taken up position in the foremost line saved the situation and repulsed the enemy tanks.

On 16th September the battalions of 96th Infantry Division and 121st Infantry Division burst into the famous park of Slutsk.[1] Scattered

[1] Now Pavlovsk.

about the extensive parkland were romantic pavilions in the French style. They belonged to the Tsars' summer residence, the famous Tsarskoye Selo which the Bolsheviks had renamed Pushkin. Now the war's fiery hand swept across this idyllic spot. Pushkin fell.

Thus the 96th, the 122nd, and the 121st Infantry Divisions were now all within 15 miles of Leningrad. Only the important industrial suburb of Kolpino, with its huge tank factories, and the heights of Pulkovo, where in 1919 the White Guards' attack against Red Leningrad was halted, were still in Russian hands. But Pulkovo was reached on 17th September and Kolpino on the 29th.

An important part of the battlefield, however, was not visible through the trench telescope on the Duderhof Hills—the battle for Schlüsselburg, the town on the western bank of Lake Ladoga, where the Neva leaves the lake and makes a wide arc towards Leningrad and the Baltic Sea. Whoever held Schlüsselburg—as, indeed, its name implies, meaning 'key fortress'—could close Leningrad's door to the east, block the waterway between the Baltic and Lake Ladoga, and thus also the system of canals linking the city with the White Sea and the Arctic Ocean.

This cornerstone in the battle of Leningrad was to be seized in a special operation. The man chosen to lead it was Harry Hoppe, the colonel commanding 424th Infantry Regiment, 126th Infantry Division. The rank and file knew him simply as "Harry," because the colonel invariably tackled all tasks and problems in a clear and simple manner which gained the troops' immediate confidence and their absolute belief in the success of every operation. Kray, one of the motor-cycle messengers, had experience of this before Schlüsselburg. The colonel was standing outside a workers' settlement on the edge of the town with a plan in his hand, and said to him, "You drive along this road right into the town, then you take the first to the right, and there you wait for me." The motor-cyclists roared off. They were quite sure that Harry would turn up.

The southern bank of the huge Lake Ladoga, with Schlüsselburg, was a strategically most important area. The Bolsheviks had utilized the lake and the lock-gates of its canals for the generation of electricity. A widely ramified system of canals had been connected with the railway network of the Leningrad hinterland, and the marsh and forest area had been cultivated.

As a result, a large area, as though designed on the drawing-board, had been developed outside Schlüsselburg, with eight large workers' settlements known as Poseloks—the Russian word for settlement. They bore the rather unimaginative names of Poselok 1, Poselok 2, Poselok 3, and so on to Poselok 8.

It was from there, from this centre of an important communications and power industry system, that the waterways from Leningrad and the Baltic to the Volkhov, via Lake Onega to the White Sea and the Arctic Ocean with Archangel and Murmansk, as well as between Leningrad and Moscow by way of the Rybinsk reservoir and the Moskva–Volga Canal, were controlled. Anyone wanting to seize Leningrad, strangle it, capture it, or starve it into submission would have to close these vital doors to the city. The key to these doors was Schlüsselburg.

It was a corner of Europe rich in history. Two hundred and thirty-five years before Harry Hoppe, Peter the Great fought a battle here in order to take from the Swedes the key to the Baltic Sea. He succeeded. For the first time the Tsar of Russia gained for his country access to Europe's most important inland sea, and to protect this conquest he founded the fortress of St Petersburg, now Leningrad. It was that fortress that was being fought for now at Schlüsselburg.

Coming from Novgorod, the 424th Infantry Regiment, 126th Infantry Division, together with Major-General Zorn's 20th Motorized Division, had been set in motion at the beginning of September along the great highway to the north, via Chudovo in the direction of Schlüsselburg. It was a good plan. The idea was that the divisions of "Group Schmidt" —XXVIII Army Corps and XXXIX Panzer Corps—under General of Panzer Troops Rudolf Schmidt, should clear up the eastern Neva bridgeheads even before the start of the general offensive against Leningrad, since Soviet formations used these to maintain contact between the Leningrad approaches and the Volkhov area.

Under cover of this flanking operation the combat groups of Colonels Count Schwerin and Harry Hoppe, with their reinforced 76th and 424th Infantry Regiments, were to reach the starting positions for an assault on Schlüsselburg by 8th September 1941, the day for which the large-scale attack on Leningrad had been fixed—Hoppe's combat group on the right and Count Schwerin's on the left.

They went into action on 6th September. At first everything went according to plan. Tanks of 12th Panzer Division supported the attack. Panzerjägers and AA batteries—including an 8·8—provided cover against enemy tank attacks. Motor-cyclists and sappers formed the vanguard.

The main weight of the attack was in the sector of Hoppe's group. The I and VIII Air Corps provided Stuka support. The troops charged over the famous railway embankment of Mga. They burst into the forest along both sides of the road to Kelkolovo. But there the Russians were waiting for them in well-camouflaged machine-gun and anti-tank positions. The attack got stuck. Infantry guns, anti-tank guns, and mortars were not much use in this wilderness.

Colonel Hoppe was crouching by the railway embankment. A runner from 3rd Battalion came scurrying over the line. "Heavy casualties at

Battalion. Three officers killed." Calls for support also came from 2nd Battalion.

"We've got to find a gap," Hoppe was thinking aloud, bent over his maps. "The Russians can't be equally strong everywhere. It's just a matter of finding their weak spot."

Hoppe's idea was either to probe the enemy's weakness by a frontal attack or to outflank him altogether. He combined in himself the dash of a First World War assault troop commander with the sound tactical instruction received in Seeckt's Reichswehr.

The runner scuttled off again. Major-General Zorn appeared at the command post. He no longer believed in the possibility of forcing a break-through in Hoppe's sector. He therefore dispatched the tanks over to Schwerin's group. That was where the main push was now to be made.

But it proved to be a case of a general proposing and a lieutenant disposing. No sooner had the tanks been withdrawn from the line than Second Lieutenant Leliveldt, with his 11th Company, discovered the looked-for gap, the weak spot in the enemy's line. He thrust into it, applied pressure to the right and left, and tore a wide breach into the front.

"Buzz over to Harry," the Second Lieutenant shouted at his runner. "We've got the gap. The front is open!"

The runner raced off. Half an hour later the entire combat group was moving. Kelkolovo fell. The notorious rail-track triangle formed by the line from Gorodok to Mga and Schlüsselburg was taken and Poselok 6 was stormed.

At 1600 hours Sinyavino with its huge stores and ammunition depots fell into the hands of 3rd Battalion. From a small hill north of the town the vast sheet of water of Lake Ladoga could be seen and a light sea-breeze felt. There was a good deal of shipping on the lake.

"Keep going," Hoppe commanded. His men took Poselok 5 and moved on as far as Poselok 1. From there the "Red Road" led to the "Red Bridge" over the canals and coastal railway-lines. This was the spinal cord of the Schlüsselburg nerve centre.

Night fell over the battlefield. From Sinyavino a gigantic fireworks display lit up the sky: some Russian ammunition dumps had been hit and were now going up. Unfortunately the vast explosions also wrecked the combat group's communications with Division.

On the following morning, 8th September, Schlüsselburg was to have been stormed. But at what time? Hoppe did not know, since Division was going to co-ordinate the time of attack with the Stuka formations. But now, with communications out of action, there was no contact with divisional headquarters. It was an awkward situation.

Over to the west, at Leningrad, the Corps launched its general attack at first light on 8th September. But in Schlüsselburg everything remained quiet. When the sun rose the town with its pointed spires and massive old ramparts lay in front of Hoppe's battalions. The shrub-grown ground favoured the attack. But there was still no contact with Division. The 9th Company made a reconnaissance in force as far as the eastern edge of the town.

At 0615 hours Sergeant Becker reported to 3rd Battalion: The eastern edge of the town is held by weak enemy forces only. Clearly the Russians were not expecting an attack at this point, from their rear. It seemed a unique chance.

Hoppe was in a quandary: should he attack or not? If he stormed the town and the Stukas did not come until his battalions were inside, the consequences were not to be imagined. But he could not just sit there waiting. To wait without doing anything was the worst thing of all—that was what the Service manual said. Better a wrong decision than no decision at all. Hoppe decided accordingly.

Shortly before 0700 hours he ordered: "The 424th Regiment will take Schlüsselburg and drive through to the 1000-yard-wide Neva river, at the point where it leaves Lake Ladoga, dividing Schlüsselburg from Sheremetyevka and the southern bank of Lake Ladoga from its western bank. Time of attack is 0700 hours." Harry had made his plan.

At 0730 hours the battalions were bursting through the weakly held eastern fringe of the town. The Russians were thrown into confusion by the unexpected attack.

At 0740 hours Sergeant Wendt hoisted the German flag over the tall steeple of the church.

Ever since the start of the attack Second Lieutenants Fuss and Pauli had been sitting in front of their walkie-talkie transmitter, trying to make contact with the nearest heavy battery, at Gorodok. It might be possible to re-establish contact with Division HQ through them.

Fuss had been talking into his microphone ceaselessly for three-quarters of an hour. Calling—switching over to receiving—calling again. Nothing happened. "Suppose we don't get through? Suppose the Stukas come?"

At last, at 0815 hours, the battery at Gorodok responded. They had been heard. "This is Group Harry. Urgently pass on to Division: Schlüsselburg already stormed. Stukas must be stopped. Have you got that?"

"Message understood."

The battery officer immediately passed on the signal. The Stukas had already taken off because Hoppe's attack had not been scheduled until 0900 hours. Most of the machines could be recalled. But one squadron had gone too far for the new order to reach it. Via the battery at Gorodok a signal was sent to Hoppe to warn him of his danger.

At 0845 exactly the JU-87s appeared in the sky. Hoppe's men waved aircraft signalling sheets. They fired white Very lights: We are here.

Would the pilots see them? Or would they think this was a trick? Their orders were to bomb Schlüsselburg.

The Stukas banked steeply—neatly, one after another. But suddenly the first one levelled out again, roared on, and dropped its bombs into the Neva. The others followed suit. At the last moment a signal from the squadron commander had reached them. Harry Hoppe and his men heaved a sigh of relief. At 1000 hours the battalions of combat group Schwerin also moved into the southern part of the town.

The conquest of Schlüsselburg meant that Leningrad was sealed off to the east. The city now was an island surrounded by troops and water. Only a narrow corridor was still open to the western shore of Lake Ladoga, because the Finns in the Karelian Isthmus were still standing by. They were waiting for the Germans to drive past Leningrad to Tikhvin. Only then did Mannerheim intend to drive along the eastern shore of Lake Ladoga, across the Svir, and thus form the eastern prong of the pincers closing around a huge pocket with Leningrad in it. That, unfortunately, proved too ambitious an objective.

The Soviet High Command was appalled at the defeat at Schlüsselburg. With every means in his power Marshal Voroshilov tried to regain this important keypoint for his eastward communications. He drove entire regiments in assault boats and landing craft across the lake from the western shore against the Schlüsselburg side. Simultaneously he ordered an attack from the landward side, from Lipki.

Colonel Hoppe's regiment was cut off at times. The Russians were bringing up more and more forces. On the German side the troops began to suspect that heavy casualties lay in store for them. And some also began to suspect that Leningrad's encirclement from the east would become illusory once Lake Ladoga froze over in winter.

The optimists laughed at such misgivings. "Winter?" they asked. "Leningrad will have fallen long before the first frost."

But Leningrad did not fall. Why not?

Because Hitler and the Wehrmacht High Command had decided not to take Leningrad before the winter, but merely to encircle it and starve it out.

Paradoxical as it sounds, this is exactly what happened. At the very moment when Leningrad's last line of defence had been broken, when the Duderhof Hills had been stormed, when Uritsk and Schlüsselburg had been taken, and the city, shaking with fright, lay right in front of the German formations, came the red light from the Fuehrer's headquarters.

General Reinhardt, commanding XLI Panzer Corps—later promoted Colonel-General—recalls the situation: "In the middle of the troops' justified victory celebrations, like a cold shower, came the news from Panzer Group on 12th September that Leningrad was not to be taken, but merely sealed off. The offensive was to be continued only as far as the Pushkin–Peterhof road. The XLI Panzer Corps was to be detached during the next few days for employment elsewhere. We just could not understand it. At the last moment the troops, who had been giving of their best, were robbed of the crown of victory."

Sergeant Fritsch merely tapped his forehead when the commander of 2nd Company Panzer Battalion 37 said to him, "We are not allowed into Leningrad. We are being pulled out of the line. I got it from a wireless operator at divisional headquarters."

"You're nuts," Fritsch said, corroborating his gesture. The rumour of the decision had leaked also to 1st Panzer Regiment, 1st Panzer Division. But the officers merely shook their heads. "It's just not possible. Surely we didn't come all the way from East Prussia to the gates of Leningrad merely to walk away now as though it had all been a mistake?" Everybody was grumbling and every conversation ended with the words: "Surely, it's not possible."

The order of Army Group was still being kept secret because Leningrad was to be surrounded as closely as possible and a number of important points on the outskirts were yet to be captured—as, for instance, Kolpino and the heights of Pulkovo. But what unit would fight with any enthusiasm if its men knew that all they were after was front-line rectifications, while the great objective was no longer to be attempted? The troops, therefore, were allowed to believe that the capture of Leningrad was the objective, and so they fought with the utmost vigour. This is shown very clearly by the following account from the diary of Second Lieutenant Stoves, commanding No. 1 Platoon, 6th Company, 1st Panzer Regiment:

On 13th September three Soviet heavy KV-1 and KV-2 tanks, fresh from the Kolpino tank factory, partly without their paintwork, came rumbling down the road from Pulkovo through the morning mist, heading for the intersection with the Pushkin–Krasnoye Selo road.

Stoves gave the action stations signal to his three tanks standing along both sides of the road to the airfield of Pushkin, ordered his own tank-driver to move behind a shed and keep his engine running, and to provide cover towards the south. He then inspected the pickets outside the village of Malaya Kabosi, together with Captain von Berckefeldt. Thick eddies of morning mist were contending with the sun. The time was 0700 hours. Sergeant Bunzel's tank, No. 612, slowly moved on to the road.

Suddenly, as if they had sprung from the ground, two enormous KV-2s stood in front of them. Stoves and Berckefeldt flung themselves into the roadside ditch. But at that moment came a crash. Bunzel had been on the alert. Once more his 5-cm. tank cannon barked. The leading Soviet tank stopped. Smoke began to issue from it. The second moved forward past it. This one was hit by Sergeant Gulich, whose

264 *Part Two: Leningrad*

tank, No. 614, stood on the far side of the road. The very first shell scored a direct hit. The crew of the KV-2 baled out.

Five more KV-2 monsters appeared. And out of the mist near Malaya Kabosi came three KV-1s heading straight for Sergeant Oehrlein's tank, No. 613. Russian infantry, who had been riding on top, jumped down and advanced in line abreast. The leading KV fired its 15-cm. gun. Direct hit on Oehrlein's tank. The sergeant was slumped over the edge of his turret, seriously wounded. Stoves ran across. Right and left of him Soviet infantry were charging. The German pickets around Malaya Kabosi were withdrawing. In the mist it was almost impossible to tell friend from foe.

Together with Oehrlein's gun-layer, Stoves first of all dragged the driver, the most seriously wounded man, over to Sergeant Gulich's tank, which was standing behind a small shed, giving covering fire with its machine-gun. Then they ran back again. They lifted Sergeant Oehrlein out of the turret. They also tried to get the badly wounded radio operator out, but this proved impossible. They could not get to him. Out of the mist, like spectres, came the Russians. Urra! Second Lieutenant Stoves quickly secured all hatches with the square-shanked key. They would get the radio operator out when they made their counter-attack later. Until then the Russians had better be kept out of the tank. At that moment the gun-layer cried out in pain. He had been hit in the arm.

"Come on, man, run," the lieutenant shouted at him. The gun-layer, a medical student, held his damaged arm with his other hand and raced off into the mist. Stoves got the unconscious Oehrlein on to his shoulder and hurried away with him.

Right and left Soviet infantrymen were charging past with fixed bayonets. Evidently they regarded the Panzer lieutenant as one of their own men—probably because of the padded Russian jacket he was wearing.

Stoves managed it. He reached his tank, which was still providing cover against the west, well camouflaged behind the shed. A Medical Corps armed infantry carrier arrived, took charge of Oehrlein, the driver, and also the gun-layer, and drove off again. The scene was still shrouded in swirling mist—it was like a witch's cauldron.

The 1st Company, 113th Rifle Regiment, meanwhile had suffered something very like an attack of panic. It withdrew from the Malaya Kabosi crossroads. The infantry guns had long left the spot, and so had the anti-tank gun. Twenty-five yards from the shed a KV-1 crawled past Second Lieutenant Stoves's tank, No. 611. It exposed its broadside. Get him! Lance-corporal Bergener, the gun-layer, got him. A second shell put the next Russian out of action. Stoves's tank was excellently camouflaged. Now it crept cautiously to the corner of the wooden shed. A third and a fourth KV were coming down the road. Their commanders were nervous and uncertain where the deadly fire was coming from.

Bergener was lying in wait. "Fire!" Too short. "Again!" The second shell hit the Russian straight on the gunshield. The fourth tank, which hurriedly tried to turn about, received a hit astern.

At that moment Stoves saw Sergeant Bunzel's tank falling back, pursued by a KV. Bunzel could not fire at him: his gun had received a hit. Stoves's gun-layer, Bergener, saved Bunzel. He shot up his pursuer. It was the fifth Soviet tank put out of action that day.

By now the Russians had located the dangerous German. Anti-tank rifles were cracking; crash-boom shells were bursting close to the shed. "We're leaving!" Stoves commanded. In a small spinney they met Bunzel's tank, No. 612. He reported: "Cannon damaged, but both m.g.s in order."

Thirty yards farther back was Gulich's tank, No. 614—somewhat the worse for wear. At the edge of a ditch near by a machine-gun party was in position. Stoves skipped across to them. He found Captain von Berckefeldt, his steel helmet askew on his head. "A fine mess," he observed drily. "To start with, my men skedaddled because of the heavy tanks. But my lieutenant is just rounding them up again. We'll be on the move again in a minute."

Stoves returned to his tank. The engine came to life with a whine. Cautiously they drove back to the crossroads, to Oehrlein's tank, to get the radio operator out.

Twenty minutes later First Lieutenant Darius, commanding the 6th Panzer Company, caught his breath sharply. Over the air came the hoarse voice of Stoves's radio operator: "Second Lieutenant Stoves has just been killed when our tank was hit."

What had happened? A KV-1 had scored a direct hit on the superstructure of tank No. 611 at a range of 400 yards. The splinters had torn open the lieutenant's head and face. Covered with blood, he had collapsed in the commander's seat. But death had not claimed him yet. Five weeks later the lieutenant was back with his regiment. But by then it was no longer outside Leningrad.

The 1st Panzer Division went on to take the suburb of Aleksandrovka, the terminus of the Leningrad tramway's south-western line, seven and a half miles from the city centre. Then, on 17th September, the Panzer Corps was withdrawn from the front—"for employment elsewhere." It was to be employed at Moscow.

The force before Leningrad had thus been deprived of its mailed fist. Although the great objective seemed within arm's reach, the infantry divisions nevertheless came to a standstill—96th and 121st Infantry Divisions in front of the legendary Pulkovo hills, where in the civil war of 1919 the White regiments had similarly been halted in their attempts to recapture Red Leningrad.

The combat-hardened 58th Infantry Division was in Uritsk, shelling targets in the centre of Leningrad with its medium artillery. The men in their trenches along the coastal road could see the smoking chimney-stacks of the Leningrad factories only four miles away. The industrial plants and shipyards were working round the clock, producing armaments—tanks, assault boats, and shells. Thirty Soviet divisions were herded together inside the city. But they were not through yet. Though quite ready now to put an end to the fighting, they were being granted a respite and time to get over their panic.

It was incredible. What was behind this incomprehensible decision?

The plan for Operation Barbarossa stipulated clearly: Following the destruction of the Soviet forces in the Minsk–Smolensk area the Panzer forces of Army Group Centre will turn to the north, where, in co-operation with Army Group North, they will destroy the Soviet forces in the Baltic areas and then take Leningrad. The directive said quite clearly: Only after the capture of Leningrad is the attack on Moscow to be continued. This plan, strategically speaking, was entirely correct and logical, especially in its pin-pointing of the centre of gravity of the campaign, and in its intention of making the Baltic available as a supply route as soon as possible and of accomplishing a link-up with the Finns.

Disregarding this clear plan, Hitler changed his mind after the fall of Smolensk. Why?

The Army High Command and the generals in the field were urging him to take advantage of the unexpectedly rapid collapse of the Soviet Central Front and to capture Moscow, the heart, brain, and transport centre of the Soviet Union. But Hitler was reluctant. For six weeks the tug-of-war continued and precious time was lost. In the end Hitler neither stuck to his plan of taking Leningrad first nor give the green light for the attack on Moscow. Instead, on 21st August 1941, he chose an entirely new objective—the oil of the Caucasus and the grain of the Ukraine. He ordered Guderian's Panzer group to drive 280 miles to the south and, jointly with Rundstedt, to fight the battle of Kiev.

That battle was won. Indeed, it was a tremendous victory, with over 665,000 prisoners and the annihilation of the bulk of the Russian forces on the Soviet Southern Front.

This victory in the Ukraine misled Hitler into assuming that the Soviet Union was on the verge of military collapse—an error which led him into further disastrous decisions. At the beginning of September he ordered the German armies in the East to attack Moscow after all—in spite of the advanced season—and to capture it. At the same time the offensive was to be continued in the south against the Caucasian oilfields and the Crimea. Leningrad, on the other hand, was to be encircled and starved into surrender.

Clausewitz, the preceptor of the Prussian General Staff, once stated that in an offensive operation one can never be too strong, either generally or at the decisive spot. Hindenburg, in a lecture at the Dresden Military Academy, paraphrased this: "A strategy without a centre of gravity is like a man without character." Hitler disregarded these axioms. He believed that with the forces available he could take Moscow as well as the Caucasus before the end of the year and force Leningrad into surrender by the stranglehold of infantry encirclement.

Since the sealing-off of Leningrad required no armoured forces, and since, on the other hand, the attack on Moscow had to be mounted quickly in view of the approaching winter, Hitler on 17th September withdrew Hoepner's Panzer Group and all bomber formations from the Leningrad front. This order came at the very moment when one last effort would have meant the capture of the city.

The decision to go over to a siege at Leningrad was no doubt largely due to the attitude of the Finns. Field-Marshal von Mannerheim, the Finnish Commander-in-Chief, had certain scruples about crossing the old Finnish frontier in the Karelian Isthmus and attacking Leningrad. True, he was prepared to drive across the Svir east of Lake Ladoga once the Germans had reached Tikhvin, but he was against any Finnish attempt to conquer Leningrad. From his memoirs it is clear that the Marshal did not wish to involve Finnish troops in the almost certain devastation of the city. Mannerheim adhered to his principle of a "war of active defence" and opposed any war of conquest.

Whatever the reasons, Hitler's decision not to take a city strategically and economically as important as Leningrad was a crime against the laws of warfare. This crime was to be heavily paid for later.

From a military point of view the fall of Leningrad and the Oranienbaum[1] pocket would have meant the disarming of nearly forty Soviet divisions. Equally important would have been Leningrad's elimination as an armaments centre. The city's tank factories, as well as its ordnance and ammunition plants, continued to turn out their products undisturbed right through the war, and to supply the Red Army with vital armaments. The fall of Leningrad, moreover, would have freed the German Eighteenth Army for other operations, whereas it was now condemned to guard duty outside Leningrad until 1944.

Finally, Leningrad would have been of inestimable value as a supply base for the German Eastern Front. Unimpeded by partisans, supplies could have been routed through the Baltic. The link-up with the Finns, moreover, would have given a different turn to the fighting in the Far North, for Petrozavodsk and for the allied supply base of Murmansk, where no progress was being made at all for the simple reason that the available forces were insufficient.

Instead of all these patent advantages the German Command gained nothing but severe drawbacks by deciding not to take Leningrad. The

[1] Now called Lomonosov.

Soviet High Command was being positively invited to try to relieve the city from outside and, simultaneously, to keep up break-out attempts from within. The desperate attempts of the Soviet Fifty-fifth and Eighth Armies to break the German ring at Kolpino and Dubrovka were the most outstanding battles in the prolonged costly fighting for the spiritual metropolis of the Red revolution. That fighting continued for more than two years.

But by far the most serious error of the German Command lay in the fact that Leningrad was encircled in the summer only. The big natural obstacles, such as lakes, river courses, and marshes, which during the summer were as good as actual parts of the German siege forces, became excellent lines of communication and huge gaps in the encircling ring in winter, the moment Lake Ladoga and the Neva froze over. Through these gaps supplies and reinforcements could be brought in right through the winter months.

Moreover, towards the east, Leningrad still had a 50-mile-wide corridor all the way to Lake Ladoga so long as the Finns did not cross their old frontier in the Karelian Isthmus. As a result, Zhdanov, the Defence Commissar, was able to build the "Road of Life" over the ice of Lake Ladoga—including a motor highway and a railway branch line connecting with the Murmansk railway. Along this lifeline on ice the city was being supplied from the lake's eastern bank. Suddenly Leningrad was no longer sealed off: the German encirclement had been breached by "General Frost."

In order to close this wintertime gap, Army Group North mounted its extensive Tikhvin operation. This aimed at including Lake Ladoga in the siege front and sealing off Leningrad east of the lake. The Finns were to drive across the Svir from the north and to link up with the German Sixteenth Army east of the lake. The XXXIX Panzer Corps under General Rudolf Schmidt was to use four mobile divisions for a thrust into the almost pathless northern Russian tayga which the German Military-Geographical Records described as "virtually uncharted."

On 15th October the Corps with 12th and 8th Panzer Divisions, as well as 18th and 20th Motorized Infantry Divisions, moved off from the Volkhov bridgeheads of 126th and 21st Infantry Divisions, crossing the big river to the east. Its first objective was Tikhvin. There the last rail connection from Vologda to Leningrad was to be cut and the advance continued as far as the Svir, where the link-up was to be effected with the Finns. That link-up would have completed the encirclement of Leningrad, including Lake Ladoga.

In the evening of 8th November the Pomeranians and Silesians of 12th Panzer Division and 18th Motorized Infantry Division entered Tikhvin after stiff and costly fighting. The two divisions organized themselves for defence—General Harpe's 12th Panzer Division west of the town and General Herrlein's 18th Motorized Infantry Division east of the town. The 18th thus represented the extreme north-eastern corner of the German front in Russia.

The first part of the operation had gone so smoothly, thanks to the employment of experienced regiments, that the Fuehrer's headquarters quite seriously asked Corps whether a drive to Vologda—i.e., another 250 miles farther east—would be possible. Two hundred and fifty miles—in winter! Major Nolte, the chief of operations of 18th Motorized Infantry Division, spoke his mind very bluntly when the question was put to him by his Corps commander.

How Utopian such an idea was was shown only two days later. The morning of 15th November brought the expected full-scale attack by a fresh Siberian division, supported by an armoured brigade with brand-new T-34s. The day began with a hurricane of fire from the very latest type of "Stalin's organ-pipes." It was a savage battle. The batteries of 18th Artillery Regiment under Colonel Berger destroyed fifty enemy tanks. For several days the Siberian rifle battalions charged against the German front line—until they were bled white. Tikhvin, though but a smouldering heap of ruins, remained in German hands.

Naturally the Soviet High Command realized that the bold German Panzer operation was aiming at a link-up with the Finns on the Svir. Stalin therefore flung further Siberian divisions into the path of the Panzer Corps. Highly critical situations arose in the area of 61st Infantry Division, which was in danger of being surrounded, and stiff fighting consumed the combat strength of the Corps. All their courage was in vain. Even the hardy Finns, familiar as they were with the climate of the North Russian tayga in winter, did not succeed in crossing the Svir. The XXXIX Panzer Corps was on its own. In desert country, in the face of unceasing attacks by Siberian operational reserves, the Corps was unable to maintain its exposed positions. General von Arnim, Schmidt's successor, therefore again withdrew his divisions to the Volkhov.

The feats of the rearguard battalions covering this retreat were unparalleled. Colonel Nolte—then Major Nolte and chief of operations of 18th Motorized Infantry Division—remarked: "Not many men make good vanguard commanders. But to command a vanguard is an easy matter compared with commanding a rearguard. The vanguard commander is backing success—the rearguard commander covers up for a failure. The former is swept ahead by the enthusiasm of thousands, the latter is weighed down by the misery and sufferings of the defeated."

In terms of military discipline and courage, the retreat from Tikhvin to the Volkhov, according to Colonel-General Halder, marked a glorious page in the history of soldierly virtue. An outstanding example was 11th and 12th Companies, 51st Infantry Regiment, under Lieutenant-Colonel Grosser, who literally sacrificed themselves—who allowed themselves to be shot, bayoneted, and battered to death in order to cover the retreat of their comrades. When the spent remnants of XXXIX Panzer Corps were brought back across the Volkhov on 22nd December 1941, in 52 degrees below zero Centigrade, they had behind them an appalling experience. The Silesian 18th Motorized Infantry Division alone had lost 9000 men. Its combat strength was down to 741. These few made their way back over the Volkhov. The Tikhvin operation, the great encirclement of Leningrad, had failed.

The fate of 3rd Battalion, 30th Motorized Infantry Regiment, demonstrates how the fighting for Tikhvin surpassed the capabilities of the units involved. On its march from Chudovo to Tikhvin, when the temperature suddenly dropped to 40 degrees below zero, the battalion lost 250 men—half its combat strength—most of them through being frozen to death. In the case of some of them the frightful discovery was made that their cerebral fluid had frozen solid because they had not worn any woollen protection under their steel helmets.

From then onward the front between Leningrad and Volkhov was to be a permanent source of danger and costly fighting for the German forces in the East.

It was the penalty for having gambled away the capture of Leningrad. The penalty for trying to do too much in too many different places. As it was, Hitler had not reached his operational objectives for 1941 either in the North or on the Central Front: Leningrad and Moscow remained unsubdued.

PART THREE: *Rostov*

1. Through the Nogay Steppe

New objectives for the Southern Front–The bridge of Berislav–Sappers tackle the lower Dnieper–Mölders's fighter aircraft intervene–The road to the Crimea is barred–Battle at the Tartar Ditch–Roundabout in the Nogay Steppe–Between Berdyansk and Mariupol.

ON 12th September 1941, when 36th Motorized Infantry Division and 1st Panzer Division were rapidly advancing towards Leningrad past the Duderhof Hills under a brilliant late-summer sky, it was raining heavily on Lake Ilmen. The headquarters staff of LVI Panzer Corps had set up their command post alongside a gutted farmhouse south-west of Demyansk. General von Manstein sat in his sodden tent with his orderly officers. They were waiting for the evening report, and until then were killing time by a rubber of bridge.

Suddenly the telephone rang. Captain Specht lifted the receiver. "The Commander-in-Chief would like to speak to the general," he said.

Manstein grunted. Telephone calls at that hour usually meant bad news. But for once this was not so. Colonel-General Busch, the Commander-in-Chief Sixteenth Army, had rung to congratulate his friend Manstein.

"Congratulate me? On what, Herr Generaloberst?" Manstein asked in surprise. Busch deliberately paused for a moment and then read out a signal he had just received from the Fuehrer's headquarters: "General von Manstein will assume command of Eleventh Army with immediate effect."

The Eleventh Army! That meant the southern end of the front—the extreme right wing of Army Group South. A few hours previously the Army commander, Ritter von Schobert, had attempted a forced landing in his Fieseler Storch aircraft and had come down in the middle of a Russian minefield. Pilot and general had been blown to pieces.

Manstein received his appointment with mixed feelings. An Army command, of course, was the crowning achievement of an officer's career—but an Army command also meant giving up the personal, active direction of troops in the field. Manstein was with all his heart a commander in the field. Yet, both as chief of staff of Rundstedt's Army Group A and later as the general commanding XXXVIII Army Corps, had he also proved himself an outstanding strategist. Indeed, the pattern of the campaign against France had been Manstein's work.

In spite of all the regret at leaving LVI Panzer Corps—the Corps he had led right up to the gates of Leningrad, the Corps with which he had overcome dangerous crises, smashed Soviet armies, and frequently borne the brunt of the campaign of Army Group North—one consideration made his departure easier for him. Because he was a gifted strategist Manstein realized the mistakes made by the High Command in the north and at the centre, and had long been unhappy about the tug-of-war between Hitler and the Army High Command on the issue of the great strategic objectives. Only that morning, on 12th September, after recording his Corps' successes in the fighting against a vastly superior Soviet force south of Lake Ilmen, he had written in his diary: "In spite of everything I lack a sense of real satisfaction at these successes."

Why did Manstein lack that sense of satisfaction? Because he saw that at the top there was no clear idea of the objective that ought to be pursued, or of the purpose which his costly operations were to promote. Bock, just as the Army High Command, wanted to head for Moscow. Leeb, sticking to Hitler's original idea, wanted to make for Leningrad. And Hitler himself? Hitler did not want to make for Leningrad or for Moscow. He was after economic objectives—grain, oil, and ores. He wanted the Ukraine and the Caucasus.

It was no accident that, at the very climax of the battle of Leningrad and at a crucial phase of Sixteenth Army's successful operation against the flanking position of Moscow's defences, Hitler dispatched his best man from the north to the south.

On the southern front, about the middle of September, Field-Marshal von Rundstedt was on the point of concluding the battle of the Kiev pocket, after an initially slow and laborious operation. Together with Guderian's Panzer Group, Rundstedt's forces destroyed the bulk of the Soviet southern armies in the Ukraine.

The Eleventh Army, mounting its offensive from Rumania, had taken no part in the battle of Kiev. Together with two Rumanian armies it was to recapture Bessarabia, which the Soviets had forced the Rumanians to surrender to them in 1940. Its re-annexation was Hitler's reward for Rumania's participation in the Eastern campaign. Following the liberation of Bessarabia, Eleventh Army was to advance to the lower reaches of the Dnieper, a huge river which ran as a colossal obstacle through the zones of operations of both Army Groups. The forcing of the Dnieper crossings marked the beginning of a dual strategic task. In the words of the order: "Eleventh Army will capture the Crimean Peninsula with some of its forces, and with the bulk of its forces will drive towards Rostov along the northern edge of the Sea of Azov."

Undoubtedly the Crimea and Rostov were both highly important strategic objectives. Rostov-on-Don with its four major railway-lines and countless road intersections towards east, west, north, and south was the gateway to the Caucasus. And whoever controlled the Crimea would control the Black Sea and could exert political pressure on neighbour countries—Turkey and Persia, for instance. Turkey, in particular, was very much in Hitler's mind. He was extremely anxious to have that country on his side, for that would mean the forging of a bridge to the Mediterranean and to the fabulously rich oilfields of the Arab world. Rommel's armies in Africa and the armies in the East might link up. Might!

The plan to seize the Crimea was, moreover, motivated by considerations of economic warfare. The peninsula was a dangerous Soviet airbase for attacks on the Rumanian oilfields of Ploesti—a permanent source of anxiety to Hitler.

By seizing the Crimea and Rostov, Eleventh Army was therefore to provide the basis for Rundstedt's conquest of the "Soviet Ruhr," the Donets basin. Stalingrad on the Volga and Astrakhan on the Caspian were the more distant objectives in Hitler's mind. In fact, they were laid down in the explanatory notes to Operation Barbarossa, and, as the A-A line, figured in the detailed schedule of war aims. The A-A line meant Astrakhan–Archangel, a gigantic line right across the Soviet Union, from the Arctic Ocean along the Northern Dvina and the Volga —a distance of roughly 1250 miles. It was Hitler's finishing-line for his operation against Stalin's empire. From this line armed patrols based on big frontier fortifications on the Volga and Northern Dvina were to contain the Soviet forces and their bases on both sides of the Urals.

Only a map in hand can fully convey the fantastic objectives pursued by the highest leaders of Germany. Yet the objectives mapped out even for Eleventh Army involved tasks which were bound to lead to a dissipation of its forces.

Manstein, the cool, sober strategist, realized at once that too much was being demanded of Eleventh Army. Even though he was taking over an excellent force he knew that the best and most self-sacrificing divisions could not be expected to do things which were far beyond their capacities.

Eleventh Army had often proved its striking power. But one of its most remarkable feats was the crossing of the Dnieper at Berislav by the 22nd Infantry Division from Lower Saxony. This classical instance of a major river crossing deserves a more detailed account—if only because it represents a glorious achievement by the sappers, so often the poor relations of military history. Unlike the armoured forces and mobile divisions, the sappers never bask in the limelight of victory, but perform their indispensable duties in the shadow of battle.

Nothing demonstrates the drama of that vital crossing of the Lower Dnieper more clearly than a factual account of the operation.

On 24th August Lieutenant-Colonel von Boddien reached the western bank of the river with an advanced formation of 22nd Infantry Division, composed of the Motorized Reconnaissance Detachment 22, the 2nd Company Panzerjäger Detachment 22, 3rd Company Engineers Battalion 22, and an AA group. The town was held by strong Soviet forces.

On the following morning Boddien attacked the town. The 16th Infantry Regiment, reinforced by 2nd Company, Engineers Battalion 22, and 2nd Battalion, 54th Artillery Regiment, were brought up on lorries. Straight from their vehicles the troops joined in the fierce street fighting that was already raging. By nightfall of 26th August Berislav had been taken and was firmly in German hands.

Now came the great moment for the sappers. The Dnieper, the second biggest river of European Russia, was 750 yards wide at that point. And on the far bank were the Soviets, knowing that the Germans were planning to force the river.

Colonel Ritter von Heigl, commanding the Engineers Regiment Headquarters 690, was in charge of the first phase of the operation, the crossing itself. Two divisional sapper battalions, Nos. 22 and 46, as well as the Motorized Army Engineers Battalion 741 and the Assault Craft Detachment 903, had the task of ferrying the first waves of assault infantry across the river under enemy fire.

On 30th August, even before daybreak, the infantrymen of 22nd Infantry Division, men from Hanover and the towns and villages of Oldenburg, had taken up positions by the water's edge. The battalions of 16th Infantry Regiment were on an island in the river, in a spot inaccessible to anyone without local knowledge. A Ukrainian fisherman had shown them how to get there. The men of 47th Infantry Regiment were awaiting the order to attack at the foot of a vineyard, in a spot almost entirely devoid of cover, pressed flat to the ground. Soviet bombers and fighter-bombers kept coming over again and again, dropping parachute flares and looking for targets. Whenever they appeared all movement had to freeze into immobility. At dawn a milky white mist began to rise from the river, a real godsend.

The time was 0427 hours. The motors of the assault craft came to life with a whine. Simultaneously, artillery and heavy infantry weapons put up a heavy barrage across the river. The Soviet river defences were being kept down. Behind the assault boats the various inflatable dinghies, small and large, were being got into the water.

From the far bank white Very lights were fired: the bank had been reached. The artillery moved its barrage farther forward. Machine-guns ticked; carbines barked. Stukas and bombers of Fourth Air Fleet roared over the river and dropped their bombs on Soviet positions on the far bank. The assault boats came back for fresh infantry and then crossed over again to the far bank.

For three hours the assault-boat men had been standing by their tillers. The river was boiling with the bursts of heavy enemy artillery. A boat was blown to bits. Others capsized through near misses. But the Russians evidently had no artillery spotter left by the river. Their fire was haphazard.

The first wave of infantry had dislodged the Soviet riverside pickets and gained a small bridgehead. Heavy infantry weapons were now ferried across on sapper ferries. The initial crossing had been successfully accomplished. The infantry extended the bridgehead. Two days later it was two and a half miles deep. The second phase, the building of a bridge for the bulk of the division and for XXX Corps, could begin.

Colonel Zimmer, commanding Mountain Engineers Regiment 620 and in control of all sapper units of XLIX Mountain Corps, was in charge of the complicated technical set-up needed for the building of an eight-ton bridge with 116 pontoons. The Engineers Battalions 46 and 240 and the Mountain Engineers Battalion 54 were employed in this task, together with the Rumanian 10th Bridge-building Company —a total of over 2500 men.

The pontoons were moored some four miles upstream from the bridging-point, well camouflaged. They were first linked in twos, to make a kind of ferry, and several of these ferries were then linked to make bridge units. In accordance with a definite plan these bridge units were called downstream and steered from both banks, into the bridging-line. In this way the bridge grew out from the two banks, until its two arms met in the middle. That was always a tense moment. Only by accurate calculations on the part of the sapper officer would the last bridging units fit together exactly to make a perfect joint.

The work began at 1800 hours on 31st August. After midnight, by 0100 hours, the two arms of the bridge were within 25 yards of each other.

By 0330 hours on 1st September the gap was closed. At 0400 hours the first group of vehicles of 22nd Infantry Division moved across to the far bank. Just then a high wind sprang up and waves of up to five feet smashed against the pontoons. The vehicles on the bridge were flung about, and a few of the pontoons sprang leaks.

Right into the middle of this difficult manœuvre burst an attack by Soviet bombers. They swooped low. A direct hit. Two ferries sank, and there were 16 dead and wounded among the sappers. Repairs in the turbulent river took two and a half hours. Then traffic resumed.

But presently the Soviet bombers and fighter-bombers returned—this time with fighter cover. There was no cover for whoever was on the bridge, and the river was over 50 feet deep. The columns could only move on, hoping for the best. Bombs came crashing down. Four pontoons were sunk.

This time the repairs took seven hours. The sappers were soaked to the skin; their hands were covered in blood and their bones were aching. The bridge built over this wind-lashed, stubbornly defended river, 750 yards wide, would make military history.

Colonel Mölders with his 51st Fighter Squadron took over the protection of the bridge which the Russians were trying at all costs to destroy. In two days Mölders and his fighters shot down seventy-seven Soviet bombers. Two Luftwaffe AA units, the 1st Battalion, 14th AA Regiment, and the 1st Battalion, 64th AA Regiment, brought down a further thirteen Russian bombers.

Nevertheless a great many sappers of 1st and 4th Mountain Divisions were killed during the next few days during their arduous work on the bridge. The bridge of Berislav exacted a heavy toll. It was probably the most fiercely contested pontoon bridge of the last war. It was the bridge across which Eleventh Army mounted its decisive attack against the Crimea and the Caucasus.

The Crimean Peninsula is separated from the mainland by the Sivash, also known as the Putrid Sea, a saline marsh impassable by infantry. The expanse was neither solid ground nor sea, and was not negotiable by water craft—not even assault boats or rubber dinghies.

There were three routes across the marsh. In the west was the Perekop Isthmus, a little over four miles wide. In the centre the railway-line crossed at Salkovo. And in the east was the corridor of Genichesk, only a few hundred yards wide. On 12th September 1941, the day that Colonel-General Ritter von Schobert was killed, XXX Army Corps and XLIX Mountain Corps were advancing rapidly east of Berislav, bypassing Antonovka on both sides. Farther south was LIV Army Corps, its vanguard being 22nd and 73rd Infantry Divisions under Lieutenant-Colonel von Boddien and Major Stiefvater. They were racing the reinforced SS Motorized Reconnaissance Detachment "Leibstandarte Adolf Hitler" under Sturmbannführer[1] Meyer, to the Perekop Isthmus. The order which had sent these units into action was the last to be issued by Schobert. An attempt was made to seize the isthmus by swift assault and thus open the western door into the Crimea.

The time was 0430 hours. Between the Dnieper and the Black Sea the Nogay Steppe was glowing bright under the rising sun. It was a fantastic display. The steppe was in flower. There was not a tree, not

---

[1] Rank in Waffen SS corresponding to major.

a hill, for the eye to rest on. The view was wide and boundless, losing itself in the misty horizon. Only the masts of the Anglo-Iranian telegraph-line, built by the German firm of Siemens about the turn of the century, stood in the silent steppe like ghostly signposts. In summer there was not a drop of water to be found. Rivulets and water-courses were dried out; deep and lifeless these 'Balkas' intersected the 12,000 square miles of desert.

The first thought to leap to a soldier's mind was: What perfect ground for armour! But the Eleventh Army had no armour, apart from the armoured scout cars of its reconnaissance detachments. Here, where they could have been put to such excellent use, there were no Panzer or armoured infantry carrier units.

The spearhead of the attack was formed by motor-cyclists and armoured scout cars of the "Leibstandarte Adolf Hitler." They were followed by an advanced formation of 73rd Infantry Division. Sturmbannführer Meyer, who was driving with his leading company, searched the horizon through his binoculars. Nothing—no movement anywhere. Forward. Von Büttner's motor-cycle platoon was moving along the coast towards Adamany, from where the ground should be visible to both sides of the Tartar Ditch. Suddenly, like ghosts, a few horsemen appeared on the horizon and instantly vanished again—Soviet scouts.

Caution was needed. "Drive in open order!" The silence was uncanny. The riflemen in the side-cars were poised to leap out. The riders were hanging over to the side so as to jump off their machines all the more quickly.

It was shortly after 0600 hours. The motor-cycle detachment under Gruppenführer[1] Westphal was carefully approaching the first houses of Preobrazhenka. The village lay close by the main road from Berislav to Perekop. A flock of sheep was coming out of the village. Westphal waved his arms at the shepherd. "Get your flock off the road, man—we're in a hurry!" But the Tartar did not seem to understand. Or perhaps he did not want to? Westphal opened his throttle till the engine screamed and drove straight into the flock. The sheep scattered wildly and scampered off in panic. The shepherd shouted and sent his dogs after them. It was no use. The sheep ran off the road. A moment later the air was rent with thunder and lightning. The sheep were being blown to smithereens. The flock had run into a minefield. As though this inferno of explosions and the bloodcurdling bleating of dying sheep were not enough, enemy artillery suddenly opened up. Shells were bursting outside and inside the village. The motor-cyclists dismounted and advanced towards Preobrazhenka along the Perekop road. Suddenly before them they saw a whole wall of fire. On the far side of the village, only a few hundred yards in front of the German

[1] Rank in Waffen SS equivalent to lieutenant-general.

spearheads, stood a Soviet armoured train: it pumped its shells and machine-gun bursts straight into Meyer's and Stiefvater's companies. The effect was terrible.

"Take cover!" The men lay pressed to the ground. Machine-gun fire swept over their heads. But this fire was not coming from the armoured train: it was coming from Russian riflemen concealed in well-camouflaged foxholes and trenches barely 50 yards in front of the Germans.

Sturmbannführer Meyer gave the order to withdraw from Preobrazhenka. His armoured scout cars opened fire at the armoured train with their 2-cm. guns, to enable the rest of the unit to withdraw under cover of smoke canisters. Meanwhile a 3·7-cm. anti-tank gun of Meyer's 2nd Company was hurriedly hauled forward and started shelling the train. But no sooner had a few rounds been fired than the gun received a direct hit. Bits of steel sailed through the air, and the crash of metal drowned the screams of the men.

Meyer meanwhile dodged through the village to its far end, accompanied by his runners. From there he could see the elaborate defences of Perekop—trenches, barbed wire, concrete pillboxes. This, he realized, was not a position to be taken by a surprise coup. Any further attempt would mean the end of his formation. Gruppenführer Westphal, who had gone forward with him, suddenly shouted for a medical orderly. A shell had torn one of his arms off. Scattered right and left were the dead and wounded of his group.

"We're getting out of here," Sturmbannführer Meyer repeated. He gave the signal for retreat. His runners passed on the order. Motorcycles came roaring up from behind and about-turned. Without stopping they snatched up their wounded or killed comrades into the side-cars and raced back. The scout cars put down a smoke-screen outside Preobrazhenka, to conceal the move from the enemy. Under cover of that smoke-screen Rottenführer[1] Helmut Balke made three more trips to the front to bring back the wounded. Meyer brought the last one back. He was Untersturmführer[2] Rehrl. A shell-splinter had torn open his back. He died in the arms of his commander.

Eleventh Army's first attempt to burst into the Crimea by a *coup de main* with advanced units of its LIV Corps had failed. An hour later Lieutenant-General Bieler, commanding 73rd Infantry Division, read a signal from Meyer and Stiefvater: "Coup against Perekop impossible. Detailed account of engagement follows."

"Panzer Meyer" and Stiefvater were right. In front of the four-mile-wide exposed Perekop approach to the Crimea a system of defences had been established in considerable depth. Its central feature was the "Tartar Ditch," a ditch 40 to 50 feet deep built in the fifteenth century,

[1] Non-commissioned rank in Waffen SS.
[2] Rank in Waffen SS equivalent to lieutenant.

in the Turkish era, to protect the peninsula against the mainland. Five hundred years later it was to become a gigantic obstacle and dangerous trap for armour. To bypass it was impossible. The fortifications extended from the saline swamp of the Sea of Azov on the one side to the Black Sea on the other. The door to the Crimean Peninsula was well barred.

On 17th September, when General von Manstein assumed command of Eleventh Army at Nikolayev, the great shipbuilding centre on the Black Sea, he instantly realized that with the forces at his disposal he could not simultaneously capture the Crimea and Rostov. One or the other objective had to be set aside. But which of the two? Manstein did not hesitate long.

The Crimea represented a permanent danger to the deep right flank of the entire German Eastern Front, since the Soviets were able to pump ever new forces into the peninsula from the south, across the sea. Moreover, in enemy hands the Crimea was also an airbase threatening the Rumanian oilfields. For that reason Manstein decided to give preference to the capture of the Crimea. On the Rostov front he merely wanted to maintain contact with the enemy forces dislodged at Antonovka.

Manstein's was a good plan. The LIV Corps under General Hansen was first of all to force the Perekop Isthmus by frontal attack. For this difficult task Hansen was assigned the entire artillery, sappers, and anti-aircraft units under Army control. In addition to his own two infantry divisions—the 73rd and the 46th—the 50th Infantry Division, a little farther to the rear, was likewise put under his command. It was a considerable striking force to tackle a defensive front only four miles wide.

Manstein, of course, was a sufficiently experienced commander to realize that with these forces he might be able to force the door to the Crimea, but not to conquer an area of 10,000 square miles, a territory nearly as large as Belgium, with its many powerful fortresses and strongpoints.

As a strategist with a regular General Staff background he therefore based the second phase of his operational plan on precision and luck. General Kübler's XLIX Mountain Corps and the SS Brigade "Leibstandarte Adolf Hitler" under Obergruppenführer[1] Dietrich were to be detached from the mainland front in the Dnieper bend the moment the break-through was accomplished and brought down in forced marches in order to advance, fan out, and occupy the whole of the Crimea.

The "Leibstandarte," magnificently equipped as it was with heavy weapons, self-propelled anti-aircraft guns, self-propelled assault guns, motor-cycles, armoured scout cars, and infantry carriers, stood a good

[1] Rank in Waffen SS equivalent to general.

chance of overtaking the retreating enemy and cutting him off from Sevastopol. It might then take this important coastal fortress in the south of the Crimea by a swift blow, before it was reinforced.

The Mountain Corps was to be employed in the Yayla mountains, which were up to 4800 ft. high; it was then to seize the Kerch Peninsula and from there, eventually, drive across the narrow space of water into the Kuban and on to the Caucasus.

This plan was not just a mirage. Manstein regarded it as realizable—provided always the enemy did not mount any surprise actions in the Nogay Steppe. That was the risky aspect of Eleventh Army's operations. In order to concentrate his forces sufficiently for the capture of the Crimea, Manstein had to reduce his mainland forces to a minimum by detaching the "Leibstandarte" and the XLIX Mountain Corps. General von Salmuth's XXX Corps, to which the 72nd and 22nd Infantry Divisions belonged, had to hold on its own the front in the Nogay Steppe, supported only by the Rumanian Third Army. Manstein took this calculated risk because he had confidence in his combat-hardened divisions.

It was 24th September 1941. Mercilessly the southern sun beat down on the featureless steppe before Perekop and lay heavily over the saline marshes of the Sivash. The Soviet 156th Rifle Division was holding its deeply staggered defences. The central approach to the Crimea was covered by 276th Rifle Division. This division belonged to the Soviet Fifty-first Army, commanded by Colonel-General F. I. Kuznetsov. His order was: "Not an inch of soil to be surrendered!"

But a general's order is valid only as long as his troops are alive. After a three days' battle the 46th and 73rd Infantry Divisions burst through the neck of land. They overcame the Tartar Ditch, took the strongly fortified village of Armyansk, and thus gained open ground again for deployment.

Colonel-General Kuznetsov threw his 40th and 42nd Cavalry Divisions as well as units of 271st and 106th Rifle Divisions into his last defences along the isthmus of Ishun. The curtain was about to rise on the last act of Manstein's plan. It was now up to the "Leibstandarte" and the Mountain Corps to complete the breakthrough and to storm the peninsula.

Victory was within reach. But for the time being the Soviet High Command was able to foil the daring plan of attack.

Farther north, in the Nogay Steppe, along the anti-tank ditch before Timoshevka, there was much cautious whispering and coming and going during the night of 23rd/24th September. The regiments of 1st and 4th Mountain Divisions were being relieved for their employment in the Crimea. Rumanian mountain troops of the 1st, 2nd, and 4th Mountain Brigades were taking over the sector. Their headquarters staffs were being briefed. One German battalion after another handed over its positions to the Rumanians and moved off to the south.

"Hurry up, men; we are off to the sunny Crimea," the NCOs were urging on the companies of 91st Mountain Regiment. The men were marching at a fast pace. By the following morning they had covered 24 miles.

Of Regimental Group 13 only one battalion of infantry and one of artillery were left in their old positions. The headquarters section of 4th Mountain Division intended to move off to the Crimea with them.

"Everything ready?" Lieutenant-Colonel Schaefer, the chief of operations of 4th Mountain Division, asked Major Eder, commanding the 2nd Battalion, 94th Mountain Artillery Regiment. "Everything ready to move off, Herr Oberstleutnant," the gunner officer replied.

"What on earth is going on over there?" Schaefer suddenly asked in surprise.

A little distance away Rumanian infantry were hurriedly pulling out of the line.

"Eder, you run across to the Rumanian Brigade HQ and ask what's happening!" Eder did not have to ask many questions. The Rumanians were busy packing. They were flinging their belongings up into their lorries and getting away as fast as they could. "Russian break-through," they assured him.

As though in confirmation, rifle-fire broke out near by. Alarm! The Russians are here!

The Soviets evidently had got wind of the relief by Rumanian formations. With newly brought up forces of their Ninth and Eighteenth Army they attacked the covering lines of the Eleventh Army just as it was regrouping. Some units of the Rumanian Third Army retreated at once. The Russians pressed on, put the entire 4th Brigade to flight, and tore a nine-mile gap in the front. Faced with this situation, Manstein was compelled to recall his Mountain Corps again and employ it at the penetration point.

To complete the disaster, the Soviets also achieved a break-through on the southern wing, at General von Salmuth's XXX Corps. A breakthrough in the sector of the Rumanian 5th Cavalry Brigade was sealed off by the combat group von Choltitz with units of 22nd Infantry Division, and the front propped up again. After that followed a penetration on the Corps' northern wing. The Rumanian 6th Cavalry Brigade retired. In order to clear up this new crisis the 170th Infantry Division, placed under the Mountain Corps, had to be stopped, and the "Leibstandarte," which was already en route for the Crimea, turned about and employed against the penetration.

Manstein's plan to break into the Crimea by surprise and take Sevastopol by a coup had failed. Instead, the Eleventh Army was now in danger of being cut off from the Crimea in the Nogay Steppe, and of being encircled and possibly destroyed in the narrow strip of land between the Dnieper line and the Black Sea.

But in large-scale operations with their changing fortunes crises frequently turn into lucky chances. The two Soviet Armies which were putting such pressure on Manstein's divisions had neglected their flank and rear cover. That was to prove their doom—and that doom was Kleist. The 1st Panzer Group under Colonel-General von Kleist had discharged its task in the gigantic battles of encirclement at Kiev by

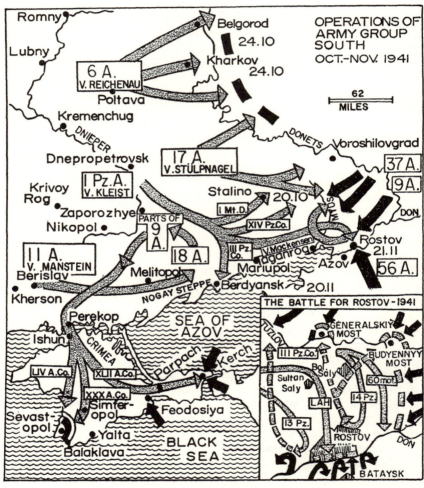

Map 13. The Donets Basin, the Crimea, and Rostov were the strategic objectives assigned by Hitler to Army Group South for 1941.

the end of September and was then available for new operations. At Dnepropetrovsk General von Mackensen's III Panzer Corps had established and held a bridgehead over the Dnieper and Samro. From this bridgehead and from Zaporozhye Kleist broke through the Soviet defences on the Dnieper, turned to the south in the direction of the Sea of Azov, and struck at the rear of the two Soviet Armies.

Before the Soviet High Command even realized what was happening its Armies, which had only just been on the point of annihilating Manstein's divisions, were themselves in the trap. Hunters became hunted, and offensive presently turned into flight. The battle of encirclement on the Sea of Azov raged across the Nogay Steppe, in the Chernigovka area, from 5th to 10th October.

The outcome was disastrous for the Soviets. The bulk of their Eighteenth Army was smashed between Mariupol and Berdyansk. The Army's Commander-in-Chief, Lieutenant-General Smirnov, was killed in action on 6th October 1941 and was found dead on the battlefield. More than 65,000 prisoners trudged off to the west. Two hundred and twelve tanks and 672 guns fell into German hands. It was a victory. But far too often during these past three weeks had the fate of Eleventh Army been balanced on a knife's edge. No doubt the German High Command took this bitter experience at the southern end of the Eastern Front as a warning that reliable victories could not be won with dissipated forces and inadequately co-ordinated operations.

At long last, therefore, Manstein received the sensible instruction to storm only the Crimea with his Eleventh Army. The capture of Rostov was assigned to Kleist's Panzer Group, to which Eleventh Army was ordered to hand over first the XLIX Mountain Corps and presently also the SS Brigade "Leibstandarte Adolf Hitler."

But the decision came three weeks too late. If this order, which at last made allowance for the actual strength of Eleventh Army, had been issued three weeks earlier the Crimea would have fallen, and Sevastopol would very probably have been taken with a surprise coup by fast formations, as envisaged in Manstein's bold plan.

Three weeks are a long time in war. And turning time to good profit was one of the outstanding skills of the Soviet High Command. As it was, Manstein and his Army were now faced with a protracted and costly battle.

2. The Battle of the Crimea

Ghost fleet between Odessa and Sevastopol–Eight-day battle for the isthmus–The Askaniya Nova collective fruit farm–Pursuit across the Crimea–"Eight girls without baskets"–First assault on Sevastopol–In the communication trenches of Fort Stalin–Russian landing at Feodosiya–Disobedience of a general–Manstein suspends the attack on Sevastopol–The Sponeck affair.

ON 16th October, while the Soviet High Command was evacuating Odessa, until then surrounded by the Rumanian Fourth Army, and transferring the evacuated units to the Crimea, General Hansen's LIV Army Corps was getting ready for the breakthrough into the peninsula north of the narrow neck of land at Ishun.

Corporal Heinrich Weseloh and Private Jan Meyer of the 2nd Battalion, 16th Infantry Regiment, 22nd Infantry Division, doubled forward, arrived at the jumping-off line for the attack. The evening of 17th October 1941 was settling over the Sivash, the saline swamp separating the Crimea from the mainland. The crater-pitted ground at Perekop and the houses of Ishun looked eerie in the falling dusk. It was cold, and there was rain in the air.

To the right of the two infantrymen a forward artillery observer was on his knees, digging himself a foxhole for the night. To the left of them were the men of their own group, also digging. Weseloh and Meyer likewise dropped down on the cold ground and began to dig a hole to give them cover for the night.

Their trenching-tools rang softly as they struck the ground. The hollow was getting deeper. They pressed themselves into it. "They say down on the coast it's still quite warm at this time of year," said Weseloh. Jan Meyer nodded. He thought of his farm back home in Hanover and cursed: "This damned war!"

"Can't go on much longer," Weseloh comforted him. "A fortnight ago, up in the steppe, we took nearly 100,000 prisoners. Three weeks ago, at Kiev, 665,000 were taken, and a little while earlier, at Uman, another 100,000. They say that some 650,000 Soviets have been taken prisoner on the Central Front to date. And at Vyazma and Bryansk they seem to have had a very good bag only a few days ago—the communiqué said 663,000 prisoners. You add that up. It makes over two million."

"I'd say there are just as many Russians about as ever," Jan Meyer grunted.

Just then a Soviet IL-15 fighter swept over their positions, firing several rounds from its cannon. Wreckage sailed through the air. The Soviets had complete air command down in the south. Even Major Gotthardt Handrick with his "Ace of Hearts" 77th Fighter Squadron was unable to do anything about it. The Soviets were vastly superior to him in numbers. In addition to ground-attack aircraft and fighter bombers they had two formations of 200 IL-15 and IL-16 fighters permanently in action. For the first time the German troops were forced to make extensive use of their trenching-tools.

Dig in—that was the first and most important commandment in the battle for the Crimea. On the entirely bare ground of the Ishun salt steppe there was no other cover than a hole in the earth. And where the Red Air Force did not strike, the Soviet artillery did. It was established in superbly camouflaged emplacements, frequently protected by reinforced concrete and armour-plating; it was fully ranged on a number of well-chosen target points and would put down sudden concentrated heavy barrages. The German artillery found it difficult to get the Russian batteries.

In these circumstances the only protection was a well-dug foxhole. And not only for the infantry: every vehicle, every gun, every horse, had likewise to be hidden several feet deep below the surface.

Night lay over Ishun—the night of 17th/18th October. In their positions between the Black Sea and the salt swamps the infantry were waiting for the dawn. The Soviets also were waiting. They knew what was going to happen and were feverishly organizing the defence of the vital peninsula. Two days earlier, on 16th October, Stalin had evacuated Odessa, which had been encircled by the Rumanian Fourth Army since early August. Major-General Y. E. Petrov's coastal Army was to help defend the Crimea. By means of hurriedly improvised naval transports Petrov's coastal Army was to be switched to Sevastopol. It was a correct move. For if Manstein succeeded in getting into the Crimea, Odessa would have lost its importance as a port and naval base on the Black Sea anyway. It was more important to hold the Crimea and, above all, Sevastopol. The quick withdrawal by sea of an entire Army from Odessa was a bold operation which few people would have expected the Soviet Union to pull off, inexperienced as it was in naval warfare.

Aboard thirty-seven large transports totalling 191,400 GRT and on a variety of large and small naval vessels the bulk of the coastal Army, some 70,000 to 80,000 troops, were embarked in a single night and, unspotted by the German Luftwaffe, shipped to Sevastopol. Admittedly, only the men were evacuated from Odessa. Horses and motor vehicles had to be left behind. The heavy guns were dumped in the harbour because there were no loading-cranes. The Soviet 57th

Artillery Regiment went on board without a single gun, without a single vehicle, and without a single piece of equipment.

Petrov's forces were then sent into action on the Ishun front in forced marches, just as they had arrived at Sevastopol—ragged and quite inadequately equipped.

For his thrust across the isthmus Manstein had lined up three divisions of LIV Army Corps. Indeed, there was no room for more formations in the four-mile-wide corridor. Reading from left to right, they were the 22nd, 73rd, and 46th Infantry Divisions and parts of 170th Infantry Division. Behind them stood XXX Corps with the 72nd, the bulk of the 170th, and the 50th Infantry Divisions. Still on the road, but later to follow the attacking Corps of Eleventh Army, was the XLII Corps with 132nd and 24th Infantry Divisions. The Fuehrer's Headquarters had made this Corps available to Manstein on condition that its divisions were moved across into the Kuban area from Kerch as quickly as possible, to advance to the Caucasus.

Manstein's six divisions were opposed by eight field divisions of the Soviet Fifty-first Army; to these must be added four cavalry divisions, as well as the fortress troops and naval brigades in Sevastopol. Moving towards the front were General Petrov's units from Odessa.

It seemed as if the night would never end. The forward observers were lying behind their trench telescopes. The riflemen were crouched in their two-men foxholes, pressed close together, shivering. Immediately behind the most forward infantry positions were the guns of the medium artillery and the smoke mortars which were to be used here for the first time in the sector of Eleventh Army. They were hidden by earth ramparts and camouflage netting. In position farther back was the heavy artillery with its 15- and 21-cm. guns. At 0500 hours a gigantic thunder-clap rent the grey dawn. The battle for the Crimea was opening with a tremendous artillery bombardment from every gun in Eleventh Army. It was an inferno of crashing noise, flashes of fire, fountains of mud, smoke, and stench. With a roar and a scream the smoke mortar rockets with their fiery tails streaked towards the enemy positions, showering the defenders of the isthmus of Ishun with a hail of iron and fire.

The time was 0530. The inferno was only 100 yards in front of the positions of the assault regiment. For a moment the barrage was silent. Then it started again, but this time the shellbursts were farther away: the guns had lengthened their range. That was the signal for the infantry. The men scrambled out of their earth-holes. "Forward!" They charged. Machine-guns gave them covering fire. Mortars were keeping enemy strongpoints quiet.

But the German artillery bombardment had not put the Soviets out of action in their long and carefully prepared positions. Russian machine-guns opened up. Soviet artillery fired well-aimed salvos and time and again forced the attackers to take cover.

Only step by step could the charging infantrymen gain ground. On the left wing Colonel Haccius of the 22nd Infantry Division from Lower Saxony made a penetration in the enemy lines with his battalions of 65th Infantry Regiment and seized the fortified ridge of high ground which blocked the passage. But heavy enemy gunfire forced them to dig in.

Things went less well in the sector of 47th Infantry Regiment. The assault companies got stuck in front of a powerful wire obstacle and were shot up by the Soviets. Those who were not killed worked their way back. The 16th Infantry Regiment, 22nd Infantry Division, had to be brought up from reserve positions; it made a flanking attack and rolled up the Soviet defences in front of 47th Infantry Division. The advance was resumed. The so-called Heroes Tumulus of Assis, a commanding earth mound in an otherwise completely flat terrain, was stormed by men of 47th Infantry Regiment. But the Russians did not surrender. They died in their foxholes and trenches.

In the sector of 73rd Infantry Division, on the right of 22nd Infantry Division, the regiments also gradually gained ground. And on the right wing units of 46th and 170th Infantry Divisions worked their way into the strongly fortified system of Soviet defences.

But these deeply staggered defences seemed to have no end to them —wire obstacles and more wire obstacles, thick minefields with wooden box mines which did not respond to the sappers' detectors, as well as emplaced and remote-controlled flame-throwers. Moreover, buried tanks and even electrically detonated sea-mines completed these "devil's plantations" which the gallant sappers had to weed.

Field position after field position had to be taken by the infantry in costly fighting through these mile-deep defences. Frequently the situation was saved only by the assault artillery employed in support of the infantry: the lumbering monsters of the Self-propelled Gun Battalion 190 breached the wire obstacles and pillbox lines for the infantry companies. The battle raged for eight days—eight times twenty-four hours. At last the entrance to the Crimea was forced at several points. Even Petrov's coastal Army had been unable to prevent this. According to Colonel P. A. Zhilin, that Army lost most of its men and equipment during the last three days of the fighting for the isthmus. Zhilin ascribed these heavy casualties to "massed German tank attacks." He is mistaken. Manstein had no tank formations at all. They were Major Vogt's two dozen self-propelled guns of Battalion 190— known as the Lions after their tactical sign—which, together with 170th Infantry Division, decisively smashed General Petrov's coastal Army.

At the same time Eleventh Army command could not help noticing that the battle strength of its own assault formations had begun to decline during these days of heavy fighting. The 25th and 26th October, in particular, had seen many crises. And on 27th October there had been some fierce engagements with Petrov's Odessa regiments before the Soviet resistance weakened.

Manstein therefore fixed 28th October as the date for the final breakthrough strike. But this blow did not connect: the Soviet Fifty-first Army had abandoned its positions under cover of darkness and withdrawn to the east. The remnants of Petrov's coastal Army were streaming south in disorder, in the direction of Sevastopol. The German breakthrough into the Crimea had come off.

Eleventh Army could now go over to the pursuit. In the office building of the Askaniya Nova collective fruit farm, just under 20 miles north-east of Perekop, runners came and went ceaselessly on 28th October. In the large conference room of Army headquarters Manstein's chief of operations, Colonel Busse, had spread out his situation maps. Arrows, lines, little circles, and flags marked the incipient flight of the Russians.

Towards noon Manstein entered the map-room together with Colonel Wöhler, the Chief of Staff of Eleventh Army. "What d'you think of the situation, Busse?" Manstein asked his chief of operations. "Are the Russians going to give up the Crimea?"

"I don't think so, Herr General," Busse replied.

"Neither do I," Manstein returned. "If they did they would lose control of the Black Sea and throw away their strong positions threatening the flanks of our Army Group South. They won't do that in a hurry. Besides, it would be rather difficult to embark two Armies and get them away."

Wöhler pointed at the map. "The Russians are certain to try to hold Sevastopol, Feodosiya, and Kerch. They will save their defeated troops by getting them into these redoubts; there they will replenish them and send them into attack again. So long as they hold the naval fortress of Sevastopol they are able to do that."

"That's just what we've got to prevent," Manstein retorted.

Busse nodded. "But how are we going to turn our infantry into mobile formations? If only we had a Panzer or motorized division! It would make things a lot easier."

Colonel Wöhler took this as his clue. "We'll amalgamate all available motorized sections of infantry divisions, from reconnaissance detachments to anti-aircraft and anti-tank guns, and send them forward as a fast combat group!" Busse wholeheartedly agreed with the idea.

"Very well," Manstein decided. "Busse, you'll see that such a combat group is formed. Colonel Ziegler is to lead it. His first objective is to be Simferopol, the main city and transport centre of the peninsula. The

In rapid succession the rocket mortars discharged their massive missiles. With a terrifying scream and a trail of fire they came down on the enemy positions.

H-hour was 0315. The first day of the war in Russia dawned. Tanks were standing by to advance. Engineers were building the first emergency bridge over the Bug.

This is what the advance looked like. The spearhead of an infantry regiment has taken a village. A motor-cycle battalion is advancing alongside a highway.

Battle is joined. An anti-tank gun in action against Soviet tanks. German artillery being hauled into firing position. Armoured carriers advancing through the blazing village of Chaussy.

Encounter in the anti-tank ditch in front of the Desna: men of 197th Infantry Division advancing to the east, prisoners from a Mongol division trudging to the west.

Railway sappers: repairing blown-up bridges and converting the Soviet broad gauge to standard gauge were also battles of vital importance.

Duel in a forest near Leningrad. A T-34 has broken through. An anti-tank gun is barring its way. They are facing each other at 20 yards. Menacingly, the tank's cannon is swinging towards its target: just then a shell from the anti-tank gun scores a direct hit on the tank's gun-shield.

This is how Russia's rivers were forced. Assault engineers of XXIV Panzer Corps racing across the Dnieper. Motor-cycle troops crossing the Berezina by ferry under enemy artillery fire. Cavalry fording the Pripet.

The Red Cross—a horse-drawn medical company during the advance.

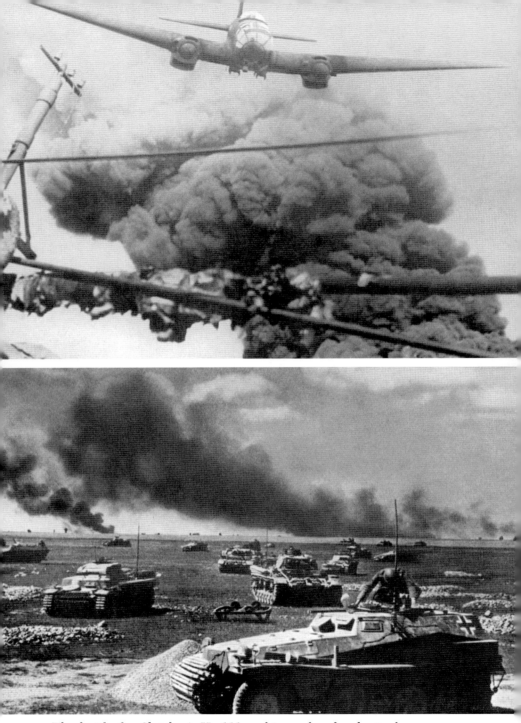

The battle for Slutsk. A He-111 making a low-level attack on enemy fuel-dumps. German tanks deploying for attack. In the background the blazing town of Slutsk.

The Dniester has been forced. Mortars digging in to provide cover for bridging operations. Roslavl, the hotly contested pivot of the central front, has fallen.

War ancient and modern. Horsemen of the East Prussian Cavalry Division on reconnaissance. Riflemen of a Panzer division advancing under cover of a Mark III tank.

The T-34, the great Russian surprise: perfect in its purposeful design, superior in fire-power, armour, and mobility to the German armoured fighting vehicles.

The golden-domed cathedral of Smolensk was packed when the German front-line troops allowed an old Russian priest to hold the first divine service.

Novgorod, the ancient trading centre north of Lake Ilmen, goes up in flames.

At the Fuehrer's Headquarters: Hitler with Keitel, Halder, and Brauchitsch.

Tank warfare: the German 8·8-cm. Flak gun was the bogey of the Russians. Units of 3rd Panzer Division advancing towards Romny for the battle of Kiev. An assault gun on picket duty on the perimeter of a pocket; the killed men covered by the tarpaulin.

The advance over Russia's roads consumed the men's strength. The first autumn rains fell.

(*Overleaf*) The offensive grinds to a halt: tanks, supplies, and horse-drawn vehicles are bogged down.

This was all that was left to them.

Hitler reaches out for Moscow. A German machine-gun providing cover 20 miles outside the city. Colonel-General Hoth (in the jeep) in conversation with Major-General Raus. Anti-tank obstacles in the main streets of Moscow. An assault gun with grenadiers on the advance.

Battle in the Arctic: supplies for the German lines being brought up by reindeer sleigh.
U-boats stalking the American convoys off Murmansk and Archangel.

The Russian winter has broken the back of the German offensive against Moscow. The retreating divisions are haunted by the spectre of Napoleon.

The Siberians are sent into action. Well equipped and experienced in winter warfare, they attacked at Kalinin, in the forests of Klin and at Mozhaysk.

The reality of winter warfare: a Soviet charge broken up by a German machine gun position. Assault guns and infantry of Ninth Army counterattacking at Rzhev.

125th Infantry Division fighting in the streets of Rostov. Field howitzers firing point-blank at stubbornly held barricades and fortified balconies.

May 1942: units of Sixth Army near Kharkov. The fortress of Sevastopol under bombardment prior to the attack of Eleventh Army.

Front line in the mountains: tanks and motor-cycle troops at the foot of the Caucasus. Mountain troops with mules crossing a stream swollen after rain.

The afternoon of 23rd August 1942: Stalingrad lies spread out before them. General Hube in his armoured command car at the head of 16th Panzer Division. With the field-glasses Colonel-General von Richthofen, C.-in-C. Fourth Air Fleet.

In the late summer of 1942 Sixth Army crossed the Don and advanced on the Volga. Four months later Colonel-General Hoth's formations were to break open the Soviet ring around Stalingrad. The spearhead of 11th Panzer Regiment was within 40 miles of its objective.

Machine-gunners in the churned-up soil of Stalingrad. A flak gun on a self-propelled chassis at the workers' settlement of the hotly contested tractor works.

Soviet assault party inside the "Red Barricade" ordnance factory.

Stalingrad's Red Square with the ruins of the department store. The graves of the dead and the dugouts of the living are close together. Field-Marshal Paulus during his first interrogation.

way to Sevastopol and the south coast lies through that town. And that way has got to be blocked."

Manstein picked up a coloured crayon. With a few quick strokes he sketched out his operational plan on the map: the XXX Army Corps with 22nd and 72nd Infantry Divisions would advance behind Ziegler's fast combat group via Simferopol and Bakhchisaray to the south coast —to Sevastopol and Yalta. The newly arrived XLII Army Corps with 46th, 73rd, and 170th Infantry Divisions would move towards Feodosiya and the Isthmus of Parpach. The LIV Corps was to drive south with 50th and 132nd Infantry Divisions, straight towards Sevastopol. Perhaps the fortress could be taken by a surprise attack after all.

That was Manstein—bold, quick of decision, and with a sure eye for the situation as a whole. His plan cut across the enemy's intentions. For General Kuznetsov was withdrawing his Soviet Fifty-first Army towards the south-east, in accordance with orders, to offer resistance at Feodosiya and Kerch.

General Petrov's coastal Army was utterly disorganized. It no longer had contact with its High Command, and hence it had no order for a withdrawal. Petrov assembled all his commanders, chiefs of staffs, and commissars of divisions and brigades at the headquarters of 95th Rifle Division in Ekibash. There was much heated discussion. Everybody was afraid to take the responsibility. Eventually it was decided to withdraw to the south, to defend Sevastopol.

That was exactly how Manstein expected the Soviets to react when he sketched out his plan at the Askaniya Nova farm. "Any questions, gentlemen?"

"None, Herr General!"

"Very well, you'll see to everything, Busse. I'm driving to XXX Corps."

Heels clicked. Outside, in the courtyard, the engine of the command car sprang into life. The radio transmitters moved off. The mobile headquarters section, the advanced command post of Eleventh Army, was moving off to the front.

As Manstein arrived at XXX Corps the message had just arrived that Major-General Wolff's 22nd Infantry Division, a former airborne division and hence somewhat better equipped with motor vehicles, had already organized its own motorized vanguard detachment from sappers, anti-tank gunners, Army anti-aircraft guns, infantry, and artillery. This force, commanded by Major Pretz, had already driven past Taganash to the road and railway junction of Dzhankoy.

On 1st November Colonel Ziegler's combat group took Simferopol. Together with the reconnaissance detachment of 22nd Infantry Division under Lieutenant-Colonel von Boddien, it then penetrated over the mountains down to the south coast at Yalta, cutting off strong

K

forces of the Soviet coastal Army still streaming back towards Sevastopol.

In the eastern part of the Crimea the 46th Infantry Division reached the isthmus of Parpach and blocked it before the bulk of the Soviet formations got there. On 3rd November the regiments of 170th Infantry Division took the town and harbour of Feodosiya. In fierce fighting the 46th and 170th Infantry Divisions burst through the isthmus of Parpach. Lieutenant-Colonel Thilo, the commander of 401st Infantry Regiment, and his adjutant, Lieutenant von Prott, were killed in front of the strongpoints and wire obstacles. Casualties were heavy. The companies were down to twenty or at the most thirty men. But the victory was complete. Only the Soviet Army headquarters personnel and some defeated formations without heavy weapons succeeded in escaping to the mainland by way of the Kerch road. On 15th November the strongly fortified town of Kerch was captured.

The advanced detachment of 22nd Infantry Division under Major Pretz was also advancing according to time-table. Bypassing Simferopol, it drove into the craggy Yayla Mountains. The men, though unused to mountain conditions, acquitted themselves admirably. In cooperation with 124th Infantry Regiment, 72nd Infantry Division, they took Alushta and surrounded a Soviet cavalry division. Yalta, the famous harbour and seaside resort, the Monte Carlo of the Black Sea, was occupied.

Lieutenant-Colonel Müller with his 105th Infantry Regiment, 72nd Infantry Division, turned along the coast road to the west, towards Sevastopol, and with a bold stroke took Balaclava, the southernmost bastion of the fortress. Everything seemed to be going according to plan.

The 50th and 132nd Infantry Divisions of LIV Corps, coming from the north, were likewise pressing against the Sevastopol approaches. But abruptly Soviet resistance stiffened. Soviet naval infantry and fortress artillery, intact crack units, including the officer cadets of 79th Officer Aspirant Brigade from Novorossiysk, intervened in the fighting. They did not yield an inch. It became obvious that with the available combat-weary German regiments Sevastopol could not be taken by surprise attack. Manstein was denied the ultimate prize of victory.

But even though Eleventh Army's pursuit lacked the crowning glory of a rapid fall of Sevastopol, the vigorous offensive spirit of its formations had nevertheless brought about the virtual annihilation of the enemy in the field. Twelve rifle divisions and four cavalry divisions had been largely destroyed. Six German infantry divisions took over 100,000 prisoners and destroyed or captured more than 700 guns and 160 tanks.

From 16th November 1941 onward Eleventh Army was faced with

the task of taking the last enemy bulwark in the Crimea, one of the strongest naval fortresses in the world, by attack from the landward side. Sevastopol had to fall—one way or another. It was not enough to bypass the huge naval fortress with its extensive port or to seal it off merely on the landward side. For that would enable Stalin, at any time he chose, to mount amphibious operations against the flank of the German Eastern Front. The time had therefore come to prepare a systematic attack against the fortress. Nothing must be left to chance. The correct employment of artillery and its supply with ammunition was one of the key problems in this operation, next to the final sealing off of the fortress on the landward side.

The kind of preliminary work done by the artillery is illustrated by a typical scene at the batteries of 22nd Infantry Division north-east of Sevastopol. Sergeant Pleyer had just emptied the rain-water from a rusty food-tin, which stood underneath a leaky spot in the roof of the ancient Russian wooden dugout, when the telephone rang.

"Dora Two," Pleyer said into the instrument. Dora Two was the headquarters of 22nd Artillery Regiment. It was situated in the notorious Belbek Valley, on Hill 304, close to the small village of Syuren. The distance to Sevastopol was 17 miles.

The voice at the other end identified itself as Albatross Three. "I'm listening," said Pleyer. And repeating every word slowly he wrote down the message: "Last night eight girls arrived without baskets. Message ends."

"Message understood. Out." Pleyer replaced the receiver and picked up a green file from the shelf by the telephone. The instrument rang again.

This time the caller was Heron Five. And Heron Five had an even more curious message for Pleyer than Albatross Three. Instead of eight girls having arrived without baskets, this time it was "Gerda bombarded with cake by organist."

Sergeant Pleyer did not laugh. Solemnly he wrote down the message, repeating, "... with cake by organist."

There was a constant string of this kind of message. They came from the forward observers of muzzle-flash and sound-ranging teams belonging to the batteries. Their messages about the Soviet gun positions located by them in the Sevastopol fortress area had to be passed in code because in the difficult mountainous area the Russians time and again succeeded in tapping the German telephone-lines. For that reason code names had been given to calibres, geographical features, battery emplacements, troop units, and the German observer posts; these then resulted in such curious combinations as "girls arriving without baskets" or "Gerda being bombarded with cake by the organist." The information was entered in the ranging-cards at the artillery command posts. Every identified gun, every observer post, and every strongpoint was carefully entered and accurately ranged. Thus the gunners were familiar with all important targets. In this way the fortress and its approaches were ceaselessly probed, reconnoitred, and plotted.

The kind of thing we have witnessed at Dora Two was going on at all command posts throughout Eleventh Army towards the end of November. They worked feverishly. Manstein wanted to take Sevastopol by Christmas. Eleventh Army had to be freed as soon as possible for its next task—the advance to the Caucasus. It could not afford to be tied down in the Crimea for months on end. For that reason Manstein concentrated all the forces he had on the attack against Sevastopol.

By difficult mountain fighting, in which Eleventh Army was now able to use also the newly arrived Rumanian 1st Mountain Brigade, the gap between the left wing of LIV Corps and XXX Corps in the Yayla Mountains was closed. But the four divisions which stood east of the fortress at the end of November were hardly enough for a final attack. As everywhere throughout the entire Russian campaign, forces too small numerically were again faced with objectives too great for them. As a result, Manstein had to accept the risk of denuding the exposed Kerch Peninsula apart from a single division—the 46th Infantry Division. That meant that the 185 miles of coastline were guarded by virtually no more than reinforced field pickets. What would happen if the Russians landed at Kerch? Manstein just had to hope for the best. He had every confidence in the commander of XLII Army Corps, Count Sponeck, an experienced and energetic general, and in his 46th Infantry Division.

On 17th December everything was ready for the attack on Sevastopol. At first light the guns of all calibres opened up along the entire 12-mile front of LIV Corps. General von Richthofen's VIII Air Corps again played its part in the operation. His ground-attack aircraft and dive-bombers attacked Soviet fortifications and gun emplacements. The first battle of Sevastopol had begun.

The town was in flames. It was to be taken from the north. The main weight of the attack was in the sector of 22nd Infantry Division, forming the right wing of LIV Corps. Alongside it were 132nd, 24th, and 50th Infantry Divisions. The grenadiers of 16th Infantry Regiment charged up the slopes of the Belbek Valley and made deep penetrations in the Soviet lines.

The 2nd Battalion penetrated as far as the notorious Kamyshly Gorge, and in a daring thrust gained the commanding height of Hill 192. Exhausted, and thinned out by heavy casualties, the platoons dropped down among the scrub. Together with units of 132nd Infantry Division, its neighbouring formation on the south, the 16th Infantry Regiment cleared the enemy out of the glacis and drove right against

The Battle of the Crimea   293

the fortified zone proper south of the Belbek Valley. The assault battalions of 132nd Infantry Division, superbly supported by mine mortars of the assault sappers, gained no more than four miles on the first day of the attack. Even the terrifying "Stukas on foot" were unable to break the tough resistance of the gallant defenders.

Farther to the right, on the ridge of high ground, the battalions of 65th Infantry Regiment were fighting their way forward through pillboxes and wire obstacles in an icy winter wind. They gained ground only slowly.

On the extreme right, in the sector of 47th Infantry Regiment and the Rumanian Motorized Regiment, the companies had been stuck for the past three days in front of the fortifications of the Kacha Valley under murderous defensive fire. Conditions were frightful.

On 21st December, in the sector of 47th Infantry Regiment, 22nd Infantry Division, Captain Winnefeld swept his company with him out of the inferno. Things could not possibly be worse: if they stayed where they were they would certainly be killed. If they charged they might possibly have a chance of surviving.

"Forward!" Into the Russian trenches! Hand-grenades—trenching-tools—machine pistols! Kill or be killed! The 3rd Battalion, 47th Infantry Regiment, likewise charged and broke into the Russian lines. On the coast the squadrons of Reconnaissance Detachment 22 and 6th Company "Brandenburg" Special Purpose Regiment seized the most forward Soviet strongpoint.

Then began a frightful tearing and clawing at the Soviet defences. At last, on 23rd December, 22nd Infantry Division reached the north-south road to the fortress with Colonel von Choltitz's 16th Infantry Regiment. The outer ring of fortifications around Sevastopol was in German hands.

But Sevastopol was tough. From the twin turrets of the underground heavy armoured battery "Maxim Gorkiy" the defenders pumped their 30·5-cm. shells into the German positions. Pillboxes and machine-gun posts were belching fire.

In this inferno the German troops spent Christmas Eve. There were no candles, no church bells, and no letters. For many there was not even a plateful of hot food.

Progress by 24th and 132nd Infantry Divisions was only step by step. Well-aimed Soviet mortar-fire was battering German reserves in the clearings among the scrub and on the road. The defenders were solidly established in earth and timber dug-outs which had to be knocked out one by one. Thus the attack was reduced to a multitude of separate actions. The battalions of 24th Infantry Division literally killed themselves fighting. The only progress made was in the sector of 22nd Infantry Division.

On 28th December, at 0700 hours, the weary men of 22nd and 24th

Infantry Divisions rallied for the final assault against the core of the fortress. The regimental commanders were sitting at their field telephones, receiving their orders.

"All-out effort," was the order. "The fortress must fall by New Year's Eve!" By New Year's Eve. So off they went.

Anyone who was in this action flinches to this day at the mere thought of it. They were appalling battles for 65th, 47th, and 16th Infantry Regiments.

Colonel von Choltitz with his 16th Regiment was in the very heart of the attack. By nightfall of 28th December his assault troops had worked their way close to the powerful Fort Stalin, which commanded the northern sector outside Sevastopol. If this fort could be smashed the road would be open to Severnaya Bay, the huge harbour of Sevastopol. And whoever commanded the bay could strangle the fortress.

At that moment, in the morning of 29th December, the disastrous news arrived at Manstein's headquarters like a bombshell: following preliminary landings at Kerch, strong Soviet invading forces had now also landed at Feodosiya, on the isthmus between the Crimea and the Kerch Peninsula. They had over-run the weak German covering lines and had taken the town. The only troops left to defend the area were the 46th Infantry Division and some weak Rumanian units. Everything else was engaged in the fighting for Sevastopol.

"What's to be done now, Herr General?" the chief of operations of Eleventh Army asked his C-in-C. What, indeed, was to be done? Should they let matters ride at Kerch and Feodosiya until after Sevastopol had fallen? Or should the battle for the fortress be suspended and the forces thus freed be switched to the threatened points in the rear of the front?

Manstein was not a man of precipitate decisions. He walked over to the school-house of the village of Sarabus, where since mid-November Eleventh Army had had its headquarters, in order to study the latest reports. He himself, as well as his chief of staff and the chief of operations, had their quarters in the old farmhouse next door, in rooms very modestly furnished. A bed, a table, a chair, a stool with a wash-bowl, and a coat-hanger—these were the entire furniture. Manstein disliked furniture being requisitioned in order, as he put it, "to create comfort which the troops have to do without."

The map in the headquarters situation room revealed the mortal danger in which the Crimean Army had been for the past five hours. A few days earlier, at Christmas, units of the Soviet Fifty-first Army had made a surprise crossing of the Strait of Kerch, only three miles wide, and, following further successful landings on 26th December 1941, had established themselves to both sides of the town.

Lieutenant-General Count von Sponeck, commanding XLII Corps, had dispatched his 73rd and 170th Infantry Divisions to Sevastopol and was now left in the peninsula with only the 46th Infantry Division. But its three regiments had succeeded, by an immediate counter-attack in a temperature of 30 degrees below zero Centigrade, in sealing off the Soviet bridgeheads and, by drawing on their last reserves, in actually mopping some of them up. Manstein had heaved a sigh of relief and had allowed the offensive operations at Sevastopol to continue. But now, on 29th December, the Russians had been inside Feodosiya since 0230 hours.

Manstein considered the red arrows on the situation map. Unless some units were quickly thrown into the path of the Soviets they would

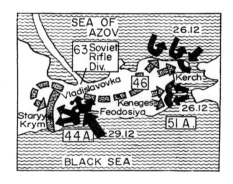

Map 14. The Soviets land on the Kerch Peninsula.

be able to seal off the isthmus of Parpach, the 12-mile-wide passage from the Crimea to the Kerch Peninsula, cut off 46th Infantry Division, and strike at the rear of the German lines before Sevastopol. Once again the German High Command's cardinal sin against the laws of modern warfare was made patent: Eleventh Army lacked mobile motorized formations as operational reserves. There was only one solution: some forces had to be detached from the Sevastopol front and switched to Feodosiya.

Anxiously the chief of staff and the chief of operations stood next to Manstein in front of the map. Was the battle for Sevastopol to be broken off at this moment? Surely that would be exactly what the Soviets hoped to achieve by their landings at Kerch?

Manstein and his staff officers weighed up the situation. Did it not look as if at Sevastopol, in the sector of 22nd Infantry Division, only one last effort was needed to break through, at least, to the vital harbour bay? If that came off a commanding position would have been gained and the attack on the town might then be suspended without any risk for a few weeks. Control over the Severnaya Bay would prevent the fortress from being reinforced from the sea. It would be firmly encircled, and the divisions thus freed could then be switched to Feodosiya and Kerch in order to throw the Soviet forces there back into the sea. Provided only that General Count Sponeck could hold out for another two or three days. Surely with all available reserves scraped together it should be possible to tie down the Russians at Feodosiya that long.

That, clearly, was the way to do it. Manstein therefore ordered: "In the northern sector before Sevastopol 22nd Infantry Division will take Fort Stalin and drive on as far as the harbour bay. From the east the attack on the fortress will be suspended; 170th Infantry Division will be withdrawn from the front at once and dispatched to Feodosiya."

Now began a race against time. Would the calculations come right? On 29th December 1941, at 1000 hours, a coded signal was received at Army headquarters from Count Sponeck's Corps. Its contents were alarming: "Corps command evacuating Kerch Peninsula. 46th Infantry Division begun to move off in direction of Parpach Isthmus."

Manstein was staggered. Some days before, at Christmas, when the Soviet 244th Rifle Division had made its landings on both sides of Kerch, Count Sponeck had suggested the evacuation of the peninsula. Manstein had firmly rejected the idea and had expressly ordered that these crucial approaches to the Crimea must be defended. Now the General commanding XLII Corps was acting without authority against this strict order.

Manstein ordered a signal to be sent back: "Withdrawal must be stopped at once."

But the signal no longer got through. Corps headquarters did not reply any more. Count Sponeck had already had his wireless station dismantled. It was the first instance of a commanding general's disobedience since the beginning of the campaign in the East. It was a symptomatic case, involving fundamental principles. Lieutenant-General Hans Count von Sponeck, the scion of a Düsseldorf family of regular officers, born in 1888, formerly an officer in the Imperial Guards, was a man of great personal courage and an excellent commander in the field. While commanding the famous 22nd Airborne Division, which in 1940 captured the "fortress of Holland" with a bold stroke, he had earned for himself the Knights Cross in the Western campaign. Subsequently, as the commander of 22nd Infantry Division, into which the Airborne Division had been converted, he also distinguished himself by outstanding gallantry during the crossing of the Dnieper.

The significance of the affair lay in the fact that Count Sponeck was the first commanding general on the Eastern Front who, when the attack of two Soviet Armies against a single German division faced him with the alternatives of hanging on and being wiped out or withdrawing, refused to choose the former alternative. He reacted to the Soviet threat not in accordance with Hitlerite principles of leadership, but according to the principles of his Prussian General Staff upbringing. This demanded of a commanding officer that he should judge each situation accurately and dispassionately, react to it flexibly, and not allow his troops to be slaughtered unless there was some compelling and inescapable reason for it. Sponeck saw no such reason.

What were the considerations which induced the Count to disregard superior orders?

Although we have no notes left by him personally, his chief of operations and his deputy chief of staff, Major Einbeck, have laid down in a memorandum the arguments of the Corps command. An instructive report is also extant from Lieutenant-Colonel von Ahlfen, the chief of staff of 617th Engineers Regiment.

This is the picture that emerges from these reports: On 28th December 1941 Lieutenant-General Himer's 46th Infantry Division, by rallying all its reserves, succeeded in smashing the Soviet bridgehead north of Kerch. The Soviets, and above all the Caucasians, had accomplished incredible feats. In spite of its being 20 degrees below zero Centigrade they had waded to the steep coast up to their necks in water, and had gained a foothold there. Without any supplies they had held out for two days. Their wounded had frozen rigid into ice-covered lumps of flesh. Frozen to death. The landings south of Kerch were likewise sealed off. But at that moment Soviet naval units attacked at Feodosiya, 60 miles behind Kerch. A heavy cruiser, two destroyers, and landing-craft entered the harbour under cover of darkness.

Of Army Coastal Artillery Battalion 147, detailed to defend Feodosiya, only four 10·5-cm. guns and the headquarters personnel had so far got to their destination. In addition, only one German and one Czech-manufactured field howitzer were in the port. The Soviet warships trained their searchlights on to the defender's gun emplacements and shelled them to smithereens with their heavy naval guns. Then the Russians disembarked.

For infantry engagements the German forces available consisted of the sapper platoon of an assault boat detachment and a Panzerjäger platoon with two 3·7-cm. anti-tank guns. Luckily the Engineers Battalion 46, en route to the west, had taken up quarters in Feodosiya for the night. Count Sponeck put Lieutenant-Colonel von Ahlfen in charge of repulsing the Soviet landing. The lieutenant-colonel mobilized every single man he could find—paymasters, workshop mechanics, the personnel of food stores and field post-offices, a road construction company, and the men of a signals unit. From this motley crew the first covering line was organized outside the town.

At 0730 hours a signal arrived at Count Sponeck's headquarters at Keneges: "Soviets are also landing north-east of Feodosiya on the open coast." An entire division was disembarking.

A few minutes later telephone connections with Army and with

Feodosiya were cut—just after Count Sponeck had received the information that Manstein was sending 170th Infantry Division from Sevastopol and two Rumanian brigades from the Yayla Mountains to Feodosiya.

What were the Soviet intentions? Their tactical aim, clearly, was to cut the narrow neck of land between the Crimea and the Kerch Peninsula, and to annihilate the trapped 46th Infantry Division. But their strategic objective, undoubtedly, was to strike swiftly into the Crimea from their foothold at Feodosiya, to occupy the traffic junctions behind the Sevastopol front, and to cut off Eleventh Army from its supplies.

That the Russians were in fact pursuing this strategic objective, and not just making local raids on the coast, was proved by the fact that their invading forces comprised two Armies—the Fifty-first under General Lvov at Kerch and the Forty-fourth under General Pervushin at Feodosiya. The Forty-fourth Army had already disembarked some 23,000 men of 63rd and 157th Rifle Divisions.

General Count Sponeck asked himself: Was 46th Infantry Division strong enough to throw the enemy forces back into the sea at Kerch and at the same time hold the Parpach Isthmus against the new landings at Feodosiya? His answer was No.

Major Einbeck records: "Corps command could only regain the initiative by immediately switching the focus of operations to the Feodosiya area. That was the place where the danger of a drive against Dzhankoy or Simferopol, now threatening Eleventh Army, might be averted. This decision involved surrendering the Kerch Peninsula as far as the Parpach line."

Count Sponeck believed that, in view of the responsibility he had for his 10,000 men, there was no time to be lost. Because of his clearer, local grasp of the situation he felt justified in acting against the order of his Army commander. He realized that he was risking his neck. He knew the iron law of military discipline. But he was also aware of a military commander's moral duty to put a meaningful order above a formal one. He did not evade the tragic dilemma which must arise whenever a man's duty to obey clashes with his personal assessment of operational necessity.

At 0800 hours on 29th December Count Sponeck ordered 46th Infantry Division to disengage itself from the enemy at Kerch, to proceed to the Parpach Isthmus by forced marches, and "to attack the enemy at Feodosiya and throw him into the sea." He sent a signal to Army informing it of his move, and then ordered his wireless station to be dismantled.

So much for Count Sponeck's strategic and tactical considerations. They made sense, they were sober and courageous. There was not a trace of cowardice, indecision, or guilty conscience.

In a temperature of 40 degrees below zero Centigrade, in an icy blizzard, the battalions of 46th Infantry Division, the anti-aircraft units, the sappers, and the gunners moved off. The distance they had to cover was 75 miles. Only occasionally was a fifteen-minute halt called to issue hot coffee to the troops. They marched for forty-six hours. Many were frost-bitten in their fingertips, toes, and noses. Most of the horses were not shod for the winter and were emaciated. They collapsed exhausted. Guns were abandoned on the icy roads.

While the regiments of 46th Infantry Division were thus withdrawing under appalling hardships but nevertheless in good order, Manstein set into motion his plan of first taking Fort Stalin at Sevastopol and then coming to Count Sponeck's aid. The companies of 16th Infantry Regiment were getting ready for the final assault. The ramparts of the fort rose steep and sinister above the tangle of barbed-wire obstacles and trenches. Noiselessly the German assault detachment had cut their way through the wire. A red flare swished upward. German artillery began to fire smoke-shells in order to unsight the Russians.

The first ramparts were stormed, the first casement was captured, the first prisoners were taken. They were worn out, utterly exhausted, and lethargic. But the battalions of 16th Infantry Regiment were down to sixty to eighty men.

In view of the situation at the Isthmus of Parpach, should this costly fighting be continued? Manstein came to the conclusion that it should not. Considering the situation at Feodosiya, he did not want to run any further risks. He ordered operations to cease. That was the last day of 1941.

Colonel von Choltitz with his 16th Infantry Regiment therefore evacuated the painfully gained ramparts of the fort and, in accordance with instructions, moved back to the crest along the Belbek Valley. The 24th Infantry Division was able to hold on to its positions. But for it too, as, indeed, for all formations of Eleventh Army on the Sevastopol front, the order of the day now was: Wait.

Five months were to elapse before the battle for this most powerful fortress of the Second World War was resumed, and five months and a half before 16th Regiment was back inside Fort Stalin.

In the morning of 31st December 1941 the leading battalions of 46th Infantry Division reached the Isthmus of Parpach. However, the forward detachments of the Soviet 63rd Rifle Division had got there before them, holding Vladislavovka, north of Feodosiya. Was the division's disengagement manœuvre to have been in vain after all?

"Attack, break through, and take Vladislavovka!" was General Himer's order to 46th Infantry Division. The troops quickly lined up for attack on the flat, snow-covered plateau. The icy wind blowing down from the Caucasus cut through their thin coats and chilled them to the marrow. The tears of impotent fury froze on their cheeks before they had run as far as their moustaches.

The exhausted regiments punched their way forward over another four miles. Then they ground to a halt. The men simply collapsed.

Under cover of darkness the battalions eventually skirted round the Russian lines on their right, pushed through the still open part of the isthmus, and presently "took up position" on the frozen ground, facing to the south and the east. The last rearguards arriving in that hurriedly improvised line belonged to 1st Company, Engineers Battalion 88.

The following noon the Russians attacked. But the German troops held them. West of Feodosiya also a tenuous covering-line was successfully established across the path of the Soviet 157th Division by 213th Infantry Regiment, 73rd Infantry Division, brought up at the very last moment, and by Rumanian formations, units of the Rumanian Mountain Corps.

When the Russians attacked with tanks the last three self-propelled guns of the "Lions Brigade" saved the critical situation. Captain Peitz had switched them to the front from Bakhchisaray, where they had provided cover against partisans. Second Lieutenant Dammann, the troop commander, managed to lead them to within 600 yards of the enemy tanks in the undulating ground south-west of Vladislavovka. Then came the first crash. Presently an infernal duel was raging. Sixteen Soviet T-26s were left on the battlefield, blazing or shattered. The armoured spearhead of the Soviet Forty-fourth Army had been broken. The danger of a Russian thrust deep into the hinterland of Sevastopol had been averted. The Russians had been stopped.

Judging by results, therefore, Count Sponeck had been justified. Or was there room for doubt? Manstein himself, in his memoirs, does not answer the question unequivocally one way or the other. He criticizes Count Sponeck for facing the Army with a *fait accompli* and making any other solution impossible.

Manstein says: "Such a precipitate withdrawal of 46th Infantry Division was not the way to maintain its combat strength. If the enemy had acted correctly at Feodosiya the division, in the condition in which it arrived at Parpach, would scarcely have been able to fight its way through to the west." If! But the enemy did not act correctly, and the outcome alone is what counts. Whichever way one judges the Sponeck affair, the general's decision sprang neither from dishonourable motives nor from cowardice. His dismissal from his command, decreed by Manstein, can be justified on grounds of principle, as an issue of obedience to superior orders. But this was not all. At the Fuehrer's Headquarters a court martial was held under the presidency of Reich Marshal Göring which sentenced Lieutenant-General Count von Sponeck, who had been summoned before it, to reduction to the ranks, forfeiture of all orders and decorations, and to death by execution.

Hitler himself must have had some misgivings about this barbarous verdict, for on appeal by the C-in-C Eleventh Army he commuted the death sentence to seven years' fortress detention. Judged by his later verdicts, this was a remarkable decision, virtually tantamount to acquittal.

But some two and a half years later, after 20th July 1944, one of Himmler's execution squads amended Hitler's clemency by brutal murder. Count von Sponeck was shot without cause and without sentence.

Count Sponeck's sentence by court martial had its repercussions on 46th Infantry Division. What Field-Marshal von Reichenau, who had meanwhile taken over Army Group South, did to the men of this division was almost as cruel as the verdict against its commanding general. Early in January 1942 its four regimental commanders were summoned to divisional headquarters. Pale and hoarse with emotion, Lieutenant-General Himer, the divisional commander, acquainted them with a teleprinter signal from Army Group. It ran: "Because of its slack reaction to the Russian landing on the Kerch Peninsula, as well as its precipitate withdrawal from the peninsula, I hereby declare 46th Division forfeit of soldierly honour. Decorations and promotions are in abeyance until countermanded. Signed: von Reichenau, Field-Marshal."

Stony silence met this death sentence upon a gallant division. What had been its crime? It had carried out an order by its commanding general. It had passed through extreme hardships and, at the end of them, had still fought bravely and prevented the enemy from breaking through to the Crimea. This now was its reward. A cruel humiliation which assumed criminal responsibility where none existed, which used exaggerated concepts of honour to conceal the excessive demands made on the troops, and which disregarded all true yardsticks.

But the verdict on an entire gallant division could not remove the real cause of the whole affair—the fact that insufficient forces were being assigned excessive tasks. This fact, dramatically illuminated by the "Sponeck affair" and the humiliation of 46th Infantry Division, was soon to reveal itself as the tragic truth—and not only in the Crimea. There and elsewhere it soon became obvious that Stalin was by no means defeated, but that, on the contrary, he was mobilizing the manpower resources of his gigantic empire in order to make good the defeats of the summer. And he succeeded because the fatal German weakness was making itself increasingly felt—too few soldiers for the difficult battles in the vast expanses of Russia.

To-day, in the age of technological wars, with mechanization and automation, the manpower potential of a numerically superior enemy can be outweighed by weapons of mass destruction. But at the time of Hitler's war in Russia these had not yet been developed. Manpower, number of divisions, still played a formidable part. With the arms supplied to the Soviets by an economically superior America this manpower potential could even become the decisive factor. That was the reason for the superiority of the Russians. After six months of unparalleled German victories the enemy, though badly battered and more than once on the point of collapse, succeeded in recovering and in scoring successes which heralded a turning point in the war. This turn of the tide was exemplified in the battles fought by Army Group South on its mainland front, to which we shall presently turn our attention. We cannot, however, leave this Crimean theatre of war with its aura of heroism, tragedy, and sombre symbolism without recording the correction made in the military annals in connection with the gallant 46th Infantry Division. At the end of January 1942 Reichenau's successor, Field-Marshal von Bock, had the following Order of the Day read out to the division: "For its outstanding performance in the defensive fighting in the Isthmus since the beginning of January I express my very special commendation to 46th Division and shall be looking forward to recommendations for promotion and decorations." The 46th Infantry Division had regained its honour.

3. In the Industrial Region of the Soviet Union

Kleist's Panzer Army takes Stalino–Sixth Army captures Kharkov–First round in the battle for Rostov–Obersturmführer Olboeter and thirty men–Rundstedt is dismissed–Ringing of the alarm bells.

HOW were things going on the remaining fronts of Army Group South?

While Manstein had burst into the Crimea the other Armies of Army Group South, fighting on the mainland, had advanced farther to the east between the Dnieper and Donets.

Kleist's Panzer Group, since promoted to First Panzer Army, had been pursuing the defeated enemy and was now lining up to attack Rostov. Between 12th and 17th October the port of Taganrog on the Sea of Azov was occupied after heavy fighting. The cost of this success is illustrated by the fact that 3rd Company of the Infantry Regiment of "Leibstandarte Adolf Hitler" came out of this operation with only seven men. The rest had been killed. But the Myus had successfully been crossed. On 20th October 1941 the German 1st Mountain Division seized Stalino from the Soviet Twelfth Army. Thus the principal armaments-making centre in the Donets area, the most important industrial region of the Soviet Union, was in German hands. According to Hitler's theory—the theory he had upheld against his General Staff and High Command, that the war would be decided by the capture of industrial centres—Stalin's defeat ought now to have been sealed.

On 28th October Colonel-General von Kleist had reached the Myus with all units of his First Panzer Army, and General von Stülpnagel's Seventeenth Army was on the Donets. Four days previously Reichenau's Sixth Army on the northern wing of the Army Group had taken the large industrial centre of Kharkov.

But then, in the south as elsewhere along the whole Eastern Front, the period of autumn mud halted all operations. The Armies were stuck. Not until 17th November, with the onset of frost, was Kleist able to resume his advance on the right wing. Forty-eight hours earlier Field-Marshal von Bock had mounted his "attack on Moscow" on the Central Front.

But the Soviets had made good use of the breathing space provided by the mud. In the Caucasus Marshal Timoshenko was raising new Divisions, Corps, and Armies. Among the members of his Military Council of the South-west Front was a man, then hardly known, who displayed great energy in raising new units and, in particular, organized partisan activities. His name was Nikita Sergeyevich Khrushchev.

While the Soviet High Command was mobilizing ever new Armies, the general shortage of all resources was making itself increasingly felt on the German side. Nowhere were there any reserves. If the Russians broke through anywhere along the front, then forces had to be withdrawn from some other spot in order to seal off the penetration. It was clear that the Eastern Front was short of at least three German Armies—one for each Army Group.

A grim illustration of the tightness of the situation and the excessive demands made on the troops was the battle waged by Army Group South for Rostov.

On 17th November General von Mackensen's III Panzer Corps had mounted its attack against this gateway to the Caucasus with 13th and 14th Panzer Divisions, 60th Motorized Infantry Division and the "Leibstandarte." The "Leibstandarte," reinforced by 4th Panzer Regiment, 13th Panzer Division, penetrated the outer fortifications at Sultan-Saly. On its left 14th Panzer Division struck at Bolshiye-Saly. General Remizov, who was defending Rostov with his Fifty-sixth Army, replied with a strong attack against the flank of 14th Panzer

Division. Mackensen thereupon employed his 60th Motorized Infantry Division in a flanking attack to the east, in order to cover his flank.

On 20th November the three fast divisions penetrated into the town, which then had 500,000 inhabitants, and pushed straight on to the Don. The 1st Battalion "Leibstandarte" stormed across the Rostov railway-bridge and captured it intact. The 60th Motorized Infantry Division meanwhile covered the exposed flank of the Corps by a dashing drive far to the east and south-east, and captured Aksayskaya, while units of 13th Panzer Division vigorously pursued the retreating enemy from the west. Rostov, the gateway to the Soviet oil paradise, was in German hands.[1]

It was a decisive victory. The Don bridges at Rostov were more than mere river crossings: they were the bridges leading to the Caucasus and to Persia. Not for nothing had Britain and the Soviet Union occupied Persia at the end of August 1941 and built a supply road from the Persian Gulf via Tabriz to the Soviet Caucasian frontier. In this way the Soviet Union had gained a direct overland link—its only one—with its rich Western Allies. The old Georgian Military Highway, the road from the Terek Valley over the Caucasian passes to Tiflis, conquered by the Russians about the middle of the nineteenth century, had acquired a new importance.

As a result, Rostov had become a kind of communications centre, a relay post between the Soviet Union and Britain for supplies shipped by the Persian Gulf. Naturally the Soviet General Staff made every possible effort to recapture Rostov from the Germans and to bar Kleist's Panzer Army from access to the Caucasus.

With his Thirty-seventh and Ninth Armies under Generals Lopatin and Kharitonov, Timoshenko now staged a very skilful operation. As a result of Mackensen turning to the south a gap had arisen between Seventeenth Army and First Panzer Army, a gap which, in view of the shortage of forces, could not be immediately closed. Here was Timoshenko's opportunity. He struck at the gap and into the rear of III Corps. It was a dangerous situation.

To meet the danger Mackensen was obliged to detach first the 13th and then also the 14th Panzer Divisions from his front and employ them at Generalskiy Most and Budennyy Most in the threatened Tuslov sector. But no sooner was the crisis in the Corps' rear more or less averted than Timoshenko pounced on Mackensen's weakened Corps along its eastern and southern flanks. The main weight of these attacks fell on 60th Motorized Infantry Division and the "Leibstandarte."

The date was 25th November 1941. The motor-cyclists of the motorized reconnaissance detachment "Leibstandarte" were holding a five-mile sector along the southern edge of Rostov, immediately on the bank of the Don, which at that point was nearly two-thirds of a mile

[1] See inset on Map 13.

wide. But the vast river had ceased to be an obstacle. It was frozen over. It was bitter weather. The men were quite inadequately protected against the cutting cold.

The alarm came at 0520 hours. Soviet regiments—units of 343rd and 31st Rifle Divisions, as well as 70th Cavalry Division—were attacking the positions along their whole breadth. Three hundred grenadiers were lying in the foremost line—a mere 300. And they were being charged by three Soviet divisions. The first assault was made by the Russian 343rd Rifle Division. For a moment the Germans were paralysed with shock: their arms linked, singing, and cheered on with shouts of "Urra," the Soviet battalions came marching up towards them on a broad front, out of the icy dawn. Their mounted bayonets were like lances projecting from a living wall. That wall now moved on to the ice of the Don. At a word of command the Russians broke into a run. Arms still linked, they came pounding over the ice.

Obersturmführer[1] Olboeter, commanding 2nd Company, was in the front line, with the heavy machine-gun of No. 3 section. "Wait for it," he said.

On the ice the first mines planted by German sappers in the snow were now exploding, tearing gaps in the charging ranks. But the great mass of them continued to advance.

"Fire!" Olboeter commanded. The machine-gun started stuttering. A fraction of a moment later other guns joined in the infernal concert.

Like a gigantic invisible scythe the first burst swept along the foremost wave of the charging Soviets, cutting them down on to the ice. The second wave was likewise mown down. To realize how Soviet infantry can charge and die one must have been on the bank of the Don at Rostov.

Over their dead and wounded the next waves charged forward. And each one got a little nearer than the last before it was mown down.

With trembling fingers Horst Schrader, the nineteen-year-old No. 2 of the machine-gun, guided the new belt into the lock. His eyes were wide with terror. The gun's barrel was steaming. As from a great distance he heard his gun commander's shout, "Change barrel! Change barrel!"

In the sector of 2nd Company the Soviet 1151st Rifle Regiment attacked with two battalions. Three waves had collapsed on the ice. The last one now, in battalion strength, was on top of the defenders.

The Russians broke into the positions and went for the machine-gun crews. They killed the grenadiers in their foxholes. Then they rallied. Unless they were thrown back by an immediate counter-attack things would be very ugly for the motor-cyclists of the reconnaissance detachment "Leibstandarte." The southern approaches to Rostov were in danger.

[1] Rank in Waffen SS equivalent to captain.

Things were also getting sticky in the 1st Company sector. Here two Soviet rifle regiments, the 177th and the 248th, were attacking. Their foremost wave was barely 20 yards in front of the German lines. Just then three German self-propelled guns, with grenadiers riding on top of them, arrived in the sector of 2nd Company for an immediate counter-attack and sealed off the Russians who had broken in. Six officers and 390 other ranks surrendered. Most of them were wounded. More than 300 Soviet killed were lying in front of the German lines.

Fierce fighting continued throughout the day. The following day the Russians came again. And the day after that.

On 28th November the Russians were inside the positions of 1st Company. They were units of the Soviet 128th Rifle Division, raised in July and brought across from Krasnodar for their first action. Obersturmführer Olboeter decided to launch an immediate counter-attack, but this time with only thirty men and two self-propelled guns. First of all, however, some one had to cut his boots off his frozen feet. He wrapped his feet and legs in gauze bandages, squares of flannel, and two horse blankets, tying it all up with string. Then he climbed aboard the leading self-propelled gun. "Off!" was all he said. "Off!"

Olboeter was an experienced tactician. With one self-propelled gun he attacked on the left wing, while he got the other to circumnavigate the enemy position until it appeared, belching fire, in the Russians' right flank. Keeping close to the self-propelled guns and firing as they ran forward, Olboeter's men broke into the Russian lines. In spite of his blanket-wrapped frozen feet, the Obersturmführer kept bobbing up to the right and left of his assault gun, directing operations, issuing orders, flinging himself down into the snow and firing his machine pistol.

The fighting lasted two hours. After that Olboeter returned with three dozen prisoners. He had rolled up the enemy position. The Soviets, taken by surprise and battle-weary, had fled across the Don. Once again a typical weakness of the Russians had been revealed: the lower commands were not sufficiently elastic to exploit local successes on a grand scale. In the recaptured position lay 300 dead Russians. But among them, also, lay most of the officers and motor-cyclists of 1st Company of Obersturmbannführer[1] Meyer's reconnaissance detachment.

But what use was a local success? The Russians came back. Impassively their massed attacks broke against the tenuously held German fighting-line. And even the greatest heroism could not offset the fact that the German formations in and around Rostov were simply too weak. Three badly mauled divisions, whose companies barely had one-third of their establishment, could not in the long run stand up to

---

[1] Rank in Waffen SS equivalent to colonel.

In the Industrial Region of the Soviet Union 307

ceaseless assault by fifteen Soviet rifle and cavalry divisions as well as several armoured brigades.

Once again the decisive German weakness was revealed—insufficient resources. The front of III Corps was 70 miles long. It could not possibly be held with the forces available. Field-Marshal von Rundstedt realized this, rang up the Chief of the Army General Staff and the Fuehrer's Headquarters, and requested permission to abandon Rostov.

But Hitler would not hear of retreat. He refused to believe that the Russians were stronger; he preached hardness when only common-sense could save the situation. Thus Rundstedt received orders to hold out where he was.

But for once Hitler had misjudged his man. The Field-Marshal refused to obey the order. Hitler thereupon relieved him of his command. Field-Marshal von Reichenau, hitherto C-in-C Sixth Army, took over Army Group South and instantly stopped the retreat which Rundstedt, with prudent anticipation, had already set in motion.

But even Reichenau could not close his eyes to harsh reality. Twenty-four hours after taking over the Army Group, at 1530 on 1st December 1941, he telephoned the Fuehrer's Headquarters: "The Russians are penetrating into the over-extended thin German line. If disaster is to be averted the front must be shortened—in other words, taken back behind the Myus. There is no other way, my Fuehrer!"

What Hitler had refused Rundstedt twenty-four hours earlier he now had to concede to Reichenau: Retreat, surrender of Rostov.

Although not a disaster, this was the first serious setback of the war. It was a skilful "elastic withdrawal." The major part of the important Donets area remained in German hands.

But nothing could disguise the fact that the German armies in the east had suffered their first major defeat. At his Army headquarters before Moscow, on Tolstoy's estate of Yasnaya Polyana, Guderian remarked glumly, "This is the first ringing of the alarm bells."

He could not know that they would ring on his own sector within six days. And not only on his sector, but along the entire Eastern Front. The blow that had fallen upon Rundstedt was only an episode by comparison with what burst upon Army Group Centre in the Moscow area six days later.

PART FOUR: *Winter Battle*

1. The Siberians are Coming

*5th December 1941–No winter clothing–Fighting for Klin
–3rd Panzer Group fights its way back–Second Panzer
Army has to give ground–Drama on the ice of the Ruza–
Brauchitsch leaves–Historic conversation at the Fuehrer's
Headquarters–Hold on at all costs–Breakthrough at
Ninth Army–The tragedy of XXIII Corps–Time-table for
"Giessen"–Guderian is dismissed.*

THE outposts in the sector of 87th Infantry Regiment had just been relieved. The time was 0500, and it was icy cold. The thermometer stood at 25 degrees below zero Centigrade. The men were trudging through the snow towards the little Yakhroma river. From the chimneys of the peasant cottages in the valley the smoke rose straight into the grey morning. Everything was quiet. The 87th Infantry Regiment belonged to 36th Motorized Infantry Division. The regiments from Rhineland–Hesse were holding the front line between the Volga reservoir south of Kalinin, also known as the Moscow Sea, and Rogachevo. The long sector could be held only in the form of separate strongpoints. For anything else the regiments were too weak. They had been bled white—and, even more so, frozen white.

In a temperature of 30 to 40 degrees below zero Centigrade no man could lie in a forward snowhole for more than an hour. Unless, of course, he was wearing a sheepskin and felt boots, a fur cap and padded gloves. But the men of 36th Motorized Infantry Division had none of these things.

They were within 30 yards of the village. Iced-up, their horse-drawn wagons stood by the stream. The shaft of the village pump rose high above the low roofs. By the pump stood some Russian women, getting water. Suddenly they all started—the pickets who had just been relieved and the Russian women. Instinctively they ducked. They scampered to the nearest cottages. And there it was—the "howling beast." There was a crash, fountains of snow rising into the air, red-hot fragments bouncing off the ground, which was frozen as hard as stone.

The shell-splinters crashed into the bath-house and into the cottages. Action stations!

The date was 5th December 1941—a Friday. A page was being turned in the history of the war. The great Russian counter-offensive before Moscow was beginning. Here, in the sector of 36th Motorized Infantry Division, in the operations zone of LVI Panzer Corps, the curtain was rising on a savage historical drama. Twenty-four hours later the great battle began also on the remaining sectors of Army Group Centre—between Ostashkov and Yelets, along a 600-mile front.

What was the situation before Moscow on that 5th December? North and west of the Soviet capital the German spearheads had got to within a few miles of the outskirts of the city. On the northern wing of Army Group Centre, Ninth Army held a 105-mile arc through Kalinin to the Moscow Sea.

The divisions of 3rd Panzer Group, which were to have outflanked Moscow in the north, had advanced as far as Dmitrov on the Moskva–Volga Canal. Farther south were the most forward units of XLI Panzer Corps, poised to cross the canal north of Lobnya. The combat group Westhoven of 1st Panzer Division, having captured Nikolskoye and Belyy Rast, had reached the western edge of Kusayevo. Adjoining on the right, 4th Panzer Group held a quadrant around Moscow, from Krasnaya Polyana to Zvenigorod; the distance to the Kremlin was nowhere more than 25 miles. The combat outposts of 2nd Panzer Division were at the first stop of the Moscow tramway. An assault detachment of Engineers Battalion 62 from Wittenberg had got closest to Stalin's lair by penetrating into the suburb of Khimki, only 5 miles from the outskirts of the city and 10 from the Kremlin.

On the southern wing of Hoepner's 4th Panzer Group, reading from left to right, were 106th and 35th Infantry Divisions, 11th and 5th Panzer Divisions, as well as the SS Motorized Infantry Division "Das Reich," and 252nd, 87th, 78th, 267th, 197th, and 7th Infantry Divisions. Next followed the divisions of Kluge's Fourth Army. They were 30 miles from Moscow, along a line running from north to south, between the Moscow motor highway and the Oka.

Next along the front came Guderian's Second Panzer Army. It had bypassed the stubbornly defended town of Tula and was holding a big eastward bulge around Stalinogorsk; its armoured spearhead, the 17th Panzer Division, pointing northward against the Oka, stood before Kashira.

On the extreme right wing the Second Army was covering the southern flank and maintaining the link with Army Group South.

This then was the 600-mile front line along which the German offensive had come to a standstill at the beginning of December—in the most literal sense frozen into inactivity. Men, beasts, engines, and weapons were in the icy grip of 45 and even 50 degrees below zero

Centigrade. In the diary of a man of 69th Rifle Regiment, 10th Panzer Division, we find the sentence: "We are waging the winter war as if this was one of our Black Forest winters back home."

That was the exact truth. Officers and troops lacked suitable special winter clothing to enable them to camp and fight on open ground at temperatures of minus 50 degrees. As a result, they clad themselves in whatever they could lay their hands on, or what they found in Russian textile mills, workshops, and stores—one garment on top of another. But this hampered the men's movements instead of making them warm. And these filthy clothes, which were never taken off, were breeding-places for lice, which got right into the skin. The men were not only cold, but also hungry. Butter arrived hard as stone and could only be sucked in small pieces as "butter ice." Bread had to be divided with the axe, and then thawed out in the fire. The result was diarrhœa. The companies were dwindling away. Their daily losses due to frost-bitten limbs and feverish intestinal troubles were higher than those from enemy action.

Like the men, the horses suffered from cold and hunger. Supplies of oats did not arrive. The frozen straw off the cottage roofs no longer satisfied their hunger, but merely made the animals sick. There was a heavy incidence of mange and colic. The animals collapsed and died by the dozen.

The engines likewise were out of action. There was not enough anti-freeze: the water in the radiators froze and engine-blocks burst. Tanks, lorries, and radio vans became immobile and useless. Weapons packed up because the oil froze in the moving parts. No one had thought of making sure of winter oil. There was likewise no special winter paste for the lenses of field-glasses, trench telescopes, and gun-sights. The optics froze over and became blind and useless.

Hardly anything was available that would have been necessary for fighting and for survival in this accursed Russian winter. The Fuehrer's Headquarters had thought that the troops would be in Moscow before the onset of the frost. The bill for this bad miscalculation in the operational time-table and the resulting lack of supplies had now to be footed by the men in the field.

Why were the needs of the hard-pressed front not met by supplies from Europe? Because what few locomotives were available likewise froze up. Instead of the twenty-six supply trains needed daily by Army Group Centre, only eight, or at most ten, arrived. And most of the JU-52 supply aircraft were unable to take off from their airstrips in Poland and Belorussia because of the biting cold and the lack of hangars.

Here is a passage from a letter by Lance-corporal Werner Burmeister of the 2nd Battery, 208th Artillery Regiment, a regiment newly arrived from France:

It's a hopeless job—you've got six horses harnessed to the gun. The front four can be led by hand, but for the two alongside the shaft some one must be mounted, because unless a man is in the saddle and jams his foot against the shaft it will hit the flanks of the animals at every step. At 30 degrees below, in those tight, nailed jackboots of ours, you get your toes frozen off before you even realize it. There isn't a man in my battery who hasn't got frostbite in his toes and heels.

That was the Russian winter—cruel in an undramatic and trivial way. The Russian troops invariably received leather boots one or two sizes too large, to enable them to be stuffed with straw or newspaper—a highly effective procedure. It was a trick well known also to the old lags in the German Army in the East. But unfortunately for them their boots were the right size.

In these conditions, was it surprising that the troops were finished? The combat strength of the regiments was down to less than half. The worst of it was that the Officers' and NCO Corps, as well as the bulk of the old experienced corporals, had been decimated by death in action, by freezing, and by disease. There were lieutenants in command of battalions, and frequently sergeants in command of companies. There were no reserves anywhere. In such conditions Army Group Centre was expected to hold a line over 600 miles long. All this must be realized in order to understand what happened next.

And what was the situation on the Soviet side? Even while the German offensive was still gaining ground the Soviet High Command had assembled a striking force south of Moscow and another north of the city. Whatever military reserves were available in the huge country were brought to Moscow. The eastern and southern frontiers were ruthlessly denuded. Siberian divisions, accustomed to winter and equipped for winter warfare, formed the nucleus of these new forces. The Soviet High Command dispatched thirty-four Siberian divisions to its Western Front; of these twenty-one were facing Army Group Centre, which had comprised seventy-eight divisions in October, but which by early December was left with the combat strength of a mere thirty-five. This greatly reduced effective fighting power was thus outweighed by the freshly arrived Siberian units alone. Their employment proved decisive.

The concentration of Soviet forces before Moscow was the result of what was probably the greatest act of treason in the Second World War. Stalin knew Japan's intention to attack not Russia but America. He knew it from his agent in Tokyo, Richard Sorge, who, as the trusted colleague of the German Ambassador and the friend of the most highly placed Japanese politicians, was acquainted with the intentions of the German and the Japanese leaders. He reported to

The Siberians are Coming 313

Stalin that Japan had refused the German Government's suggestion that she should attack Russia. He reported that the Japanese military were preparing for war against America in the Pacific. Since Sorge's reports about the German offensive intentions in spring 1941 had been so fully confirmed by events, Stalin this time believed the reports from Tokyo and withdrew his entire forces from the Far East to Moscow, even though the Japanese Kwantung Army was poised ready to strike in Manchuria.

But for these dispositions Moscow could not have been saved. The ultimate proof of the tie-up between Sorge's information and military events is provided by the fact that Stalin opened his offensive on the very day when the Japanese warships set course for Pearl Harbour to start the war against the USA. Even this top-secret date, the date of Japan's attack on America, had been passed on to Stalin by Sorge. And as soon as Soviet reconnaissance aircraft confirmed the Japanese naval deployment Stalin, mistrustful though he was, felt sure that Dr Sorge's information was reliable. He could now safely employ his Siberians outside Moscow.

At the beginning of December 1941 the Soviet High Command had concentrated altogether 17½ Armies for an attack against the German Army Group Centre. Three of these—the First, Tenth, and Twentieth Armies—consisted of Siberian and Asian divisions which had been newly raised. The other Armies, according to the reliable military historian Samsonov, had been "trebled or quadrupled by the inclusion of reserves."

Russian military writers, who are fond of playing down their own numbers while invariably overestimating the German forces, quote the ratio between German and Soviet strength at the beginning of the counter-offensive as 1·5 to 1 in favour of the Soviets. And this Soviet superiority became more marked with each week that passed.

Throughout December the German Army Group Centre received not a single fresh division. The Russian "Western Front," on the other hand, which was facing it, was reinforced during that same period by thirty-three divisions and thirty-nine brigades. These figures speak for themselves. Germany's resources were inadequate. She was waging a war beyond her capabilities.

What were the Soviet High Command's plans for its counter-offensive? Even without official Soviet sources the answer would be easy. It sprang from the situation itself. The first task was to smash the two powerful German armoured wedges threatening Moscow from the north and south.

Whether—as is nowadays claimed by Soviet military writers—the Red High Command had been planning from the very start to follow up this first objective by that of encircling the entire German Army Group Centre must remain a matter for speculation. It does not seem very plausible. But if this was indeed the plan from the outset, then it was badly conceived.

We shall presently see why.

The Soviet counter-offensive started north of Moscow with a battle for the Klin bulge. This projecting arc of the front of 3rd Panzer Group was the most serious threat to the Red capital.

At the very heart of this battle were the German XLI and LVI Panzer Corps with 36th and 14th Motorized Infantry Divisions, 6th and 7th Panzer Divisions, as well as—since 7th December—1st Panzer Division. General Schaal, formerly the commander of the well-tried 10th Panzer Division, was now in command of LVI Panzer Corps. There exists a report of his which, together with the operation reports of the separate divisions, provides an impressive and historically interesting picture of the dramatic happenings. They show, by the example of the Klin bulge, how the fate of the northern wing of Army Group Centre early in December 1941 frequently hung by a thread. They also show under what difficult conditions, with what inadequate forces, and with what heroic efforts the troops and their officers were meeting the danger.

In the grey dawn of 5th December, as the initial Russian artillery bombardment made the relieved pickets of 87th Infantry Regiment run for cover by the Yakhroma, Soviet regiments were already charging the forward lines of 36th and, next to it, 14th Motorized Infantry Divisions between Rogachevo and the southern edge of the Volga reservoir. A Soviet ski battalion broke through in the sector of 36th Motorized Infantry Division and thrust towards the West. The Russians were imitating German Blitzkrieg tactics.

At noon on 7th December—i.e., forty-eight hours later—the Soviets appeared in front of General Schaal's Corps headquarters at Bolshoye Shchapovo, four miles north-east of Klin. Staff officers, runners, and clerks snatched up their carbines. Three armoured cars, a few 2-cm. self-propelled AA guns, and two anti-tank guns of the Corps' escort party fired round after round. The general himself was lying behind a lorry with his carbine, firing aimed single rounds. The chief of operations led an AA combat detail into action and sealed off the northern entrance to the village with two machine-gun sections. In the evening a tattered company of 14th Motorized Infantry Division arrived from the punctured front line and immediately took up position to check the Russians. Shortly afterwards Colonel Westhoven, the commander of 1st Rifle Regiment, arrived on the scene, having hurried ahead with his combat section; soon after midnight he was followed by the bulk of 2nd Battalion, 1st Rifle Regiment, coming from Belyy Rast.

The following morning at 0830 hours the Russians attacked with tanks. Had the last hour of Corps headquarters struck? The first tank broke into the German lines on the northern edge, coming from

Selchino. Two regiments of infantry, supported by strong artillery units, moving towards the south-west, bypassed Shchapovo. Just then the noise of battle came from the left flank: Colonel Westhoven was attacking with units of 1st Panzer Division. The foremost tanks of 25th Panzer Regiment, 7th Panzer Division, also arrived in the nick of time, and, led by Lieutenant Ohrloff, struck at the enemy's flank. The Russians were caught off balance. Their infantry fell back and suffered heavy casualties. Corps headquarters were moved to Klin.

Map 15. The defensive battle of Klin.

At Klin General Schaal received more bad news. The enemy had succeeded in making a deep penetration at the juncture between 36th and 14th Infantry Divisions. Strong formations had pushed through the breach, bypassing Klin in the north; they had blocked the Corps' supply route and turned towards Klin via Yamuga. Only one road was left to 3rd Panzer Group—and that was seriously threatened. If the enemy succeeded in blocking it the whole Panzer Group would be threatened with disaster. It would be cut off. The men would have to try to break through on foot, leaving all vehicles and heavy weapons behind. About noon on 8th December this danger became acute. The Soviets took Spas-Zaulok, and subsequently Yamuga, five miles north of Klin.

The Thuringian 1st and the Viennese 2nd Panzer Divisions—two of the three founder members of the German armoured forces, the third being the Berlin 3rd Panzer Division, then still on Guderian's part of the front—were Schaal's last hope. They were to save the situation and keep open the vital road of retreat towards 3rd Panzer Group's new interception line, the Lama position. The 1st Panzer Division, most of which was still holding a switch line around Nikolskoye on the Rogachevo–Moscow road on the morning of 7th December, was pulled out by General Reinhardt and dispatched to Klin. There was a growing danger that this traffic junction, so vital to the withdrawal of the motorized formations, might be lost even before the division got there, but this was averted on 8th December by an attack with hurriedly assembled emergency formations under the command of Colonel Kopp. The Panzer Engineers Battalion 37 seized and held Maydanovo on the northern edge of the town. In this way the worst danger had been met for the moment. The defence of the northern edge of Klin was immediately organized and continued to be strengthened, to make sure the town was held until the arrival of the first units of 1st Panzer Division. Major-General Siry was in charge of these operations.

It was a difficult task because the Soviets knew quite well what was going on. General Schaal reports: "Encouraged by the German retreat, excited by the picture they found along the routes of the German withdrawal, and urged on by the orders of the Soviet High Command, the Russians fought very stubbornly and bitterly. Moreover, in their advance—some of them on skis, but the great mass on foot or with light vehicles, supported by T-34s—the Russians had nearly all the advantages of the terrain on their side. In this heavily wooded and difficult ground the heavy and clumsy German motorized units were confined almost entirely to hard roads. More and more the fighting turned into a series of close combat engagements in which the normally so successful co-operation of the different German service branches was no longer possible, so that, as a result, the Russians were usually superior to us."

In spite of these difficulties the combat groups of the reinforced 1st Panzer Division, together with combat groups of 5th and 2nd Panzer Divisions, succeeded in keeping open and covering the road of retreat from Klin to the west, in dislodging such enemy forces as had made penetrations, and in ensuring—though only with supreme efforts and at heavy cost—the withdrawal of four mobile divisions and parts of several infantry divisions. At this point General Schaal decided that an end had to be put to these desperate stopgap operations. He therefore conceived a daring plan designed to gain some breathing space for the Corps and Panzer Group, and to enable them to recapture the initiative.

The idea was to foil the enemy's intentions by several swift counter-attacks. Colonel Hauser, commanding 25th Panzer Regiment and known throughout 7th Panzer Division as a man of verve and initiative, was to be given every tank available in the Corps area, as well as the fifty-odd tanks promised by Army Group, with orders to break out of the switchline east of Klin, mop up the enemy divisional headquarters identified and located by radio reconnaissance between Yamuga, Spas-Zaulok, and Birevo, attack the Soviet artillery from the rear and put the guns out of action, and then, having spread chaos and confusion, to return inside the German defensive ring.

Everything was got ready for this counter-attack.

Meanwhile two combat groups of 1st Panzer Division mounted a relief attack towards the north. First of all the combat group Westhoven dislodged fairly strong enemy forces south of Kirevo. Next, on 9th December, towards 1030 hours, the combat group von Wietersheim, with Motorcycle Battalion 1, half a dozen Mark III tanks under Second Lieutenant Stoves, and supported by Artillery Battalion Born, made a thrust along the Kalinin highway towards Yamuga. At first the operation made good progress. In spite of their numerical superiority the Russians did not stand firm, but gave ground. They left 180 dead, 790 prisoners, and a large quantity of heavy weapons, including three T-34s, behind on the battlefields around Yamuga. The village itself, however, could not be retaken by the Germans.

Towards evening 1st Panzer Division took back the combat group to the northern edge of Klin and there organized itself for defence. Enemy formations following up the movement were repulsed in hand-to-hand fighting. During the night of 9th December 1st Panzer Division headquarters were given the overall command over the defence for Klin. General Krüger defended the town until 14th December. The moment had now come for "Operation Hauser."

Everything was ready—the last tanks of 1st and 7th Panzer Divisions, a tank company of 2nd Panzer Division, and about twenty-five tanks of 5th Panzer Division. At that moment a report came in from the right wing, to the effect that the enemy had broken through in the area of 4th Panzer Group, in the sector of 23rd Infantry Division. General Kuznetsov's First Striking Army had launched the southern prong of its attack against Klin. It became obvious that the Soviet First and Thirtieth Armies intended to link up west of Klin and to trap 3rd Panzer Group and any other operational troops inside the Klin bulge.

Action stations! Only a vigorous and immediate counter-attack with armour could save the situation. Bitter though the decision was, Schaal had to switch the combat group Hauser to the south-east in order to avert the imminent danger.

In the early morning of 12th December the German tanks moved off to the south-east. A sudden break in the weather had caused the thermometer to shoot up to within a few degrees of freezing-point. The winter sunshine on the road together with the worn tank-tracks made it a very slippery journey. Nevertheless the force succeeded in intercepting the Russians, relieving scattered German formations, and bringing back various groups still holding out in the broken front line into the safe shelter of 3rd Panzer Group's switchline around Klin.

The defence of the town of Klin proper, where thousands of wounded still remained in spite of continuous evacuation, was in the hands of three *ad hoc* formations. Initially the town had been kept open for the withdrawing divisions by hurriedly organized emergency formations under Colonel Kopp and Lieutenant-Colonel Knopf, with sappers, road-building details, a few anti-tank and anti-aircraft sections, three self-propelled guns, Luftwaffe ground staff, workshop mechanics, and a few repaired tanks. During the next few days, however, every available man was roped in for active defence—including along the northern outskirts twenty-five drummers of the band of 25th Panzer Regiment, employed as infantry under their band-leader. Presently the mixed combat groups Westhoven, von Wietersheim, and Caspar were fitted in along the north-eastern and north-western edges of the town. Klin by then was under Soviet artillery bombardment, and fires were burning everywhere.

On 13th December Panzer Group, with Hitler's approval, ordered the abandonment of the positions east of Klin. Everything into reverse —all along a single road, the road through Klin.

Since the night of the 13th the eastern edge of the town had been held by the reinforced 14th Motorized Infantry Division with combat groups of 2nd Panzer Division and Colonel Hauser's group. Just outside the northern part of Klin, west of the town, 1st Panzer Division was covering the great withdrawal route against furious Soviet attacks from the north. Time and again it cleared the road, and thus ensured the removal of the last few thousand wounded and of the heavy material. Under cover of these operations the Klin bulge was evacuated about noon on 14th December. But while the fighting troops in the line were making superhuman efforts, the retreat of the supply troops and scattered units became a veritable tragedy.

General Schaal records his personal recollections as follows:

> Discipline began to crack. There were more and more soldiers making their own way back to the west, without any weapons, leading a calf on a rope, or drawing a sledge with potatoes behind them —just trudging westward with no one in command. Men killed by aerial bombardment were no longer buried. Supply units, frequently without officers, had the decisive say on the roads, while the fighting troops of all branches, including anti-aircraft artillery, were desperately holding out in the front line. The entire supply train—except where units were firmly led—was streaming back in wild flight. Supply units were in the grip of psychosis, almost of panic— probably because in the past they had only been used to headlong advance. Without food, shivering with cold, in utter confusion, the men moved west. Among them were wounded whom it had been impossible to send back to base in time. Crews of motor vehicles unwilling to wait in the open for the traffic jams to clear just went into the nearest villages. It was the most difficult time the Panzer Corps ever had.

How was that possible? How could panic exist so closely behind the disciplined and indeed heroically fighting troops in the forward lines?

The answer is simple enough. The German Wehrmacht had never learned the principles and methods of retreat. The German soldier regarded retreat not as a special type of operation, to be bent to his will, but as a disaster imposed by the enemy upon him.

Even in Reichswehr days the practising of withdrawals had been looked upon askance. Somewhat contemptuously it used to be said: One does not practise withdrawals; it merely teaches the men to run away.

Later, after 1936, even elastic resistance was deleted from the training programme. 'Attack' and 'holding' were the only two techniques taught to the German soldier. As far as fighting retreats were concerned, the Wehrmacht went into the war unprepared. The cost of this omission was heavy. At Klin it had to be met for the first time.

On 14th December at 1300 hours a Russian second lieutenant with a white flag appeared in front of Captain Hingst, commanding 8th Company, 3rd Panzer Regiment, who was employed with units of 2nd Rifle Regiment along the south-eastern edge of the town as part of the combat group Hauser. The Soviet officer carried a letter signed "Colonel Yukhvin" and demanding the surrender of Klin. "The position of the defenders is hopeless," the Soviet colonel wrote. It was the first Soviet invitation to surrender presented under a white flag on the Eastern Front.

Captain Hingst treated the Russian very courteously and, having reported to Colonel Hauser and asked for instructions, sent him back at 1400 with the answer that the colonel was mistaken, the situation was by no means hopeless for the defenders.

Hingst was right. The disengagement of LVI Panzer Corps had meanwhile progressed according to plan. At 16.30 hours, when the road had been cleared, the 1st Panzer Division with its motor-cycle battalion moved off to the west. By 15th December all units had reached the interception line of 2nd Panzer Division at Nekrasino. On the southern edge of the town Colonel Hauser withdrew his forces over the small Sestra river into the western part of the town itself. As soon as the last tank was across, the bridge was blown up. Headquarters and combat groups of 53rd Motorized Infantry Regiment, as well as the Tank Company Veiel of 2nd Panzer Division, held Klin, by then in flames, until 2100 hours. Then this rearguard likewise moved off to the west. The Russians infiltrated into the town.

Klin was lost. The front of 3rd Panzer Group had been pushed in. The German armoured wedge aiming at Moscow from the north had been smashed. The two Soviet Armies, the Thirtieth and the First, had succeeded in eliminating the dangerous threat to Moscow. On the other hand, the Soviets had not succeeded in annihilating 3rd Panzer Group. Thanks to the bravery of the fighting forces and the skilful handling of 1st Panzer Division, the divisions of two Panzer Corps and units of V Army Corps had been successfully saved from encirclement, and the men as well as a large proportion of the weapons and material withdrawn to the Lama position 56 miles farther back.

But what was the situation like at the other focal points of the Moscow front—with 4th Panzer Group west of the city and with Guderian's Second Panzer Army down in the south?

Moscow is situated on the 37th meridian. On 5th December the two wings of Guderian's Panzer Army, which were to have enveloped the Soviet capital from the south, stood with 17th Panzer Division before Kashira, about 37 miles north of Tula, with 10th Motorized Infantry Division at Mikhaylov, and with 29th Motorized Infantry Division north-west of Mikhaylov. Mikhaylov, however, is on the 39th meridian. In other words, Guderian was well behind the Soviet metropolis. The Kremlin, in a sense, had already been overtaken. As a result, Guderian's thrust, though still 75 miles south of Moscow, was every bit as dangerous as the armoured wedge in the north which had got within some 20 miles of the Kremlin. For that reason Guderian's front, the area from the southern bank of the Oka via Tula to Stalinogorsk, became the second focus of the Soviet counter-offensive.

The Soviet High Command employed three Armies and a Guards Cavalry Corps in a two-pronged operation designed to encircle Guderian's much feared striking divisions and annihilate them. The Soviet Fiftieth Army formed the right jaw of the pincers, and their Tenth Army the left jaw. General Zhukov—the leading brain of the Soviet High Command, who was personally in charge of the Soviet counter-blow at Moscow—tried to apply the German recipe here in the south, just as he had done with General Kuznetsov's formations in the north, at Klin. He tried to pinch off the protruding front line of Second Panzer Army, and to do it so quickly that the German divisions were left no time to withdraw.

It was a good plan. But Guderian's strategic perception was even better. On 5th December Guderian's attempt to achieve a link-up north of Tula between 4th Panzer Division and 31st Infantry Division, with a view to encircling the town finally, had failed. As a result, the Second Panzer Army was tied down in heavy defensive fighting. During the night of 5th/6th December, the night preceding the Soviet offensive, Guderian therefore ordered the withdrawal of his exhausted forward formations to the Don-Shat-Upa line. This movement was in progress when, on 6th and 7th December, the Russians charged against LIII Army Corps and XLVII Panzer Corps at Mikhaylov. They encountered only the rearguards, which offered delaying resistance and covered the withdrawal already in full swing.

Even so, things were bad enough. The retreat in icy wind through waist-high snow, over mirror-smooth roads, was hell. Not infrequently the formations, as they painfully struggled along the roads, were involved in skirmishes with fast Siberian ski battalions. Like ghosts the Siberians appeared in their white camouflage smocks. Soundlessly they approached the road on skis through the deep snow. They fired their rifles. They flung their hand-grenades. And instantly they vanished again. They blew up bridges. They blocked important crossroads. They raided supply columns and killed the men and the horses.

But Guderian's battle-hardened divisions were no inexperienced rabbits. The 3rd Panzer Division, for instance, was withdrawing with its vehicles from the area north of Tula, from one sector to the next, through an icy blizzard. Fast mixed rifle companies with their armoured infantry carriers, anti-tank guns, and self-propelled anti-aircraft guns formed their rearguard or even acted as assault reserves for 3rd and 394th Rifle Regiments. When the bulk of their units had detached themselves from the enemy these mixed companies launched swift counter-attacks or, by continuously changing their position and rapidly firing all their automatic weapons, produced the impression and the sound effects of strong formations. On suitable occasions they would even launch lightning-like counter-blows over distances of three to six miles.

One such occasion arose near Panino on 14th December. The regiments had to cross the bridge over the Shat. The Russians were putting on the pressure with tanks and ski battalions. The villages in front of the bridge had to be burnt down to gain a clear field of fire and to deny cover to the Soviets.

Second Lieutenant Eckart with his 2nd Company, 3rd Rifle Regiment, was covering the vital road fork. The Russians came in battalion strength—Uzbeks of a Rifle Regiment of Fiftieth Army. With them, in their foremost line, were anti-tank guns and heavy mortars. Eckart sent a signal to Second Lieutenant Lohse, commanding 1st Company, in which all the armoured troop carriers of the rifle regiments were grouped: "I require support." Lohse quickly collected four tanks to add to his half-dozen armoured troop carriers and drove off. Behind the smouldering ruins of a village he cautiously crept up against the flank of the attacking Russians.

Now! They broke cover. Three Soviet anti-tank guns were over-run. The Russians were driven into the fire of 2nd Company. Those who were not killed pretended to be dead—a favourite Soviet trick.

Lohse's command carrier was the last to cross the bridge. The squat silhouette of a T-34 appeared on the horizon. It fired. But its aim was poor. The Germans succeeded in blowing up the bridge.

Lohse's armoured troop carrier company had lost one vehicle carrying an anti-tank gun. Sergeant Hofmann was wounded, and one man was missing. But an entire Soviet battalion had been destroyed.

Men like Lohse and Eckart—second lieutenants and sergeants,

L

captains as much as machine-gunners or the drivers of tanks, of armoured infantry carriers, gun tractors, lorries or horse-drawn vehicles of every kind—these were the men who tackled the situations which were frequently critical for entire combat groups or divisions. The dramatic retreat through blizzard and fire made the German front-line soldier hard; it produced that tough, patiently suffering, self-reliant, and enterprising individual fighter without whom the German armies in the East could not have survived the winter before Moscow.

The cruelty of the winter, its savagery towards German and Russian alike, was gruesomely illustrated for a rearguard of 3rd Rifle Regiment on the fourth Sunday in Advent of 1941. It happened at Ozarovo. Through his binoculars the second lieutenant spotted a group of horses and troops standing on a gentle slope in the deep snow. Cautiously the German troops approached. There was a strange silence. The Soviet group seemed terrifyingly motionless in the flickering light of the snowy waste. And suddenly the lieutenant grasped the incredible —horses and men, pressed closely together and standing waist-deep in the snow, were dead. They were standing there, just as they had been ordered to halt for a rest, frozen to death and stiff, a shocking monument to the war.

Over on one side was a soldier, leaning against the flank of his horse. Next to him a wounded man in the saddle, one leg in a splint, his eyes wide open under iced-up eyebrows, his right hand still gripping the dishevelled mane of his mount. The second lieutenant and the sergeant slumped forward in their saddles, their clenched fists still gripping their reins. Wedged in between two horses were three soldiers: evidently they had tried to keep warm against the animals' bodies. The horses themselves were like the horses on the plinths of equestrian statues—heads held high, eyes closed, their skin covered with ice, their tails whipped by the wind, but frozen into immobility. The frozen breath of eternity.

When Lance-corporal Tietz tried to photograph the shocking monument the view-finder froze over with his tears, and the shutter refused to work. The shutter release was frozen up. The god of war was holding his hand over the infernal picture: it was not to become a memento for others.

Like 3rd Panzer Division, the other divisions of Guderian's two Panzer Corps likewise withdrew from the frontal arc north-east of Tula, fighting all the time against the attacking Soviet Fiftieth, Forty-ninth and Tenth Striking Armies, and thus evaded the pincer operation, the bear-hug in which Zhukov wanted to clasp the Second Panzer Army.

At Mikhaylov, where on 8th December Zhukov's Striking Army made a surprise attack, the 10th Motorized Infantry Division, engaged in delaying defence, suffered considerable losses. At XXIV Panzer

Corps the 17th Panzer Division halted the first Soviet thrust from the direction of Kashira. South-east of Tula the "Grossdeutschland" Regiment stood up stubbornly against fierce Soviet attacks from inside the city, and thus defended the Corps' left switch-line covering the withdrawal towards the Don-Shat-Upa line. Under cover of these engagements the bulk of the Army retreated. Stalinogorsk was evacuated. Yepifan was abandoned after heavy defensive fighting, in accordance with orders, by the 10th Motorized Infantry Division, which had fought its way into the town on its retreat. The defensive line along the Don-Shat-Upa line was reached on 11th December.

However, Guderian's hope of holding out here proved unrealizable. Units of the Soviet Thirteenth Army broke into the front of General Schmidt's Second Army on both sides of Yelets, south of Guderian's Panzer Army. On 13th December Yefremov was abandoned by Second Army. The 134th and 45th Infantry Divisions—some of whose units were encircled at Livny for a few days—offered desperate resistance, but were forced to give ground and march for their lives. Lance-corporal Walter Kern of 446th Infantry Regiment, 134th Infantry Division, reports: "Whenever we moved into a village in the evening we first had to eject the Russians. And when we got ready to move off again in the morning their machine-guns were already stuttering behind our backs. Our killed comrades, whom we could not take along, lined the roads together with the dead bodies of horses, or remained lying in the ravines where we stopped to offer resistance, but which often turned out to be dangerous death-traps."

The situation was similar in the sector of 45th Infantry Division—the former Austrian 4th Division—in action south of 134th Infantry Division. Cut off at one moment, bursting through the enemy at the next, their supply columns destroyed, their operational and withdrawal directives dropped to them from the air, the combat groups of this gallant division fought their way out of the Livny pocket to the south-west.

With the front line of his right-hand neighbour withdrawn from the Yelets–Livny area to the south-west, Guderian's right wing in the Don-Shat-Upa position was left hanging in mid-air. Guderian, therefore, had to withdraw once more, taking his line a further 50 miles to the west, to the Plava.

Presently, when the Russians broke through with twenty-two rifle divisions between Yelets and Livny, Guderian was compelled to move his line back even farther. In the course of this move the connection was lost between Second Panzer Army and Fourth Army, so that a gap of 20 to 25 miles appeared in the front line between Kaluga and Belev.

The Soviet High Command seized its opportunity and launched the I Guards Cavalry Corps through the vast gap which was forming in the German lines. General Belov's Cavalry Regiment, supported by combat troops on skis and motor sledges, chased westward towards Sukhinichi and north-west towards Yukhnov. Matters were moving to a climax.

The gap in the front became the nightmare of the German High Command. From now onward there was a danger that the southern wing of Fourth Army might be enveloped. Indeed, if the Russians succeeded in breaking through via Kaluga to Vyazma on the Moscow motor highway they might even strike at the rear of Fourth Army and cut it off. A single thrust from the north could thus close the huge pocket.

It was obvious that this was also the aim of the Soviet Command. This bold strategic operation was positively asking to be executed. The spectre of defeat, distant as yet, began to haunt the badly mauled German forces of Army Group Centre.

The Soviet High Command set about its plan correctly. Kluge's Fourth Army, in the middle of Army Group Centre, was at first exposed only to tying-down attacks. In this way the Soviets tried to prevent Kluge from switching some of his forces to the wings of Army Group, or even withdrawing his Army and employing the large formations thus freed against the Soviet offensive in the south and north. Kluge was to be pinned down in the middle of the central sector until the two jaws formed by the northern and southern Russian Army Groups had smashed the wings of the German front.

It was in just this way that Field-Marshal von Bock had dealt with the Soviets at Bialystok and Minsk, Hoth at Smolensk, Rundstedt at Kiev, Guderian at Bryansk—and Kluge at Vyazma, which was the finest example of a battle of encirclement in military history. Was Zhukov now going to do the same, this time with a Russian victory at Vyazma?

It could happen, provided the Soviets succeeded in breaking through to the west, also north of Fourth Army, and in wheeling round to the south, towards the Moscow–Smolensk motor highway.

What meanwhile was the situation on the front of 4th Panzer Group? The VII and IX Army Corps had ground to a standstill along the Moskva–Volga Canal at the beginning of November. The IX Army Corps made one last attempt to improve its positions. One of the participants in this attempt—in these last pulse beats of the German frontal attack on Moscow by Hoepner's 4th Panzer Group along the motor highway—was Lieutenant Hans Brämer with his 14th Company, 487th Infantry Regiment, fighting under 267th Infantry Division. That was on 2nd December 1941.

The 267th Infantry Division from Hanover was to make one last attempt to break open the Soviet barrier west of Kubinka by means of

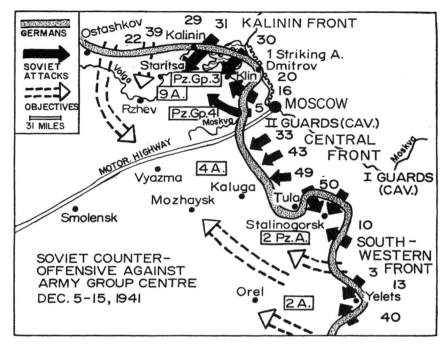

Map 16. Turning-point before Moscow. In a grand offensive the Soviet High Command hoped to surround and annihilate the exhausted Armies of Army Group Centre in the approaches to Moscow.

an enveloping attack across the frozen Moskva river. In a temperature 34 degrees below zero Centigrade it took hours to start all the vehicles needed to get the men and the heavy weapons into the deployment area. The artillery, on the other hand, put down a massive barrage as in the good old days. But, in spite of it, the move did not come off. The Russians had fresh Siberian regiments in magnificently camouflaged and well-built positions in the woods. As a result, the normally so useful 3·7-cm. anti-tank guns of Brämer's 14th Panzerjäger Company were not much help, even though two troops with six guns had been attached to the assault battalions of Lieutenant-Colonel Maier's combat group. The gun crews were killed. The guns were lost. That was the end. The men had to withdraw again. They simply could not get anywhere.

The 267th Infantry Division thereupon took up its prepared winter positions a few miles farther to the north, on the western bank of the Moskva, now in the role of left-wing division of VII Corps. Beyond it to the north stood IX Corps with 78th, 87th, and 252nd Infantry Divisions. In this way the line as far as Istra was reasonably well held.

But over on the right, along the Moskva river, 267th Infantry Division, supposed to hold a sector of roughly four miles with the remnants of the weakened 497th Infantry Regiment, could do no more than man a few strongpoints, scarcely more than reinforced field pickets. It was asking for trouble.

During the next few days the Russians continually attacked across the Moskva—sometimes with small formations and sometimes in strength. Evidently they were trying to discover the weak spots along the join between Hoepner's 4th Panzer Group and Kluge's Fourth Army. Supposing they discovered the virtually undefended gap along the Moskva on the right flank of 267th Infantry Division? The defenders kept pointing out their weakness—but without effect, since neither Corps nor Army had any reserves left which they could have dispatched to reinforce the line.

On 11th December towards 1000 hours a runner, Corporal Dohrendorf, excitedly burst into Brämer's dugout: "Herr Oberleutnant, over there on the right there are columns on skis moving to the west. I believe they are Russians!"

"Blast!" Brämer leapt to his feet and ran out. His field-glasses went up. A suppressed curse. Brämer scurried back to the telephone. Report to Regiment: "Soviet forces, several big columns in battalion strength, are passing through the front line to the west on skis." Action stations.

Corporal Dohrendorf and Lieutenant Brämer had observed correctly. Soviet Cossack battalions and ski combat groups had wiped out the German pickets along the thinly held Moskva strip and were now simply bypassing the well-fortified strongpoints of 467th and 487th Infantry Regiments, where the bulk of divisional artillery was also emplaced.

In vain did General Martinek, commanding 267th Infantry Division, try to stop the gap with the battered companies of his 497th Infantry Regiment. He did not succeed. The Russians enlarged their penetration. Contact was lost between divisional headquarters and the two northern regiments.

The division next in the line, the 78th, which was now being threatened in its flank and rear by the Soviet penetration, flung in units of its 215th Infantry Regiment. Colonel Merker, commanding 215th Infantry Regiment, assumed command in the penetration area, and from units of 467th and 487th Infantry Regiments, 267th Artillery Regiment, and with the support of battalions and batteries of Army artillery, built up a new defensive front.

But the Russians operated very skilfully, cunningly, and boldly in the wooded country. That was hardly surprising: the units were part of the Soviet 20th Cavalry Division—a crack formation of Major-General Dovator's famous Cossack Corps, given Guards status by

Stalin on 2nd December 1941 and now proudly bearing the name of II Guards Cavalry Corps.

After their break-through the Cossack regiments rallied at various key points, formed themselves into combat groups, and made surprise attacks on headquarters and supply depots in the hinterland. They blocked roads, destroyed communications, blew up bridges and viaducts, and time and again raided supply columns and wiped them out.

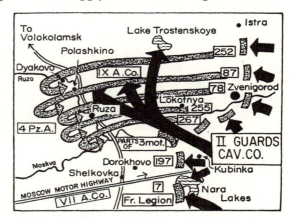

Map 17. Dovator's Cossack Corps strikes at the rear of IX Army Corps.

Thus on 13th December squadrons of 22nd Cossack Regiment overran an artillery group of 78th Infantry Division 12 miles behind the front line. They threatened Lokotnya, a supply base and road junction. Other squadrons thrust north behind 78th and 87th Divisions.

The entire front of IX Corps now hung in the air. The forward positions of the divisions were intact, but their rearward communications had been cut off. Supplies of ammunition and food did not get through. And there were several thousand wounded in the forward fighting area.

On 14th December, at 1635, a Cossack squadron attacked the 10th Battery, 78th Infantry Division, 16 miles behind the front, while the German battery was on the move to new positions farther back. The Cossacks attacked with drawn sabres. They cut up the surprised artillerymen and slaughtered men and horses.

The Russians likewise tried to break through along the Moscow highway and the old postal road, where 197th Infantry Division was guarding the supply routes. But the 197th was on the alert. Wherever the Russians made penetrations with tanks they were pinned down by concentrated fire and dislodged by immediate counter-attacks. Thus it went on day after day. At 0300 the Soviets would come out of the villages, where they had been keeping themselves warm, and in the evenings they would go back. They would take their wounded with them, but leave their dead behind.

During the night of 13th/14th December a supply column of the Cossacks, consisting of forty lorries, tried to get past the positions of 229th Artillery Regiment, 197th Infantry Division.

The temperature was 36 degrees below zero Centigrade. The recoil devices of many of the guns were frozen up. The lenses of the gunsights were frosted over and blind. The gunners scored hits nevertheless just by firing by eye. The Cossack column was smashed up at point-blank range.

But neither the determined resistance of the line along the motor highway nor the gallantry of the grenadier and artillery battalions of 197th Infantry Division, nor yet the tough and successful defensive fighting of 7th Infantry Division, in whose ranks the French volunteers also fought, was able to avert the general disaster triggered off for VII and IX Army Corps by the Cossack Corps' penetration north of the highway. There was only one thing to do—pull back the front along the entire right wing of 4th Panzer Group. The new main line for the defensive fighting was to be the Ruza line, 25 miles behind the present front line. In exceedingly hard fighting the 197th Infantry Division, together with rapidly brought-up units of 3rd Motorized Infantry Division, held open the highway at the now famous, or infamous, Shelkovka–Dorokhovo crossroads for the withdrawal of the heavy equipment and the divisions of 4th Panzer Group.

The situation is well illustrated by the Order issued by 78th Division to its regiments for the withdrawal: "The main thing is to break through the enemy's barrier behind our front line. If necessary, vehicles are to be left behind and only the troops saved."

In this manner they fought their way back: the Swabians of 78th Infantry Division, whose tactical sign was Ulm Cathedral and the iron hand of Götz von Berlichingen, the Thuringians of 87th Division, the Silesians of 252nd Division, the men from Rhineland and Hesse of 197th Infantry Division, and the battalions of 255th Infantry Division. The French legionaries trudged along next to the Bavarians of 7th Infantry Division, and their words of command in the language of Napoleon echoed eerily through the frosty nights and the icy blizzards —just as 129 years before.

Lieutenant Brämer of 267th Infantry Division and his men were now, in December, retreating along the road by which they had advanced earlier in the autumn. They were carrying their wounded with them: two infantrymen were leading a captured Cossack horse on which sat a corporal whose leg had been torn off by a shell-splinter as far as the knee. The wound was frozen up, and in this way bleeding had been stopped. Only willpower kept the man in the saddle. He wanted to live. And in order to live one had to make one's way westward.

Who was the man who won these victories between Zvenigorod and

The Siberians are Coming 329

Istra? Who was the man in command of the Cossacks who had broken through along the flank of VII Corps, dislodging IX Corps and forcing its divisions to retreat? His name was Major-General Dovator. This Cossack general must have been a superb cavalry commander. Within the framework of the Soviet Fifth Army he led his Corps with extraordinary skill, daring, and dash. He led his fast cavalry formations in the manner of a tank commander—and, after all, armour was merely the mechanized successor of the old cavalry.

"A commander must be in front," was Dovator's motto. And he led his troops from the front. He and his headquarters squadron were invariably in the front line. More than once Major-General Dovator was cited in the Soviet High Command communiqué for personal bravery.

Soviet military sources say nothing about his origins, which rather suggests that they were not proletarian but bourgeois. He probably came from the Officers' Corps of the old Tsarist Army—one of those men of middle-class origin who embraced the military profession and became unreserved supporters of the Bolshevik regime.

252nd Infantry Division, the "Oak Leaf" Division from Silesia, was among the divisions which had to fight their way back to the Ruza line. It was this division which took its revenge on Dovator's Cossacks and made the general pay the supreme price for his victory—his life.

An account of this episode reflects both the gallantry of the German troops and that of an outstanding Russian general who knew how to fight and how to die.

On 17th December 1941 the reinforced 461st Infantry Regiment hurled itself against Dovator's forward formations which were trying to block the route of 252nd Division near Lake Trostenskoye. The danger was averted. All the units of the division reached the Ruza, even though it was a constant race against Dovator's cavalry regiments.

On 19th December the 252nd crossed the Ruza river north of the town of the same name. But Dovator had got there too. He did not want to let 252nd Division escape his clutches. The Ruza was frozen over. The general prepared to mount a flank attack. From the right wing he intended to strike at the Silesians across the ice of the river. The clash came near two small villages, Dyakovo and Polashkino.

Lieutenant Prigann was in position outside Dyakovo on the higher western bank of the river with the remnants of 2nd Battalion, 472nd Infantry Regiment, and 9th Battery, 252nd Artillery Regiment. On the right, in Polashkino, Major Hoffer had taken up position with the 3rd Battalion, 7th Infantry Regiment, from Schweidnitz. They were good commanding positions. Hoffer and Prigann were determined to make good use of them.

The day was grey and cold. In the late morning snow began to fall—light, dry December snow which was blown by the wind over the fields and the frozen Ruza. The corpses of horses, gutted motor vehicles, and men killed in action and frozen rigid were covered by its shroud.

From the edge of the forest General Dovator was watching his vanguards riding down to the river. He could hear the exchange of fire in the distance. The Cossacks dismounted.

Dovator turned to the commander of his point regiment, Major Linnika: "Attack to the right of the vanguard!"

The major saluted and drew his sword. He issued his command. The 1st Squadron burst out of the wood. It was like a phantom chase. Past the village of Tolbuzino, down towards the river. At that moment the German machine-guns opened up.

The squadron instantly fanned out, dismounted, and flopped into the snow. The charge had not come off.

General Dovator was annoyed. With Major Linnika, the regimental commander, he rode down the path to the north, as far as the highroad from Ruza to Volokolamsk. This was where the spearheads of his 20th Division stood. The 14th Mounted Artillery Battalion was just moving through the forest. The time was noon.

From the edge of the forest there was a good view of Polashkino. On the roads to the west were the baggage trains of 252nd Infantry Division.

"Colonel Tavliyev," the general called out. The commander of the 20th Division leaned forward. "We'll cross the river, bypassing the village of Polashkino on the right. Then we'll strike at the rear and the flank of those columns. I'm coming with you."

The squadrons moved off at a gallop. But no sooner were they out of the wood than they were caught in very heavy machine-gun fire.

"Deploy, colonel," the general shouted. "Drive the fascists out of the village."

With his staff, Dovator galloped down to a hut by the river. He leapt from his horse and patted the animal's neck. The chestnut's name was Kazbek. He was nervous. "Steady, Kazbek," the general calmed him. He threw the reins to Akopyan, his groom. "Walk him up and down a bit, otherwise he'll get cold."

Dovator watched the fighting through his binoculars. On the right Dyakovo was in flames. It was being shelled by Russian artillery. But the dismounted men of the Soviet 22nd Cavalry Regiment were pinned down.

Now the 103rd Regiment was galloping out of the woods, deploying, but a moment later was likewise forced to dismount. The cavalrymen advanced on foot. They reached the ice of the river. There they were kept down by continuous German machine-gun fire.

"We've got to snatch the men up from the ice," the general shouted. He pulled his pistol out of its holster, cocked it, and with long, loping strides ran down to the river himself. His ADC, the political officer of the headquarters squadron, the duty officer, and the headquarters guard followed him.

Less than 20 yards lay between the general and the line of men pinned down in the middle of the river. At that moment a German machine-gun burst swept across the ice from the right-hand edge of the village. Dovator stopped short as though something had frightened him. Then he fell over heavily into a drift of powdered snow which the wind had piled up on the ice.

His ADC ran up to him. But the machine-gun was still stuttering. The German lance-corporal did not take his finger off the trigger. The little spurts of snow showed him the exact position of his bursts. They cut down also the adjutant who bore the German name of Teichman. The bursts also caught Colonel Tavliyev and dropped him down by the side of his general.

"Dogs!" screamed Karasov, the political officer. "Dogs!" His coat flying behind him in the wind, he raced over the ice to Dovator and picked him up. But just then the trail of bullets danced up through the snow and mowed him down too. Dead, he collapsed on to the ice.

At last Lieutenant Kulikov and Second Lieutenant Sokirkov succeeded in crawling over to the general. Under heavy machine-gun fire they dragged their general over the ice and carried him behind the hut.

Kazbek, the stallion, reared as his dead master was brought back. At Polashkino the machine-guns were still stuttering. The infantrymen from Schweidnitz were resisting furious attacks by Shamyakin's cavalry regiment, out to avenge Dovator.

Defeats invariably need their scapegoats. On the day General Dovator was killed in action on the Ruza the first political storm swept through the German Generals Corps. Adolf Hitler dismissed Field-Marshal von Brauchitsch, the Commander-in-Chief Army, and personally assumed the command of all land forces. Field-Marshal von Bock, Commander-in-Chief of the hard-pressed Army Group Centre, was given "sick leave." He was succeeded by Field-Marshal von Kluge. And Kluge was succeeded by General Heinrici as Commander-in-Chief Fourth Army.

On 20th December 1941 a very worried Guderian flew to East Prussia to see Hitler at his headquarters. He wanted to persuade him to take the German front line back to more favourable positions, if necessary over a considerable distance.

The five-hour interview was of historic importance. It showed the Fuehrer irritable, tormented by anxiety, but resolved to fight fanatically; it revealed a powerless and obsequious High Command, resembling courtiers in uniforms; and it showed Guderian, alone but courageous, passionately arguing his case and fearlessly giving Hitler his frank opinion on the situation at the front.

The first time the word retreat was mentioned Hitler exploded. The word seemed to sting him like the bite of an adder. It conjured up for him the spectre of the Napoleonic disaster of 1812. Anything but retreat!

Passionately Hitler tried to convince Guderian: "Once I've authorized a retreat there won't be any holding them. The troops will just run. And with the frost and the deep snow and the icy roads that means that the heavy weapons will be the first to be abandoned, and the light ones next, and then the rifles will be thrown away, and in the end there'll be nothing left. No. The defensive positions must be held. Transport junctions and supply centres must be defended like fortresses. The troops must dig their nails into the ground; they must dig in, and not yield an inch."

Guderian rejoined: "My Fuehrer, the ground in Russia at present is frozen solid to a depth of four feet. No one can dig in there."

"Then you must get the mortars to fire at the ground to make shell-craters," Hitler retorted. "That's what we did in Flanders in the first war."

Guderian again had to put Hitler right on his facts. "In Flanders the ground was soft. But in Russia the shells now produce holes no more than four inches deep and the size of a wash-basin—the soil is as hard as iron. Besides, the divisions have neither enough mortars nor, what's more important, any shells to spare for that kind of experiment. I myself have only four heavy howitzers left to each division, and none of them has more than 50 rounds. And that is for a front sector of 20 miles."

Before Hitler could interrupt him Guderian continued: "Positional warfare in this unsuitable terrain will lead to battles of material as in the First World War. We shall lose the flower of our Officers Corps and NCOs Corps; we shall suffer gigantic losses without gaining any advantage. And these losses will be irreplaceable."

There was deathly silence in the Fuehrer's bunker at the Wolfsschanze. Hitler too was silent. Then he stepped up close to Guderian and in an imploring voice said, "Do you believe Frederick the Great's grenadiers died gladly? And yet the King was justified in demanding of them the sacrifice of their lives. I too consider myself justified in demanding of each German soldier that he should sacrifice his life."

Guderian realized at once that with this bombastic comparison Hitler was merely trying to evade the issue. What Guderian was talking about was not sacrifice as such, but useless sacrifice. He therefore said calmly, "Our soldiers have proved that they are prepared to sacrifice their lives. But this sacrifice ought only to be demanded when the end justifies it. And I see no such justification, my Fuehrer!"

The Siberians are Coming 333

From the horrified expressions on the faces of the officers present it was clear that they expected Hitler to explode. But he did not. He said almost softly, "I know all about your personal effort, and how you lead your troops from in front. But for this reason you are in danger of seeing things too much at close quarters. You are hamstrung by too much compassion for your men. Things look clearer from a greater distance. In order to hold the front no sacrifice can be too great. For if we do not hold it the Armies of Army Group Centre are lost."

The argument continued for several hours. When Guderian left the situation room in the Fuehrer's bunker late at night he overheard Hitler saying to Keitel, "There goes a man whom I have not been able to convince."

That was no more than the truth. Hitler had been unable to convince the man who had moulded the German armoured forces. To Guderian Hitler's operational principle of holding on at all costs, and, what was more, under the worst possible conditions, was an insult to the traditional and long-tested strategic thinking of the Prussian General Staff. In a hopeless situation one withdraws in order to avoid needless casualties and in order to regain freedom of movement for new operations. One does not hold on merely to be killed.

At the same time one cannot entirely dismiss the argument that permission to retreat in the wastes of the Russian winter and under pressure by a victory-intoxicated and fanatical Red Army might have turned the retreat of the mauled German troops into a rout.

And what would happen once the troops were on the run? In a retreat panic spreads quickly, turning a withdrawal into disorderly flight. And nothing is more difficult than to halt a flight of panicking units.

These considerations had induced Hitler to turn down Guderian's argument with an uncompromising No. He even countermanded the authorization he had given during the early days of the Soviet offensive for the shortening of front-line sectors and withdrawals to rearward lines, and instead issued his hold-out order which has since been so hotly disputed by military historians: "Commanders and officers must, by way of personal participation in the fighting, compel the troops to offer fanatical resistance in their positions, regardless of enemy break-throughs on the flank or in the rear. Only after well-prepared shortened rearward positions have been manned by reserves may withdrawal to such positions be considered."

This order has been a subject of contention to this day. On the one side it is said that the order was a piece of lunacy in that it resulted in the substance of the German forces in the East being needlessly sacrificed. The troops, it is argued, would have been quite capable of orderly retreat. Favourable defensive positions, as, for instance, along the high ground of Smolensk, would have compelled the Soviet High

Command to launch costly attacks which would have decimated the Soviet divisions instead of the German troops.

No doubt this argument holds good for certain sectors of the front. But there are also many commanders in the field, General Staff officers and Army Commanders-in-Chief who take the view that a general withdrawal under pressure from the Siberian assault divisions, with their superiority in winter warfare, would have led to chaos at many spots and to the collapse of substantial sectors of the front. Gaps would have arisen which no commander could have closed again. Holes would have been punched into the line, and the Soviet Armies would have simply raced straight through them, pursuing and overtaking the retreating Germans. And behind Smolensk the Soviets could then have closed the trap round the whole of Army Group Centre.

Perhaps this theory credits the Soviets with rather too much strength and skill. But it cannot be denied that—from a purely military point of view—Hitler's simple and Draconic hold-out order probably offered the only real chance of averting the terrible danger of collapse. Subsequent events entirely justified Hitler. The chronicler concerned only with military history must accept this fact. Political, moral, and philosophical considerations, needless to say, are an entirely different matter. The mental conflicts to which this hold-out order gave rise among commanders in the field, the tragedies it led to, and also the unparalleled heroism and self-sacrifice shown in obeying it are illustrated by the operations of Ninth Army on the northern wing of Army Group Centre, in the Kalinin-Rzhev area, and in the defensive battles of the neighbouring Sixteenth Army of Army Group North between Lake Seliger and Lake Ilmen.

Colonel-General Strauss's Ninth Army had been holding the line between the Moscow Sea and Lake Seliger with three Army Corps since the end of October. The line ran from Kalinin to Lake Volgo— the source of the Volga—and like a big barrier blocked the Volga bend, on the southern leg of which was the town of Rzhev.

Since mid-December 1941 the Ninth Army had been retreating, step by step, from Kalinin to the south-west.

The first attacks by the Soviet Thirty-first and Twenty-ninth Armies were directed against General Wäger's XXVII Corps in the area southeast of Kalinin. The temperature was 20 degrees below zero Centigrade. Deep snow covered the frozen ground. Artillery preparation was moderate. Only a few tanks accompanied the Soviet infantry over the ice of the Volga. On the right wing of the Corps, at Lieutenant-General Witthöft's Westphalian 86th Infantry Division on the Volga reservoir, the Soviet infantry attack collapsed in the German machine-gun fire.

In the adjoining sector on the left, held by the Pomeranian 162nd Infantry Division, however, the Russians punched a hole through the

The Siberians are Coming 335

line with the aid of a few T-34s, widened their penetration, and struck through with Siberian ski battalions. In spite of this threat, General Förster's VI Corps, on the left of XXVII Corps, held its sector against furious Soviet attacks. In the sector of 26th Infantry Division the combat-tested 39th Infantry Regiment under Colonel Wiese was down to two battalions—3rd Battalion having been divided up in order to replenish the depleted companies—and the similarly weakened Westphalian 6th Infantry Division had to hold a line of 16 miles. But the Russians did not get through.

Against the 110th Infantry Division, on the other hand, on the left wing of XXVII Army Corps, the Soviets succeeded in getting across to the southern bank of the Volga. From there they were now threatening VI Corps' only supply route, the road from Staritsa to Kalinin. At the same time the town of Kalinin was beginning to be outflanked.

The 3rd Battalion of the Westphalian 18th Infantry Regiment, the Corps reserve of 6th Infantry Division, was ordered to throw the Russians, who had penetrated with 200 men, back over the Volga again. The Westphalians prepared for action. The thermometer stood at 40 degrees below zero Centigrade. Their line of attack was through knee-deep snow. They tried three times.

But the Russians were across the river in regimental strength. It was impossible to dislodge them. True, the battalion took 100 prisoners, but it also lost 22 of its own men killed and 45 wounded, and, moreover, had 55 of its men affected by severe frost-bite.

At least they had stopped any further Soviet advance. The important supply road was cleared again and covered, and the threatening encirclement of Kalinin had been prevented. As a result, the Corps gained time to withdraw the units fighting in Kalinin. On 15th December 1941 the town was abandoned. On 16th December Soviet troops under Generals Shvetsov and Yushkevich moved into Kalinin.

The Soviet penetration into the German front on the Volga and the capture of Kalinin were a heavy blow. The eastern wing of Ninth Army had to be taken back. The Soviet High Command had thus gained the prerequisites for striking deep into the flank of the German Army.

Colonel-General Strauss had seen the danger approaching. He intended—like Guderian in the south, following Zhukov's breakthrough towards Stalinogorsk—to abandon the front bulge at Kalinin and to swing his Corps back to a greatly shortened line with Lake Seliger as the pivot; the line he intended to hold was a flat arc running from Lake Volgo to Gzhatsk on the Moscow motor highway. Rzhev was to be the centre and the core of the arc. The code name for this winter position was "Königsberg."

The disengagement was to be carried out in small, swift moves through a number of accurately defined intermediate positions, all of them bearing the names of towns as codes—Augsburg, Bremen, Coburg, Dresden, Essen, Frankfurt, Giessen, Hanau, Ilmenau, and Königsberg. However, the timetable functioned only as far as the "Giessen" stop. There the 'train' came to a halt.

Thanks to the gallantry of the fighting rearguards, the divisions had managed to get as far as "Giessen" more or less intact. In spite of the deep snow they had even managed to take most of their heavy weapons with them. For two weeks they had succeeded in holding off the strong enemy and preserving the cohesion of the front line.

The troops accomplished superhuman feats. Frequently the vehicles could be started only after twelve to fifteen hours of extremely hard work. Small fires had to be lit underneath the motor vehicles to thaw out the frozen gear-boxes and transmissions. Even then nearly all the vehicles had to be towed by human labour.

Covering lines organized by the fighting formations kept the pursuing Soviets at bay while the rest of the troops got the withdrawal going. The key role was played by individual fighters. In deep snowdrifts they lay behind their machine-guns, opposing the furious Soviet attacks. Their thin gloves were not enough to prevent their fingers from freezing off. They therefore wrapped their hands in rags and pieces of cloth. This, of course, made them too clumsy to work the triggers of their machine-guns or machine pistols. They therefore wedged little sticks, twigs, or chips of wood from the charred beams of peasant cottages between the rags enveloping their fists, and with these worked the triggers of their weapons.

Thus the Corps on the right wing of Ninth Army "travelled" via "Augsburg" and "Bremen," "Coburg," "Dresden," "Essen," and "Frankfurt," until Hitler's hold-on order halted their systematic withdrawal long before "Königsberg."

The divisions of 3rd and 4th Panzer Groups had already stopped their withdrawal at the Ruza line. For that reason Ninth Army was now ordered to hold the continuation of this front line as far as the Volga.

Field-Marshal von Kluge, the new C-in-C Army Group Centre, demanded strict observation of this order. He instructed Ninth Army: "Everybody must hold on wherever he stands. Anyone failing to do so tears a hole in the line, a hole which can no longer be sealed."

The only faint ray of light in the order was the passage: "Disengagement from the enemy can be useful or purposeful only when it results in more favourable fighting conditions, and if possible in the formation of reserves." But the Field-Marshal immediately restricted this concession: "Any disengagement of units from division upward requires my personal authorization."

On 19th December 1941 Colonel-General Strauss arrived at the

Page i, top: Guderian, the expert on tank battles. *Bottom:* Prisoners taken in the Vyazma pocket

Page ii, top: Tanks in the Don steppe. *Bottom:* Village street on a frosty day

Page iii: War-time Madonna

Page iv, top: A Village near Rzhev: women carrying a child in an open coffin to the burial ground. *Bottom:* On the fringe of the battle of Kiev

Below: A main road in the Tula area during the muddy period

Opposite, top: Flak on the bank of the Desna near Bryansk. *Bottom:* Advance in southern Russia during the summer of 1942

Below: A tarpaulin and a wooden cross

Opposite, top: Scorched earth: horses reverted to wild state. In the great Don loop, 1942. *Bottom:* Village north of Bryansk

Below: In a mechanized war men do not die among the flowers of the field with shouts of urrah on their lips. *Overleaf:* Stalino, the Soviet steel centre in the Donets Basin—now Donetsk

Left and below: Kharkov, the industrial and spiritual metropolis of the Ukraine: cathedral and administration blocks

Opposite: Over the passes of the Caucasus

Overleaf: Taking off for Stalingrad: the last supply flight into the pocket

Military cemetery in front of the cathedral at Zhizdra.

headquarters of General Schubert's XXIII Corps, which comprised 251st, 256th, 206th, 102nd, and 253rd Infantry Divisions, with a new order: "Not another step back."

Three days later the assault regiments of General Maslennikov's Soviet Thirty-ninth Army struck at the Corps' right wing with T-34s and tried to break through the line of the 256th Infantry Division from Saxony. Maslennikov wanted to reach Rzhev.

The Saxon regiments of 256th Infantry Division resisted desperately. They allowed the Russian tanks to roll past them, and from their holes in the snow shot up the Soviet infantry. Tank demolition squads of the artillery then tackled the T-34s.

There was Second Lieutenant Falck of 1st Battalion, 256th Artillery Regiment, lying behind a snowdrift. For camouflage he had slipped on a home-made snow smock. A Soviet tank rumbled past him, spraying the ground with machine-gun fire.

That was Falck's moment. He leapt at the tank and swung himself up on its stern. Hanging by his fingers, he wriggled round the turret. He pulled the string of two egg-shaped hand-grenades, holding on to the gun-barrel with his right hand. He then leaned well forward and with his left slipped the grenades down the barrel with a vigorous push. He quickly dropped off the tank, into the soft, two-foot-deep snow. The first bang came at once, followed by a succession of explosions as of fireworks. The hand-grenades had done their job. The tank's ammunition was going up.

The line of 256th Infantry Division held on 22nd and 23rd December. It still held on Christmas Eve, and on Christmas Day and Boxing Day. The temperature was 25 to 30 degrees below zero Centigrade. The sky was dark and cloudy, and the ground was shrouded by light flurries of snow. Visibility was less than 100 yards.

Out of this backdrop of a "Napoleonic winter" Russian tanks kept emerging like phantoms. German Panzerjägers often engaged the T-34s with their 3·7-cm. anti-tank guns at no more than six yards' range. If the tanks survived, then the anti-tank gunners were crushed. Frequently the 8-cm. AA combat groups of the Luftwaffe or the explosive charges used by daring individual fighters like Second Lieutenant Falck were the only salvation against T-34s.

By 29th December the men of 256th Infantry Division had been resisting a ten times superior enemy for seven days. By then they were holding only minor strongpoints—at road-forks, in forest clearings, on the edge of villages.

The Russians attacked also in the sector of the neighbouring division. Three Armies of Colonel-General Konev's "Kalinin Front" were battering at the German lines along the Volga bend. It was becoming increasingly obvious that Konev intended to strike via Rzhev to the Moscow motor highway, in order to link up with Zhukov's southern prong in the rear of the German Army Group. Rzhev became a keypoint in the destinies of the Eastern Front.

By 31st December, the last day of 1941, the main fighting line of 256th Infantry Division had been torn open all over the place, in spite of support by VIII Air Corps. The Russians were infiltrating. The 206th Infantry Division, too, was finished. Its 301st Infantry Regiment was down to a few hundred men. On the same day the cohesion of Ninth Army's line was lost west of Staritsa. In the sector of 26th Infantry Division, north-west of the now burning town of Staritsa, the two battalions of 18th Infantry Regiment and what was left of 84th Infantry Regiment, together with 2nd Battalion of the Divisional Artillery Regiment, were holding out against an enemy coming at them from all directions.

The railway station of Staro-Novoye was also blazing fiercely. Christmas parcels, special Christmas rations, and the division's winter clothing, which had at long last arrived, were all going up in flames. All the troops were able to save was a dump of Swiss cheese. Everywhere, in all the peasant huts along the sector, there were piles of the large round cheeses. As the relieved men came in from outside they would carve themselves large chunks with their bayonets.

But twenty-four hours later the peasant huts and the cheeses had to be abandoned. The regiment had to establish a new switchline against enemy groups which had broken through six miles farther to the south-west, at Klimovo.

Aerial reconnaissance had reported a strong enemy column on the right wing of 256th Infantry Division, outside Mologino. Mologino was 19 miles from Rzhev. And in Rzhev were 3000 wounded.

Division received an order by radio from XXIII Corps to reinforce its right wing and to "hold on at all costs." The remnants of 476th and 481st Infantry Regiments flung themselves into the path of the Russians along the road.

Hitler's order "Ninth Army will not retreat another step" nailed down the Corps to the line reached on 3rd January 1942, outside Latoshino, east of Yeltsy.

On 31st December at 1300 hours Colonel-General Strauss had turned up at General Schubert's Corps Headquarters in Rzhev with the order: "Mologino must be defended to the last man." What other order could he give? Twenty minutes later, at 1325 to be exact, Lieutenant-General Kauffmann, commanding 256th Infantry Division, entered the room. He came straight from Mologino. He was as white as a sheet and half frozen. In a voice trembling with emotion he reported to his Army Commander-in-Chief: "Herr Generaloberst, my division is down to the combat strength of a regiment and is surrounded by Soviet ski troops. The men are at the end of their tether. They are just dropping with fatigue. They flop into the snow and die from exhaustion. What they are expected to do is sheer suicide. The young soldiers are turning on their officers, screaming at them: 'Why don't you just go ahead and kill us—it makes no odds to us who does us in.' Mologino is lost already."

Colonel-General Strauss stood petrified. Then he said slowly, "It is the Fuehrer's express order that we hold out. There is no other way than to hold on or to die." And turning to General Kauffmann he added, "You'd better drive to the fighting-line, to your men, Herr General— that's where your place is now." The General saluted without a word and left the room.

In point of fact, the situation at Mologino was not quite as desperate as Kauffmann had made out. In the afternoon of 31st December the remnants of the reinforced 1st Battalion, 476th Infantry Regiment, had been rushed into the town, which was still being stubbornly defended by Reconnaissance Detachment 256 under Major (Reserve) Mummert. The remaining units of the regiment were assigned places in the defensive positions planned west of the town. However, by nightfall Siberian ski troops had occupied the forest between Mologino and the intended line of defence. What was left to be held? It was now merely a case of defending Mologino for as long as possible, thereby tying down the Russian forces and preventing them from interfering with the Corps' withdrawal. In fierce fighting the men of the Reconnaissance Detachment and of 1st Battalion repulsed the attacks of the Siberians. Frequently they would hold on to just a few isolated houses in the middle of the township. Then they would gain a little breathing space again by immediate counter-attacks.

Radio communication with Division had been cut off since 2nd January. Contact with the neighbouring unit on the left was maintained by patrols shuttling between them. Nevertheless Major Mummert was determined to hold Mologino.

During the night of 2nd/3rd January the signals section succeeded in restoring contact with Division. At Division there was much surprise that Mologino was still being held, and the order was given to evacuate the place at once and to rejoin Division. Towards 0600 hours Mummert evacuated Mologino. The heavy equipment was left behind. Through the black night and the Siberian lines, moving silently in single file along a trail normally used by patrols, the men made their way to the neighbouring unit.

Once more 206th Infantry Division succeeded in sealing the gap torn in the front of Ninth Army, but on 4th January 1942 the front cracked for good. Between VI and XXIII Corps a yawning gap appeared, between 9 and 12 miles wide. Through this opening strong Soviet forces now thrust across the Volga. The Soviet Twenty-ninth Army wheeled round towards Rzhev and tried to take the town from the south-west. Major Disselkamp, chief of the supply services of

6th Division, an energetic officer, intercepted the first thrust with hurriedly scraped-together emergency units. These consisted of drivers and other supply personnel, a few self-propelled and anti-tank guns, units of a repair company, and above all the veterinary surgeons and orderlies of Veterinary Company, 6th Infantry Division. With these men Disselkamp halted the Soviets. In this manner VI Corps was given a chance to establish a new defensive front with 26th and 6th Infantry Divisions. Rzhev was held as a cornerstone for future counter-actions. The Soviet Thirty-ninth Army and General Gorin's Soviet Cavalry Corps, however, bypassed the town in the west and drove on southward via Sychevka towards Vyazma.

Although the front was in flames along its entire length, the key sectors and the long-term objectives of the Russians were beginning to emerge clearly. Colonel-General Konev, having torn a hole in XXIII Corps on the northern wing of Army Group Centre, intended to envelop and wipe out the Ninth Army. On the southern wing Marshal Zhukov was racing through the gap between Second Panzer Army and Fourth Army, aiming at Vyazma, and at the same time hoping to strike at the flank of Second Panzer Army.

General Golikov with his Soviet Tenth Army was already encircling the town of Sukhinichi. But General Gilsa's 4000-strong combat group refused to give way and turned the town into a breakwater against the Soviet storm. Gilsa's group held out for four weeks. It was a chapter that will have to be mentioned later.

The year 1941, which had begun with such confidence in victory, drew to its close in an atmosphere of gloom and anxiety. On Boxing Day Hitler had seized the occasion of a complaint by the new Commander-in-Chief Army Group Centre to rid himself of Guderian, whose constant warnings he had found irksome. Field-Marshal von Kluge had accused the Colonel-General of disobedience: he had had some major differences with him earlier in December. Hitler thereupon deposed Guderian. The troops in the field were dumbfounded. What was to become of them if they were deprived of the best military brains?

Full of forebodings, Guderian concluded his farewell order to his combat-tested Second Panzer Army: "My thoughts will be with you in your difficult task."

And a difficult task it was. Nowhere could the German Command raise sufficient reserves to halt the Russian penetrations. Red cavalry formations were already pressing against the weak covering lines north of Yukhnov and threatening the vital supply routes to Smolensk. Soviet airborne troops were being put down behind the German lines. The partisans became a major threat.

Hitler was ranting against fate, cursing the Russian winter, railing against God and his generals. His wrath fell also upon the well-tried

Commander-in-Chief 4th Panzer Group. Early in January, when Colonel-General Hoepner withdrew his 4th Panzer Group—since the New Year raised to Fourth Panzer Army—without, as it was thought at the Fuehrer's Headquarters, asking for permission, Hitler seized on this 'disobedience' in order to make an example. Hoepner was deposed, degraded, and dismissed the service with ignominy. After Guderian, the troops in the field thus lost their second outstanding tank leader.

That there was no factual justification whatever for this measure is attested by Major-General Negendanck, then signals chief of the Panzer Army, who was a direct witness of the incident. This is the account he has given to the author: "A few of us were having lunch with Colonel-General Hoepner, who had just come back from the front. At table he expounded his view to his chief of staff, Colonel Châles de Beaulieu, that the right wing of Fourth Panzer Army should be taken back since the adjoining wing of Fourth Army would not be able to stand up to a Russian thrust into the gap which had arisen there. This view was presently conveyed by telephone to the Chief of Staff of Army Group Centre. We were still at table when Field-Marshal von Kluge rang through to discuss the matter with Colonel-General Hoepner, and I remember clearly how Colonel-General Hoepner repeated at the end of the conversation, 'Very well, then, Herr Feldmarschall, we'll only withdraw the heavy artillery and baggage trains to begin with, to make sure we don't lose them. You will explain to the Fuehrer the need for this measure and request his authorization.' When the colonel-general rejoined us after his conversation he said, 'All right, Beaulieu, you will make all the necessary preparations.' At midnight, suddenly, came the dumbfounding news of Hoepner's dismissal. I was shortly afterwards told by a General Staff officer of Army Group Centre that Field-Marshal von Kluge had reported the affair to Hitler not on the lines agreed with Colonel-General Hoepner, but describing the withdrawal of the front as an accomplished fact. Hitler immediately exploded and ordered the dismissal of our outstanding and universally revered Army commander."

Thus far Major-General Negendanck's report. It represents an exceedingly interesting and valuable contribution to military history, on the much-discussed question of Hoepner's dismissal and on the equivocal role played by Field-Marshal von Kluge in Hoepner's as well as in Guderian's recall from their commands.

342   Part Four: Winter Battle

2. South of Lake Ilmen

The fishing village of Vzvad–Charge across the frozen lake–Four Soviet Armies over-run one German Division–Staraya Russa–The Valday Hills–Yeremenko has a conversation with Stalin in the Kremlin shelter–The Guards are starving–Toropets and Andreapol–The tragedy of 189th Infantry Regiment.

TO return to the hard-pressed front. Having struck at the middle and at the northern and southern flanks of Army Group Centre, the Soviet High Command now also struck at the right wing of Army Group North. South of Lake Ilmen, where the 290th Infantry Division from Northern Germany was in position, the great Soviet breakthrough battle opened early in January.

Viktor Nikolayevich was an experienced fisherman on Lake Ilmen. He wore a goatee beard and was known by his fellow villagers as "The Counsellor." He was the leader of Vzvad's eighty-strong village guard against partisan attacks. Viktor Nikolayevich and his friends simply wanted to be left in peace. At the beginning of September 1941 the Germans had arrived with their pleasant Lieutenant-Colonel Iffland and his Panzerjäger Battalion 290. They had settled down at the northernmost point of the strategically important neck of land between Lakes Seliger and Ilmen. That point marked the end of the only road leading up from Staraya Russa through 10 miles of steppe, forest, and swamp to the lake and the Lovat estuary.

The fishing village of Vzvad was therefore a strongpoint, a roadside fortress, and the terminal point of the front between Lakes Seliger and Ilmen, on the wing of 290th Infantry Division. In the autumn the Panzerjägers had left. What point was there in guarding swamps and marshes? But towards the end of December units of the battalion had returned. For what in the summer had been impassable territory—except for those with local knowledge—might become an easy passage through the German front in the winter, once the swamp was frozen solid. Scouts and supply columns for the partisans tried their luck there. Patrols sent out by the Russian units holding a line through the forests north of Sinetskiy Bay were crossing the marshes and frozen lakes and ponds on skis.

A major Soviet breakthrough in the direction of the traffic junction of Staraya Russa would have been a mortal threat to the two Corps holding the line between Lakes Ilmen and Seliger. The Russians had

South of Lake Ilmen 343

tried before—and often successfully—to unhinge entire sectors of the German front by capturing their rearward supply bases.

The temperature on 6th January 1942 was 41 degrees below zero Centigrade. The ice-cover on the lake and on other waters was over two feet thick. The depth of the snow was nearly two feet. German patrols were constantly on the move, searching the area for tracks. They found nothing.

In the early afternoon "The Counsellor" came to see Captain Pröhl, the commander of the Panzerjägers in Vzvad and Lieutenant-Colonel Iffland's representative there. "There's talk in the village that the liberation battle for Staraya Russa is starting to-day, our Russian Christmas Day," he said.

The captain knew that Viktor Nikolayevich was not given to gossip. He also knew that no amount of patrolling could prevent secret contacts between the fishermen on both sides of the fighting-line. He immediately sent out two ski patrols.

Two hours later the first patrol was back. "Numerous ski tracks by the Lovat river," they reported.

The second patrol brought back three prisoners—two Soviet infantrymen and a suspect civilian.

Night fell. Pröhl put his men on immediate alert. On the far side, over the Russian lines, red and green flares rose into the sky.

The icy night passed quietly and uneventfully. Not a shot was fired along the snow-bound front between Lakes Ilmen and Seliger.

The prisoners were interrogated through an interpreter. The civilian claimed to come from a near-by village. He had been forced by the two soldiers, he said, to show them the way to Vzvad. His shaven head, however, suggested that he too was a soldier, probably on an Intelligence assignment. Captain Pröhl had him locked up in a sauna.

The interrogation of the two uniformed men yielded some interesting facts. Both belonged to the Soviet 71st Ski Battalion. They reported that their battalion had been freshly sent up to the front and was equipped with snow-ploughs and motor-sledges. Food was poor, they complained. All the supplies consisted of were weapons and ammunition.

Was there any talk about attack, the interpreter asked. For a moment the prisoners hesitated, but presently they started talking. "Yes, it's said that the balloon's going up to-morrow."

Pröhl received the interrogation protocol with caution. Surely they would notice soon enough when the artillery started its preliminary bombardment. That was the invariable sign of an impending attack.

On the morning of 7th January Pröhl informed Division. Then he sent out more patrols. The icy eastern wind was freshening, developing into a blizzard and obliterating tracks, paths, and even the road to

Staraya Russa. The thermometer outside "The Counsellor's" cottage showed 45 degrees below zero.

At dusk the sound of aircraft was heard. The lighthouse at Zhelezno was blinking all the time, no doubt acting as a beacon to the Soviet aircraft. Strangely enough, not a single machine came anywhere near the front. Not a shot was fired. Not a gun opened up.

At 2120 hours the telephone rang. Second Lieutenant Richter reported from strongpoint "Hochstand 5," two miles south-east of Vzvad: "Strong enemy movements. Motor-sledges and ski troops bypassing us."

A runner arrived from the observation post on top of the church tower of Vzvad: "Columns of motor vehicles approaching from the south-east with headlights full on."

Two strong patrols left at once. Panting, runner after runner came back: "Scrub-covered ground at 'Hochstand 5' held by enemy." "Enemy ski troops near the hamlet of Podborovka—*i.e.*, south-west of Vzvad, on the road to Staraya Russa. They are providing cover for snow-ploughs employed on road-clearing."

What could it mean? The Soviets, unobtrusively and well camouflaged, had clearly broken through the German front, which was held only intermittently, by separate strongpoints. They had moved without artillery preparation.

Action stations! The telephone-line to "Hochstand 5" was still intact. Pröhl rang Second Lieutenant Richter: "Pack up at once and get through to Vzvad with your men."

"We'll try," Richter replied.

A continuous procession of Soviet columns was moving past the German strongpoint. Richter and his twelve men pulled their snow-smocks firmly over their uniforms. Then they infiltrated into the Soviet columns. At a suitable spot they dropped out again and reached Vzvad unmolested.

At 0300 hours the Russians attacked the German strongpoint. Telephone connection with Division was abruptly cut.

But Captain Pröhl knew even without orders from above that the strongpoint of Vzvad had to be held as a 'breakwater.' Meanwhile the 6th Company, 1st Luftwaffe Signals Regiment, units of Motorcycle Battalion 38, under 18th Motorized Infantry Division, and 2nd Company Local Defence Battalion 615 had moved into Vzvad to avoid being over-run by the Russians. As a result, Pröhl now had 543 men under his command.

These 543 men held the isolated strongpoint on Lake Ilmen, far ahead of the main German fighting-line, for thirteen days—an island in the enemy flood.

Furiously the Russians tried to wipe out Vzvad, the strongpoint controlling the road. They employed ski battalions. They tried

South of Lake Ilmen 345

"Stalin's organ-pipes." They brought on fighter bombers. And eventually they came with tanks. But Vzvad held out.

The Soviets fired incendiary shells into the village in order to destroy the troops' quarters: each of the shells contained twenty to thirty phosphorus sets. The wooden houses blazed like torches. Hospital quarters and dressing stations were consumed by the flames. Twenty-eight wounded had to be laid out in the open on mattresses and blankets, behind the ruins of buildings, in the snow. In 35 degrees below.

The operations diary and the radio signals sent to 18th Motorized Infantry Division in Staraya Russa, under which Pröhl was placed after contact had been lost with 290th Infantry Division, are profoundly moving in their unadorned, matter-of-fact account, and compel the admiration of every reader.

12th January. Ceaseless enemy artillery bombardment. A German aircraft dropped ammunition. But instead of HE shells the container was full of AA shells, which were useless. In another container was the Knights Cross for Captain Pröhl. Five Iron Crosses 1st Class and 20 Iron Crosses 2nd Class were, moreover, awarded by radio signal from Division.

By 1640 hours ammunition and bandages were getting low. An urgent signal was sent to Division asking for supplies. A request was added that these should be dropped from greater height; the day before all four ammunition containers had exploded on hitting the ground.

At 1900 hours Pröhl urgently repeated his demand for ammunition and food supplies. The wounded horses were slaughtered, and one day's ration gained as a result. But there were no potatoes or bread whatever.

By 2000 hours five men had been killed and thirty-two wounded.

14th January. The commander of the Soviet 140th Rifle Regiment sent a horseman under a white flag. He demanded capitulation. He was sent back, and a salvo was fired by anti-tank and infantry guns against Podborovka, where the Russian regimental headquarters were situated.

During the night the Russians came with tanks. A T-26 broke through and stopped right outside Pröhl's command post. Inside, the men waited calmly to see if the Russian would open his turret. He did not. They flung explosive charges at the tank. The crash of the grenades seemed to worry the Russians. The tank withdrew to the southern end of the village. There it passed right in front of Sergeant Schlünz's anti-tank gun. Two shots rang out. Both were direct hits. The tank went up in flames.

A Fieseler Storch aircraft arrived with a medical officer, Dr Günther, and medical supplies. A signal from Hitler commended the defenders and simultaneously informed them that relief was impossible. Pröhl was given permission to evacuate Vzvad if the garrison was threatened with annihilation.

This *carte blanche* faced Pröhl with a difficult mental conflict: were they already threatened with annihilation, or not just yet? The strongpoint had already been bypassed by the Soviets to a depth of 10 miles. Should Pröhl evacuate it? At that moment of doubt came a signal from 18th Motorized Infantry Division: "Staraya Russa is holding out, in spite of being encircled."

Pröhl realized that these island fortresses tied down the enemy and broke the momentum of his headlong advance. Vzvad too would hold out.

18th January, the 11th day of encirclement. The thermometer had dropped to 51 degrees below. Minus 51 degrees Centigrade! At night patrols went out and stripped the felt boots off the Soviet dead in front of the lines. They collected their fur caps and cut the fur coats off the bodies frozen into rigid postures.

19th January. Soviet large-scale attack during the night. Penetrations. Hand-to-hand fighting in the flickering light of blazing houses. Savage battles for the sauna and the collective farm store. Four tanks knocked out with hand-grenades in close combat.

The fighting lasted eight hours. The Soviets were repulsed. German casualties totalled seventeen killed and seventy-two wounded. "One more such attack and we're finished," Lieutenant Baechle reported to Captain Pröhl in a calm voice on the following morning, the morning of 20th January.

Pröhl nodded. He had already made his decision. "To-night's our last chance. After the losses they've suffered the Russians will be regrouping. That's when we must act."

Officers, platoon leaders, and the commander of the local civilian guard were summoned to a meeting. It was decided to break out across the ice of Lake Ilmen. The objective was Ushin on the western shore of Tuleblskiy Bay. It meant marching 12 miles over the piled-up ice of the lake and through chest-deep snow.

The dead were buried by the "House Olga," which had burnt down, and where, as a result, the ground was still thawed from the blaze. A mass grave was blasted and dug there. Sixty-two wounded men, incapable of marching, were put on sledges and the last remaining healthy horses harnessed to them. Snow was falling. There was a haze. On the other hand, it was not quite as cold as the day before—a mere 30 degrees below.

At nightfall they moved off. A patrol with local guides went in front, trampling a firm path for the rest. The men of Motorcycle Battalion 38 were waist-deep in snow. The lead group had to be relieved every half-hour: that was as much as even the strongest man could stand.

The separate marching units followed at ten-minute intervals, in close order. The civilian guard of Vzvad moved off with them, with "The Counsellor" Viktor Nikolayevich at their head. Not one of them dared stay behind. It would have meant death.

The last signal to 18th Motorized Infantry Division read: "Breakthrough beginning. Our identification signal: Flares in sequence green —white—red."

Second Lieutenant Richter, left behind with two platoons as the rearguard, continued for the next two hours to put up as much harassing fire as possible, simulating positions held in strength. Then Sergeant Steves moved off with the sapper platoon. No. 3 Platoon of the Reconnaissance Squadron stayed behind for another thirty minutes, keeping the machine-guns stuttering. After that one gun after another fell silent. A strange stillness fell over Vzvad—now utterly

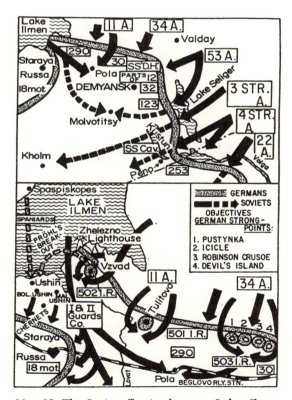

Map 18. The Soviet offensive between Lakes Ilmen and Seliger in January 1942 and the operations area of 290th Infantry Division and 18th Motorized Infantry Division.

gutted and wrecked. Sergeant Willich was the last to move out—past "House Olga," where the dead were resting.

It was a bad journey. First they moved over the ice of the Lovat to the north, as far as the lighthouse, then in a north-westerly direction on to the ice of the lake, and finally south-west towards the shore. The temperature was 40 degrees below, and on the lake as low as 50 degrees. The men were like moving icicles. The horses were reeling. Some of them collapsed. A quick *coup de grâce*, and the men moved on again.

Their compass needles froze up. They had been on the move for six hours. First Lieutenant Mundt stopped and let his group march past him. "Everything all right?" he asked Second Lieutenant Voss as his platoon moved past. "Everything in order."

But when No. 2 Platoon came past, First Lieutenant Beisinghof was not at its head. It was being led by Sergeant Matzen, Beisinghof and Dr Wiebel, the M.O., were with a man who refused to go on. He had sat down in the snow and wanted a rest. "Only half an hour—until the next group comes," he begged. But it would have been certain death. They pulled him to his feet; they argued with him; they ordered him. The lieutenant and the medical officer supported him one on each side. A hundred yards behind their platoon they slowly made their uncertain way forward.

Beisinghof once more moved to the head of his column. That was what they all did—Captain Pröhl, Second Lieutenant Matthis, Second Lieutenant Güle, and Dr Günther, the M.O., with the main body of the force, and Sergeant Feuer with the vanguard and Second Lieutenant Richter with the rearguard. Like sheep-dogs they moved forward and backward along their columns, seeing to it that no man was left behind or had thrown himself despairingly into the snow. Dog-tired themselves, they covered the distance twice and three times over.

After fourteen hours' march they made it. At 0800 hours Sergeant Feuer caught sight of men with German steel helmets, wrapped up to the tips of their noses. He called out to them, stumbled over to them, and caught hold of the nearest one: "Kamerad, Kamerad!"

They embraced. But what on earth was the man saying? Feuer understood only the words "Santa Maria" and "Camarada." But he guessed that "Bienvenido" meant "Welcome." The German combat group had encountered a Spanish unit. Spaniards, volunteers of 269th Infantry Regiment of the Blue Division, employed on the Eastern Front, north of Lake Ilmen, as the 250th Infantry Division.

On 10th January the Spanish ski company under Captain Ordás had left the northern shore of Lake Ilmen with 205 men in order to reinforce their German comrades in Vzvad. But the ice barriers on the lake had made the 20 miles as the crow flies into 40 miles as the men had to march. The Spaniards' radio equipment broke down and their compasses froze up.

When Captain Ordás reached the southern shore of Lake Ilmen a long way west of Vzvad half his men were suffering from severe frostbite. On their further move they were attacked by Siberian assault detachments. The Spaniards fought excellently and even took some prisoners. They recaptured Chernets and, together with a platoon from a police company, repulsed furious Soviet counter-attacks.

On 21st January only thirty-four men were still alive of the 205 men of the Spanish ski company. Hence the demonstrative way in which they welcomed the German garrison of Vzvad, four miles east of Ushin. Two days later they mounted a counter-attack against the lost strongpoints of Malyy Ushin and Bolshoy Ushin side by side with German infantrymen, in the sector of 81st Infantry Division, which had only just arrived from France. Twelve Spanish soldiers survived—twelve out of 205.

The combat group from Vzvad had lost five men on its journey over the lake. They had fallen victim to the cold. Exhausted and lethargic, they had dropped into the snow, unnoticed, and had gone to sleep for good in the boundless waste.

As the survivors staggered into their cold quarters in Ushin they could hear the distant rumble of the front and see the fires of Staraya Russa. That magnificent ancient town, the old trading centre on Lake Ilmen, was once more in flames. Many a battle had been fought for its possession throughout the centuries. It had been captured, and it had been destroyed. In this winter of 1941/42 Staraya Russa had become a traffic junction, a supply base, and the heart of the supply services for the German front between Lake Ilmen and Lake Seliger. If it fell the whole front would fall.

It was Major-General Herrlein's 18th Motorized Infantry Division from Liegnitz which experienced at Staraya Russa a kind of 'super-Vzvad' by desperately defending the town from all directions against the divisions of the Soviet Eleventh Army. The 18th was under the command of Colonel Werner von Erdmannsdorff, deputizing for his severely ill general. By its resistance in Staraya Russa the division was to foil the grand plan of operations of General Morozov's Eleventh Army.

What was that plan? Morozov intended to move around Lake Ilmen, and then, in co-operation with a strong Army Group operating north of the lake against the Volkhov, to strike at Colonel-General von Küchler's Eighteenth Army east of Leningrad, and thus to start the liberation of that city. It was a good plan. The units employed for this operation against Staraya Russa on the western wing of the Soviet Eleventh Striking Army were excellent crack formations—the I and II Guards Corps. This showed the importance attached by the Soviet

High Command to the task. After all, the successful conclusion of the operation would mean two significant successes—the gaining of the elbow-room necessary for further operations, now barred by Staraya Russa, and the seizure of the German Sixteenth Army's huge supply depots and stores of war material. These would be a valuable prize for the poorly supplied Soviet Corps, a prize of particular value since these Corps would presently be operating behind the German lines.

During the first week Morozov's Guards penetrated five times into the town centre—in fact, right into the Army's supply depot. But each time they were driven back again with heavy casualties. The ammunition dumps blew up. Whatever ancient and historical buildings of Staraya Russa had survived the battles of the summer were now reduced to rubble by shells and fire. But the human wall around Staraya Russa held. Staraya Russa was the rock in the crashing breakers of the battle, the point of crystallization at which Corps time and again re-established the shattered flank of Sixteenth Army. The credit for this must go not only to the men but also to the staff of 18th Motorized Infantry Division. This division's defensive success was a good example of skilful leadership and experience of Russian conditions applied at divisional level. The background is worth mentioning briefly.

The 18th Motorized Infantry Division, badly mauled in the fighting at Tikhvin, had been dispatched by Colonel-General Busch to the Staraya Russa area, as an Army reserve, and spread out over the villages. Reconnaissance, experience of Soviet tactics, and a well-developed 'nose' for the situation led the divisional staff to conclude that the Russians would cross the frozen Lake Ilmen to strike at Staraya Russa. For that reason the division's acting commander, Colonel von Erdmannsdorff, and his chief of operations kept urging Corps, and eventually even Army, to concentrate the division—with all its units, including the baggage train—and to deploy it in an already reconnoitred position along the lake.

Corps would not hear of such precautionary measures, and regarded any anxiety about a Soviet attack across the ice of Lake Ilmen as "unrealistic." Colonel-General Busch, however, thought there might be something in Erdmannsdorff's hunch and let him have his way. On 4th January Army therefore issued appropriate orders. But not until 4th January. During the night of 7th/8th January—seventy-two hours later—the Russians came across the ice.

After the very first report from the fighting area Corps and Army immediately realized that the Soviet attack on the northern wing of the line between the two lakes was no local operation. That much was clear from the signals from forward strongpoints and patrols. The offensive, for once, had not started with the customary and traditional artillery bombardment. It had started in complete silence in order to deceive the Germans about the extent of the operation.

By the time the Soviet artillery opened up to support the frontal attack against 290th Infantry Division, aimed principally at Tulitovo and Pustynka, strong Russian forces had already driven through the wide gap in the front, reaching the Lovat estuary across the frozen ponds and marshes and, even more serious, getting into the rear of 290th Infantry Division across the ice of Lake Ilmen.

Freight-carrying gliders and transport aircraft fitted with skids had landed on the frozen lake and unloaded ski battalions and rifle brigades. Soviet armoured brigades were crossing the lake with heavy tanks, making for the penetration points. Like nightmarish monsters 52-ton KVs came crawling over the ice. Noisy snow-ploughs were moving ahead of Soviet infantry and tank battalions, clearing the way for them. Motor-sledges packed with infantry roared through the landscape, throwing up huge sprays of snow.

No German eye had ever seen anything like it. No German staff officer had ever witnessed this sort of thing at manœuvres.

Consequently the first reports produced a good deal of surprise and incredulous shaking of heads at Corps and Army headquarters. But very soon there was no doubt that a large-scale Soviet offensive had been launched across the lake, and that its first objective was Staraya Russa, the transport junction of the German front on Lake Ilmen.

Colonel von Erdmannsdorff had arrived from the Shimsk area during the night at the head of his 18th Motorized Infantry Division. General Hansen put him in command of the garrison units in the town, the baggage trains, the rearward services, and the construction battalions. With these units Erdmannsdorff succeeded in establishing a defensive front outside the town and in stabilizing the situation.

Against this unshakable bulwark of Silesian infantry regiments and the units subordinated to them the first part of the Soviet plan came to naught. The Soviet Eleventh Army had to bypass Staraya Russa. It had to turn to its second task—to strike south along the Lovat river in order to get behind the divisions of the German X Corps. In this attempt General Morozov came up against the North German regiments of 290th Infantry Division under Lieutenant-General von Wrede.

As the defenders of Vzvad had done, so the companies of 290th Infantry Division held on with their inadequate numbers, even though their positions had been bypassed on both sides, and in this way formed breakwaters against Soviet attacks.

At Tulitovo the 2nd Battalion, 502nd Infantry Regiment, held out for nearly five weeks. After that it was over-run. At Pustynka Second Lieutenant Becker with his 1st Company 503rd Infantry Regiment resisted for exactly twenty-six days, tying down strong enemy forces. The strongpoints named "Devil's Island," "Icicle," and "Robinson

Crusoe Island" were defended by companies of 503rd Infantry Regiment under Eckhardt and Wetthauer, although the men had had no rations for several days.

The situation is illustrated by an exchange of signals between 290th Infantry Division and X Army Corps. The division radioed: "Most urgently request ammunition."

Corps replied: "According to our calculations too much ammunition used."

290th Infantry Division rejoined: "Your calculations of no consequence."

In this way weakened regiments mounting about 130 small-arms to the mile were holding entire divisions of the Soviet Eleventh Army. The Russians were prevented from making a decisive frontal breakthrough. But it was impossible to prevent outflanking attacks by two Soviet crack units. The II Guards Corps captured Parfino, a railway station on the important Leningrad–Staraya Russa–Moscow line, and the I Guards Corps with an even wider sweep struck at the rear of 290th Infantry Division and finally succeeded in infiltrating.

At that crucial moment the Soviet Thirty-fourth Army made a penetration to the right of 290th Infantry Division, in the sector of the 30th Infantry Division from Schleswig–Holstein, severed the link between the two divisions, likewise turned against the rear of 290th Infantry Division, and at Pola, along the river of the same name, joined up with II Guards Corps forming the other jaw of the pincers.

The trap was closed round 290th Infantry Division. The left wing of the German front on Lake Ilmen was outmanœuvred. The Soviets had cut X Corps in two and faced it with an exceedingly dangerous situation.

And what meanwhile was the situation on the right wing, in the area of Count Brockdorff-Ahlefeldt's II Army Corps?

On 9th January there erupted across Lake Seliger a Soviet large-scale attack which, in concentration of forces and momentum, surpassed anything previously known. Four Russian Armies, the Twenty-second and Fifty-third Armies, as well as the Third and Fourth Striking Armies, charged across the frozen lake with approximately twenty divisions and several dozen independent armoured and ski brigades.

They pounced upon the weakly held 50-mile sector of a single German division, the 123rd Infantry Division from Brandenburg, and its right-hand neighbour, the 253rd Infantry Division, the wing division of Army Group Centre.

The main impact of the blow fell upon the 123rd and shattered the front of the Brandenburg regiments. In vain did its left-hand neighbour, the 32nd Infantry Division from Pomerania, come to its aid with whatever forces it could spare. It was of no use: the 123rd Infantry Division was swept aside.

South of Lake Ilmen 353

What were the Russians after? The two Striking Armies had not been sent into battle in order to operate on the narrow strip of land between the two lakes: they had different strategic objectives, going far beyond the two German Corps on the Lake Ilmen front.

The attack of the Soviet Fifty-third Army, on the other hand, was directed exclusively against the German lines between the two lakes. Having broken through, the Soviet forces swiftly wheeled towards the north-west in order to link up with units of the Soviet Eleventh Army coming down from the north, thus encircling most of the German X Corps and all of II Corps.

At the centre of the pocket which was thus taking shape, on the commanding Valday Hills, stood the little town of Demyansk, until then unknown and unimportant. This little town was to go down in military history as the site of one of the strategically most important battles of encirclement—the "battle of the Demyansk pocket."

For more than a year—until the spring of 1943—fierce and savage fighting continued for the virgin forests, swamps, and miserable villages of the Valday Hills, the region where Volga, Dvina, and Dnieper have their sources, the watershed of European Russia. Under the command of General Count Brockdorff-Ahlefeldt six German infantry divisions of II Corps resisted a vastly superior enemy force—cut off though they were from the main German front, relying entirely on themselves, and most of the time only scantily supplied from the air. They prevented the Soviets from breaking through to the south and west, and thus saved Army Group North from annihilation.

What then was the task of the remaining three Soviet Armies which, on that 9th January, had likewise swept over the crushed remnants of the German 123rd Infantry Division on Lake Seliger? What was their strategic objective? What was the purpose pursued by the Soviet High Command with its offensive? The aim of the operation was bold and far-reaching. The Third and Fourth Striking Armies and the Twenty-second Army were to drive deep into the hinterland of the German front and cause the whole of Army Group Centre to collapse. The offensive, therefore, had been conceived as the strategic consummation of the Soviet winter battle.

The man who was to accomplish this grand project was Colonel-General Andrey Ivanovich Yeremenko, Commander-in-Chief of the Soviet Fourth Striking Army. He was the same man whom Stalin had repeatedly employed at crucial points of the Central Front during the German summer offensive, as a daring improviser and a saviour in critical situations. Now Yeremenko was to have his revenge for his defeats.

His task was to break through at the most sensitive point of the German Eastern Front—the junction between Army Groups Centre and North—to separate the two Army Groups, and to destroy the

M

German Central Front, which was already reeling under the Soviets' heavy blows. The Vitebsk area, 175 miles from Yeremenko's starting-line on Lake Seliger, was the strategic objective.

The plan of the Soviet High Command sprang from Stalin's confident belief that the earlier winter battles south and north of Moscow had so badly mauled the German Armies that only a *coup de grâce* was needed now.

General Yeremenko, to-day a much-decorated Marshal of the Soviet Union, is the first Soviet commander in the field to have published an exceedingly interesting, and at times astonishingly critical, account of his campaign, including the operation of his Fourth Striking Army, under the title *Towards the West*. It will be of interest to view this decisive phase of the war in the East also through the eyes of a commander of the other side.

In mid-October 1941 General Yeremenko had been caught by a German fighter bomber in the battle of the Bryansk pocket. Just before he could dive into a forester's hut he had been hit by a few bomb-splinters. Severely wounded, he had been flown out of the pocket. Until the middle of December he had been in a military hospital in Kuybyshev. On 24th December he had been summoned to Stalin. The Generalissimo had received him in his underground headquarters in the Kremlin. This is how Yeremenko describes the interview:

"Tell me, Comrade Yeremenko, are you very touchy?" Stalin asked.
"No, not particularly," I replied.
"You won't be offended if I put you temporarily under Comrades who until recently were your subordinates?"
I replied that I was prepared to take over a Corps or any other post if the Party considered this necessary and if I could serve the mother country in this way.
Stalin nodded. He said this measure was necessary in order to solve a most important task. He considered me the right man for it.

In the course of this conversation Stalin then explained to Yeremenko what it was all about. As an experienced commander in the field the colonel-general was to take over the newly raised Fourth Striking Army, a crack unit, a kind of Guards Army, which, moreover, enjoyed the same privileges as Guards units. The officers received pay and a half and the men double pay, and they had better rations than other formations.

Nothing can show more clearly the importance which Stalin attached to the tasks of the Fourth Striking Army than the fact that he entrusted with its command one of the best Soviet military leaders, a man with the rank of a colonel-general, although Kurochkin, the Commander-in-Chief of the North-western Front, under whom the Fourth Striking Army would come, was only a lieutenant-general.

South of Lake Ilmen   355

Yeremenko was given all conceivable powers for raising, equipping, and supplying his Army. When Stalin dismissed his general in the Kremlin bunker he left him in no doubt that the operation of the Fourth Striking Army was to become the "culmination of the Russian winter offensive." The hopes of the Supreme Commander, the General Staff, and the mother country rested upon Yeremenko's shoulders.

To capture booty is an old-established and legitimate practice of war. Anything belonging to the enemy forces becomes the prize of the victor. More than once Field-Marshal Rommel had made his Afrika Corps mobile by means of lorries and fuel captured from British stores. English corned beef from Field-Marshal Sir Claude Auchinleck's well-assorted desert food-dumps often provided a welcome change from blood sausage and pork, just as the fragrant Virginia tobacco of the 10,000,000 navy-cut cigarettes captured in Tobruk's storehouses acted as a great morale booster for the German "desert foxes."

But to base a decisive offensive on the assumption that the troops' grumbling stomachs must be filled with food captured in enemy dumps was something quite unprecedented.

It was Colonel-General Yeremenko's contribution to military history. In his account of the operation of his Fourth Striking Army he writes:

> An efficient preparation of our offensive by the rearward services would have required the stockpiling of major quantities of food in the immediate neighbourhood of our zone of operations. Instead, the staff of the North-western Front 'relieved' us of whatever supplies we had laboriously obtained. We had to share our foodstuffs with our right-hand neighbour, the Third Striking Army, which had practically no rations whatever.

That was bad enough. But worse was to come.

"After ten days," Yeremenko writes, "our own supplies had been used up." Some divisions did not even have one day's rations at the beginning of the offensive. The 360th Rifle Division, for instance, was one of these. In its war diary we find the following entry under 8th January: "Division has no rations." The identical entry is found in the diary of 332nd Rifle Division on the following day. On the day of the offensive itself, on 9th January, the men of nearly all divisions went without breakfast. They went into battle on empty stomachs. The 360th Rifle Division was eventually given the dry bread earmarked for 358th Division, so that the troops should at least have a mouthful of bread on the evening of the first day's fighting.

How was this disastrous food-supply situation to be coped with? How could entire armies fight and conquer in 40 degrees below if they had nothing to eat? Even Guards Corps and crack divisions needed bread and could not live by slogans alone. Yeremenko hit on the solution. He instructed his divisions: "Get your rations from the Germans!" The capture of field kitchens, supply columns, and food-stores became the most important military task. War reverted to its archaic form.

The extent to which the taking of booty influenced strategic decisions is made clear by Yeremenko:

> From the interrogation of prisoners and the reports of our scouts behind the German lines we knew that there were large supply-dumps with huge quantities of foodstuffs in Toropets, since that town was a major supply base of Army Group Centre. This fact was of decisive importance to us. Here we had a chance of obtaining food supplies both for my own Army and our neighbour Army.
>
> Major-General Tarasov, commanding 249th Rifle Division, was given the task of taking Toropets by a swift encircling action and seizing the dumps undamaged. The plan came off.
>
> We captured approximately forty food stores containing butter and other fats, canned meat and fish, various concentrates, flour, groats, sugar, dried fruit, chocolate, and much else besides. The dumps simply became our own army stores: only the personnel changed. These supplies fed our Army throughout a whole month. The success of the Toropets action was of great importance to our operations. I felt very proud when I reported it to General Headquarters.

Yeremenko's details about his valuable booty are quite correct—but his account of the fighting shows a little doctoring. Toropets was not taken by just one division. Yeremenko employed the 249th Rifle Division, two rifle brigades—the 48th and the 39th—as well as units of 360th Rifle Division, to overcome the German defenders of the town. These consisted of 1200 field security troops, one regiment of 403rd Local Defence Division, one company of cyclists, and one platoon of Panzerjägers of 207th Local Defence Division. These forces were joined, in the course of the fighting, by the remnants of the smashed 416th and 189th Infantry Regiments, as well as by a few dozen men from Fegelein's overrun SS Cavalry Brigade. This pitiful force was, of course, unable to stand up to Yeremenko's powerful steam-roller, nor was there any time to destroy the huge dumps at Toropets.

Almost equally sensational as the information about the supply situation of the Soviet Armies during the winter offensive of 1941–42 is the glimpse we are allowed by Colonel-General Yeremenko of the military preparations and the training of the force which was to win the palm of victory on the Central Front:

> The information which the staff of our Army Group had sent me about the enemy's position seemed to me unreliable. I thought it doubtful that the Germans—as our Army Group maintained—should still possess a second defensive system with strongpoints and fortified field positions in deep echelon. I established that during the past two months not a single prisoner had been brought in from the sector of the German 123rd Infantry Division west of Lake

Seliger. Immediately upon my arrival at Army headquarters I therefore gave orders to the 249th Rifle Division to carry out reconnaissance in strength and to bring in prisoners. The division discharged this task in an exemplary fashion. Within five days I possessed data about the enemy's system of defences and his units. No second line of defence was found to exist over a depth of 8 to 12 miles.

This instance illustrates the importance of information given by prisoners. The Soviets were past masters at getting what they wanted even from German soldiers determined to say nothing. The old-established right of a prisoner to refuse information had lost all practical significance in the German-Soviet war—on both sides.

Yeremenko attached great importance to having his formations undergo really tough training for winter warfare in forest country. To this end he had thought out a truly Draconian but effective method. He would order his divisions, complete with their commanders and officers, just as he found them, to spend four days in the dense forests in midwinter—without accommodation, without field kitchens, and without food supplies. They were not allowed to light any fires, even though the temperature was 30 to 40 degrees below. During the day there were military exercises under realistic conditions, and in the evening there were lectures. Melted snow and two handfuls of dried millet were the troops' daily rations.

No other Army in the world could make this kind of demand on its soldiers. But through the centuries this has always been one of the secrets of the Russian Army. It is unequalled in the endurance of hardships, and is still capable of fighting under such primitive conditions as would mean certain disaster for any Western Army. Naturally, the winter weather with its extreme frost was no kinder to the Soviet weapons and equipment than to those of the Germans, but the Russians were good at improvising. They made themselves independent of their frozen-up technical equipment.

When their wireless-sets packed up because of the hard frost, communications officers were appointed in each Soviet unit who would see to it that orders and reports were carried from unit to unit by the quickest possible route—on horseback, by horse-drawn sleigh, or on skis. Moreover, an aerial communications group was organized, equipped with old-fashioned but sturdy light aircraft. In the difficult wooded country these proved an important aid to orientation.

Finally, there was propaganda. The Soviets spent more time and effort on propaganda than on food supplies for their troops. Up to the very last minute before an attack political officers belaboured the hearts and minds of the Red Army men with stirring slogans. The rousing slogans took the place of the issue of brandy of former days. The combined effect of slogans and spirits was often terrible.

Yeremenko writes:

Part Four: Winter Battle

In order to stiffen our formations, hundreds of Communist Party and Komsomol members were attached to them from rearward formations. Workers from the Sverdlovsk and Chelyabinsk areas visited the troops in their starting-lines at the front. Sitting in their trenches and positions, the men from Sverdlovsk and Chelyabinsk chatted with the troops, telling them about their successes on the industrial front. They gave a pledge to the soldiers to produce even more and to supply the fighting front with whatever it needed for victory over the enemy. The soldiers and officers, in turn, solemnly promised to fight bravely and fearlessly, to crush the enemy and honourably to discharge their duty.

Party and Komsomol meetings were held in all formations and units at the front. Communist Party and Komsomol members undertook solemn obligations to set an example to the rest in the impending battle, not to spare themselves, but to be an inspiration to every one. In this way the representatives of the Party among the troops in the field created the prerequisites for a vigorous and successful discharge of the military tasks on each sector.

The scale on which the political fanaticism of the Communist Party was mobilized in the military machine is revealed very clearly in Yeremenko's account. The Marshal records: "249th Rifle Division included in its ranks 567 members and 463 probationary members of the Communist Party, as well as 1096 Komsomol members." That was a quarter of the Division's combat strength.

"Assignments which require the greatest sense of responsibility," Yeremenko further reports, "were given to Komsomol members. Thus in the 1195th Rifle Regiment, 360th Rifle Division, all the No. 1 machine-gunners, all sub-machine gunners, and all the scouts were Komsomol members."

Yeremenko's offensive erupted on 9th January 1942. "Attack," the Marshal writes, "is an ordinary, everyday word to the soldier. At that time, however, in the winter of 1941–42, it had a solemn ring. That word contained our hope of smashing the enemy, of liberating our native land, of saving our near and dear ones and all our fellow countrymen who had fallen under fascist servitude; it contained our hopes of revenge against a perfidious enemy and our dream of peaceful life and peaceful work."

A little bombastically Yeremenko concludes: "And every soldier, from the supply-column driver to the assault-unit man, was dreaming of attack as the most wonderful and important thing in his life."

This is what the "most wonderful and important thing," which, according to Yeremenko, every Red Army man was dreaming of, looked like in reality. Two hours of artillery bombardment; infantry attack with two divisions through breast-deep snow towards the town of Peno; charge over the ice straight into the machine-gun fire of the German front.

South of Lake Ilmen 359

Peno was taken on the second day of the offensive after heavy and costly fighting. The reconnaissance detachment of Fegelein's SS Cavalry Brigade was overrun. The first breach had been punched for Yeremenko's break-through.

But the two wings of the Soviet Army did not succeed in making any real progress in spite of their colossal superiority. The Russian 360th Rifle Division came to a halt in front of the positions of the 416th Infantry Regiment from Brandenburg. On the left wing, on Lake Volgo near Bor and Selishche, the Russian 334th Rifle Division was badly mauled by the Westphalian 253rd Infantry Division and thrown back again.

But at the centre of the attack the Russian 249th Rifle Division made further progress. It was a crack unit, shortly afterwards to be raised by Stalin to the rank of 16th Guards Division and decorated with the Order of Lenin. Major-General Tarasov swept on with his division towards Andreapol. His objective was a break-through towards Toropets, the traffic junction and German supply base, Yeremenko's coveted "bread-basket." His road to the food-dumps was blocked by the Silesian 189th Infantry Regiment under Colonel Hohmeyer, which had been rushed post-haste to Andreapol. The regiment belonged to 81st Infantry Division and was reinforced by 2nd Battalion, 181st Artillery Regiment, as well as a sapper company and a few supply units.

Yeremenko repeatedly pays tribute to the feats and self-sacrifice of this German regiment. It caused a lot of difficulties for his Army, at the centre of its attack, and resisted two Soviet crack divisions literally to the last man, inflicting serious casualties upon the foremost divisions of the Fourth Striking Army.

The tragedy of the Silesians and Sudeten Germans of 189th Infantry Regiment was enacted between the railway station of Okhvat and the villages of Lugi, Velichkovo, and Lauga. Only a few men survived the ferocious battle against Yeremenko's Guards in three feet of snow and a temperature of 46 degrees below. One of the few survivors of the 189th, in a position to describe the extinction of his regiment, is First Lieutenant Erich Schlösser, who participated in the fighting before Andreapol as an NCO in the 3rd Company.

The 81st Infantry Division, to which the 189th Infantry Regiment belonged, had gone through the campaign in France without appreciable losses. Just before Christmas 1941 the division was in quarters along the Atlantic coast, enjoying a distinctly cushy billet.

But they were not to enjoy any Christmas festivities there on the Atlantic coast. On 22nd December 1941 came the order: Prepare for departure. On 23rd December the companies clambered aboard their train. Where were they off to? It did not look like a long journey. They had not been issued with any special food or with winter clothing.

They had received no new weapons and no equipment of any kind. No one believed the rumour which was slowly making its way through the train from the regimental staff: We are off to the Eastern Front— to Russia!

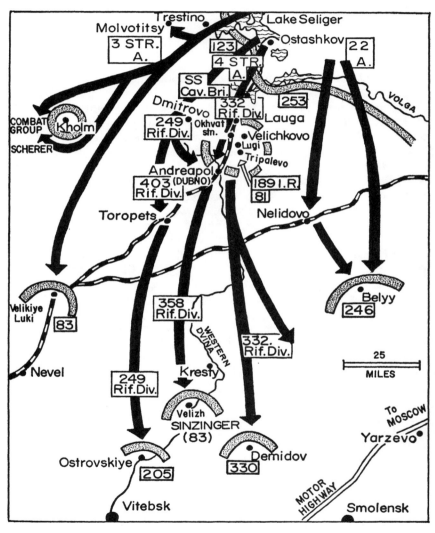

Map 19. At the beginning of January 1942 the German front was ripped open also along the junction between Army Group Centre and Army Group North. The Soviet Fourth Striking Army is aiming at Vitebsk and Smolensk.

Monotonously the wheels clanked over the rails all the way across France. The men spent Christmas Eve in the straw of their goods wagons. They were beginning to shiver in their light-weight coats. On they went, through Germany. Then through Poland. In Warsaw they were issued with food. When they drew their next issue they were well inside Belorussia—at Minsk. The temperature was 25 degrees below, and the cold was seeping through the sides of the wagons. The primitive stoves were red-hot. But the men were miserably cold.

After thirteen days of uninterrupted journey the companies clambered out of their train on 5th January 1942. They stood at the station of Andreapol, in three-foot-deep snow and a temperature 30 degrees below. There was not a single winter greatcoat between them. There were no balaclava helmets and no ear-muffs. Before they knew what had happened to them many men had their toes and ears frozen off.

The war diary of II Corps records: "The regiment's total lack of winter requirements defies description." But before it was possible to supply the regiment, which had a mobilization strength of barely 3000 men, with even the most urgent necessities it was ordered into action against Yeremenko's Guards Regiments of 249th Rifle Division, who were pouring through the breach at Peno and south-west towards Andreapol. Soviet ski battalions were already racing across Lake Okhvat.

Colonel Hohmeyer flung his battalions into their path. The 3rd Engineers Company 181 was put under his command.

The 1st Battalion, 189th Infantry Regiment, reinforced by a battery of 181st Artillery Regiment, arrived in the village and at the station of Okhvat at exactly the same time as the Russian vanguards. The Russians seized the eastern edge of the small town, while Captain Lindenthal's 3rd Company hung on grimly to its western edge. The Soviet 249th Rifle Division sent its 925th Regiment into action—Siberians who charged across the frozen lake with shouts of Urra. Hohmeyer also switched his 3rd Battalion to Okhvat.

By the railway embankment Captain Neumann was trying to ward off the Russian attacks with his 11th Company and to relieve the 1st Battalion at Okhvat. The Russians had to be halted—at least long enough for a stop-gap line of defence to be established in the wide breach between Dvina and Volga. Unless that was done the Soviet divisions would drive on to their objectives—Vitebsk, Smolensk, and the motor highway—in order to link up with the Soviet Armies attacking from the south and to close the trap around Army Group Centre.

Sergeant Maziol with his platoon was in position at the south-western edge of Okhvat. "Tanks!" Corporal Gustav Praxa suddenly shouted into the peasant hut. Everybody out! From the entrance to the village came the first tank, a light T-60. In line behind followed another, then a second and a third—eight in all. They were a combat group of the Soviet 141st Tank Battalion.

The tanks fired into the houses. They tore the thatched roofs to shreds. Clearly they intended to wreck anything that could serve the Germans for accommodation. It was a typical Russian fighting method.

Maziol, together with Praxa and Sergeant Müller, who led the 1st section, were lying behind the corner of a house. An enemy tank on the far side of the wide village street was spraying the ground with machine-gun fire, churning up the snow and pinning down the three men.

"If they get past us they'll shoot up our supply vehicles and keep on to Andreapol," Maziol observed in his unmistakable Silesian accent. Then he added in a matter-of-fact way: "We've got to finish them off with grenades."

Müller and Praxa understood. With numb fingers they got their hand-grenades ready. Already the first of the T-60s came rumbling past the corner.

That was Müller's moment. He leapt to his feet, ran alongside the tank, and swung himself up on its stern. He grabbed the hatch-handle. He tore open the hatch. He held it open with his left hand while his right clutched his egg grenade. With his teeth he pulled the detonator ring, calmly waited two seconds, and then dropped the egg into the tank. He flung himself down. A crash. A sheet of flame.

The second tank stopped. Its hatch opened. The Russian wanted to have a quick look to see what was happening. It was just long enough for Maziol to get him into the sights of his machine pistol. A burst spat from his barrel. The Russian dropped backward into the turret. And already Müller was on top of the tank, dropping a stick grenade into the still open turret-hatch.

The two tanks were enveloped in black smoke which blanketed out the road. Like a phantom the third tank emerged through the smoke. Abruptly it tried to reverse, but got stuck in the snow.

Corporal Praxa leapt on to its turret, but could not open the hatch. But the Russian gunner was just opening it from inside. He wanted to have a quick look round. On catching sight of Praxa he immediately ducked again. But the hand-grenade rolled in just before the hatch closed.

Seeing the disaster that had befallen the spearhead of their combat group, the remaining five Soviet tanks careered around wildly in the deep snow. Eventually they about-turned in the wide village street and retreated.

At dusk the Siberians of 925th Rifle Regiment came again. To support them General Tarasov this time employed the 1117th and 1119th Rifle Regiments, 332nd Rifle Division. Lieutenant-Colonel Proske's 1st Battalion was badly mauled. Captain Neumann's 11th Company, fighting by the railway embankment, also had to give ground.

During the night of 12th/13th January the thermometer dropped to 42 degrees below. In each company some twenty to thirty men were out of action because of severe frostbite.

By the morning the average combat strength of the German companies was reduced to fifty to sixty small-arms. In the sector of 1st Battalion there were only three peasants' houses left where the men could warm themselves a little. The horses stood out in the open. Their eyes were feverish, and they were shaking with cold.

Yeremenko was angry to find a single German regiment holding up his advance to Andreapol and Toropets and denying him access to the coveted supply-dumps. For that reason he now employed the 249th and 332nd Rifle Divisions in an outflanking operation. On 14th January the Russians struck at the rear of 189th Regiment. They smashed supply columns in the Lugi and Velichkovo area. They blocked the supply routes. They overran the dressing stations and field hospitals. They closed the trap.

At 1800 hours Colonel Hohmeyer ordered the break-out from the encirclement. In a sudden concentrated bombardment of Velichkovo and Lugi the artillery spent its last shells. Then the companies charged. The date was 15th January. Since 11th January the men had had no proper sleep and on only two occasions hot food.

Lugi was retaken by 1st Battalion. Soviet counter-attacks with tanks were stopped by Second Lieutenant Klausing at the edge of the village. Only in the church did a Soviet machine-gun post hold out. Its fire blocked the road. One of its victims was Second Lieutenant Gebhardt. His platoon was shot up.

A lance-corporal, anonymous to this day, worked his way through the ruined nave of the church and by climbing up to the organ-loft finished off the machine-gun with three hand-grenades.

But it proved impossible to retake Velichkovo. The reinforced 2nd Battalion was pinned down in the centre of the village and slowly wiped out.

By 16th January only a few remnants survived of 189th Infantry Regiment. The Russians once more burst into Lugi with five tanks, overran the regiment's sledge column, blocked the railway embankment in their rear, and presently stood before Andreapol.

Colonel Hohmeyer gave the battalions *carte blanche* to fight their way back to Toropets through the woods. It meant a march of over 30 miles. The colonel himself rode out on horseback to reconnoitre. It was a ride into eternity. He did not return; he died somewhere in the snowy wastes outside Andreapol, like most of the men of his regiment. Hohmeyer was posthumously promoted major-general.

Lieutenant-Colonel Proske also set out on horseback with two officers in order to reconnoitre a way to slip through. None of them returned.

With small combat groups the officers and NCOs tried to penetrate through the deep snow of the forests. But only one detachment of 1st Battalion succeeded in completing the frightful trek to Toropets. They had set out with 160 men. Forty of them reached their destination on 18th January.

"The German 189th Infantry Regiment left behind on the battlefield 1100 dead," Yeremenko reports. One thousand one hundred dead.

With Colonel Hohmeyer's units smashed, the road was open for Yeremenko to his first objective—the huge supply-dumps in Toropets. The rearward German formations of 403 Local Defence Division with their few captured enemy tanks and police units were unable to hold the town. Five Soviet crack regiments were making an encircling attack. On 21st January General Tarasov seized the Toropets supply-dumps undamaged. For the first time since the beginning of their offensive Yeremenko's soldiers had adequate supplies of food.

After the break-through at Toropets there was no continuous German front left along an 80-mile stretch, between Velikiye Luki and Rzhev. It was the most humiliating and the most dangerous moment experienced by Army Group Centre since 6th December 1941. Three Soviet Armies—with Yeremenko's Fourth Striking Army well in front with four rifle divisions, two rifle brigades, and three ski battalions—were reaching out for the great victory which, Stalin hoped, would bring the destruction of the German Army Group Centre and hence the turning-point of the war.

In this situation General von der Chevallerie, commanding LIX Corps, was ordered to seal the Vitebsk gap with three divisions. It was an easy order to give—but of the three divisions not one had arrived in Russia in its entirety. The bulk of all these divisions was still en route from France to the Eastern Front—the 83rd Infantry Division from Northern Germany, the 330th Infantry Division from Württemberg, and the 205th Infantry Division from Baden. The only units within reach were remnants of the 416th Infantry Regiment of the Berlin–Brandenburg 123rd Infantry Division who had gone through the hell of Lake Seliger.

General von der Chevallerie and the advanced personnel of his Corps headquarters in Vitebsk had been working feverishly since 20th January to get his units into Russia. It was a race against time.

General Yeremenko's 249th Rifle Division and units of 358th Rifle

South of Lake Ilmen 365

Division were meanwhile advancing from Toropets towards Ostrovskiye and Velizh, both of them important road junctions on the Dvina and the last obstacles on the road to Vitebsk, the main supply and food base of Army Group Centre.

Lieutenant-General Kurt von der Chevallerie could do nothing else but send his units into action in driblets, as it were, as they arrived in the East, straight from their trains, to halt Tarasov's regiments. The feats performed by these German battalions, who were hurled straight from the mild French winter into temperatures of 40 to 50 degrees below and expected to avert the disaster threatening Army Group Centre—which, in fact, they did avert in months of fierce fighting—surpass all comprehension.

With his combat groups Chevallerie defended the crucial points in the gap between Ninth and Sixteenth Armies until, at the end of January 1942, Third Panzer Army took over behind him. The names of the villages have become savage memorials to the winter battles—Demidov, Velizh, Kresty, Surazh, and Rudnya. The men from Northern Germany, Swabia, Baden, and Brandenburg made these gutted villages into breakwaters against which Yeremenko's waves crashed and were held back.

The fiercest fighting was for Velizh and Kresty. There a combat group under Colonel Sinzinger, commanding 257th Infantry Regiment, with units of 83rd Infantry Division, was offering stubborn resistance to the Russians. The men from the Lüneburg Heath, from Schleswig-Holstein, from Hamburg and Bremen, spent the nights in their tents at 25 to 40 degrees below, without straw and without camp-fires. In daytime they worked their way through chest-deep snow. They were cut off. They counter-attacked and fought their way out. They fought their way forward and they fought their way back again. But they did not cease resisting.

Facing them were four Soviet divisions and units of three rifle brigades, trying at all costs to get via the road junction of Rudnya to the Minsk–Smolensk–Moscow motor highway in order to sever the lifeline of Army Group Centre.

They did not succeed. The Soviet offensive petered out in the face of the unexpected opposition of LIX Corps. Yeremenko undisguisedly names the reasons for the failure of his full-scale offensive: the Soviet High Command had underestimated the German troops' powers of resistance in winter warfare under Siberian conditions. It had thought the German divisions to be utterly exhausted. Stalin thus made the same mistake as Hitler had made before Moscow. The Soviet High Command had underrated its opponent and overrated its own strength.

Inadequate supplies of ammunition, fuel, and foodstuffs, the shortage of officers, the poor training of the troops, and unexpectedly heavy casualties had made the Soviet troops battle-weary. Yeremenko's Guards, the 249th Rifle Division, numbered only 1400 men at the end of January 1942, according to his own figures. On 9th January the division had joined action with 8000 men.

Even the most stringent orders from the Soviet High Command were unable to drive Yeremenko's Fourth Striking Army to its envisaged strategic objective—Vitebsk. It just could not make it.

The two Armies on Yeremenko's flanks, the Third Striking Army on the west and the Soviet Twenty-second Army to the east, likewise failed to reach their objectives of Velikiye Luki and Yarzevo on the Smolensk–Moscow motor highway. General Purkayev's Third Striking Army was stuck before Kholm, where the German combat group Scherer was holding out in all-round defence, halting the Russian divisions. General Vostrukhov's Twenty-second Army did not get past Belyy, where units of the Hessian 246th Infantry Division were holding out unshakably.

Thus the most dangerous thrust of the Soviet winter offensive against the German Army Group Centre, the drive into the rear of its Ninth Army, had failed. The outer prong of the Russian pincers, designed to bite deep behind the German front, had been broken.

3. Model Takes Over

The supply-dumps of Sychevka–"What have you brought with you, Herr General?"–A regiment holds the Volga bend–"I am the only one left from my company"–Stalin's offensive gets stuck–Sukhinichi, or the mouse in the elephant's trunk–A padre and a cavalry sergeant–Spotlight on the other side: two Russian diaries and a farewell letter.

HOWEVER, disaster was still threatening north and south of the motor highway, at Rzhev and Sukhinichi. The inner prong of the Soviet offensive was a direct threat to the front-line units of the German Ninth and Fourth Armies.

Rzhev, more than anything, was the objective of the Soviet attacks. The Russians wanted to take this cornerstone of the German Central Front at all costs. If they succeeded it would mean the outflanking and encirclement of the Ninth Army.

When enemy tanks suddenly rumble past the front door of an Army headquarters and the front is only half a mile away, these are sure signs of incipient disaster. On 12th January 1942, in the late afternoon, the German Ninth Army was faced with just such a disaster.

The time was 1600 hours. In the map room of Army headquarters in Sychevka, in front of the situation map, stood Lieutenant-Colonel Blaurock, the Army chief of operations, with Major-General Krüger, commanding 1st Panzer Division from Thuringia. Also present, to acquaint themselves with the situation, were Lieutenant-Colonel Wenck, the division's chief of operations, Lieutenant-Colonel von Wietersheim, commanding 113th Rifle Regiment, and Lieutenant-Colonel Holste, commanding 73rd Artillery Regiment. The most forward combat group of 1st Panzer Division had just arrived at Sychevka. A week ago this small town with its huge railway goods yard had been a quiet spot behind the lines—an Army headquarters and rearward supply base, a paradise for supply officials and paymasters. For two days now it had been the front line.

The stuttering of machine-guns and the dull thud of mortars could be heard in the room. "If I may acquaint you, sir, with this bloody mess of a situation," the Army chief of operations said to Krüger. "Since 9th January the Russians have been keeping up their full-scale attack from the Ostashkov area against the left wing of our cut-off XXIII Army Corps, and have pushed it down to the south. At the same time there have been fairly strong attacks against the left wing of VI Army Corps—here." Blaurock stabbed the map with his finger. "Our request for permission to take the front back to the Gzhatsk–Volga line was turned down. Since 11th January there have been strong enemy attacks from the north-west, striking towards the south and west of Sychevka, with the most forward enemy units on the town's western outskirts." Blaurock placed his hand on Sychevka and said imploringly, "Hold Sychevka for us, Herr General—it must not be lost."

The commander and officers of 1st Panzer Division nodded. They understood the difficult situation they saw before them. What surprised them was that Colonel-General Strauss, the Army C-in-C, was not personally present at this conference. The chief of operations explained: "Colonel-General Strauss's health is finished. The chief of staff, too, has to go on sick leave. We're expecting the new Army Commander-in-Chief any day now—General Model." There were surprised faces all round.

So Model was the new C-in-C Ninth Army. His had been a meteoric rise. Three months previously he had still been commanding a division—the famous 3rd Panzer Division.

The short, wiry man from Genthin, born in 1891, was well known at the various headquarters throughout Army Group Centre. He was known even better to the men of 1st Panzer Division who had fought under his command within the framework of XLI Panzer Corps ever since Kalinin. He was popular with his troops, much though he differed from his predecessor, Colonel-General Reinhardt. Everybody knew that where Model was in command the good fortune of war was present: the most daring enterprises came off and the most critical situations were retrieved. Nowhere was a man of his type needed more urgently at that moment than with Ninth Army.

Blaurock once more stepped up to the large map. "The situation has really become more than extremely critical during the past forty-eight hours," he said with typical General Staff detachment.

He pointed to the heavy red arrows. "Here, to the west of Rzhev, the Russians have punched a nine-mile hole into our front. Two Soviet Armies, the Twenty-ninth and the Thirty-ninth, have for the past two days been pouring through this hole to the south, with armour, infantry, and columns of sledges. Approximately nine divisions have already got through. Our XXIII Corps is severed, encircled, and can be supplied only by air. The VI Corps, thank heaven, has succeeded in

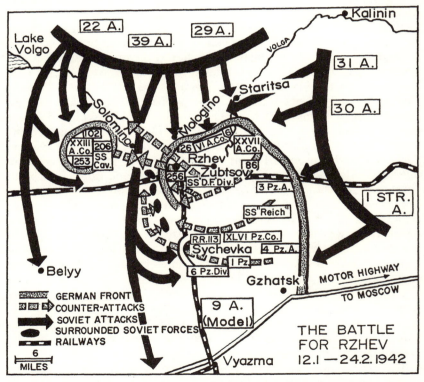

Map 20. At the beginning of January 1942 the Soviets broke through the front of the Ninth Army and drove deep into the rear of Army Group Centre. A most critical situation had arisen.

establishing and holding a new defensive front west and south-west of Rzhev."

Blaurock traced the red arrows with his hand. "The Soviet spearheads—cavalry, mind you—are already west of Vyazma on the motor highway, the lifeline of the whole Army Group Centre. But so far these forces do not appear to be very strong, and they do not represent the main problem. Far more difficult at the moment is the situation here." And Blaurock pointed to a tangle of red rings and arrows about 30 miles south-west of Rzhev.

"As you see," he continued, "strong Soviet forces have swung round towards Sychevka behind the most forward units of Twenty-ninth and Thirty-ninth Armies. The Russians quite clearly intend to take the town, to wheel northward, and to encircle the Army. At the moment they are fighting for the railway-line to Rzhev. If they capture it the supplies of our entire Army will be cut off. Our entire supplies and reinforcements hang by this one line. If Sychevka falls we shall be outmanœuvred. The Russians, gentlemen, are already on our doorstep. Their spearheads have already penetrated into the railway-yards, but are fortunately busy looting. I say fortunately, because the edge of the town is held only by emergency units hurriedly scraped together of runners and supply drivers, led by Colonel Kruse, the chief of Army artillery. On the edge of the goods yard is your 6th Company, 1st Rifle Regiment, which arrived just before midnight."

Major-General Krüger, an easy-going Saxon who was not readily ruffled, uttered a much-used trooper's word. Wietersheim nodded and muttered, "That's putting it mildly."

Half an hour after this conversation at Army headquarters in Sychevka the forward units of combat group von Wietersheim were moving into action to engage the Soviets, who had already established themselves in the railway-yards and in the sheds of the extensive supply depot. The troops consisted of a few armoured infantry carriers and a reinforced company of the Motorcycle Battalion, 1st Division—men from Langensalza and Sondershausen, led by First Lieutenant Pätzold.

Pätzold, who had worked his way forward to a shed with his runner, was watching the goods station of Sychevka-North through his binoculars. "It's as busy as a Michaelmas Fair," he uttered in surprise.

The motor-cyclists beat their arms round their bodies to keep themselves warm. The lieutenant came back and mounted his motor-cycle. "Forward!"

The scene among the sheds and store huts of the goods station really looked like a fair-ground. The Russians were dragging crates and cases of foodstuffs from the sheds and were beside themselves with delight at the things they found. The special rations for airmen and tank crews, in particular, met with their approval—chocolate, biscuits, and dried fruit. But the legs of pork in aspic, the liver sausage, and the fish conserves were also acknowledged with shrieks of delight. They opened the tins with their bayonets and ceaselessly tried one after another. "Papushka, look at this—try it."

Then there were the cigarettes! "Just have a whiff of this—not at all like our makhorka smelling of printer's ink and *Pravda* newsprint."

But by far the greatest magnetic attraction was exercised by the French cognac. The soldiers knocked off the necks of the bottles and quaffed the magnificent liquor, which, by comparison with their rough vodka, seemed as wonderfully mild as sweetened tea.

The Russians were getting merry. They no longer felt the 40 degrees below. They were forgetting the accursed war. They burst into song. They cheered. They embraced and kissed each other. No warning was given by any sentry. Not a single rifle was fired. But suddenly the bursts of machine-gun fire from Pätzold's motor-cyclists swept among the sheds. Hand-grenades exploded. Sub-machine-guns spat their bullets from infantry carriers. In wild panic the Soviets ran away. They did not get far. They fell in the machine-gun fire and died among the canned food and the cigarettes, among the Hennessey cognac and the tins of biscuits.

If one wanted to put it frivolously one would be entitled to say that the first German victory at Sychevka was won by chocolate and French cognac. Only because the Russians were so busy with their precious booty could weak units of 1st Panzer Division succeed in snatching the vital railway-yards from a greatly superior enemy. It was a situation that was by no means unique.

General Infantes, the Commander of the Spanish Blue Division, for instance, makes the following point in an impressive study of the Spanish volunteers' operations in Russia: "We found not infrequently that, after successful local attacks, the Russian troops would forget their tasks and waste precious time. By mounting immediate counter-attacks we would often catch them searching our dug-outs for food, or emptying tins of jam or bottles of brandy. This weakness of theirs was always fatal, because they rarely got away alive. Sometimes we would overcome them in our counter-attacks because they had lost their way in the labyrinth of our trench system. It is true that the Red Army men will advance unflinchingly towards any objective they have been given. What makes them dangerous is not only their up-to-date weapons, but also the vodka issued to them, which turns them into savage fighters. Their well-prepared large-scale mass attacks are undoubtedly very dangerous, since the 'Russian steamroller' crushes anything that opposes its progress. All one can do then is face the attackers with cold steel. But a well-organized counter-action will always take the Russians by surprise."

During the next two days further parts of 1st Panzer Division arrived. Together with 337th Infantry Regiment, air-lifted from France to Russia, they cleared the enemy from the immediate surroundings of Sychevka and restored connections with the airstrip at Novo-Dugino, south of the town, where the Luftwaffe had been holding out in all-round defence for a number of days. Surrounded bakery companies, dug in around their huge baking-ovens, and a hard-pressed Army signals company were relieved. The men of an Army horse hospital were freed from their encirclement. Immediate Soviet counter-attacks were successfully repulsed.

A few days after their first conference at Ninth Army headquarters the commander and chief of operations of 1st Panzer Division had again called on the Ninth Army chief of operations to acquaint themselves with Army's further intentions with regard to the fighting for Rzhev and Sychevka. Greetings had only just been exchanged when a door was heard slamming outside—the door of a German jeep. Words of command were shouted. An orderly entered and announced: "General Model."

In a three-quarter-length greatcoat, with old-fashioned but practical earflaps over his ears, with soft high boots, the indispensable monocle in his right eye, the new Commander-in-Chief stepped into the room. The man radiated energy and fearlessness. He shook hands with the officers. He flung his coat, cap, and ear-flaps on a chair. He polished his monocle, which had steamed up in the warm room. Then he stepped up to the situation map. "Rather a mess," he said drily, and briefly studied the latest entries.

"I have already informed the gentlemen in rough outline about the main problems," Blaurock reported. "The first thing Ninth Army has to do is to establish the situation around Sychevka and to secure the Rzhev–Sychevka–Vyazma railway-line. Following a stabilization at Sychevka itself, through 1st Panzer Division, the forward units of the 'Reich' SS Motorized Infantry Division are at present arriving."

General of Panzer Troops Model, a dashing commander in the field as well as a coolly calculating staff officer, nodded. "And then the first thing to do will be to close the gap up here." He ran his hand over the wide red arrows indicating the Russian penetrations west of Rzhev between Nikolskoye and Solomino. "We've got to turn off the supply-tap of those Russian divisions which have broken through. And from down here"—Model put his hand on Sychevka—"we shall then strike at the Russian flank and catch them in a stranglehold."

Krüger and Wenck were amazed at so much optimism. Blaurock summed up their astonishment in the cautious question: "And what, Herr General, have you brought us for this operation?"

Model calmly regarded his Army chief of operations and said, "Myself." Then he burst out laughing. With a great sense of relief they all joined in the laughter. It was the first time in many days that loud and happy laughter was heard in the map room of Ninth Army headquarters in Sychevka. A new spirit had moved in.

It was a strange thing, but the moment Model assumed command of the Army the regiments seemed to gain strength. It was not only the crisp precision of the new C-in-C's orders—but he also turned up everywhere in person. While Colonel Krebs, his chief of staff, was in Sychevka, looking after staff affairs, Model was at the front. He would suddenly jump out of his command jeep outside a battalion headquarters, or appear on horseback through the deep snow in the foremost line, encouraging, commending, criticizing, and occasionally even charging against enemy penetrations at the head of a battalion, pistol in hand. This live-wire general was everywhere. And even where he was not his presence was felt.

It was largely that presence which decided the impending battle. To understand it one must know what led up to it.

As early as 8th January Colonel-General Strauss had tried to close the breach in the north. Units of the replenished SS Cavalry Brigade Fegelein under the command of Obersturmbannführer[1] Zehender had been switched east from the Nelidovo area and had mounted the attack via Olenino. Units of VI Corps from Rzhev had thrust westward to meet them. But the Russians were much too strong in the penetration area, and the German forces too weak. The counter-attack of the combat group Zehender was utterly paralysed for several hours by a frightful blizzard, and subsequently unable to succeed in the face of several Soviet brigades. East of Olenino the attack ground to a halt. The attempt to close the gap had failed.

In order to repeat the attempt with stronger forces, Army Group Centre had withdrawn 1st Panzer Division from the Ruza line and dispatched it to Rzhev. It was a lucky move. For as a result the division could now be quickly redirected and switched to Sychevka in order to redeem the critical situation there.

But mere defence in the areas held did not lead anywhere. "Attack, regain the initiative, impose your will on the enemy." That was Model's recipe. The Thuringian 1st Panzer Division from the ancient central German towns of Weimar, Erfurt, Eisenach, Jena, Sondershausen, and Kassel made a virtue of necessity: since they lacked tanks, the tank crews transformed themselves into infantrymen on skis.

Lieutenant Darius, whom we met earlier on the Duderhof Hills, was now in charge of a noiselessly operating "ski company." By daring thrusts and patrol operations his men gave cover to the railway engineering detachments who were continually busy repairing the track between Sychevka and Rzhev, a favourite target of Russian sabotage units.

[1] Rank in SS equivalent to colonel.

But it was rather a long stretch of line. Major Richter, commanding 2nd Battalion, 4th Flak Regiment, therefore thought up an unconventional method of protecting the vital railway traffic to Rzhev. He got his men in Rzhev to build a kind of mobile "AA battery": on a number of flat-cars two 8·8-cm. AA guns, four machine-guns, and two light 2-cm. AA guns were installed, the wagons were hitched to an engine, and the home-made "armoured train" was manned by a crew of forty under the command of First Lieutenant Langhammer.

This train ran a shuttle service between Rzhev and Sychevka. First of all, at the urgent request of the duty transport officer, it steamed to the south to pick up an ammunition train. When he received his first assignment Lieutenant Langhammer is quoted as having asked doubtfully, "You don't think a U-boat would be more suitable?" But the AA gunners of the "armoured Flak train," which soon became famous throughout the sector, discharged their task admirably.

Physically, service on board the unprotected "armoured train" was torture. The headwind caused the temperature on the surface of the weapons and on the open flat-cars to drop to 50 and even 58 degrees below zero. The muffled lookouts on the locomotive wore leather masks on their faces because otherwise their noses and cheeks would have frozen off within minutes. In front of it, the locomotive pushed several goods wagons to act as "mine detectors."

Time and again the "armoured" train dispersed strong enemy sabotage detachments which had made their way up to the railway embankment. Moreover, the battery on wheels brought up the supply trains to Rzhev, following in convoy behind it, and thus ensured vital supplies during the first, most difficult days.

Things were by no means rosy for the Soviets who had broken into the German lines. This is shown by a glimpse at the other side of the front.

Sergey Kambulin, a twenty-six-year-old lieutenant in command of the machine-pistol company of a rifle regiment in 381st Rifle Division, was hustling his men onward. "Davay," he shouted; "get a move on, don't dawdle!"

Grumbling, the men put their shoulders to the wheels and pushed two captured German infantry guns forward. The horses had died of hunger and cold. As for the men of the company, two, three, or sometimes four and even more would drop out every day.

They were advancing along a wide snow-track packed hard by the tanks. The caterpillar tracks had made the snow as firm as concrete. But they had also made it as smooth as a skating-rink in a Leningrad park. Painfully the men struggled forward. One of them asked, "What's the name of that village over there, Comrade Lieutenant?"

Kambulin looked at his map. "Solomino," he said. With thumb and forefinger he measured the distances on the map. "We're already 20 miles west of Rzhev, moving in a southerly direction. You know what that means? It means we are striking at the fascists' rear!"

At Solomino was the westernmost breakthrough point of the big gap through which Kambulin's company was advancing to the south. The penetration point was covered by anti-tank guns and heavy 15·2-cm. field howitzers. A hundred yards on the company's right a horse-drawn supply column was moving along the road. The field kitchens were steaming. Longingly Kambulin's men looked across. They had not had any hot food since the previous evening. The time was 1100 hours.

The day before, on 21st January, Second Lieutenant Kambulin had at last received a pair of felt boots. He had refused to accept a pair until every single man in his company had been issued with them. The thermometer stood at 45 degrees below zero Centigrade.

"They say the Germans are still running about in tight leather boots —some of them even in cloth boots," remarked one of the soldiers, a young village schoolmaster. "I hope the bastards freeze to death," Kambulin grunted.

"Enemy aircraft!" a man shouted. Everybody scattered and flung themselves into the snow. A German fighter-bomber was already opening up at them with its cannon. In the distance German aircraft were wreaking havoc among the Soviet supply column.

Shortly afterwards Soviet fighters appeared. But German fighters arrived almost simultaneously and chased off the Soviet machines.

From the west came the thunder of German artillery. The shell-bursts were a little short of Kambulin's company, but presently they got nearer, straddled the platoons, and continued to creep forward, to the east. The worst was over.

Kambulin straightened up. What on earth was happening? The supply column was hastily retreating. Machine-guns rattled. From the west came infantry, in line abreast, wearing snow smocks. Between them lumbered massive tanks without cupolas.

"Those are German assault guns—German self-propelled guns," Kambulin realized. The village schoolmaster too was shouting: "Those are Germans, Comrade Lieutenant!"

Calmly Second Lieutenant Kambulin made his dispositions. The sections dispersed. And already the first machine-pistol salvos swept over the enemy. The two light guns which they had captured from the Germans were barking.

The Germans on the other side flopped into the snow. They were seen waving and signalling to their rear. They were calling up their infantry guns. Model's battle for the big Soviet penetration area west of Rzhev had begun.

The new C-in-C Ninth Army had launched the second phase of his operation against the Soviet Armies which had broken through the

German front. He had done so in 45 degrees below zero, a temperature which froze a man's breath.

Regimental and divisional commanders had asked Model to postpone the date of attack because of the frightful cold. Model's reply had been: "Why, gentlemen? To-morrow or the day after won't be any warmer. The Russians aren't stopping their operations."

Attack—that was Model's element. His great achievement in January 1942 consisted in leading Ninth Army from a hopeless situation of desperate all-round defence all along the front into a liberating counter-offensive with clearly defined centres of gravity.

Model's plan was simple. From Sychevka he made the reinforced 1st Panzer Division and units of the newly brought up "Reich" SS Division drive towards the north-west, in the direction of Osuyskoye, in order to strike at the flank of the most forward Soviet formations.

Twenty-four hours later, on 22nd January, Model ordered VI Corps to attack from the area west of Rzhev, striking in a westerly direction at the Soviet break-through zone, the main weight of this operation being borne by 256th Infantry Division, reinforced by battalions of four other divisions, by artillery, Panzerjägers, and AA guns.

Simultaneously XXIII Corps—cut off at Olenino—attacked from the west with 206th Infantry Division, the SS Cavalry Brigade Fegelein, and Assault Gun Battalion 189, in order to break through and link up with the formations of VI Corps coming from the east. The men who were thus unexpectedly facing Second Lieutenant Kambulin belonged to the SS Combat Group Zehender: in fact, horsemen employed as infantry, together with some self-propelled guns of the "Ritter Adler Brigade"—the 189th Brigade. In vain did Kambulin try to stop them. Two days later a German patrol found him dead in the snow, surrounded by his shot-up company.

Kambulin, gravely wounded, had frozen to death. Shortly before he died he had made a last entry in his diary: "The German assault guns are a deadly weapon. We've got no defence against them."

The German two-pronged thrust against the Soviet penetration area between Nikolskoye and Solomino, an operation mounted with the very last ounce of strength, had succeeded. The VIII Air Corps under Air Force General Wolfram von Richthofen smashed Soviet AA and artillery positions in the penetration area. Heavy mortars shattered the Soviet anti-tank guns. At 1245 hours on 23rd January the spearheads of XXIII Corps and of Combat Group Recke of VI Corps were shaking hands.

XXIII Corps was able to restore physical communications with Ninth Army, even though, for the time being, only across a narrow strip of ground. The two "snow roads" laid by the Soviets across the Volga had been severed, and the Soviet Corps belonging to Twenty-ninth and Thirty-ninth Armies had been cut off from their rearward communications and from all supplies.

It was a great hour for Model. He had regained the initiative on the battlefield between Sychevka and the Volga, and he had no intention of surrendering it again. The first thing the new C-in-C did was to reinforce the newly gained land connection between VI and XXIII Corps. For naturally the Soviets tried desperately to break through the barrier again and to restore communications with their nine divisions which had made the original penetration. That had to be prevented.

For this task Model chose the best man. As always when he had a particularly difficult operational assignment, Model succeeded in picking the best man for the job—in this case Obersturmbannführer[1] Otto Kumm, commanding the "Der Führer" Regiment of the "Das Reich" SS Division. With his regiment Kumm was dispatched to the Volga, to the exact spot where the Soviet Twenty-ninth Army had crossed the frozen river.

"Hold on at all costs," had been Model's order to Kumm. "At all costs," the general had repeated emphatically.

Kumm saluted. "Jawohl, Herr General!" Would he be able to hold on, with just one regiment?

On 28th January, while he was reinforcing his barrier in the north, Model launched his encircling attack in the south against the Soviet units which had broken through. The attack was made from the Osuga–Sychevka area with all available troops: 1st Panzer Division, 86th Infantry Division, the bulk of the "Reich" SS Division and of 5th Panzer Division, as well as 309th Infantry Regiment and the Combat Group Decker of 2nd Panzer Division, had all been united in XLVI Corps under the command of General von Vietinghoff and were pressing towards the north-west. The Russians knew what was at stake and resisted desperately.

There was much bitter fighting. In the deep snow of the forests every wooden shack became a fortress; in the villages every wrecked house was an inferno.

On 26th January came the expected large-scale attack against the northern front of 256th Infantry Division and the right-wing of XXIII Corps, where 206th Infantry Division was employed. There were many highly critical situations, retrieved only by supreme efforts of the dog-tired men.

In daytime Model would spend about an hour over his maps and the remaining ten hours with his troops. Wherever he appeared he had the effect of a battery recharging the spent energies of the unit commanders.

The unaccustomed temperature fluctuations caused the German

[1] Rank in the Waffen SS corresponding to colonel.

troops extreme hardships. With the milder weather came blizzards. Then, abruptly, the thermometer dropped again to 52 degrees below zero Centigrade. The men cursed the Russian winter.

Nevertheless the Soviets were repulsed, compressed, and split up along the Rzhev–Olenino railway line. The Russian commanders sacrificed entire battalions in pointless counter-attacks.

On 4th February the Westphalian 86th Infantry Division took the keypoint of Osuyskoye. Forty-eight hours later Thuringian grenadiers of 1st Panzer Division, riding in armoured infantry carriers, broke through to the railway line at Chertolino. There the foremost units of Combat Group Wietersheim linked up with the spearheads of Combat Group Zehender. The ring around nine Soviet divisions, representing the bulk of two Armies, was closed.

Kumm and his 650-strong regiment had meanwhile built themselves an improvised but serviceable position along the frozen Volga. Holes had been blasted into the ground with blasting cartridges and mines. Machine-gun positions and infantry dug-outs had been set up at regular intervals of 100 to 200 yards. It was a thin line, and Kumm had no reserves.

The Russians attacked ceaselessly. Day after day their formations grew more numerous. They were intent on getting through, on restoring contact with the cut-off divisions. It was at that point that the battle of Rzhev was being decided.

Kumm's headquarters were only half a mile behind the fighting line of 3rd Battalion. Every day Model called by Fieseler Storch, landing on the ice of the Volga. Or else he would come by jeep. On one occasion, when the vehicle had got stuck, he arrived on horseback.

On 28th January, just as Model was at Kumm's headquarters, men of 1st Battalion brought in a Red Army prisoner. He was a signaller from the headquarters of the Soviet Thirty-ninth Army. Such men had rarity value. They knew more than many a commander in the field.

The loquacious Russian reported that a large-scale attack was planned for the next day. He claimed that several Russian rifle and armoured brigades were all lined up for it. The break-through was to be achieved regardless of the cost, and the encircled Corps liberated.

Model left the headquarters, a worried man. "Obersturmbannführer, I'm relying on you," were his parting words to Kumm. And with a grin he added, "But maybe that Russian was leading us on."

The prisoner had not been leading them on. On the following morning the full-scale attack began. It came exactly at the earlier penetration point of the Soviet Twenty-ninth Army, where the wide tank-tracks marked out the road across the ice.

Kumm's regiment, though numerically small, was well equipped. In the foremost line was an 8·8-cm. AA gun. The Panzerjäger Company had 5-cm. anti-tank guns. The Heavy Company comprised a heavy and a light troop with infantry guns and two more troops equipped with 3·7-cm. anti-tank guns. Moreover, in the course of the fighting the Motorcycle Battalion of the "Reich" Division was placed under the regiment, as well as a battery of Assault Gun Battalion 189. Even so it was still a modest force compared with the mass of the attackers.

The Russians kept up their charge ceaselessly—by day and by night, throughout three weeks. But they committed a tactical mistake, a typical Russian mistake: they failed to concentrate their strength on a single major break-through. They omitted to form a centre of gravity. They flung in battalion after battalion, then regiment after regiment, and eventually brigade after brigade.

Anti-tank cover for the group resisting at Klepenino was provided by two Panzerjäger troops of Panzerjäger Battalion 561. The thirteen 5-cm. anti-tank guns under Second Lieutenant Petermann had destroyed twenty T-34s by 3rd February. On 5th February Second Lieutenant Hofer took over the anti-tank troop from the wounded Petermann. The ferocity of the fighting is shown by the fact that the crew of the gun outside Klepenino had to be changed three times within five hours. Two dozen shot-up enemy tanks lay in front of the position. The neighbouring gun had been crushed by a T-34. The infantrymen had to tackle the colossus with mines and demolition charges.

On the sixth day the Russians appeared in front of 10th Company with thirty light tanks. They advanced to within 50 yards of the positions. They halted. And then the whole armada opened fire at the infantry dug-outs and machine-gun posts. They continued pasting them from all barrels for a full thirty minutes. Then they drove back into the forest. Silence and brittle cold hung over the plain. Two hours later a man crawled out of the shattered position of 10th Company back to battalion headquarters. He was helped in. He was Rottenführer[1] Wagner. Seriously wounded, with frost-bitten hands, he tried to stand up in front of Bollert, the battalion commander, to make his report. But he collapsed, and reported lying on the floor: "Hauptsturmführer, I'm the only one left from the company. They're all dead." A tremor ran through him. A moment later there was no survivor of 10th Company.

There was now a gap in the front about two-thirds of a mile wide. The VI Army Corps rushed 120 men into the line—drivers, cooks, bootmakers, and tailors. Paymasters were in charge of platoons. Fine men, but wholly inexperienced in this kind of fighting. They moved into the positions of 10th Company. The Russians, after a sudden concentrated mortar bombardment, charged with shouts of "Urra." That was too much for the nerves of the men of the supply services. They simply took to their heels. They were picked off one by one like rabbits.

[1] Rank in the Waffen SS equivalent to lance-corporal.

When dusk fell the Soviets were within 50 yards of Kumm's regimental headquarters at Klepenino. The small village originally had thirty houses, but only eight were left.

Hauptsturmführer[1] Holzer, the regiment's adjutant, had cut deep holes under the floor and sawn firing slits into the lower beams which formed the wall. From the regimental commander down to the drivers each man stood in his firing-pit, with carbine, machine-pistol or machine-gun. They were supported by an anti-tank gun and by the Panzerjäger Battalion 561, now fighting as infantry.

No matter how often they attacked, the Soviets never got closer than 15 yards. The words found in the operational reports were not a figure of speech, but the most literal appalling truth: "Outside Klepenino the dead were piled high in huge heaps."

Corps sent aid in the shape of an infantry regiment. But it was shot up by the Soviets while making a counter-attack. Its remnants were shared out among Kumm's battalions or else employed for flank cover. During the night of 7th/8th February the Russians eventually broke into 2nd Company's positions in battalion strength. Ferocious hand-to-hand fighting continued for four hours. The 2nd Company of the "Der Führer" Regiment was killed to the last man.

At that moment the Motorcycle Battalion of the "Das Reich" Division arrived at Klepenino. In addition, units of Assault Gun Battalion 189 and the Reconnaissance Detachment, 14th Motorized Infantry Division, under Major Mummert were rushed to Kumm's front.

A 21-cm. mortar was got into position in a patch of woodland, and the enemy who had broken into the "Russian grove" was pounded with it. That grove changed hands ten times. After the eleventh charge it remained firmly in the hands of Major Mummert's Reconnaissance Detachment 14.

Kumm's front on the northern edge of the great pocket held firmly. Relief brigades of the Soviet Thirty-ninth Army did not succeed in crossing the Volga. They bled to death. The killed were lying in their thousands in front of the German lines by the Volga bend.

In the meantime operations against the Soviet divisions encircled south and west of Rzhev continued. On 17th February the Combat Group von Wietersheim penetrated into the core of the last major Soviet pocket—in wooded country near Monchalovo—with tanks, Panzer sappers, and armoured infantry carriers of the reinforced 1st Panzer Division. The last desperate break-out attempt by 500 Soviets under the personal leadership of a general collapsed in the fire of the German combat group.

The battle was drawing to its close. The Soviet Twenty-ninth Army and major parts of the Thirty-ninth had been destroyed. Model, promoted Colonel-General on 1st February, had brought about a turn of

[1] Rank in Waffen SS equivalent to major.

the tide in the winter battles on the German Central Front. The ferocity of the fighting is revealed by two figures: 5000 Russians were taken prisoner; 27,000 lay dead on the battlefield. Six enemy rifle divisions had bled to death, four had been smashed, and nine more, as well as five armoured brigades, had taken heavy knocks.

German casualties, too, had been heavy. On 18th February, when Obersturmbannführer Otto Kumm reported at his divisional headquarters, Model happened to be there. He said to Kumm, "I know what your regiment has been through—but I still can't do without it. What is its present strength?"

Kumm gestured towards the window. "Herr Generaloberst, my regiment is on parade outside." Model glanced through the window. Outside thirty-five men had fallen in.

Heavy and indeed appalling though the price was which Ninth Army had to pay for smashing the Soviet Armies which had broken through between Sychevka and the Volga bend, it was not too high if one considers that the fate of the whole of Army Group Centre was at stake. The deadly danger of encirclement which had threatened it from the north had now been averted. But what was the situation on the southern wing of the Army Group, where the divisions of the Soviet Tenth Army had driven through the breached German front between Belev and Kaluga and had already bypassed Sukhinichi in their attempt to reach the motor highway east of Smolensk, deep in the rear of Fourth Army, and thus cut the lifeline of Army Group Centre?

The stables and cattle-sheds of the Voin collective farm were deep in snow in the wide plain between Orel and Mtsensk. Major-General Nehring had established there the headquarters of his 18th Panzer Division, and Lieutenant Winter, in charge of his headquarters, had placed the tractors and combine harvesters, old Soviet lorries and German armoured infantry carriers, between the sentries and buildings of this former imperial estate in such a way that a veritable fortress had been created, a divisional headquarters 'hedgehog.'

This was a necessary measure because the winter war, with its swift and dangerous Russian penetrations and attacks by partisans, had made even the higher command headquarters potential front-line positions. These therefore constituted a system of fortified strongpoints between the thin German main fighting-line and the rearward areas.

Major-General Nehring had just returned from a visit to the front. His chief of operations, Major Estor, met him with the words: "The Commander-in-Chief urgently asks you to telephone him. Something's up. He wants you to ring him at once."

Nehring had himself connected with Colonel-General Schmidt, Guderian's successor as C-in-C Second Panzer Army. It was a short conversation.

Schmidt said, "We need you. Would you come over to-morrow morning, please. It's an important matter."

Major Estor's log entry of the telephone conversation is dated Tuesday, 6th January 1942.

On the following morning Nehring drove to Orel, a bustling base in the hinterland which had become a front-line town overnight. Colonel-General Schmidt was not there. He had driven over to General Kübler, who had been in command of Fourth Army since Christmas and who now found himself very hard pressed by the enemy.

Nehring was received by the Chief of Staff, Colonel von Liebenstein. First of all the colonel served him some heated-up chicken broth straight from the tin—a detail the general remembers to this day. It was most welcome after the drive through the frosty winter waste.

Without any preliminaries Liebenstein came to the point: "The situation in the gap between Belev and Kaluga is getting more and more critical. Unless something is done Fourth Army will be in serious danger." He pointed to the map. "These strong Soviet forces are already deep in Kübler's rear. Fourth Army headquarters in Yukhnov has already become the front line. We've got no reserves. True, General von Gilsa's 216th Infantry Division was switched by the High Command from France to Sukhinichi towards the end of December, because until then the first Soviet assault was being held by hurriedly scraped-up forces. But now units of Gilsa's division have been encircled by the Soviet Tenth Army. Gilsa is resisting desperately. His men are very well equipped and are a brave lot—but they are unaccustomed to such winter conditions and can only inadequately be supplied from the air. Gilsa already reports some thousand wounded. But if this last breakwater is washed away it will mean disaster."

Nehring stood in front of the situation map, studying the red arrows and loops in the notorious 50-mile breach in the front between Kaluga and Belev—the breach which had been the nightmare of all headquarters personnel for the past fortnight.

"And what is to happen?" Nehring asked.

Liebenstein answered, "We've no other choice but to detach units from our Orel front, hard-pressed though we are ourselves, in order to stabilize the situation in the gap. We must link up with Gilsa again and strengthen his defensive front. And that is where you come in with your well-tried 18th. Needless to say, you'll be given some further forces, to be put under your command. We have in mind 12th Rifle Regiment, 4th Panzer Division, and Major-General von Scheele's 208th Infantry Division, which has just arrived from France and is at present employed as flank cover south of Belev. Admittedly, we've taken 309th and 337th Infantry Regiments away from the division because they were urgently needed by Ninth Army in the Sychevka area."

Nehring, an experienced commander in many critical situations, was not exactly pleased with the assignment. But he realized the need for the action.

The regiments took ten days to cover the roughly 125-mile journey from their sector via Orel, Bryansk, and Ordzhonikidzegrad to the assembly area near Zhizdra. Their journey in a temperature of 40 degrees below zero, through three-foot-deep snow and mountainous drifts, was sheer hell.

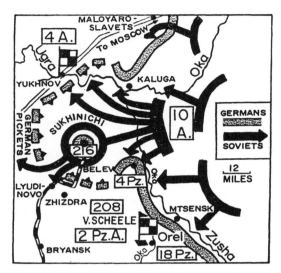

Map 21. The Soviet breakthrough between Fourth Army and Second Panzer Army in January 1942. Sukhinichi acted as a breakwater behind the breached front.

Captain Oskar Schaub from Vienna, a battalion commander in 12th Rifle Regiment, has described the way the units struggled through open country. The narrow wheels of guns and supply vehicles, he recalls, sank into the snow up to their axles. Lorries kept getting stuck. On the whole, the horse-drawn units managed best. The tough little farm-horses averaged about three miles an hour with their carts or sledges. The motorized troops with their tracked and wheeled vehicles managed barely more than a mile—half as much as a pedestrian would cover under normal conditions. The horse, in fact, was greatly superior to motor vehicles and armour in these conditions. The result was that all Panzer divisions were equipping themselves with horses during the winter.

On 16th and 17th January the reinforced 18th Panzer Division got ready to move off from Zhizdra. Its left wing was covered by 12th Rifle Regiment under Colonel Smilo von Lüttwitz, while the right flank was protected against enemy surprise attacks by units of 208th Infantry Division. Strong patrols on skis screened the area of the advance. Makeshift snow-ploughs cleared the road for the marching columns.

Operation Sukhinichi got going. It was one of the most extraordinary, harebrained, and risky operations of the winter war.

Nehring's present comment is: "A piece of strategic impertinence." He was right. In the Sukhinichi area there were no fewer than thirty Soviet rifle divisions operating, as well as six rifle brigades, four armoured brigades, two air-borne brigades, and four cavalry divisions. That was a truly gigantic force—an elephant about to be attacked by a mouse.

However, cunning, skill, and daring succeeded in outmanœuvring the Russians. The mouse slipped into the elephant's trunk.

Colonel Kuzmany, the one-armed Austrian commander of 338th Infantry Regiment, was standing up on his little peasant sledge. He was driving at the head of his combat group. Under him were three battalions, reinforced by tanks and artillery. His thrust was in the middle of the attack, via Bukan and Slobodka towards Sukhinichi. Colonel Jolasse, commanding 52nd Rifle Regiment, cleared some elbow-room for Kuzmany on the left flank and in the rear, and with his combat group attacked the strongly defended township of Lyudinovo. His force consisted of two battalions, the Panzer Company von Stünzner, the 2nd Company Panzerjäger Battalion 88, and a battery of 208th Artillery Regiment. The Russians were completely taken by surprise: they had not expected an attack so far from the fighting-line by German forces emerging like phantoms from the snowy wastes.

The companies of Jolasse's combat group dislodged the enemy from Lyudinovo and chased him into the forests and the snow-covered lake district. In ferocious street fighting against Soviet emergency units the battalions Wolter and Aschen cleared a road through the town. This first engagement yielded a considerable quantity of captured weapons, 150 prisoners, and over 500 killed.

Kuzmany meanwhile also battered his way through a surprised enemy. Wherever the Russians attempted to offer resistance they were smashed by concentrated fire from all weapons.

First Lieutenant Klauke, commanding 2nd Battery, 208th Artillery Regiment, stood on a sledge directing the fire of field howitzers. Attacking battalions were dispersed and machine-gun and mortar positions smashed at point-blank range.

There was no time for the gunners to aim their guns by calculation. "A quick look through the barrel, and you knew the direction was all right" is how Corporal Werner Burmeister, a gun-layer in 2nd Battery, recalls the situation. Colonel von Lüttwitz meanwhile with his reinforced 12th Rifle Regiment was working his way forward in the deep western flank of Nehring's formations. Captain Schaub records in his report of engagement: "The wheels kept getting stuck in the chest-deep snow. Working till late at night, the 2nd Company shovelled their way through to a signal-box on the Ordzhonikidzegrad–Sukhinichi line. The temperature was 40 degrees below. Rifles and machine-guns had to be wrapped up as carefully as the men's noses and hands, or else the oil would freeze—and that could be fatal."

Every yard of their way had to be shovelled clear. And at any moment the enemy might appear from the right, from the left, from behind or from in front. Out of this situation Lüttwitz evolved a novel fighting method. Schaub describes it as follows: "The point company would struggle through the deep snow along both sides of the road to the nearest village and attack the enemy in this narrow deep formation, working like an assault detachment. The attack would be opened with concentrated mortar-fire. After that the hand-grenade was the principal weapon—or, at close quarters, the trenching-tool. Meanwhile the remaining companies would shovel the road clear for the motor vehicles. Thus our combat group resembled a slowly moving hedgehog."

The front line was everywhere. Even the divisional headquarters personnel had to fight for their lives on 20th January, when, late in the evening, a Russian battalion charged Slobodka in the snowy light. They were saved by a 2-cm. AA gun of the headquarters security detachment until the sapper battalion hurried to the scene to retrieve the situation.

Thanks to this bold improvisation and the continuous alternation of attack and defence, advance, flank-screening, and rear cover, "Operation Sukhinichi" succeeded. Two weak divisions had slashed a 40-mile-long corridor right through an enemy Army, to reach a besieged strongpoint.

On 24th January at 1230 hours Colonel Kuzmany shook hands with a battle outpost of the combat group under Freiherr von und zu Gilsa. A bridge had been built to the cut-off 216th Infantry Division and the formations subordinated to it. It was a narrow bridge, but it held.

The following morning Nehring drove into the town to discuss the situation with Gilsa. A thousand wounded were lying in the cellars of the ruined houses: to get them out was one of the most urgent tasks. This, like everything else about this operation, was done in an unconventional way. Five hundred local sledges with Russian peasants and prisoners as drivers were available in Lyudinovo. Each sledge could accommodate only one wounded man. Every driver, therefore, had to make the 40-mile trip through no-man's-land four times. But not a single one dropped out. Every one of them stood up to the tremendous demands made by the night-time sleigh-ride through biting frost, blizzard, and enemy patrols.

In command of this mercy fleet was a corporal—a village priest in civilian life. His assistant was a cavalry sergeant. Their helpers were 500 Russians. The two Iron Crosses which Nehring had earmarked for them were never awarded: the two good Samaritans disappeared in the turmoil of battle. Their names are not known.

The importance of Operation Sukhinichi in the general situation is reflected by the fact that Hitler paid tribute to it in a special announcement. This was his way of demonstrating that surrounded units which obeyed his order and held on as breakwaters regardless of being by-passed by the enemy would not be forgotten. This demonstration was an important prerequisite for the troops' perseverance at other points of the front where major or lesser formations were encircled, as, for example, at Kholm and Demyansk.

"They'll get us out": this unshakable belief of surrounded troops and their officers was justified time and again during that winter of 1941–42. It is a point that must be remembered by all those who to-day shake their heads in uncomprehending amazement at the blind faith shown a year later by the encircled German Sixth Army at Stalingrad.

Sukhinichi was a decisive strategic success. Nevertheless, in a sound appreciation of the situation, Lieutenant-General Freiherr von Langermann-Erlenkamp, commanding XXIV Panzer Corps, decided to evacuate the exposed town of Sukhinichi itself. This move made it possible to establish a more favourable defensive front across that notorious breach which was now being closed again. The nightmare of the German High Command was at an end. Conditions had been created for smashing the southern prong of the Soviet offensive.

After weeks of heavy fighting, lasting well into the spring, the bulk of the divisions of the Soviet Tenth and Thirty-third Armies, the I Guards Cavalry Corps, and the 4th Parachute Commando which had penetrated, were annihilated south-east of Vyazma. That was the great battle in the Ugra bend, with its focal points at Ukhnov, Kirov, and Zhizdra. In this battle downright superhuman feats were performed by the divisions from Brandenburg and Bavaria, from Schleswig-Holstein and Mecklenburg, the Upper Palatinate and Hanover, from Hesse and Saxony, as well as by the independent 4th "Death's Head" SS Regiment and the Parachute Assault Regiment Meindl.

What this savage fighting looked like from the other side is revealed by two impressive documents—both of them captured diaries of Soviet officers. They afford a glimpse of the morale of the Russian front-line troops in the Sukhinichi–Yukhnov–Rzhev area.

The first diary is that of Lieutenant Goncharov, a company commander and temporary battalion commander in 616th Rifle Regiment. He was killed in action north-west of Yukhnov on 9th February 1942.

The second diary is that of a lieutenant of 385th Rifle Division whose name had been proposed for the title of "Hero of the Soviet Union." Since it is not certain whether he is alive, his name shall not be given here. Both diaries are extant in the original and come from the archives of the Intelligence Officer of the German XL Panzer Corps.

Goncharov's notes reveal a simple soul—a man who believed in Stalin's political slogans, was annoyed with his superiors, and passed on all kinds of front-line gossip. The entries are revealing in many ways:

> 2nd January 1942. When the 4th Battalion withdrew from Yerdenovo we had to leave our dead and wounded behind. The wounded were killed by the Germans.
>
> 5th January 1942. I talked to the civilian population about the Germans. Generally speaking, they all tell the same tale—looting, executions, rape. But it strikes me that they relate these fascist atrocities without revulsion. They talk about them just as though they are reporting a lecture by the collective farm chairman. And yet, how revolting is everything about that Aryan race! No sense of decency. They strip themselves naked in the presence of women and kill their lice. We have always regarded the Aryans as people of culture. Now it is clear that these Aryans are dull, stupid, shameless bourgeois.
>
> 10th January 1942. I have read Molotov's note to-day about the German atrocities. One's hair stands on end when one reads the few examples listed. In my opinion the world has not room enough for retribution for all that this Nordic race has inflicted on us. But we shall have our revenge—we shall have it of their whole race—in spite of our very humane and moderate leader Stalin. To hell with international considerations! Sooner or later we shall have to fight England as well.
>
> 14th January 1942. At Shanskiy Zavod I slept at the home of a woman partisan. Nearly half the village collaborated with the Germans. The partisans were not only not supported, but betrayed and opposed. I had to get up at three in the morning to get back to the front. It was difficult. I had slept on a warm stove. It was covered with white tiles which I had never come across before. I must confess it looked very good.
>
> 23rd January 1942. The Germans are in the village of Agroshevo, nine miles behind us. It's freezing hard. I had to rub my nose with snow several times—otherwise it would have frozen off. Some 50 per cent. of my men have frost-bitten noses; in some of them gangrene is beginning to develop. By nightfall it was clear that we were surrounded. No supplies coming through. We are hungry.
>
> 25th January. "You know, Comrade Lieutenant," one of my men said to me yesterday, "when one gets really cold one becomes indifferent to freezing to death or being shot. One only has one wish—to die as quickly as possible." That's the exact truth. The cold drains the men of the will to fight.
>
> 26th January 1942. At midnight a breakout attempt was mounted towards Rubikhonov. The 4th Company tried to envelop the enemy from the left, with one machine-gun and three mortars. One mortar-shell hit the machine-gun crew of my No. 1 Platoon. Three men were wounded and three men killed. One of the wounded screamed, cried, and begged to be carried out of the fire. Another implored me to shoot him dead. In the open the frost was so hard that I could

not bandage the man, because I would have had to undress him. I only had the choice between letting him freeze to death or bleed to death. The battalion is down to 100 men, including headquarters staff and supplies. In the main fighting-line only 40 to 50 men are left. Our strength is ebbing away. These damned Germans fight like the devil.

1st February 1942. Zass is no longer our regimental commander. I'm very pleased. The man was perpetually drunk. While drunk he would make idiotic decisions, and as a result of these we have lost a lot of men.

6th February 1942. Several men of the Ski Regiment were shot to-day because of pilfering, going absent without leave, and offences while on sentry duty.

8th February 1942. The Germans are attacking.

That is the last entry. Goncharov was killed in action near Papayevo, six miles north-west of Yukhnov, on 9th February 1942, fighting against formations of the Hessian 34th Infantry Division. His regiment was smashed.

The author of the second diary, Second Lieutenant V., was a very different man from Goncharov. He too was employed south-east of Vyazma. Facing him were units of the German 19th Panzer Division and the 3rd Motorized Infantry Division. Fanatical, ambitious, yet astonishingly clear-sighted, this young "Hero of the Soviet Union" is interesting source material for the study of the lower commands of the Red Army. Clearly this lieutenant had on several occasions averted dangerous situations for his regiment.

"The company's morale," he wrote on 7th February 1942, "is excellent—if it were not for the disastrous food situation. If only the men had enough to eat one could win any battle with them."

On 10th February we read:

I have become weak from this nomadic life of the past year. Yesterday I had an upset stomach from the bad bread and frozen potatoes. But I've got to remain in harness. Last time I was away from the company for twelve days all discipline went to hell. If only these damned supplies would get through! The men would charge into a rain of bullets without batting an eyelid. But they are hungry. They are losing their strength. The weapons, too, are getting rusty: we've got no lubricating oil. Yesterday I organized a meeting in a barn, together with the political commissar. I explained why fascist prisoners must be allowed to remain alive—as a source of information.

I intended to submit my application for Party membership since I believed we would go into action to-day. I shall only hand it in just before a battle, so that no one should suspect me of seeking personal advantage.

19th February 1942. Yesterday I handed in my application for Party membership, to become a Bolshevik. In the evening I was ordered to go out with only my sub-machine-gunners and take the forest which 3rd Company had been unable to take. It's a formalistic plan. But orders are orders. I'm off in an hour.

The entry after the attack reads:

The men stood up well, and the attack was successful. Ten Germans were killed and five taken prisoner. Four of these we had to shoot because they refused to come back with us. My thirty men received spirits and several tins of cigarettes, biscuits, sausage, and butter from the CO's personal-disposal supplies.

24th February 1942. As from to-day I am a probationer for Party membership. I'm beginning to get enthusiastic about the war. I would have made quite a good partisan.

25th February 1942. To-day sapper-instructor B. arrived. When he was told that he was in the front-line he got very dejected and asked Gladev to play a funeral march to him on his guitar. Gladev obliged. Fifteen minutes later B. was killed. Fate? Or do the bullets seek out the cowards?

The bulk of our active officers went to Officers' Training Courses only for the sake of the uniforms and the gold collar-flashes. They are better at marching than tactical knowledge, and better at making reports than carrying out orders. At best they know how to charge and be killed.

26th February 1942. I had my photograph taken to-day for the Party papers. We are being sent on a special mission. My company is ready for action. I hope they are not sending us in in daylight: that would be dangerous and stupid. The Red Army must fight at night.

27th February 1942. I had to carry out a court-martial sentence—an execution. In response to my question three men volunteered at once. The two culprits had hidden themselves while out on patrol and had shirked action. Silly fools! They thought they could escape honourable death and now had to die ignominiously.

4th March 1942. At last a letter from my wife: I keep re-reading it. Letters from home bring joy and pain.

This ambiguous sentence is the last entry. It dates from the time when Soviet hopes of a victory on the Central Front were evaporating.

From this time also dates the letter of a young Russian. Addressed to a friend, it was found in his pocket, unfinished, when he was killed near Dorogobuzh. It is quoted here for all those, wherever they may be, who lost friends in the war. It may even be that its reproduction decades after it was written will enable it at last to get to its addressee. It reads:

Greetings, dear friend! Greetings and also farewell—because I am no longer alive. This letter will be sent to you only in the event of my death. But that day is not distant—I can feel it. I do not know how long this letter will remain in my pocket, crumpled, but sooner or later it will reach you to remind you for a last time of your schoolmate.

Model Takes Over 389

I feel an urge to say a great deal to you on this final occasion—a great deal. I want to pour out my whole sadness of unfulfilled hopes, to communicate to you my fear of this unknown death. Yes, dear friend, fear—for I am afraid of what comes after death.

I do not know how and where I shall die, whether I shall be hit by the bullet of a German machine-gunner, torn apart by an aerial bomb, or killed by a shell-burst—but each one of these possibilities terrifies me equally. I have seen hundreds of men killed. I have repeatedly heard the rattle of death in the throats of comrades with whom I had cheerfully eaten a meal out of the same mess-tin a little while before.

I have met death face to face many times. Once a shell-splinter tore my cap off my head. Another time a bullet went through my mess-tin so that my soup ran out and I was left hungry. But never before have I been so afraid as now.

Look about you—spring is coming. The six letters of this word keep disturbing me. For this is no ordinary spring—this spring I shall be twenty. Twenty years old—almost a man. And to die just when nature is smiling upon you, when your heart pounds with joy because of a bird's song or the gentle caress of the moist spring breeze....

It was spring when this letter was written. But at the time when the German gunner Burmeister was aiming his light field howitzer at Lyudinovo, when the unending columns of sledges were evacuating the wounded from Sukhinichi, when the machine pistols stuttered at the Ugra bend, and when all along the front shouts were going up: "Russian tanks breaking through!" snow was still lying several feet deep on the battlefields of the Eastern Front.

But the winter battle had already been decided. True, the men in the front line did not yet realize this. They were still engaged in exceedingly heavy defensive fighting. But the maps at the Army headquarters were already revealing the truth: the great crisis at Army Group Centre was over.

Mobile Soviet cavalry formations continued to advance as far as Dorogobuzh, east of Smolensk, during the next few weeks, but these were the last waves of a drive that had lost its momentum. The Soviets had reached the end of their offensive strength. They had failed to reach the strategic objective of their winter battle—to destroy Army Group Centre and thus cause the entire German Central Front to collapse.

The turning-point had been due to two things. First, the Soviet High Command had bitten off too much. Operational leadership, conditions, and supplies for their offensive Armies had not been up to such far-reaching aims.

The second reason was the outstanding achievements of the German formations in denying the Soviets their success and preventing disaster.

In terms of discipline, gallantry, hardships, and self-sacrifice, officers and men had surpassed anything previously known. Military organization had remained intact and capable of functioning in spite of the troops being over-extended, in spite of hunger and inadequate clothing. That was how the situation was saved at the threatened Central Front in the winter of 1941–42. That alone ensured the success of Hitler's hold-on order and of the tactics of hanging on to vital reinforced positions.

Thus Rzhev was saved and Sukhinichi relieved. At the last moment and with the last ounce of strength the enemy was kept off the Smolensk–Moscow motor highway. The encirclement of Army Group Centre was prevented. The great crisis on the Central Front was, on the whole, overcome.

But what was the situation like in the area of Army Group North? How did the troops on the Leningrad Front and on the Volkhov river survive the Soviet winter offensive?

4. Assault on the Valday Hills

The Soviet 57th Striking Brigade charges across the Volkhov–Rendezvous: Clearing Erika–Two Soviet Armies in the bag–Demyansk: 100,000 Germans surrounded–An unusual Order of the Day by Count Brockdorff-Ahlefeldt–A pocket is supplied from the air–Operation "Bridge-building"–Kholm, a fortress without guns.

THE point where the Tigoda river runs into the Volkhov was the junction between the German 61st and 21st Infantry Divisions. These junctions were always vulnerable and a favourite target for Russian attacks. The Russians had found from experience that the overlapping command pattern along these junctions made it more difficult to clear up any penetrations.

Who was responsible for sealing off a penetration along a junction? No commander ever was particularly anxious to do so himself; he would much rather leave the responsibility to his neighbour. For that reason special "junction reserves" had been a customary thing in the First World War. But the weak German Eastern Front was only rarely able to afford such a luxury during the winter of 1941–42.

"That damned junction, of course," Colonel Lohmeyer cursed as, on 3rd January 1942, he received the curt and clear-cut order from the command of 291st Infantry Division: "Enemy forces which have penetrated between 61st and 21st Infantry Divisions at the mouth of the Tigoda must be thrown back and the main fighting-line restored."

The East Prussians of the Elk Division had been withdrawn from the line only a few days previously for rest and replenishment. But what was the use?

With his well-tried 505th Infantry Regiment and part of an SS Battalion of the 9th ("Death's Head") SS Regiment, newly arrived from Finland, Lohmeyer flung himself against the Soviet ski battalions which had penetrated at the mouth of the Tigoda. It was a fantastic kind of fighting in chest-deep snow, 42 degrees below, in thick forest matted with impenetrable undergrowth.

In the early evening of 4th January Colonel Lohmeyer, the hero of Liepaja, was killed in a forest by an enemy mortar-bomb. The news spread like wildfire and was a profound shock to the regiment. Colonel Hesse took over command and with the angry battalions of 505th Infantry Regiment cleared up the Soviet penetration.

But the attack they had repulsed at the mouth of the Tigoda was not the expected Soviet full-scale attack. During those first few days of the New Year there was fierce local fighting everywhere between Kirishi and Novgorod. The Russians were prodding the Volkhov front to find its weak points; they were carrying out reconnaissance in force to identify German positions and units; they were searching for a gap through which they could thrust. The old-timers could feel it in their bones: something was in the air, a large-scale attack was imminent. When would it come? And where? Those were the agonizing questions.

Major Rüdiger, commanding the Intelligence detachment of 126th Infantry Division, heaved a sigh of relief when a monitoring company NCO knocked at his door late in the evening of 12th January and brought him an intercepted and decoded signal from the Soviet Fifty-second Army to its 327th Rifle Division: "Positions to be held at all costs. Offensive postponed. Continue feinting attacks.'

So there was not going to be an offensive after all—at least not in this sector, Rüdiger concluded. He immediately rang up Lieutenant-General Laux, his C.O. Laux, an experienced officer, thanked him for the information, but added, "I wouldn't trust those fellows too far."

The contents of the intercepted radio signal soon spread about. When, therefore, Soviet artillery started shelling the German positions over a broad front at 0800 hours on the following morning, 13th January, the troops did not take it too seriously.

But after a while things began to look suspicious. The heavy bombardment did not look like a blind. The guns then lengthened their range beyond the German lines. The time was 0930. Under the massive artillery umbrella numerous packs of infantry emerged from the haze of the dawning winter day, and ski detachments came gliding over the ice of the Volkhov. "The Russians are coming!"

The radio signal of the night before had been a Soviet ruse to deceive the German command. The battle of the Volkhov had begun: it had started north of Novgorod, at the junction between 126th and 215th Infantry Divisions.

By 1030 hours the Soviets had established their first bridgehead across the Volkhov at Gorka, in the sector of 422nd Infantry Regiment, and had broken into the German main fighting-line.

Colonel Harry Hoppe, the conqueror of Schlüsselburg, mounted an immediate counter-attack with units of 424th Infantry Regiment and sealed off the penetration. But he did not succeed in regaining the old main fighting-line.

In the morning of 14th January the enemy attacked again, and strong formations succeeded in infiltrating through the snow-bound forests into the rear of the German positions. By nightfall the spearheads of fast Soviet ski battalions stood in front of the gun emplacements of the divisional artillery. The German gunners defended themselves with trenching-tools, carbines, and pistols and repulsed the Soviets. But for how long?

While Division and Corps were still convinced that the main weight of the Soviet attacks was in the sector of 422nd Infantry Regiment, a far greater disaster was unrolling farther north, in the Yamno-Arefino area. It was there, at the junction between 126th and 215th Infantry Divisions, where the sectors of the two wing regiments, 426th and 435th Infantry Regiments, abutted, that the Soviets had concentrated their main effort.

On a very narrow front, therefore, the crack Soviet 327th Rifle Division and the superbly equipped independent 57th Striking Brigade charged across the Volkhov against the positions of the three weakened battalions of a single German regiment—426th Infantry Regiment under Lieutenant-Colonel Schmidt.

Simultaneous Soviet attacks on 435th Infantry Regiment, lying on the left of the 426th, prevented help being sent from there. Skilfully taking advantage of deep hollows in the ground in front of the German main fighting-line, the Soviets punched their way into the German positions, cracked the line of strongpoints, and, with the bulk of XIII Cavalry Corps of the Second Striking Army, swept like a tidal wave through the burst dike into the hinterland. Ceaselessly the Russians pumped reinforcements through the two-to-three-mile gap and pressed on towards the Novgorod–Chudovo road.

In a cutting cold of 50 degrees below zero the dispersed German companies hung on to clearings in the woods and to high snowdrifts and made the Soviets pay heavily for their slow advance. It took the Russians four days to cover the five-mile distance to the road. And when they did reach the road they had not gained much, for three German strongpoints continued to hold out firmly along it like pillars against the battering waves—Mostki, Spasskaya Polist, and Zemtitsy.

Surrounded by the enemy, these strongpoints held on for weeks in the rear of the Soviet flood. They became focal points in the fighting for the vital road, the north-south link of the Volkhov front.

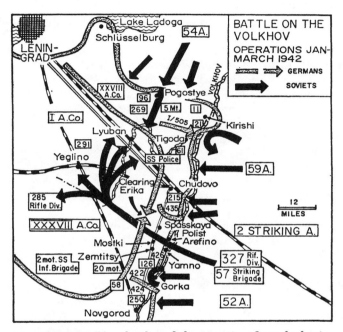

Map 22. The Volkhov battle and the operations from the beginning of January to the end of March 1942. The bulk of the Soviet Second Striking Army, having broken through the German front, was pinched off at the clearing "Erika."

By 24th January the Russians had pumped sufficient forces into the penetration to launch their drive in depth. With cavalry, armour, and ski battalions they raced boldly through the narrow—in fact, all too narrow—bottleneck towards the north-west. It was a perfect breakthrough. But its basis was dangerously narrow.

What were the Russians after? Were their operations aimed against Leningrad, or did they have other, more far-reaching intentions? That was the question agitating the German Staff. They did not have to worry their heads very long. After eight days the spearheads of the Soviet assault regiments were already 55 miles behind the German front. If they were aiming at Leningrad they had covered half the distance.

On 28th January Russian forward detachments attacked Yeglino. The direction of the attack, therefore, was towards the north-west, bypassing Leningrad in the south, towards the Soviet–Estonian frontier. We know to-day that the big drive outflanking Leningrad was in fact originally intended to go as far as Kingisepp—a more than optimistic idea. But then the Russians suddenly stopped at Yeglino and, instead of continuing westward, swung north-east towards Lyuban, on the Chudovo–Leningrad road. Were they aiming at Leningrad after all?

General of Cavalry Lindemann, who had succeeded to the command of Eighteenth Army when Field-Marshal von Küchler took over Army Group North on 15th January, needed only one glance at the large situation map in order to read the Russian intentions. Their penetration area, the bottleneck through which they were pouring, was too narrow and their exposed flanks were too long. To advance farther would have been foolhardy.

Since the Soviet Fifty-fourth Army was just then attacking the German 269th Infantry Division at Pogostye, south of Lake Ladoga, the Russian intentions were clearly revealed by the map: to begin with, the German I Corps was to be annihilated in a pincer operation.

"We must be prepared for anything and not lose our nerve," remarked General of Infantry von Both, commanding the East Prussian I Corps in Lyuban, as the commander of the Corps headquarters started issuing carbines and machine pistols to the officers and clerks. Not to lose one's nerve—that was the problem.

The 126th Infantry Division has been severely blamed for allowing the Russians to break through in its sector. That is unfair. No regiment of any other weakened division on the Eastern Front could have held this concentrated Soviet attack. In judging the 126th Infantry Division one should remember not so much the Soviet break-through as the fact that this division continued to hold the flanks and cornerstones of this barely 20-mile-wide gap day after day against the onslaught of Soviet combat groups, and by doing so prevented the penetration from being widened.

In spite of continuous costly attacks the Russians did not succeed in widening their narrow corridor. They left some 15,000 dead in front of the positions of 126th Infantry Division. This circumstance presently had dramatic consequences.

It was Colonel Harry Hoppe, the hero of Schlüsselburg, and his 424th Infantry Regiment who enabled a new main fighting-line to be established along the southern edge of the penetration.

The northern edge was held with admirable steadfastness by the 215th Infantry Division under Lieutenant-General Kniess. A vital contribution was made by the stubborn defence of the strongpoints

Assault on the Valday Hills 395

Mostki, Spasskaya Polist, and Zemtitsy. There the "Brigade Köchling," an *ad hoc* collection of units from fifteen different divisions, defended the strongpoints throughout many weeks.

Exemplary resistance was offered in Zemtitsy by Captain Klossek with his 3rd Battalion, 422nd Infantry Regiment. If any proof was needed that Hitler's blunt and uncompromising hold-on order and its self-sacrificing observance by the men could in certain conditions avert disaster and create the prerequisite for future successful operations, then this proof was supplied by the battle on the Volkhov.

Some 125 miles away from the Volkhov front, meanwhile, the 58th Infantry Division from Lower Saxony was still holding the Leningrad suburb of Uritsk—the same division which nearly six months previously had reached the first tram-stop of Leningrad and thus, virtually, the cradle of the Red revolution.

"All unit commanders to report to Divisional HQ at 1100 hours for a conference!" Telephones rang. "Conference with the General," the telephonists informed their friends at battalion and company level. "Something's up," they added.

The date was 1st March 1942. General Altrichter, the OC 58th Infantry Division, greeted his officers. They all suspected that their division was to be once again employed on some special mission.

"Bound to be Volkhov," the officers muttered to each other. A fortnight previously Lieutenant Strasser with his 9th Battery, 158th Artillery Regiment, had moved off towards Volkhov together with the Emergency Battalion Lörges as a "ski battery." A week later further batteries had been dispatched towards Novgorod.

Their surmise was quickly confirmed. "Gentlemen," Altrichter opened the conference, "we have been assigned a task which will have a vital effect on the overall situation."

"Volkhov after all," Colonel Kreipe, commanding 209th Infantry Regiment, said softly to his neighbour, Lieutenant-Colonel Neumann.

General Altrichter had heard him. He nodded and continued: "58th Division has been chosen to be the striking division for sealing off the Volkhov penetration from the south and encircling the enemy forces which have broken through."

Friedrich Altrichter, a doctor of philosophy and the author of interesting essays on military education, formerly on the staff of the Dresden Military College, was good at explaining strategic problems. Many a German officer had passed through his class. He died in Soviet captivity in 1949.

Altrichter stepped up to the large situation map and began his lecture: "You see what the situation is: the enemy has already driven deep into our lines, and in strength. Frontal engagement can no longer lead to success since we don't have the inexhaustible reserves that would be necessary for that kind of operation. Our only chance is to strike at the Russians at the basis of their operation, at the breakthrough point—to pinch them off and thus to isolate the forces which have penetrated. Fortunately, the 126th and 215th Infantry Divisions have once more established firm fronts along the edges of the bottleneck, and we shall be able to assemble under their cover. We shall strike at the gap from the south. The SS Police Division will attack from the north. Rendezvous is the clearing known as Erika. The regiments of 126th Infantry Division and all other formations employed there, including the battalions of the Spanish Blue Division, which have fought splendidly so far, will be subordinated to us. With these forces we should be able to manage it. And we've got to manage it—otherwise Eighteenth Army is lost. But if we succeed in closing the trap, then we shall have the bulk of two Soviet Armies in the bag."

There was silence in the small room. Then a clicking of heels. Outside it was still bitterly cold.

"Volkhov," the whisper ran through all the unit offices. Volkhov. The men looked it up on their maps. "Some 125 miles to the south of us. And in this weather," they grumbled.

Preparations were complete by 15th March. About 15th March, in Western and Central Europe, people began to think of the spring. But on the Volkhov the temperature was still 50 degrees below zero. The snow lay four feet deep in the thick forests.

The Russians knew what was at stake at their penetration point, and had therefore strengthened it as much as possible. Along the stretch of road controlled by them flame-throwers had been installed and thick minefields laid across all negotiable clearings.

The 220th Infantry Regiment charged the Soviet barriers with its captured enemy tanks and its sapper assault detachments. Fighter-bombers and Stukas of I Air Corps dropped their bombs on Soviet positions and strongpoints. But the deep snow cushioned the bomb-bursts, and the blast effect was slight. The Russians held their block, and 220th Infantry Regiment did not get through.

Things went better two miles farther west. There the battalions of 209th Infantry Regiment fought their way forward step by step through the clearings of an almost impenetrable snow-bound forest. The 154th Infantry Regiment likewise struggled through the undergrowth, with self-propelled guns and sappers clearing a path for them. Savage small-scale fighting raged throughout the forest.

Time and again the mortars failed in the severe cold: ice kept forming inside the barrels so that the bombs no longer fitted. The gunners had shells bursting inside their barrels because the rifling kept icing up. Machine-guns packed up because the oil got stiff and sticky. The most reliable weapons were hand-grenades, trenching-tools, and bayonets.

Towards 1645 hours on 19th March the most forward units of 2nd

Assault on the Valday Hills 397

Battalion, 209th Infantry Regiment, under Major Materne made their way across the clearing marked on their maps with "E"—known to them as clearing Erika. It is a name well remembered by every one who fought on the Volkhov. It marks a patch of dismal, hotly contested forest. Along the wooden road constructed across this clearing and serving as a supply route a trooper had erected a noticeboard with the words "Here begins the arse-hole of the world." For many months it remained in the same spot where, on that 19th March, the spearheads of Materne's battalion were lurking.

A machine-gun was stuttering from the far end of the clearing. "That's a German m.-g.," one of the men said. "Better be careful, boys," warned Major Materne.

A white flare soared up on the far side. "Answer it!" Materne commanded.

A white fiery ball, like a miniature sun, hissed over the clearing. At the far end a figure wrapped up to the tip of his nose and wearing a German steel helmet emerged from behind a bush, and waved. "Ours!" The men were beside themselves with joy.

Through the snow they raced towards each other, slapped each other's backs, fished out cigarettes, and pushed them between each other's lips. "What do you know," they said to their pals from the forward assault detachments of the SS Police Division. "What do you know—we made it!"

They had made it. The breach was sealed. They had linked up at the clearing Erika, and they had cut the supply route of the Soviet Second Striking Army.

Two Soviet Armies were in the bag. As at Rzhev and at Sukhinichi, the unshakable resistance offered by individual units had created the prerequisites for retrieving a seemingly hopeless situation by means of bold counter-blows and for snatching the initiative from the Russian Armies, grown overconfident and careless through a taste of victory. Once more the hunters became the hunted and the pursued became the pursuers.

Thus the Soviet assault against the Volkhov was halted and their attempt to relieve Leningrad foiled. But what was happening in the meantime on the strip of land between Lakes Ilmen and Seliger, where five Soviet Armies had broken through and torn a wide gap in the front between Army Group Centre and Army Group North?

On this strip of land only two German barriers were left to stem the Soviet flood—Demyansk and Kholm—and everything now depended on them. If these two strongpoints were over-run or swept away the Soviet Armies would have a clear road into the deep, virtually undefended hinterland of the German front. In the Demyansk area there were six German divisions barring the way to the Russians. Other formations—such as units of 290th Infantry Division, which had held their ground when the Russians broke through at Lake Ilmen and had subsequently been unable to withdraw from their partial encirclement towards Staraya Russa—later broke through in a south-easterly direction to reach II Corps and thus reinforced the defenders of the Demyansk pocket. The II Army Corps under General Count Brockdorff-Ahlefeldt drew the bulk of the Soviet attacking forces of five Armies and tied them down.

The Soviet Third Striking Army, advancing farther south, was likewise unable to make any progress in the face of the second immovable barrier across the otherwise empty gap between Demyansk and Velikiye Luki, a barrier blocking the road into the rear of Sixteenth Army—Kholm.

Demyansk and Kholm became crucial for the turn of the tide on the northern wing of the German Armies in the East. The defenders of

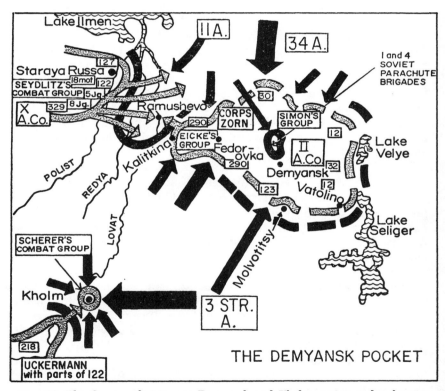

Map 23. The German divisions at Demyansk and Kholm, acting as breakwaters against the Soviet flood, succeeded in halting three Soviet Armies. The heavy fighting continued until the spring of 1942, when the two German pockets were liberated.

Assault on the Valday Hills 399

Demyansk and Kholm wrested the victory from the Soviets by sheer stubborn resistance.

The history of the battle of the Demyansk pocket—a battle lasting twelve and a half months and hence ranking as the longest battle of encirclement on the Eastern Front—began on 8th February 1942. Count Brockdorff-Ahlefeldt, the general commanding II Army Corps, was on the telephone to Sixteenth Army. "We'll try to keep communications open with you at all costs," Colonel-General Busch was saying.

At that moment there was a click in the receiver. The two generals heard the cool voice of the operator at the Corps exchange cut in: "I'm disconnecting now: the enemy's in the line."

The general replaced the receiver. He looked at his ADC and said, "That was probably the last talk we've had with Army for some time to come."

"That means the ring is closed?" the ADC asked.

"Yes," replied the general. After a short pause he added, "At least we know where we are. We'll just have to wait and see."

"Waiting and seeing" turned out to be twelve months and eighteen days of ferocious fighting. The Corps was encircled in an area of 1200 square miles.

The reasons why this battle had to be fought, why a pocket in the middle of the Valday Hills around the dismal dump of Demyansk had to be held against the Soviet assault, were explained by the general in an unusual Order of the Day. Unusual because it not only ordered the officers and men to do a certain thing, but also explained the circumstances and reasons for the order.

On 20th February, therefore, twelve days after they had first been encircled, Count Brockdorff-Ahlefeldt had the following order read to all his formations in the pocket:

"Taking advantage of the coldest winter months, the enemy has crossed the ice of Lake Ilmen, the normally marshy delta of the Lovat river, and the shallow valleys of the Pola, Redya, and Polist, as well as numerous lesser watercourses, and placed himself behind II Corps and its rearward communications. These river-valleys form part of an extensive area of swampy lowlands which get flooded and entirely impassable, even on foot, the moment the ice and snow begin to melt. Any enemy transport, especially the bringing up of supplies on any scale, will then be entirely impossible.

"Russian supplies during the wet spring-time would be possible only along the major hard roads. The intersections of these roads, however —Kholm, Staraya Russa and Demyansk—are firmly in German hands. Moreover, the Corps with its battle-tried six divisions commands the only real piece of high ground in the area. It is therefore impossible for the Russians with their numerous troops to hold out in the wet lowlands without supplies in spring.

"What matters, therefore, is to hold these road junctions and the high ground around Demyansk until the spring thaw. Sooner or later the Russians will have to give in and abandon that ground, especially as strong German forces will be attacking them from the west."

The officers and men listened and nodded. They understood. They were determined to hold "the county," as they called their pocket in an allusion to their general's title.

The battle began. It was the first major battle of encirclement in which the German troops were the encircled. For the first time in military history an entire Corps of six divisions with roughly 100,000 men—virtually a whole Army—was successfully supplied by air. It was on the Valday Hills in Russia that the first airlift in military history went into operation.

Some 500 transport aircraft supplied the 100,000 men of II Corps with everything they needed for survival and fighting—day in, day out. The machines flew regardless of blizzards and frosts, of fog or winter thunderstorms, and regardless of the furious anti-aircraft fire of the Soviets.

About 100 aircraft had to make the flight into the pocket and back each day. On certain days there were as many as 150. This meant that during every hour of the short winter days 10 to 15 aircraft had to land and take off from the two makeshift airstrips.

The feats of these transport units under Colonel Morzik, the Chief of Air Transport, were unparalleled at the time. The scope of the operation is illustrated by two figures: 64,844 tons of cargo were flown into the pocket and 35,400 men—wounded or transferred—were flown out.

The airlift was a decisive contribution to success. But it also mortgaged future German strategy: the German transport squadrons were decimated. Many pilots were killed.

More disastrous still was the fact that the success at Demyansk confirmed Hitler in his decision, nine months later, to hang on to Stalingrad, because he believed he would be similarly able to keep the encircled Sixth Army with its approximately 300,000 men supplied from the air.

Major Ivan Yevstifeyev, born in 1907, commanded the famous 57th Brigade, the spearhead of the Soviet Second Striking Army, on the Volkhov. He was an outstanding officer, courageous, a skilful leader in the field, and with the full background of a Soviet General Staff training.

When he was taken prisoner he remarked, "It was bound to happen like this, with so much stupidity in our Supreme Command." The report which he wrote shows that the news of the Second Striking Army being pinched off at the bottleneck of the clearing Erika had had a disastrous effect in Moscow. It meant the shattering of Stalin's hopes of liberating Leningrad and annihilating the German Army Group North. Stalin was looking for a scapegoat.

And just as he had dismissed General Sokolovskiy, the Army Commander-in-Chief, during the first week of January for attacking on a broad front and not breaking through, so General Klykov and his Chief of Staff were now sacked because they broke through on too narrow a front. But who was to save the situation now? Who could blow out the bung from the bottleneck and free the bulk of the two Armies encircled in the huge pocket?

Stalin's choice turned out to be a man who was then one of the stars of the Soviet Generals' Corps—Andrey Andreyevich Vlasov. In the late summer of 1941 Vlasov had courageously defended Kiev for two months, and subsequently, as Commander-in-Chief Twentieth Army, thrown back the northern wing of the German offensive against Moscow at Solnechnogorsk and Volokolamsk. He had been rewarded with orders, praise, fame, and the position of a deputy Commander-in-Chief of the Army Group Volkhov. Now he was to go into action again to prove his generalship.

Vlasov was the son of a small peasant, born in 1901. At considerable sacrifice his father sent him to a priests' seminary. Lenin's revolution decided him to become a Communist, to join the Red Army, to become a regular officer and eventually a general. Owing to the fact that during the thirties he was in China as Chiang Kai-shek's military adviser, he escaped the great purges which claimed Marshal Tukhachevskiy and most of his friends as victims. When Vlasov eventually returned to Russia there was no limit to his career. Very soon his fame as an organizer was proclaimed throughout the Soviet military Press.

The man who had once changed the 99th Rifle Division from the most notorious riffraff in the Red Army into a crack unit had now been chosen to save the two surrounded Armies.

Before dawn on 21st March Vlasov was flown into the Volkhov pocket and assumed command of the seventeen divisions and eight brigades in the forests between Chudovo and Lyuban. He immediately set about battering down the bolted door from inside.

At the same hour as Vlasov had summoned his unit commanders to a forester's cabin east of Finev-Lug to discuss the bursting of the German ring of encirclement around the Soviet Second Striking Army on the Volkhav, some hundred miles to the south-east the German Colonel Ilgen reported at Lieutenant-General Zorn's headquarters in Fedorovka in the Demyansk pocket to learn the plan for breaking through the Russian ring around the six German divisions in the Demyansk pocket. It was a strange parallel.

"Today's the first day of spring, Herr General," Colonel Ilgen said with a smile. General Zorn was standing outside the crooked wooden building which served as the Corps Group's headquarters. "Spring indeed," he grunted, "with two feet of snow and 30 degrees below."

"Well, that's what spring's like in Demyansk," Ilgen grinned. "You're right," Zorn nodded. "But joking apart, Ilgen, I hope the frost continues for a while. Once the thaw comes the mud will be terrible—not a wheel will be able to turn. And Seydlitz has got to get here before then."

The crimson disc of the morning sun broke through the haze of the new day. From the distance, from the sector of 290th Infantry Division at Kalitkino and of the Combat Group Eicke at the extreme western point of the pocket, came the flashes of Soviet heavy artillery. Zorn glanced at his watch. "0730," he said. "Now Seydlitz is mounting his attack."

At that moment, 25 miles away, south-east of Staraya Russa, the guns opened up along a six-mile front. They put down a heavy barrage. Stukas roared and screamed above the Russian lines. And then the regiments of General von Seydlitz-Kurzbach's Corps Group charged ahead—just as in the old days of the summer offensive. "Operation Bridge-building," the German offensive to relieve II Corps in the Demyansk pocket, had begun. The ring around Count Brockdorff's Corps had been closed for forty-one days. Only 25 miles separated "the county" from the German main fighting-line. The encircled six divisions had to defend a front line of nearly 190 miles: they were clearly not strong enough to hold a continuous line everywhere, and many sectors were defended only by intermittent strongpoints.

Apart from their numerical weakness, the defenders were suffering from the exceedingly tight food situation. The 96,000 men and roughly 20,000 horses had to be supplied by air. Rations had been reduced nearly by half.

Obviously the supply planes could not carry any hay or straw for the horses. Thus, in spite of the ingenuity of their attendants, the animals were getting thinner and thinner. The rotten straw of the wrecked peasant shacks was no adequate substitute for fodder. True, the animals were supplied with tree bark, pine branches and needles, reeds, and beans, but this did not assuage their hunger. They ate the sand and died of sand colic. They went down with mange, founder, and other diseases. The veterinary surgeons fought for every animal's life, but frequently the *coup de grâce* was all they could prescribe. Thus the horses performed their last service in the field kitchens. The Russian civilian population inside the pocket used to come and collect the bones and entrails. Nothing was left, except the hooves.

All this was now to end. General von Seydlitz-Kurzbach was mounting his attack from Staraya Russa, in the strength of four divisions, in order to blast a corridor through to the Demyansk pocket and reunite the cut-off divisions with the main fighting-line.

Meanwhile in the western salient of the pocket the "Corps Group Zorn" had been formed: at the appropriate moment it was to launch a break-out—"Operation Gangway"—in order to meet Seydlitz's divisions half-way. The spearhead of this operation was to be provided by Colonel Ilgen's regiment, which had been put together from various battalions of the surrounded divisions.

"What then is the general plan, Herr General, and when is zero hour?" Colonel Ilgen asked.

With his walking-stick General Zorn drew the outline of the Demyansk pocket in the snow. Left of it he drew an arc indicating the main front at Staraya Russa. "Count Brockdorff has informed me that Seydlitz's Group is mounting its attack from the Staraya Russa front with four divisions." And General Zorn added four arrows to his sketch-map in the snow.

"Here"—he indicated the two arrows in the middle—"are the Silesian 8th and the Württemberg 5th Jäger Divisions, carrying the main weight of the attack. Both these divisions were brought to the Eastern Front from France at the beginning of the year, and both have experience of active service. The Ulm Jägers, in particular, fought superbly well in the heavy battles at Staraya Russa at the beginning of February." Zorn pointed his stick to the right and to the left. "The flanks of the relief attack will be covered by 329th Infantry Division on the right and 122nd Infantry Division on the left. Seydlitz's Jäger Divisions will be aiming directly at this point, the westernmost point of our pocket, at Kalitkino and Vasilkovo. The moment he gets to the Lovat crossing at Ramushevo, along the Staraya Russa–Demyansk road—in other words, when he is still 8 miles away—we attack. Your task, Ilgen, will be to tear open the Russian positions outside our pocket and to reach the Lovat at Ramushevo."

Ilgen nodded. The snow was sparkling under the morning sun. From afar came the rumble of guns. An orderly came running from the building: "Telephone, Herr General."

At first everything went according to plan. After the preliminary artillery bombardment and the concentrated use of Stukas, Seydlitz's offensive moved smoothly, just as during the first few weeks of the Blitzkrieg. Presently, however, the difficulties began. The wintry forest and scrub east of Staraya Russa slowed down the momentum of the advance. Russian defences in depth, consisting of five systems of positions, had to be pierced. Progress was only a step at a time—and that required courage and cunning, and blood and tears. No quarter was given. The fighting continued for four weeks. It began in a temperature of 30 degrees below, over swamps frozen as hard as stone. A few days later the thermometer rose to freezing-point. The thaw came. Everything sank into the morass.

By the end of March the temperature was down again to 20 degrees below. In daytime there were heavy snowfalls, and at night the swamps and forests were swept by icy spring gales which instantly froze any living thing that had not sought shelter in a cave, in a hut, in a hole in the ground, or under hurriedly felled tree-trunks.

In April the weather broke for good. Snow and ice melted. The water was knee-deep on the roads. The men waded waist-deep through the icy swamps and marshes. Rafts had to be built for the heavy machine-guns from branches of trees and bushes, or otherwise they would disappear in the mud.

The wounded had to be laid on stretchers made of branches, or else they would have drowned. Anything that weighed anything—rifles, horses, and men—sank into the swamp. The men's uniforms were sodden. And in the thick scrub the enemy was lurking. The Russians too suffered greatly from the mud. Their heavy tanks were unable to intervene, and their artillery was immobilized.

On 12th April Seydlitz's spearheads caught sight of the shattered towers of Ramushevo, rising like a mirage through the haze and smoke. They had reached their objective. "Operation Bridgebuilding," they knew, would stand or fall with the possession of Ramushevo. For this small town controlled the road and the crossing over the Lovat, which, now the spring thaw had set in, was once more a major obstacle.

The following day Colonel Ilgen was severely wounded while reconnoitring in preparation for his attack. His regiment, for whose attack he had made the most painstaking preparations, was taken over by Lieutenant-Colonel von Borries. The attack was mounted at dawn on 14th April.

Six days later, on 20th April, the battalions reached the first houses of Ramushevo on the eastern bank of the Lovat at nightfall. The forward units consisted of the reinforced SS Panzerjäger Battalion of the "Death's Head" Regiment under Hauptsturmführer[1] Bockman.

On the far bank the western part of the town was blazing. Tracer bullets streaked through the night. The noise of battle came over the river. The river was in full spate and more than a thousand yards wide. It was impossible to see what was happening on the far side: the smoke, the dust, and the glare of the fires reduced visibility to nil. It was the same throughout the next day.

Seydlitz's companies were fighting furiously for a stretch of riverbank. At nightfall Borries's men caught sight of figures with German steel helmets waving from the other side. "They're here! They're here!"

[1] Rank in Waffen SS equivalent to major.

Assault on the Valday Hills 405

The time was 1830 hours on 21st April 1942. Only the turbulent Lovat now separated the corridor which would reunite the encircled regiments with the main fighting-line. Demyansk, the powerful breakwater on the Valday Hills, had performed its duty. For several months six German divisions had barred the way to the Soviet Armies. Now they were once more part of a continuous front.

And what was the situation at Kholm, 55 miles farther south?

For the past hundred days General Scherer's combat group with about 5000 men had been holding the road junction in the middle of the vast swampy area, the strongpoint and crossing on the upper Lovat, commanding both the river and the hinterland. Kholm, the only solid point in the torn front line between Velikiye Luki and Demyansk, the bolt on the back-door to the Sixteenth Army, was halting the Soviet drive to the west just as Demyansk was holding the Soviet wedge driving to the south.

The small provincial town with its 12,000 inhabitants had become a front-line town overnight. Supply troops and scattered units of divisions of the line had been organized for its defence. Major-General Scherer, the commander of 281st Local Defence Division, had been appointed Fortress Commandant. His instructions were to hold Kholm at all costs.

Kholm was held. Its defence has gone down in military history as the story of a highly creditable performance, a story of courage, military improvisation, and soldierly bearing.

The Combat Group Scherer was a motley crowd. It consisted of units of 123rd Infantry Division, of the 218th Infantry Division, which had only just been transferred to the Eastern Front from Denmark, and of the 553rd Infantry Regiment, 329th Infantry Division. There were also Mountain Jägers from Carinthia and Styria organized as Commando 8, then the 3rd Battalion, 1st Luftwaffe Field Regiment, and the Reserve Police Battalion 65 of 285th Local Defence Division. There was even a naval motor-transport unit. From these groups of varying sizes Colonel Manitius, commanding 386th Infantry Regiment belonging to 218th Infantry Division, the Operations Commandant, moulded an efficient fighting force of which he himself was the soul and inspiration.

On 28th January Kholm was fully encircled. Parts of the Machine-gun Battalion 10 managed to get inside the pocket a little later, but behind them the trap sprang shut for good.

The fortress area measured barely a square mile, and later shrank to only about half a square mile. It was held by 5000 to 5500 men. In parts, the front of the pocket ran right through the middle of the town. The men knew every house, every ruin, every tree, and every bomb-crater between the northern cemetery, the hairpin bend, the

GPU prison, and the police-station ravine. These were the four most notorious strongpoints of the fortress. The town was being besieged by three Soviet rifle divisions, who charged day after day.

The defenders could be supplied only from the air. In a field outside the front of the pocket, in no-man's land, sappers built a makeshift airstrip measuring 70 by 25 yards. Every single landing was an adventure. And most of the JU-52s were damaged in the attempt. Before long the field was dotted with wrecks of aircraft. The Luftwaffe therefore switched over to sending personnel and heavy equipment by freight-carrying gliders and dropping foodstuffs and ammunition in containers.

Anxious minutes invariably followed the appearance of a JU or two with gliders from over the edge of the forest in the west. If the gliders were cast off only a few seconds too soon they would land among the Russians. Even if a glider touched down in the correct spot an ever-ready assault detachment had to secure the precious consignment as quickly as possible. For, needless to say, the Soviets were also lying in wait for these prizes. Frequently the two sides would race each other to the point where the freighter had crash-landed.

Eighty freight gliders landed in the Kholm pocket. Twenty-seven JUs were lost in the supply runs. But being entirely supplied by air was not the most typical or most unusual thing about Kholm. Far stranger still was the fact that Kholm was a fortress without artillery.

A few 8-cm. mortars, some 3·7-cm. anti-tank guns, and one 5-cm., as well as two light infantry guns—that was all the heavy equipment inside the pocket. There were no heavy guns or howitzers. How then was the fortress able to hold out against a powerful enemy attacking it with guns and tanks? How could a fortress hold out even for a few days without its most vital weapon—the fortress artillery?

In Kholm this problem was solved in a manner that is probably unique in military history. The fortress artillery was emplaced outside the fortress, but its fire was directed from within.

The guns which pounded the Soviet concentrations by day and night, the artillery barrages which came down to protect the German lines whenever the enemy charged—all these came from heavy batteries standing at the end of a narrow corridor which General von Uckermann's combat group had driven to within 6 miles of Kholm, straight through enemy country.

Against all the rules of military practice these batteries of 218th Artillery Regiment and Heavy Artillery Battalion 536 were emplaced at the end of this corridor, as if on a proffered dish, firing for all they were worth.

In Kholm itself First Lieutenant Feist and Second Lieutenant Dettmann acted as forward observers and directed the fire by manually keyed telegraphy. On some days more than 1000 heavy-calibre shells would roar overhead, across Kholm, into the Russian positions or into charging enemy formations.

The forward observers in Kholm, the artillery officers and their signallers, gradually developed such skill that even individual attacking Russian tanks were shelled and knocked out by direct hits.

The Soviets were determined to take Kholm before the onset of the thaw. They wanted to get past this accursed barrier which was holding up an entire Army. On some days, therefore, they launched as many as eight attacks. They would make penetrations. They would be thrown back again in fierce hand-to-hand fighting. They would come again. They would occupy a ruin here, a snowdrift there, or a bomb-crater. An immediate counter-attack would be launched with hand-grenades and flame-throwers. Thus it went on day after day. Among the German combat groups there were specialists in the lobbing of hand-grenades who could hit pin-point targets, and there were tank-wreckers who performed their task with the unruffled expertise of spider-men working on tall buildings.

The incipient thaw with its mud and slush imposed a temporary halt on the Soviet attacks, but it also made life inside the pocket unbearable. In the cellars of the ruined houses 1500 wounded were lying on the bare floors; a few lucky ones were bedded on planks. Their lot was far worse than that of the 700 wounded who were flown out of the pocket.

Doctors and medical orderlies under the command of Dr Ocker worked until they dropped. Dr Huck, the surgeon, risked the boldest operations in an attempt to save the lives of badly wounded men. But the dirt and the lice were almost more dangerous enemies than the Russian shells and exacted a heavy toll. The situation is illustrated by two figures: there were 2200 wounded and 1550 dead.

On 12th March Second Lieutenant Hofstetter made the following entry in his diary: "The first case of typhus." There it was, the spectre of all besieged fortresses—typhus.

Vaccine was dropped. Medical personnel and doctors spent every free minute inoculating the men. War was declared on the lice. The race against time began. Then came the news: hold on—divisions under General von Arnim are on the way to get you out. But on 1st May it seemed as though all the heroism and all the sacrifices had been in vain. The Russians mounted their full-scale attack. First an artillery barrage—then the tanks. And after them the infantry. "Urra!"

The eastern end of the front snapped. The Russians were within a hundred yards of the Lovat. If they succeeded in covering these hundred yards that would be the end, for they would hold the higher river-bank and would be able to crack the fortress from inside. But they did not manage it. They were stopped by the Stukas, by Uckermann's guns, and by the sangfroid of Scherer's men.

One of these men was Sergeant Behle with his 5-cm. anti-tank gun at the southern cemetery. His optical sight was out of action. Five Soviet tanks were approaching. Behle aimed by looking down the barrel. He shoved in the shell. He slammed the breech shut. He fired. A loud crash—a direct hit. Behle fired twenty rounds. Not one of the five tanks got past him. The fifth was stopped only 40 yards in front of his gun.

On the airstrip Sergeant Bock was lurking behind cover with his anti-tank rifle. Standing, his barrel supported on a piece of masonry, he knocked out four Russian light tanks.

All was quiet on 2nd May. But on 3rd May action began again at 0300 hours. It was a rainy, hazy day. The German aircraft could not take off. But the men in the pocket could hear the approaching noise of battle from the south-west. "Those must be our chaps," they called out to each other. The hope lent them new strength. They must not give in now.

On 4th May they saw German Stukas dropping their bombs just outside their pocket. They were blazing a trail for the relieving units.

On 5th May the weather was again hazy. A runner came bursting into Captain Waldow's command post. "Herr Rittmeister, two German assault guns!" A moment later the soldiers could hear the clank of the tracks. Doubled forward, the grenadiers were advancing alongside the steel monsters camouflaged with twigs and branches. "Are they really our men? Or is this a Russian trick? Careful!" But now they were here, the steel-helmeted men and the self-propelled guns of the "Greif" Battalion under Lieutenant Tornau and the sappers of Lieutenant-Colonel Tromm's 411th Infantry Regiment, a unit of 122nd Infantry Division. Second Lieutenant Dettmann, describing the scene, says, "The men were received with wonderment, like visitors from another planet."

The relief came at the eleventh hour. There were only 1200 men left in the firing-pits, the trenches, and the ruins. Some 1500 wounded were cowering in their miserable quarters. About the same number of dead lay buried between the positions. Lieutenant-Colonel Tromm joined them at the last moment: he was hit by a Soviet shell.

Kholm was once more part of the German front, part of the main fighting-line in the stabilized area south of Lake Ilmen. From then onward that line held until 1944.

5. General Vlasov

A crack Soviet Army in the swamp–The wooden road across the clearing Erika–A merciless battle–Engineers Battalion 158–Break-out from hell–The disaster to the Soviet Second Striking Army–"Don't shoot; I am General Vlasov"–Maps buried under a river.

WHAT had happened in the meantime on the Volkhov, where Vlasov's Second Striking Army had been cut off in the forests? Here, too, spring had come. The snow had melted away, and the ice on rivers and swamps had thawed. In the dug-outs and trenches the water stood waist-high. In the dead forests thousands of millions of midges woke to new life. Where columns on sledges and skis had darted swiftly only a short while before there were now water-courses and throbbing swamps. And in the very midst of this hell was General Vlasov with his fourteen Soviet rifle divisions, his three cavalry divisions, his seven rifle brigades and one armoured brigade—an Army in the swamp.

Vlasov was an energetic general. On 27th March he had burst open the German barrier in the clearing Erika by striking at it with Siberian assault brigades and tanks from the west. Admittedly, the gap they punched was only a mile wide, but nevertheless it was a gap through which the encircled units could be supplied. In vain did the battalions of the German 58th Infantry Division and the SS Police Division try to drive Vlasov's Siberians out of the clearing. They did not succeed. They were too weak themselves, and the swampy and wooded ground on both sides of the clearing Erika was too difficult to allow the bringing up of sufficiently strong formations to seal off the Soviet pocket completely. As a result, the 58th Infantry Division and the group holding the area north of it had to stand up to continuous Soviet attacks throughout six difficult weeks.

At last, at the beginning of May 1942, a second, carefully prepared attack by the 58th Infantry Division, which had by then been reinforced, resulted in success—*i.e.*, in a solid link-up with the Police Division employed to the north of the clearing Erika.

At that stage Vlasov decided to break out of the hell of the Volkhov swamps.

But by then his regiments were no longer able to cross frozen marshes or make their way through the thick forests. The marshy forest floor and the swamps forced them to keep to roads and paths. And there was only one path open to them—the wooden causeway across the clearing Erika.

On 20th May General of Cavalry Lindemann issued an Order of the Day to his Eighteenth Army. It opened with the words: "The Russians are pulling out of the Volkhov pocket." It was like a signal for the German men fighting on the Volkhov. On that 20th May they once more sealed the gap across the clearing Erika. Throughout the heavy fighting of these months the infantrymen, the gunners, and Panzerjägers were helped by the sappers of Engineers Battalion 158 under Captain Heinz. Day and night, under the most difficult conditions, the men were in action. Their casualties were heavy. At the end of the battle, when the gallant battalion was pulled out, the combat strength of its three sapper companies was down to three officers, three NCOs, and thirty-three men. The commander of the 2nd Company Engineers Battalion 158, Second Lieutenant Duncker, a parish priest in civilian life, was awarded the Knights Cross in recognition of the part he played in the successful fighting at the clearing Erika.

By the end of May 1942 the savage battle on the Volkhov had been won by the German troops. Those units of Vlasov's Army which had not succeeded in slipping out were now irretrievably caught in the trap. They were nine rifle divisions, six rifle brigades, and units of an armoured brigade. The end of the Soviet Second Striking Army was at hand.

That end was frightful. Only 32,000 men survived the battle, and they were taken prisoner. Several tens of thousands were lying in the forests and swamps—drowned, starved to death, bled to death. It was an appalling field of slaughter. Huge swarms of flies buzzed over the marshes and over the corpses which were sticking out of the swamp. A terrible stench hung over the clearing. It was hell itself.

Through that hell roamed General Andrey Andreyevich Vlasov with his staff. The Germans were after him. Suddenly the general had vanished. Where was he? Had he been killed? Had he shot himself? Or was he hiding out?

Vlasov's description, complete with photograph, was dropped by German aircraft in many thousands of leaflets over the villages of the Volkhov pocket. High rewards and special leave were promised for his capture. Naturally, from that moment onward reports were coming in every day: Vlasov has been seen; Vlasov has been found dead; Vlasov has been taken prisoner. On being followed up the reports invariably proved to be based on mistakes, empty boasts, and misunderstandings.

On 11th July yet another report arrived at XXVIII Corps to the effect that Vlasov had been found dead. Captain von Schwerdtner, the Intelligence Officer, set out at once. He found a dead officer covered with a general's greatcoat. He was about six foot two tall—about Vlasov's height.

But it was no longer possible to identify the body or to establish any similarity. Schwerdtner gave orders for the body to be taken to Corps and himself drove back.

In the next village the Russian headman stopped him and reported, "I've got a man locked up in my shed who looks like a partisan. There's a woman with him too, maybe an agent. Would you like to see them?"

Schwerdtner told him to lead him to the shed. The village headman unlocked it. Schwerdtner's interpreter and escort party held their machine pistols at the ready.

"*Vykhodi!*" the Russian shouted. "Come out!"

Out of the dark into the blinding light of day stepped a giant of a man, covered in dirt, with a beard, in an officer's tunic with leather shoulder-strap and muddy leather boots. He blinked his eyes behind the thick black horn-rimmed glasses. He saw the machine pistols, raised his hands, and said in broken German, "Don't shoot; I am General Vlasov."

The sun was high in the sky. The flies were buzzing; otherwise there was complete silence.

Out of that shed in a Volkhov village history pushed a man—one of the best produced by Bolshevik Russia—a man whom the carnage and the corpses on the Volkhov had turned into a mortal enemy of Stalin. Vlasov was Russia. If anyone could defeat Stalin it was he.

The battle in the forests by the Volkhov was one of the most frightful battles ever. The mere fact that one of the best and politically most reliable Soviet generals emerged from it as an opponent of Stalin and Bolshevism merely confirms the horror of the hell through which Vlasov's Second Striking Army had passed. From that date onward Vlasov remained a political factor of considerable importance in the background of the duel between Hitler and Stalin.

But the battle also yielded another prize of supreme military importance, even though perhaps less spectacular and known at the time only to a few experts. Preliminary interrogation of captured staff officers had revealed that the Soviet offensive on the Volkhov was superbly equipped in every respect—and that included the map material of a large cartographic office specially set up for this offensive. But where were the maps? The vast battlefield was closely searched, but no trace was found.

Eventually a second lieutenant was tracked down who had been on the staff of the cartographic office. The lieutenant talked. He led the German experts to a small river and told them to divert the water at a certain point. And there, buried in the river-bed, were the maps of the Soviet cartographic office. Just as the Western Goths once buried their King Alaric under a river, so the Soviet head of the cartographic office had hidden three lorry-loads of exceedingly valuable maps in the river-bed and then ordered the waters to be led back to their original course again. It was the most important cartographic find made by the German forces in the whole war. The cache contained Russian maps from the western frontier of the Soviet Union to well beyond the Urals. The prize was sent to Berlin, and before long the troops on all fronts were supplied with the latest Soviet maps.

PART FIVE:
The Ports on the Arctic Ocean

1. "Operation Platinum Fox"

The Murmansk railway–Offensive on the edge of the world–General Dietl reaches out for Murmansk–Across the Titovka and Litsa–No roads in the tundra–An error costs the Finns their victory–Mountain Jägers in the Litsa bridgehead.

THE very first drafts for "Operation Barbarossa" list a surprising objective—Murmansk. This little-known place was named alongside the great strategic objectives like Moscow, Leningrad, Kiev, and Rostov. What was so important about Murmansk? It was a port and a railway station on the wind-swept roof-top of Europe, on the Arctic Ocean north of the Arctic Circle, in the same latitude as the vast glaciers of Greenland, some 600 miles away from civilization.

In the summer of 1941 Murmansk had 100,000 inhabitants. For three months of the year there was scorching summer, and for eight months there was deep winter and polar night. All around was desolate tundra, without a tree or a shrub. Why then was this godforsaken town listed alongside the great objectives in the secret drafts for "Operation Barbarossa"? Why was Murmansk named in the same breath as the capital of the Communist empire, or Leningrad, or the industrial Donets region, or the Ukrainian grain area, or the Caucasian oilfields —all of them objectives aimed at by entire Army Groups, Air Fleets, and Panzer Armies, and considered worthy of the most savage battles in history?

"Under every sleeper of the Murmansk railway a German lies buried," the Lapps used to say. Like all legends, this one should not be taken too literally—although it is not so very far from the truth.

Between 1915 and 1917 some 70,000 German and Austrian prisoners-of-war were employed in these virgin forests, swamps, and Arctic tundra between St Petersburg and Murmansk in the building of a railway originally started in 1914 with convict labour. The hardships of the prisoners-of-war defied description. During the short scorching summer they were mown down by typhoid, and during the eight months of the Arctic winter they were killed by cold and hunger. Within twenty-four months 25,000 men died. Every mile of the 850-mile-long line cost twenty-nine dead.

When Adolf Hitler received General of Mountain Troops Eduard Dietl at the Reich Chancellory in Berlin on 21st April 1941 he did not show him the balance-sheet of the lives lost in the construction of the Murmansk railway, but calculations showing the number of freight trains carrying goods, armaments, and troops along the Kirov railway—as the Soviets had named the line—between Moscow and the Arctic Ocean.

General Dietl, the hero of Narvik, the general commanding the "Mountain Corps Norway," had known about Directive No. 21, "Operation Barbarossa," since the end of December. Like most of the generals, he too had been taken aback when he first saw the secret paper. But being an obedient soldier he had got down to work to prepare for his tasks for D-Day. These tasks, according to the directive, were as follows: "The 'Mountain Corps Norway' will firstly secure the Petsamo[1] area with its ore-mines, as well as the Arctic Ocean road, and subsequently, in conjunction with Finnish forces, advance to the Murmansk railway and cut off overland supplies to the Murmansk area."

For three and a half months Dietl and his four staff officers, whom he had taken into his confidence, had been working on these tasks. Now, on 21st April, one day after the celebration of his fifty-second birthday, Hitler was anxious to know how the plans for this part of the operation were progressing. At that time neither he nor Dietl had any inkling of the importance the Murmansk railway was to gain for the Soviet war economy in later years. They did not suspect that American convoys would sail into the Arctic Ocean once Germany was at war with Russia, to unload their supplies of military aid at Murmansk.

To Hitler at that time the railway was a line of communications along which Stalin was rapidly able to switch major contingents of troops, artillery, aircraft, and tanks from Central Russia to the Soviet-Finnish frontier on the Arctic Ocean in order to snatch from Germany the vital nickel-mines of Petsamo and the ores of Narvik.

That was Hitler's nightmare. That there was another, far greater, possibly decisive danger lurking behind the Murmansk railway he did not see at the time—or certainly not in its full implications. Yet he, or any of his strategists, should have been able to predict it.

When the Tsar hurried the construction of the railway in the First World War he did so not in order to conquer Norway or seize the nickel of Petsamo, but in order to put to use the only ice-free port of

[1] Now Pechenga.

his empire, the only port from which Russia had unrestricted contact with the world's oceans. In the whole of the giant Russian empire Murmansk is the only ice-free port with free access to the Atlantic.

True, Archangel on the White Sea also has a port which communicates with the open sea, but although it is situated farther south than Murmansk, it is closed by ice for three months in the year. And Vladivostok, the "ruler of the east," as its name suggests, is likewise subject to freezing up for about a hundred days. Besides, it is situated at Russia's back-door, and 4350 miles away by rail from European Russia. The ports on the Black Sea are blocked by the Bosphorus, and those on the Baltic by the strait between Denmark and Sweden. Murmansk therefore is Russia's only open gateway to the world. The remote town owes its importance to a freak of nature—the Gulf Stream. Some of its warm waters wash through the 750-mile-wide gap between Greenland and Norway, where the Arctic and Atlantic Oceans meet. These warm water masses from the Gulf Stream prevent the Norwegian fjords from freezing up, and the very last scrap of warmth from the sun-drenched Gulf of Mexico makes sure, before being swallowed up by the Arctic Ocean, that the Kola Bay does not freeze up even in the severest Arctic winters with temperatures of 40 or 50 degrees below zero Centigrade.

That was why the Tsar built a railway from St Petersburg to the fishing village of Murmansk. And in 1917, when America entered the war against Germany on the side of Russia, the Murmansk railway became the shortest and the most important all-the-year-round supply-line between the USA and Russia.

The map-room at the Reich Chancellory was flooded with April sunshine. The large windows into the garden were open. From this very room the sovereigns of the old Europe had once gazed out on the greenery of the fine old trees. For Hitler's map-room in the old Reich Chancellory was the same salon where in 1878 Bismarck's Congress of Berlin was sitting to curb Russia's hegemony in the Balkan Peninsula.

As General Dietl entered the room on 21st April General Jodl had just submitted to Hitler the draft of the High Command communiqué. Hitler was wearing his old-fashioned nickel-gilt glasses. He read through the text and made one or two alterations. Victories—everywhere victories. In Greece the German divisions were in headlong advance to the south via Larissa; at Metsovon mountain troops were crossing the Pindus Mountains, chasing the retreating British. In North Africa Rommel's regiments had breached the Tobruk defences at Ras-el-Madaur and were now fighting at the Halfaya Pass, half-way to Cairo. Three days earlier the Yugoslav Army had surrendered. Yugoslavia had been over-run in only eleven days. In Greece the end was imminent. Nothing was impossible for the German soldier!

Hitler took off his spectacles and welcomed General Dietl. He was fond of this plain Bavarian hero of the Mountain Troops, the popular conqueror of Narvik. Major Engel, the High Command ADC, spread out a 1:1,000,000 map of the Finnish–Norwegian area.

"Have you made good progress with your preparations?" Hitler asked. "We haven't got much time left." Without waiting for a reply he walked over to the map-table, put his glasses on again, and bent over the map. With complete confidence, as though he had never done anything else but plan major military operations, he began to lecture:

"Murmansk is the most dangerous deployment centre of the Russians in the extreme north. The harbour and railway have a considerable capacity, and the town and its airfields are probably held in strength. It would take Stalin only a very short time to dispatch a few additional divisions to Murmansk and mount an attack against the West. Murmansk hasn't been extended for nothing. In 1920 it was a dump with 2600 inhabitants, and to-day it has 100,000. Our aerial reconnaissance has revealed gigantic railway installations, enormous quays, factories, exit roads—in short, a modern fortified centre, a dangerous strongpoint in the thinly populated territory along the Arctic Ocean."

Hitler had warmed to his subject. He placed the index finger of his right hand at Murmansk and that of his left on Petsamo. "The distance to the nickel-mines is only 60 miles."

He stabbed another point on the map. "And from Petsamo to Kirkenes on the Varanger Fjord is only another 30 miles. To have the Russians in this area would be disastrous. Not only should we lose the nickel ore which is indispensable to our steel manufacture, but it would also be a heavy strategic blow to the whole of our Eastern campaign. The Russians would be on the Arctic Ocean road, the transport lifeline of Northern Finland. It leads deep into the rear of the Finnish front and right to Sweden's back-door. To have the Russians on the Varanger Fjord would mean a most serious threat to the Arctic Ocean and to our ports in Northern Norway."

Hitler straightened up, took off his glasses, and looked at Dietl. "All that depends on your Mountain Corps, Dietl. We must eliminate this danger at the very beginning of our Eastern campaign. Not by waiting, but by attacking. You've got to manage those ridiculous 60 miles from Petsamo to Murmansk with your Mountain Jägers, and thus put an end to the threat."

"Ridiculous 60 miles," had been Hitler's words. They are fully attested. But then who can blame him for his optimism, considering the High Command communiqué he had just signed?

General Eduard Dietl was amazed, not for the first time, at the way this former corporal managed to outline grand strategic designs and operational problems. But he was far from happy. To him the 60 miles from Petsamo to Murmansk did not look ridiculous at all. In his forthright way he voiced his opinion to Hitler. Step by step he took him through the results of the research work done by his staff officers.

"My Fuehrer," he said in his engagingly simple manner, "the landscape up there in the tundra outside Murmansk is just as it was after the Creation. There's not a tree, not a shrub, not a human settlement. No roads and no paths. Nothing but rock and scree. There are countless torrents, lakes, and fast-flowing rivers with rapids and waterfalls. In summer there's swamp—and in winter there's ice, snow, and it's 40 to 50 degrees below. Icy gales rage throughout the eight months of Arctic night. This 60 miles of tundra belt surrounding Murmansk like a protective armour is one big wilderness. War has never before been waged in this tundra, since the pathless stony desert is virtually impenetrable for formations. Unless, of course, roads are first built, or at least cart-tracks, so that the men and—what's more difficult—the beasts of burden can be kept supplied. But if I am to do all that with my own forces, then it must be at the expense of my fighting formations —and my two mountain divisions are not all that well equipped technically as it is. In fact, my officers use the term 'economy kit.' We haven't got enough tractors, we haven't got enough mules, we haven't got enough mobile artillery, and each division has only two regiments."

Anyone else talking to Hitler like that would have fared badly. But Dietl could afford to do so. He argued soberly and without bombast, and now and again interlarded his language with Bavarian sayings. The purpose of his arguments—which had been supplied to him by his efficient chief of staff, Lieutenant-Colonel von Le Suire—was to get Hitler to drop his idea of attacking the town and fortress of Murmansk and persuade him to defend the strategically and economically vital Petsamo area instead. In Dietl's opinion the Murmansk railway should be cut farther south, in country more favourable for military operations.

"It may well be that the Russians will attack," Dietl continued his line of thought, his finger moving on the map from Murmansk to Petsamo. "For them an attack is easier than for us. Their supply base lies directly behind their front, and their railway runs practically into the battle zone, whereas we have to bring up every shell, every loaf of bread, every bundle of hay, and every sack of oats either by the enormous sea-route from Hamburg and the Baltic ports, via Kirkenes, or from Rovaniemi along the 375-mile-long Arctic Ocean road to Petsamo, and from there first by lorry, then by horse-drawn cart, then by mules, and eventually by human carriers. But if we can cut the Russians' railway at any point at all they will be just as badly off as ourselves."

Hitler was much impressed by Dietl's exposition. He realized that the elimination of Murmansk did not necessarily entail a direct attack. Its lifeline could be severed at any other point of the 850-mile railway. In that case its terminus at Murmansk would simply wither away. And the fine ice-free port would become worthless because it would have lost its rearward exit.

"Leave me your papers," Hitler said thoughtfully. "I'll think it over." The matter was left open when General Dietl took his leave. Hopefully he reported the outcome of his interview to his staff.

Three weeks later, on 7th May 1941, Hitler's decision came by courier via the Army Commander-in-Chief Norway, Colonel-General von Falkenhorst. The decision was neither one thing nor another, but a poor compromise. Hitler was ordering the Army in Norway, now put in charge also of operations in Northern Finland, to attack the Murmansk railway at three points: Dietl's Mountain Corps was to move with two divisions from Petsamo against the town and port of Murmansk; XXXVI Corps was to strike at the same time with two infantry divisions via Salla towards Kandalaksha, some 220 miles farther south, and cut the railway there. Finally, another 93 miles farther south, the Finnish III Corps was to advance via Kestenga towards Loukhi, also with two divisions, and seize the railway there. Six divisions were being employed at three different points.

The main effort was to be made by Dietl's Austrian Mountain Jägers of the 2nd and 3rd Mountain Divisions. On the day the campaign began they would have to cross the Finnish frontier from Kirkenes in Norway and occupy the Petsamo area. Seven days later "Operation Platinum Fox" was to start—the attack through the tundra against the town and harbour of Murmansk.

It is not known who persuaded Hitler to disregard Dietl's weighty arguments. The only lasting effect they produced was the dispatch of twelve efficient and terribly hardworking detachments of the Reich Labour Service into Dietl's zone of operations. They were the Reich Labour Service groups K363 and K376 under the command of Chief Labour Leader Welser. The courier who brought Dietl Hitler's orders on 7th May also brought with him the necessary maps. On these maps things no longer looked quite so bad. Only a small border strip of the zone of operations was shown as lacking all roads or tracks. A few miles inside the country, however, roads and tracks began to be marked—one from the bridge over the Titovka frontier river over to the Litsa, and another farther south from Lake Chapr to Motovskiy; indeed, from there there was another road running north again to Zapadnaya Litsa. All these roads connected with the main roads leading into Murmansk. Things were looking more hopeful.

The date was 22nd June. The time was 0200 hours, but the sun, veiled by a light haze, hung over the horizon like a large pale full moon, steeping the country in a watery light. Right across the continent, from the Baltic to the Black Sea 3,000,000 soldiers along a 1250-mile front were waiting at that moment for the order to start a great war. But up north, before Murmansk, under the midnight sun, the element of surprise had to be waived from the start. There was Finnish territory between the German troops' starting-lines in Norway and the Soviet frontier. In Petsamo, moreover, there was a Soviet Consul. He would have noticed any occupation prior to 22nd June and would have alerted Moscow. The surprise element of the whole Operation Barbarossa might have been jeopardized.

For that reason, with the consent of the Finns, a single company of sappers had crossed into Finnish territory during the night of 20th/21st June, in small groups and wearing civilian clothes, in order to prepare for the crossing of the Petsamo river.

In outward appearances the Finns acted most correctly. The Finnish frontier guards waited with bureaucratic pedantry for the hands of their watches to move to 0231 hours eastern time. All right now: the war against Russia had started a minute before. The barrier went up. The men from Styria, the Tyrol, and Salzburg moved off— into their great adventure north of the Arctic Circle.

By 24th June the ground had been reconnoitred as far as the frontier.

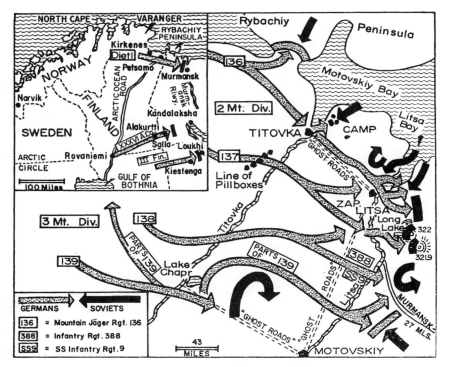

Map 24. Outline sketch of the front in the Far North and situation map of the operations of 2nd and 3rd Mountain Divisions of the Mountain Corps Norway between 29th June and 18th September 1941.

Local Finnish guides conducted the German patrols across rock and scree, over glistening red granite, through streams and snow-drifts.

The first major obstacle was the Titovka, an icy mountain river. Near its estuary, on its eastern bank, close by the little town of the same name, the Soviets had an Army camp with units of a NKVD frontier regiment. Finnish scouts had also established the existence of an airfield.

A particular problem was the Rybachiy Peninsula—or Fishermen's Peninsula. It was not known whether it was held in strength. Major-General Schlemmer was instructed to cut it off quickly at its narrow neck with units of his 2nd Mountain Division, in order to protect the Corps' flank against surprise attack. Simultaneously, battalions of his 136th Mountain Jäger Regiment were to take the Titovka bridge hard by the river's mouth into the fjord.

At first all went well. The 136th sealed off the Rybachiy Peninsula. They captured the bridge and crossed the river. The Army camp and the airfield were deserted. The battalions of the 137th Mountain Jäger Regiment found things more difficult. They came up against a well-prepared Soviet line of defences along the frontiers. Fortunately for them fog descended. While preventing the German artillery and Stukas from supporting the infantry in their attack on the pillboxes, it also enabled them to break through the positions without appreciable casualties. The pillboxes were bypassed, to be subsequently subdued with Stuka and AA gun support.

The resistance offered by the Siberian and Mongolian pillbox crews was a foretaste of what the attackers might come up against later. The defenders did not yield an inch. Even flame-throwers did not induce them to surrender. They fought until they were shot dead, beaten dead, or burnt to death. Only a hundred prisoners were taken.

There was little Soviet air activity. The Russians had left their hundred Rata machines unprotected and uncamouflaged on their two airfields near Murmansk, even after 22nd June. Attacks by a German bomber squadron on the airfields resulted in the destruction of most of the Soviet fighters.

In the evening of 30th June the most forward units of Major-General Schlemmer's 2nd Mountain Division were on the Litsa river. The regiments of Major-General Kreysing's 3rd Mountain Division laboriously struggled forward past Lake Chapr, looking for the road to Motovskiy which was shown on their map. If all went well they should presently link up with the 1st Company of Major von Burstin's Special Purpose Tank Battalion 40, which was equipped with captured French tanks, and advance to Murmansk along the new Russian road.

But everything did not go well. There was no trace of a road to be found. There was much excitement and to-ing and fro-ing of runners. A signal was sent to Corps headquarters. The Luftwaffe was ordered to investigate. Presently aerial reconnaissance confirmed that there was no road to Motovskiy—not even a track or mule-path. Very soon afterwards 2nd Mountain Division also realized that in its sector there was no road from the Titovka to Zapadnaya Litsa, and no road from there down to Motovskiy.

The German map analysts in the High Command of the Armed Forces had taken the conventional signs to mean the same as they did in Central Europe: they had interpreted the dotted double lines on the Russian maps as tracks. In actual fact these were the routes of telegraph-lines and the approximate movements of the tundra nomads, the Lapps, during the winter.

That was the end of the originally planned employment of the 3rd Mountain Division. Without a road there could be no advance. Admittedly, penetrations of 5 to 10 miles could be made into pathless country if necessary, but it was not possible to hold out, let alone advance, in such country unless cart-tracks were built for the most urgently needed supplies.

Everything therefore had to be regrouped. Without respite, the young Labour Service men, hardly more than boys, were building cart- and mule-tracks.

On 3rd July the 1st Battalion, 137th Mountain Regiment, reached the fishing village of Zapadnaya Litsa on the western bank of the Litsa river, just above its estuary. In inflated rubber dinghies the Jägers crossed the river. They came to an abandoned Army camp, where they found hard bread, groats, makhorka tobacco, and, unexpectedly, 150 lorries. There was great astonishment: where there were lorries a road could not be far off. A moment later a great cheer went up. Down along the valley floor ran a magnificent modern road— the road to Murmansk.

Anxiously the Jägers waited for supplies, for ammunition, for artillery. At last, on 6th July, the attack across the Litsa was mounted on a broad front. The 3rd Mountain Division crossed over in inflated dinghies. The sappers of Lieutenant-Colonel Klatt's Mountain Engineers Battalion 83 tirelessly paddled the dinghies from bank to bank. Now and again they had to pick up their carbines to ward off Russian attacks. The Soviets were shelling the crossing-point. Worse still, they were employing ground-attack aircraft. The German Luftwaffe was absent. The units of Fifth Air Fleet had been withdrawn in order to support the second prong of the attack against Salla and the Murmansk railway 250 miles farther south.

The road to Murmansk was almost within an arm's reach of the 138th Mountain Jäger Regiment. If only they had had a dozen Stukas, a dozen tanks, and some heavy guns they could have burst through the Soviet barrier. But as it was they failed. They were defeated by the terrain: the horse-drawn guns could not get through. The two mountain batteries which had got to the front were down to forty shells. Infantry support did not materialize either. Two-thirds of the division had to be employed on supplies, leaving only one-third to do the fighting. The Russians, on the other hand, brought up their reinforcements in long columns of lorries right to the battlefield. Battalion after battalion was unloaded and deployed for counter-attack to protect the road.

At this tense moment a further piece of alarming news arrived at Dietl's Corps headquarters in Titovka: the Russians were using naval units to land three battalions in the Litsa Bay in the flank and rear of 2nd Mountain Division. The landing was repulsed—but only at the cost of reducing the fighting strength of Major-General Kreysing's Mountain Jäger Regiments, which were by now greatly over-extended.

But the men from Styria and Carinthia did not give in. A flanking attack against the dominating high ground was to gain them breathing-space. With this *tour de force* Dietl wanted to gain access to the road. Meanwhile the German 6th Destroyer Flotilla under Lieutenant-Commander Schulze-Hinrichs was to hold the Soviet forces in the Litsa Bay in check. It was an excellent plan.

Detailed orders were sent down from Corps to the regiments. Those for 136th Regiment were carried by the regiment's motor-cycle orderly. But among the rocks and scree he missed regimental headquarters. The German sentry called out after him, yelled, and finally fired his rifle into the air to warn him. It was no use: the noise of the engine drowned it all. At six miles an hour, the dispatch-rider struggled forward until, suddenly, he found himself faced by Russians. He whipped his machine round. One of the Russians fired. The orderly was hit. Three Russians dragged him into a Soviet dugout. The Germans mounted an immediate counter-attack, but it came too late. The orderly and the Russians had gone. The plans for the attack were in Russian hands.

On 13th July Dietl tried another plan. A penetration was made into the Soviet positions, but it was not a break-through. The Soviets just did not budge from the strongly fortified commanding Hills 322 and 321.9 by the "Long Lake." Those infuriating hills were less than a thousand feet high, but the Germans could not take them. They lacked artillery, they lacked dive-bombers, they lacked reserves.

Headquarters personnel, Labour Service men, and the mule attendants worked ceaselessly and practically without sleep. To take one wounded man to the rearward dressing station two relays of four men each were needed, since he had to be carried for up to ten hours. Entire battalions were used up in this way.

In the evening of 17th July Dietl decided with a heavy heart to suspend his attack and go over to the defensive. He was only 28 miles from Murmansk. The chronicler to-day must shake his head in disbelief: why was a job which clearly required a steam-hammer tackled with a bare fist? Dietl, after all, had put Hitler in the picture. And Hitler himself had spoken so heatedly about the importance of Murmansk. Why then was the operation conducted with insufficient weight behind it? Why were three separate sectors of the front attacked with two divisions each, and why was the Luftwaffe switched first one way and then another, instead of all available ground and air forces being concentrated at one focal point?

The answer to the question is that the Finns had miscalculated and had badly advised the Germans. Field-Marshal Mannerheim's High Command had declared that for reasons of terrain it was not possible at any point of the Lapland Front to employ and supply more than two divisions. That had been the reason behind Hitler's plan to attack in three sectors with two divisions each. But the result was that no penetration was made at any of the three sectors.

The two divisions of XXXVI Corps under General of Cavalry Feige, the 169th Infantry Division and the SS Combat Group "North," which mounted their attack 250 miles south of Dietl on 1st July, with the objective of reaching the Murmansk railway at Kandalaksha, admittedly got as far as Alakurtti, having fought their way through Salla, 22 miles from their objective, but at that point their strength gave out, and there they got stuck.

Major-General Siilasvuo's Finnish III Corps with its 6th and 3rd Divisions similarly got no farther than Ukhta and Kestenga, and got stuck about 43 miles from the railway.

The Finns had advised the Germans badly. Their view had been based on their own military capabilities and their own equipment. But it is clear in retrospect that it would have been eminently possible to mount an operation with a pronounced and clear main effort either towards Murmansk itself or, better still, from Salla towards Kandalaksha, in which case the railway from Rovaniemi to the front would have been available. Admittedly, such an offensive in the strength of four to six divisions would have required revolutionary methods of supply, possibly supply by air during the attack, as well as the large-scale employment of road-building labour equipped with machinery.

But the German High Command was unwilling or unable to make an effort on such a scale. The importance of the objective was dimly realized, but the operations planned for its capture were nevertheless regarded as being in a secondary theatre of war. As for those "ridiculous 60 miles," the heroism of a crack force and the proved skill of an outstanding general were thought sufficient to cope with them.

The High Command would not admit that operations in the Arctic tundra were not possible in the way planned. Orders therefore came for another attempt to be made. On 8th September, the day when

General Hoepner's Panzer Divisions mounted their attack against Leningrad and when Guderian's divisions moved off to reduce the Kiev pocket, Dietl's Mountain Jägers once more grabbed the reins of their mules, picked up their ammunition-boxes, and put their shoulders to their mountain guns in another attempt to defeat the tundra and the Soviets and to capture Murmansk.

It had become clear in the meantime that in addition to their 14th and 52nd Rifle Divisions the Russians had moved further crack units into their defensive front. Nevertheless all the reinforcements the Mountain Corps received were two regiments—the 9th SS "Death's Head" Infantry Regiment and 388th Infantry Regiment. Neither had any experience of mountain warfare.

Things happened as they were bound to happen. The cleverly conceived flanking attack, after a promising beginning, ground to a halt against the last Soviet defences among the maze of lakes and patches of swamp outside Murmansk.

Unceasingly the Stukas whined overhead, bombing the Russian positions. Units of 3rd Mountain Division got as far as the new road to Murmansk. On the left wing regiments of 2nd Mountain Division dislodged the Soviet 58th Rifle Regiment from the high ground along the "Long Lake." Presently, however, came the Soviet counter-attacks, nourished time and again from their near-by supply bases. The Siberians attacked again and again. They were lurking behind boulders, leaping up out of caves in rocks and dips in the ground. They would collapse in the German fire, but more would come. Every step, every yard, of the advance took hours and demanded a high toll in dead and wounded.

On 19th September Dietl's regiments were compelled to withdraw behind the Litsa, that fateful river of the Arctic tundra. The third attempt to get across had failed. That accursed river had already cost the Germans 2211 dead, 7854 wounded, and 425 missing.

While the scorching sun was beating down in the Kiev pocket upon unending columns of 665,000 Soviet prisoners the first snow fell at Murmansk on 23rd September. The Arctic winter with permanent night and ice was beginning. It was a mere 30 miles to Murmansk—but in the Arctic winter night this was an infinite distance. Yet must not the attempt be made again—in spite of everything?

Day after day Murmansk was increasingly revealing its true significance. The cranes were busy on the piers. In all corners of the fjords lay ships with British and American names. The great stream of Western aid had begun to flow. And since Archangel was frozen up from November onward, supplies for the desperately fighting forces outside Moscow and Leningrad had to come via Murmansk. It was an endless stream, a stream which was not to cease again, but to grow in volume, a stream which ultimately decided the German-Russian war.

Here are a few figures to prove the point. During the first year of the Soviet aid programme the following supplies were delivered along the northern sea route alone—*i.e.*, through Murmansk and Archangel —in nineteen convoys:

3052 aircraft: Germany entered the war in the East with 1830 aircraft.

4048 tanks: the German forces on 22nd June 1941 had 3580 armoured fighting vehicles.

520,000 motor vehicles of all types: Germany had entered the war with altogether 600,000 vehicles.

That door on the Arctic Ocean was getting more and more dangerous every day. Must it not be closed?

2. Battle in the Arctic Night

From Athens to Lapland–1400 horses must die–The Petsamojoki river–Supply crisis–Nightmare trek along the Arctic Ocean road–The Soviet 10th Rifle Division celebrates the October Revolution–An anniversary attack –Fighting at Hand-grenade Rock–Convoy PQ 17–The Soviet 155th Rifle Division freezes to death–The front in the Far North becomes icebound.

D AMN this snow! Damn this whole country!"
The howling gale drowned the men's curses and carried them away into nothingness. Visibility was barely ten paces. For the past twenty-four hours an Arctic blizzard had been sweeping over the tundra, whipping fine powdered snow through the air and turning the half-light of the Arctic winter day into an icy inferno. They could feel the gale on their skins. They could feel it stinging their eyes like needles. It felt as if it was going right into their brains.

Hans Riederer stumbled. His rucksack rode up to the back of his neck. Was the gale mocking him?

In long single file they were trudging through the powdered snow, which did not get compressed under their boots, but slipped away like flour, offering no footholds. Suddenly, like a ghost, a heavily swathed sentry appeared in front of the marching column. He directed the company over to the right, down a small track branching off the Arctic Ocean road.

First Lieutenant Eichhorn was now able to make out the outlines of the bridge over the Petsamojoki, the bridge leading to the fighting-line, into the tundra, to the sector they were taking over. "Half right!" the lieutenant shouted behind him. The Jägers passed it on. The long file swung over to the right, to the edge of the road by the ramp of the bridge.

A column was coming over from the far side. Men heavily wrapped up. Most of them had beards. They were leaning over forward as they marched, weighed down by their heavy burdens.

"Who are you?" they called out over the storm.

"6th Mountain Division—come to relieve you," Eichhorn's men replied. The men on the bridge gave a tired wave.

"Haven't you come from Greece?"

"Yes."

"God, what a swap for you!" And on they moved. The fragments of a few curses were drowned by the blizzard. Like ghosts the men moved past Eichhorn's company. Then came four walking casualties, the dressings on their faces heavily encrusted, their hands in thick bandages. Immediately behind them six men were hauling an akja, a reindeer sledge. On it was a long bundle tied up in tarpaulin.

They stopped, beating their arms round their bodies. "Do you belong to the 6th?"

"Yes. And you?"

"138th Jäger Regiment." That meant they were part of 3rd Mountain Division.

The lance-corporal in front of the sledge noticed the officer's epaulettes on Eichhorn's greatcoat. He brought his hand to his cap and ordered his men: "Carry on!"

They moved on. On the sledge, trussed up in the sheet of tarpaulin, was their lieutenant. He had been killed five days before.

"He must have a proper grave," the lance-corporal had said. "We can't leave him here in this damned wilderness." So they had carried him down from the hill where they had been entrenched—down granite-strewn slopes, then through moss and past the five stunted birches to the first fir-tree. That was where their akja stood. They had now been hauling him for four hours. They had another two hours to go before they reached Parkkina with its military cemetery.

The date was 9th October. The day before they had finished their "Prince Eugene" bridge over the Petsamojoki. They had barely driven in the last nail before the Arctic blizzard began. It marked the beginning of the Arctic winter. Fifty hours later all transport to the front came to a standstill. The Arctic Ocean road was blocked by snowdrifts, and the newly built tracks in the forward area had vanished under deep snow.

In the forward lines the battalions of 2nd and 3rd Mountain Divisions had been waiting for the past ten days for their relief, for supplies, for ammunition and mail—also for a little tobacco and perhaps even a flask of brandy.

But ever since 28th September supplies had been getting through only in small driblets. This was due to a strange mishap which, in the afternoon of 28th September, completely destroyed the 100-yard-long wooden bridge over the Petsamojoki at Parkkina.

A few Soviet bombs had dropped on the river-bank just below the bridge. A few minutes later, as if pushed by some invisible giant hand, the whole bank had begun to move. About 4,000,000 cubic yards of soil started slipping. The 500-yard-wide shelf between river and Arctic Ocean road fell into the river-valley over a length of some 800 yards.

Entire patches of birch-woods were pushed into the river-bed. The waters of the Petsamojoki, a fair-sized river, piled up, overspilt their bank, and flooded the Arctic Ocean road.

The bridge at Parkkina was crushed by the masses of earth as if it were built of matchsticks. Telephone-poles along the road were snapped and, together with the wires, disappeared in the landslide. Abruptly the entire landscape was changed.

Worst of all, connection with the front across the river was severed. Urgent messages were sent to headquarters. There the officers gazed anxiously at the Arctic Ocean road, the lifeline of the front. Was it really cut?

What had happened? Had the Russians been up to some gigantic devilry? Nothing of the sort: the Soviets had just been very lucky. The landslide by the Petsamojoki was due to a curious geological phenomenon.

The river had carved its 25- to 30-feet-deep bed into a soft layer of clay which at one time had been sea-floor and therefore consisted of marine sediment. Following the raising of this layer through geological forces, these deposits hung along both sides of the river as a 500-yard-wide shelf of clay between the masses of granite.

When the half-dozen 500-lb. bombs—aimed at the bridge—hit the bank, one next to the other, the soft ground lost its adhesion and developed a huge crack almost 1000 yards long. The adjacent strata pressed on it, and like a gigantic bulldozer pushed the mass of earth into the river-valley, which was about 25 feet deep and 160 yards wide.

There is no other recorded case in military history of supplies for an entire front of two divisions being interrupted in so curious and dramatic a fashion. Suddenly 10,000 to 15,000 men, as well as 7000 horses and mules, were cut off from all rearward communications.

Major-General Schörner immediately made all units and headquarters staff of his 6th Mountain Division already in the area available for coping with this natural disaster. Sappers dug wide channels through the masses of earth which had slipped into the river-bed to allow the blocked water to flow away. In twelve hours of ceaseless night work, jointly with the headquarters personnel, supply drivers, and emergency units, they built a double foot-bridge from both banks. Columns of porters were organized; parties of a hundred men at a time, relieved every two hours, carried foodstuffs, fodder, ammunition, fuel, building materials, and charcoal from hurriedly organized stores on the western bank across to the eastern bank. They shifted 150 tons a day.

Simultaneously, the sappers of 6th Mountain Division started building a new bridge. Up there, on the edge of the world, even the construction of a bridge was an undertaking of almost unimaginable difficulty and hazard.

For that new bridge at Parkkina the men of Mountain Engineers Battalion 91 had to get their heavy beams from a newly set-up sawmill 125 miles away. The lighter planks were brought by ship from Kirkenes to Petsamo. Some 25,000 lengths of round timber were picked up by the sappers from the timber store of the nickel-mines.

Meanwhile the battalions of the 2nd and 3rd Mountain Divisions were in the front line, unrelieved and without adequate food-supplies. Would they be able to hold out? Would it be possible under such conditions to hang on to the forward winter positions? The battalions, which had been in action there since June, were exhausted and bled white. The men were finished, physically as well as psychologically. With a heavy heart the German High Command therefore decided to withdraw the two divisions from the front and replace them by Major-General Schörner's reinforced 6th Mountain Division. At the time the decision was taken Schörner's men from Innsbruck were still in Greece. In the spring of 1941 they had burst through the "Metaxas Line," overcome Greek resistance along the Mount Olympus range, stormed Larissa in conjunction with the Viennese 2nd Panzer Division, taken Athens, and finally fought in Crete.

These men had then been switched from the Mediterranean to the extreme north, into the winter positions in the Litsa bridgehead. In the autumn of 1941 Schörner's Austrian Mountain Jägers would have been more than welcome before Leningrad or Moscow. The fact that Hitler dispatched them not to these sectors, but to the northernmost corner of the Eastern Front, proves the German Command's determination not to yield an inch of ground outside Murmansk. On that sector there could be, there must be, no retreat. The enormous volume of American aid to the Soviet Union, which had since begun flowing, lent Murmansk new and vital importance.

Whereas at the beginning of the war Hitler had viewed the capture of Murmansk merely as the elimination of a threat to the vital ore-mines and the Arctic Ocean road, it was now a case of seizing a port of vital importance to the outcome of the war, and the railway-line serving that port. The German starting-line, the springboard for a new offensive against Murmansk, must therefore be held.

On 8th October the new bridge at Parkkina was finished—two days ahead of schedule. It was named the "Prince Eugene Bridge," after Eugene of Savoy, as a tribute to the Austrian Mountain troops who made up the bulk of General Dietl's Mountain Corps.

Long columns of supply vehicles had been held up on the Arctic Ocean road for weeks. Now they could start moving again.

But it seemed as if there was a jinx on that sector of the front. Winter came surprisingly early—as, indeed, everywhere else on the Eastern Front. Only up there it started with a frightful Arctic gale. By the evening of 9th October all movement to the front had come to a standstill. Drivers who tried to defy the weather and get their lorries through were buried under snowdrifts and suffocated by their exhaust gases. Columns of porters lost their way and froze to death. Even the reindeer refused to budge. First Lieutenant Eichhorn's Company was stuck in front of the bridge.

In the front line the failure of supplies to arrive had terrible consequences. The men were starved, they were cold, and they were out of ammunition. The condition of the wounded was frightful. There were not enough bearers to take them out of the line quickly. Horses and mules were also badly affected.

Major Hess, the Quartermaster of the Mountain Corps Norway, reports in his book *Arctic Ocean Front, 1941* that the draught horses of 388th Infantry Regiment and those of 1st Battalion, 214th Artillery Regiment, in particular, were not up to the hardships. Within a few weeks 1400 horses died. Of the small Greek mules brought by Schörner's newly arrived division not a single one survived the hell of the tundra.

Nevertheless the Litsa front held out. The Austrian Jägers stood up to the Arctic winter, which hit their front eight weeks earlier than it did the divisions before Moscow. At long last their relief came. The companies of Schörner's 6th Mountain Division, which had taken over from 2nd and 3rd Mountain Divisions at the end of October, moved into the positions in the Litsa bridgehead and along the Titovka.

To hand over this difficult sector of the front, right up in the polar night, to a unit which had only just come from the sunny south and had no idea about living and fighting conditions in the extreme north was one of the most risky experiments of the whole war.

In long columns, moving in single file, the companies trudged through the snow between the lakes and up on to the granite plateau. The snow was a foot deep. The temperature was already 10 degrees below zero.

Near the front the men encountered heavily wrapped figures—NCOs detailed to take the new units to their positions. There was much waving of arms and subdued shouting. Careful there—the Russians are only a few hundred yards farther on. Now and again a Russian flare rose into the air and a few bursts of machine-gun fire swept over the ground.

"Follow me!" Behind the NCOs the platoons moved off in different directions, and presently split up into sections. Thus the whole marching column vanished into nothingness. Where were they being taken?

Lance-corporal Sailer, with eight men from Innsbruck, was trudging through the snow behind his guide. "Where the hell is that man taking us?" he muttered. The guide merely grunted and a moment later stopped. "Here we are."

A massive granite boulder with a machine-gun on top. Behind it a few miserable caves made from piled-up stones, lined with moss and roofed over with pine-branches, with smaller stones on top and a frozen tarpaulin over the entrance. These were their fighting positions and living quarters for the winter.

The Jägers were speechless. No dug-out, no pill-box, no trench, no continuous front line. And their cave was not high enough for a man to stand upright, but only just big enough for the men to cower close to one another. That was the winter line in the Litsa bridgehead.

That then was journey's end. They had come from sun-drenched Athens, from the Acropolis, from the market-place of humanity; they had driven right across Europe, sailed through the Gulf of Bothnia, and marched along the 400 miles of Arctic Ocean road from Rovaniemi.

Others had come by ship up the coast of Northern Norway, until the British had caught them off Hammerfest and chased them into the fjords. From there they had marched to Kirkenes, along road No. 50, foot-slogging it for over 300 miles. And now they were in the tundra before Murmansk, swallowed up by the polar night.

With a few whispered words of advice the emaciated figures of 2nd and 3rd Mountain Divisions handed over to them. "You're sure to get some material for living quarters and better dug-outs," they comforted them. Then they packed their rucksacks and with a sigh of relief moved off, into the night. Many of them, especially the battalions of 3rd Mountain Division, went back along the same long road, the Arctic Ocean road south to Rovaniemi, which their comrades of the 6th had travelled in the opposite direction. Only the reinforced 139th Mountain Jäger Regiment remained behind in the area of Mountain Corps Norway, as an Army reserve. Thus it was spared the nightmarish journey along the Arctic Ocean road to the south. For by then winter had started in earnest and the trek to Southern Lapland was torture.

The Arctic Ocean road was the lifeline of the fighting front. All traffic moving away from the front had to give way to traffic moving up. As a result, only one battalion moved south each day, always with predetermined destinations and bivouacs. Everything moved on foot:

Battle in the Arctic Night 431

only the baggage went by vehicle. The guns, taken to pieces, infantry weapons, and ammunition stayed with the marching men and horses.

General Klatt, then Lieutenant-Colonel Klatt and in command of 138th Mountain Jäger Regiment, gives the following account of this trek in the divisional records of 3rd Mountain Division: "Once we had reached the tree-line the worst was over. Now at last each day ended at a blazing camp-fire. These were a great help also to our emaciated animals, the first of which began to collapse after about ten days. What was to be done? We lifted their shivering bodies off the ground, collected enough wood to light a fire, and supported the animal's weakened flanks until it was warm again and could stand on its own feet. If it then returned to its accustomed place among the other animals we knew that we had outwitted death this time. We succeeded quite often, but by no means always, and it was invariably touch and go—a matter of a few minutes—whether we could save our dumb, shaggy friends." Beyond Ivalot came the first Lapp settlements, and then Finnish farmhouses. The soldiers down from the Arctic saw the first electric light again, children playing, reindeer sleighs, and the Rovaniemi railway. A last forced march, and the Gulf of Bothnia came into sight. They had reached their objective.

The 24th anniversary of the Socialist October Revolution—which, under the Gregorian calendar since introduced, falls on 7th November —was marked in Moscow by the German assault on the capital. The Soviet metropolis was starving and shivering. Marauders roamed the streets. Special courts were sitting.

The 8th and 9th November had been declared ordinary working days, in view of the situation. Only on the 7th were short celebrations to be held. The traditional mass rally of the Moscow City Soviet on the eve of the anniversary had been transferred below ground: at the lowest level of the Mayakovskiy station of the Moscow Metro, Stalin addressed the Party and the Red Army. He invoked victory and demanded loyalty and obedience.

In the morning of 7th November military formations on their way to the front marched across the snow-covered Red Square, past Stalin. Stalin was standing on top of the Lenin Mausoleum—where later his own embalmed body was to lie for seven years—and saluted the Army units which were trudging in silence through the falling snow. All round the square countless AA guns had been emplaced for fear of a German air-raid. But Goering's Luftwaffe did not show up.

Some 1600 miles away from Moscow, on the icy front before Murmansk, the commander of the Soviet 10th Rifle Division decided to mark the 24th anniversary of the Revolution in a very special way: he wanted to make Stalin a present of a victory.

During the night of 6th/7th November Corporal Andreas Brandner in strongpoint K3 put his hand to his ear. The easterly wind was carrying across the noise of singing and hilarious revelry. From the Soviet positions the strains of the Internationale wafted across time and again.

The corporal made a "special incident" report to Company. The company commander telephoned Battalion. Suspicion and caution were called for: when the Russians had vodka to drink it usually boded nothing good. Were they merely celebrating, or was this the prelude to an attack?

By 0400 hours the men knew the answer: the Soviets were coming. With shouts of "Urra" a regiment charged against K3, and another against K4. The Siberians fought fanatically. They got inside the German artillery barrage. They gained a foothold on two unoccupied commanding heights immediately in front of the German positions.

By immediate counter-attacks the Russians were dislodged—except from one conical rock in front of K3. There an assault-party and hand-grenade battle was fought for the next few weeks, a kind of operation typical of this sector. It was reminiscent of the assault-party operations at Verdun and in the Dolomites during the First World War.

The Siberians were established under cover of an overhanging slab just below the summit of the rock, in the dead angle barely 10 yards below the German defenders. It was impossible to get at them with small arms or artillery. The hand-grenade was the only effective weapon in the circumstances.

Time and again, like cats, the Siberians would scramble up the overhang on their side and appear right in front of the small German strongpoint. They would fire their sub-machine-guns and charge the defenders. There would be hand-to-hand fighting, with rifle-butt, trenching-tool, and bayonet.

This kind of fighting continued for five days. The outcrop of granite was soon known as "hand-grenade rock." The men of 2nd Battalion, 143rd Jäger Regiment, climbing up on the German side to relieve their colleagues, would ask themselves anxiously: Shall I be walking down again on my own two feet, or shall I be carried on a stretcher? During that short period the German defenders threw 5000 hand-grenades and the Russians left 350 killed in front of their lines.

After that the period of hibernation set in also on the Litsa bridgehead until mid-December 1941. Similarly, at the neck of the Rybachiy Peninsula, where the Machine-gun Battalions 13 and 14, as well as companies of the 388th Infantry Regiment, 214th Infantry Division, were in position, nothing much happened. The Arctic winter, now at its height, did not permit any major operations. The snow was lying many feet deep, and an icy blizzard swept over rocks and through the valleys. Only patrols kept the war alive.

The German troops cut down the telegraph-poles of the Russian line to Murmansk and burnt them in the simple stoves which had since arrived. The Russians retaliated by attacks on sentries and columns of porters.

On 21st December, three days before Christmas Eve, the Soviet winter offensive which had been in full swing on the main front for the past fortnight opened also in the extreme north. The Soviet 10th Rifle Division, since promoted to Guards status for its attacks on the Revolution anniversary, as well as the 3rd and 12th Naval Brigades, once more charged against K3, and presently also against K4 and K5. That was on the sector of 143rd Jäger Regiment.

Its sister regiment, the 141st Jägers, holding the southern part of the bridgehead, was not at first affected by the attacks, and could thus be drawn upon for counter-attacks to clear up enemy penetrations.

One regiment of the Soviet 12th Naval Brigade which had broken through the German lines was pounced upon and routed by the 3rd Battalion, 143rd Jäger Regiment, which had been held in reserve, in a counter-attack launched in hard frost and a blinding blizzard on Hill 263.5. Those who escaped ran straight into the fire of the German artillery.

Along the Arctic Ocean Stalin's winter offensive did not gain an inch of ground. This failure of an attack mounted by numerically superior, superbly equipped, and winter-trained troops proved that the terrain and climate represented an almost insuperable obstacle to any attacker faced by a determined opponent.

But Stalin was no more willing to accept this fact than Hitler. The danger threatening his Murmansk lifeline seemed too great to him. Its severance would have been a fatal blow to the entire Soviet war effort.

By employing all available forces Stalin therefore tried to eliminate that threat and annihilate the German Mountain Corps. No price could be too high provided only Murmansk was held.

In the battle of Kiev in the autumn of 1941, the greatest German battle of encirclement in the Eastern campaign, the German Armies, after weeks of fighting, captured or destroyed approximately 900 tanks, 3000 guns, and about 10,000 to 15,000 motor vehicles. In the subsequent battle of the Vyazma and Bryansk pockets, the greatest battle of annihilation in the Eastern campaign, the Soviets lost 1250 tanks. That was when Hitler authorized his Reich Press Chief to announce that "the enemy will never recover from this blow."

In fact, the American armament supplies during 1942 almost completely made good the material losses of the Red Army. The decisive effect of American aid on the destinies of the war could not be revealed more clearly than by this fact.

The Western Powers soon discovered how to protect their convoys against the German U-boats in the Arctic Ocean and the German aircraft operating from airfields in Northern Norway and Northern

Finland. Powerful naval forces would escort the huge convoys of thirty, forty, or even more merchant-ships right into Murmansk or into the White Sea. But they paid a heavy price for that lesson by the disaster which befell convoy PQ 17.

This famous convoy, at the same time, was a warning to the German High Command of the colossal volume of American aid that was being shipped to Russia's northern ports. In that sense PQ 17 was an important milestone in the war—for both sides.

Early in July 1942 a convoy of thirty-three transports, twenty-two of them American, steamed into the Northern Ocean. Almost the same number of naval units—cruisers, destroyers, corvettes, anti-aircraft vessels, submarines, and minesweepers—escorted the merchant armada, which was sailing in close order; its distant cover was provided by the British Home Fleet, with two battleships, one aircraft carrier, two cruisers, and fourteen destroyers.

On 4th July, as the convoy rounded Jan Mayen Island to turn into the Barents Sea, the British Admiralty in London received an urgent signal from an agent: "German surface units—the battleship *Tirpitz*, the armoured cruiser *Admiral Scheer*, and the heavy cruiser *Hipper*, as well as seven destroyers and three torpedo-boats—have put to sea from Altenfjord in Northern Norway."

That could only mean a full-scale attack on PQ 17 with greatly superior forces. The Home Fleet was too far away to arrive at the spot in time. The escort units were therefore ordered to take evasive action and order the convoy to disperse. The merchantmen were to try to reach their destination singly.

That decision was a fatal mistake. The German High Seas Fleet had no intention of attacking PQ 17; indeed, fearing enemy aircraft carriers, the units presently returned to port.

The scattered convoy, however, abandoned by its shepherds, was presently attacked by Admiral Dönitz's pack of U-boats and by bomber squadrons and torpedo aircraft under the "Air Chief Kirkenes," and in a dramatic battle lasting several days utterly destroyed. Twenty-four transports and rescue vessels were sunk.

The true weight of this blow can be judged only if one knows what lay in the holds of the sunken transports. War material lost included 3350 motor vehicles, 430 tanks, 210 aircraft, and 100,000 tons of other cargo. That was the equivalent of the booty taken in a medium-sized battle of annihilation, like the one of Uman.

The Allies learnt their lesson from this disaster. Never again did they send out their convoys without maximum cover by naval units and aircraft carriers. The result was that of 16·5 million tons of total American supplies dispatched to the Soviet Union 15 million tons reached their destination—most of it via Murmansk. These supplies included 13,000 tanks, 135,000 machine-guns, 100 million yards of uniform cloth, and 11 million pairs of Army boots.

But to return to the fighting for Murmansk. Towards the end of April 1942, after a lull of several months, Lieutenant-General Frolov, the C-in-C of the Soviet "Karelian Front," mounted his large-scale offensive with his Fourteenth Army. This offensive was intended to be decisive and to annihilate the German Mountain Corps which had been under the command of Lieutenant-General Schörner since January 1942. By means of a boldly conceived combined land and naval operation the Soviets wanted to crush the 6th Mountain Division in a big two-pronged pincer operation, reach Kirkenes and the ore-mines, and occupy Northern Finland.

The prelude was a frontal attack by the Soviet 10th Guards and 14th Rifle Divisions in the Litsa bridgehead. Concentrated artillery-fire, followed by a charge: at 0300 hours, in the milky light of the polar night, the Russians attacked in unending waves. At first they came in silence, then with shouts of "Urra."

Pounded by heavy shell-fire, smothered by a blizzard reducing visibility to a bare 10 yards, the men of the Austrian 143rd and 141st Jäger Regiments stood in their strongpoints, without yielding an inch. Whenever the Russians succeeded in penetrating the machine-gun and carbine fire and breaking into a strongpoint, they were overcome in hand-to-hand fighting.

For three days the Soviet 14th and 10th Divisions ran amok—then their strength was spent. They had not gained an inch of ground.

But General Frolov did not give up. He had another trump card. On 1st May six ski brigades, including the famous 31st Reindeer Brigade, circumvented the southern wing of the German line of strongpoints and made an enveloping attack against the rear of the 6th Mountain Division.

Simultaneously the replenished and reinforced Soviet 12th Naval Brigade with 10,000 to 12,000 men landed on the western coast of Motovskiy Bay. Under cover of gunfire by Soviet torpedo-boats the naval infantry—or marines—charged ashore, burst through the weak German covering line held by only two companies, and drove on against the Parkkina–Zapadnaya Litsa supply route. "Revenge for 28th December" was their slogan. And it looked as though they would get it.

The situation was extremely critical. General Schörner personally brought up rearward units, supply formations, and headquarters personnel to the threatened supply route. There he flung himself down alongside his Jägers, firing his carbine, directing the counter-attacks, and ceaselessly urging his men: "Hold on! We've got to gain time!"

He succeeded. Enough time was gained for the hurriedly summoned battalions of 2nd Mountain Division to be brought up from Kirkenes. On 3rd May, just before midnight, they moved into action—units of 136th and 143rd Jäger Regiments.

The heavy, costly fighting continued until 10th May, when General Frolov's marines were forced to withdraw. The Soviet naval units in Motovskiy Bay evacuated the remnants. The Soviet northern prong had been smashed.

The southern prong, with the 31st Reindeer Brigade at its centre, encountered the picket lines of 139th Jäger Regiment, along the Titovka river. The strongpoints of these experienced troops, who had been through the fighting for Narvik, held out. But the Soviets seeped through the front and with their reindeer formations threatened the Arctic Ocean road, the airfield, and the nickel-mines.

Schörner mounted a successful counterblow. Battalions of 137th and 141st Jäger Regiments, together with a mixed combat group of Reconnaissance Detachment 112 and Engineers Battalion 91, halted the attack and smashed the enemy.

But the Soviet High Command had another card up its sleeve—a dangerous card at that. But it was prevented from playing it. The fortunes of war intervened in Schörner's favour in a most terrible way.

Along the entirely unprotected southern flank of the German front, in the worst wilderness of the tundra, General Frolov had employed the Soviet 155th Rifle Division. It was to have given the German Mountain Corps its *coup de grâce*. But the Russians too were extended to capacity.

The 155th Division had not received their winter equipment in time. By entire companies the Red Army men froze to death in the tundra. Under vast heaps of snow, along the lines of their advance, they lay dead and buried. It was an appalling repetition of Napoleon's tragedy: of 6000 Russians only 500 reached the combat zone. They were so emaciated that the smallest German picket groups were able to smash them.

But in spite of all defensive successes the general balance-sheet of the campaign in the Far North was shattering. For lack of strength three offensive wedges of the German–Finnish Armies had ground to a standstill in the vastnesses between Finland's eastern frontier and the Murmansk railway.

The offensive of the Mountain Corps Norway had come to a halt in the bridgehead east of the Litsa.

General Feige's XXXVI Army Corps succeeded in taking Salla, in smashing the Soviet XLVI Corps, and in capturing the high ground of Voytya and Lysaya. But after that its offensive vigour was likewise spent.

The front of the Finnish III Corps under General Siilasvuo seized up west of Ukhta in a bridgehead east of the narrow neck of land between

Lakes Topozero and Pya. The great objective, the Murmansk railway, though within arm's reach, was never attained.

One question inevitably arises: if it was impossible to seize Russia's vital lifeline from the Arctic Ocean to the Leningrad and Moscow fronts, why on earth were not the railway, the bridges, and the Murmansk transhipment installations put out of action by air-raids? The answer can be found in the records of the German Luftwaffe Command, and it is significant of the war in the East as a whole. The Luftwaffe was able to score only partial successes. Any prolonged interruption of the railway or extensive destruction of engineering works and power stations proved impossible. Why? Simply because the Luftwaffe lacked adequate forces. The Fifth Air Fleet operating on the Northern Front was being dissipated by the need to support too many operations at the same time: hence it was unable to make any concentrated major effort that might have promised success.

The fronts in the Far North had frozen rigid. Murmansk, the objective of the campaign, had not been reached. And Archangel, the finishing-point laid down in the plan for the war in the East, was a long way off.

PART SIX:
The Caucasus and the Oilfields

1. Prelude to Stalingrad

Halder drives to Hitler's Headquarters–Anxieties of the Chief of the General Staff–The Izyum bend–Balakleya and Slavyansk–Fuehrer Directive No. 41–"Case Blue"–Curtain up over the Crimea–Failure of a Russian Dunkirk–Mid-May south of Kharkov–"Fridericus" will not take place–Kleist's one-pronged armoured pincers–The road to death–239,000 prisoners.

COLONEL-GENERAL Halder's car swung out of the Mauer Forest in East Prussia, where OKH, the High Command of Land Forces, was situated in a well-camouflaged spot, on to the road to Rastenburg. A spring gale was sweeping through the branches of the ancient beeches. It was whipping up the surface of the Mauer Lake into white caps, and it was driving the clouds so low over the ground that one almost expected to see them slit open by the tall stone cross on the hill where the military cemetery of Lötzen was situated.

It was the afternoon of 28th March 1942. Colonel-General Halder, Chief of the Army General Staff, was driving over to Hitler's Headquarters at the "Wolfsschanze," hidden in the forests of Rastenburg.

On the lap of his orderly officer lay a brief-case—at that moment perhaps the most valuable brief-case in the world. It contained the German General Staff's operational plan for 1942.

In his mind Halder once more rehearsed his proposals. The ideas, thoughts, and wishes voiced by Hitler as Commander-in-Chief of the Army and Supreme Commander of the Armed Forces at the daily situation conferences had been laboriously written up by Halder into a carefully considered draft. The main feature of the plan of campaign for 1942 was a full-scale attack in the southern sector towards the Caucasus; its objective was the destruction of the bulk of the Russian forces between Donets and Don, the gaining of the passes across the Caucasus, and eventually the seizure of the vast oilfields by the Caspian.

The Chief of the General Staff was not happy about the plan. He was beset by doubts whether a major German offensive was justified after the heavy drain of strength during the winter. Many a dangerous crisis which the reader of the preceding chapters will already have seen liquidated was then, at the end of March, still causing anxiety to the German High Command and the General Staff of the Army. At that time, too, General Vlasov's Army on the Volkhov had not yet been finally defeated. Count Brockdorff-Ahlefeldt with the divisions of his II Corps was still surrounded in the Demyansk pocket. "Operation Bridge-building" had begun, but had not yet been successfully accomplished. In Kholm, Scherer's combat group had not yet been relieved.

Even in the Dorogobuzh-Yelnya area, only 25 miles east of Smolensk, the situation was still critical at the end of March. There the Soviets were operating with units of their Thirty-third Army, the I Guards Cavalry, and the IV Airborne Corps. Farther north the Soviet Thirty-ninth Army and the XI Cavalry Corps were still holding the dangerous front-line prominence west of Sychevka.

But these were by no means all the worries in the mind of the Chief of the General Staff at the end of March. In the Crimea Manstein with his Eleventh Army was still immobilized before Sevastopol, and the Kerch Peninsula had even been recaptured by the Russians in January. But the situation was most critical of all at Kharkov, where heavy fighting had been going on since mid-January. The Soviet High Command was making a supreme effort to nip Kharkov off in a pincer movement. The northern jaw of the pincers had been held at Belgorod and Volchansk, but the southern jaw, the Soviet Fifty-seventh Army, had breached the German Donets front on both sides of Izyum over a width of 50 miles. The Soviet divisions had already established a bridgehead 60 miles deep. The spearheads of the attack were threatening Dnepropetrovsk, the supply centre of Army Group South. Whether this Soviet penetration in the Izyum area would develop into a dam-burst of incalculable consequences depended on whether the two cornerstones north and south of the penetration point, Balakleya and Slavyansk, could be held. For the past few weeks these points had been held by the battalions of two German infantry divisions in what had already become a heroic saga of defensive fighting. On its outcome depended the entire future of the southern front. Slavyansk was held by the 257th Infantry Division from Berlin, and Balakleya by the 44th Infantry Division from Vienna.

In savage and costly engagements the Berlin regiments under General Sachs, and later under Colonel Rüchler, defended the southern edge of the Izyum bend. A combat group under Colonel Drabbe, commanding 457th Infantry Regiment, displayed such skill, valour, and self-sacrifice in fighting for miserable villages, collective farms, and small homesteads that even the Soviet operation reports—as a rule very reticent about German feats of arms—were full of admiration. The village of Cherkasskaya was typical of this kind of fighting. In eleven days Drabbe's group lost there nearly half of its 1000 men. Some 600 defenders were manning an all-round front of 8½ miles. Soviet dead actually counted in front of this village numbered 1100. Eventually the Russians took the village, but it had by then consumed the strength of five regiments.

Before leaving his headquarters for the "Wolfsschanze" in the afternoon of 28th March, Colonel-General Halder had asked for the operations report of 257th Infantry Division about the battle that had now been raging for seventy days. He wanted the division to be cited in the High Command communiqué: to date its regiments had repulsed 180 enemy attacks, and 12,500 Soviet dead had been counted in front of its lines. Three Soviet rifle divisions and a cavalry division had been mauled, and four more rifle divisions and an armoured brigade had taken a heavy toll. The German casualties, admittedly, also testified to the severity of the fighting: they amounted to 652 killed, 1663 wounded, 1689 frost-bitten, and 296 missing—a total of 4300 men, half the total losses suffered by the division in ten months' fighting in Russia. That was Slavyansk.

The northern edge of the Izyum penetration, the Balakleya area, was held by the Viennese 44th Infantry Division, whose 134th Infantry Regiment was the successor of the famous ancient Hoch-und-Deutschmeister Regiment. The division under Colonel Debois held a front from Andreyevka through Balakleya and Yakovenkovo to Volokhov Yar. That was a line of 60 miles. And along those 60 miles an entire Soviet Corps was attacking, reinforced by armour and rocket batteries.

Here, too, the combat groups and their commanders were the inspiration of the resistance. The exploits of the combat group under Colonel Boje, commanding 134th Infantry Regiment, in the vital sectors of Yakovenkovo and Volokhov Yar, on the high bank of the Balakleya, whipped by icy winds, are among the most remarkable chapters of the war in the East.

The fighting hinged on villages and farms—in other words, on the troops' shelter quarters. In a temperature of 50 degrees below zero a house and a hot stove, offering as they did the chance of an hour's sleep, became a matter of life and death. The Germans clung to the villages, and the Russians tried to drive them out of them, because they too wanted to get away from their snowy fortifications, behind which they assembled for attack, to find a roof, a warm corner, and a few hours' sleep without the fear of freezing to death.

Once again the war centred around the most elementary requirements of life. Germans and Russians alike spent their last ounce of strength. For once the interests of the men in the line and those of the General Staffs coincided: both wanted Balakleya and the villages to the north which were being held by 44th Infantry Division—the former for the sake of the houses and the latter for reasons of strategy. If the cornerpost of Balakleya and the high ground commanding the road to the west were to be lost Timoshenko would be able to convert his penetration at Izyum into a large-scale strategic break-through to Kharkov.

But Balakleya held on, stubbornly defended by 131st Infantry Regiment under Colonel Poppinga. But, north of it, Strongpoint 5 of 1st Battalion, 134th Infantry Regiment, was attacked in such strength that it could not be held. The battalion resisted the Soviet tank attacks to the last man. Among those killed in action was First Lieutenant von Hammerstein, a nephew of the former Chief of the Army Directorate, von Hammerstein-Equord. Young officers like Hammerstein, ready for any kind of action or sacrifice, were typical of these terrible defensive battles. Together with the hard-boiled fearless old sergeants and corporals, they formed small fighting parties which almost invariably pulled off incredible exploits.

Thus First Lieutenant Vormann with remnants of 2nd Company smashed an entire Soviet battalion in a savage night engagement.

First Lieutenant Jordan, commanding 13th Company, personally lay in front of the Russian lines at Yakovenkovo night after night, directing the fire of his infantry guns against enemy machine-gun positions on the hill, and one after the other putting them out of action. Vormann and Jordan are both buried at Stalingrad.

The ferocity of the fighting in the Balakleya sector is also shown by the fact that Colonel Boje himself and his staff were forced more than once to join in hand-to-hand fighting with pistol and hand-grenade. A Soviet ski battalion eventually reached the crucial Balakleya–Yakovenkovo road on the southern flank of the combat group and established itself in huge ricks of straw. Boje threw in his last reserves to save his combat group from the mortal danger of encirclement. The Russians did not yield an inch. Even when their straw ricks had been set ablaze by Stukas they still fired their guns and defended themselves to the end.

The interesting thing about that terrible fighting was that the decisive rôle was invariably played by the individual. Altogether, the successful defensive battles fought by the German troops in the winter and spring of 1942 hinged very largely upon the individual soldier. He was—certainly still at that time—superior to his Russian opposite number in experience and fighting morale. This fact alone explains the astonishing feats performed by German troops, frequently depending entirely on themselves, all along the front from Schlüsselburg down to Sevastopol against a numerically and materially superior enemy.

Prelude to Stalingrad 443

The following instance of courage, sangfroid, and practical skill from the hotly contested Kharkov area is typical of many.

In March 1942 the 3rd Panzer Division was employed in that sector in the rôle of a fire-brigade on a continually threatened front. In the Nepokrytaya sector Sergeant Erwin Dreger with fifteen men of 1st Company, 3rd Rifle Regiment, was holding a front of a little over a mile. That, of course, was possible only because of the particular tactics thought up by Dreger and thanks to his men's nerves of steel—all of them Eastern Front veterans. From stocks of captured material Dreger had armed every one of them with a machine-gun, keeping three more guns in reserve, just in case. Inside the village, and along the edge of it as well as all over the ground outside, piles of captured machine-gun ammunition had been laid out so that the guns could be operated without a No. 2 to look after the ammunition-belts.

Dreger's men were spaced out in a wide arc, facing the corner of a patch of woodland from where the Russians regularly mounted their attacks. It was at that point—though, of course, neither Dreger nor 3rd Panzer Division realized it—that the Soviet forward detachments intended to launch their break-through. The date chosen by them was 17th March. Towards 1030 hours the Russians charged in battalion strength. Dreger with his machine-gun was in the middle—*i.e.*, at the most rearward point—of the arc. The Russians were getting closer and closer without a single round being fired. Dreger had given his men strict orders: "Wait for me to open fire." The spearheads of the enemy attack were within 50 yards when Dreger's "lead machine-gun" at last opened up. Although they were a hard-boiled lot, Dreger's men felt greatly relieved when at last they were permitted to fire. Since the enemy assault had aimed almost exactly at the centre of their defensive line it was possible to deal with it almost as if by envelopment. Under the effective fire from both flanks the Russian assault collapsed after twenty minutes. Whereas the Russians suffered very heavy casualties, Dreger's group had not lost a single man.

After an hour's pause the Russians directed pin-point artillery-fire at the machine-gun positions they had identified. But Dreger and his men merely laughed: needless to say, they had long abandoned their former positions and established themselves in new ones.

The Russians made five attacks within fourteen hours. Five times their battalion collapsed in the concentrated machine-gun fire of Dreger's group. Sixteen determined men were opposing an enemy who was about a hundred times superior to them in numbers.

Three days later, however, the war exacted the supreme price from Dreger too. He and his platoon had in the end been forced out of the village. But in a temperature of 30 degrees below the men needed a house, a room, or a cellar for a few hours during the night, to protect them from the icy wind. Dreger intended to recapture a collective farm by a surprise coup. He was caught by a burst from a machine pistol. His men dragged him behind a straw rick and made him comfortable. Dreger was tapping his icy fingers against each other as if to warm them. And while doing so he seemed to be listening into the still icy night. Softly the normally so undemonstrative man said to his comrades, "Listen—that's death knocking at the door." With that he died.

Colonel-General Halder did not know the story of Sergeant Dreger —but he was well acquainted on that 28th March with the operational reports sent in by 44th Infantry Division about the fighting at Balakleya.

On 13th February he had passed them on to Jodl for inclusion in the High Command communiqué. On the 14th the Viennese were cited for the first time. Six weeks had passed since then. The force of the Soviet assault had on the whole been broken by the astonishing resilience of the troops at the cornerposts of the Izyum bend. But even with optimism and confidence that this crisis, like all the rest, would soon be liquidated, there still remained the justified question: would it not be better to make a pause along the entire Eastern Front, including Army Group South, and to let the Russians attack and wear themselves out against the German defences until their reserves were gradually spent?

That was the question which Halder had been asking himself and his officers time and again while planning the 1942 campaign.

But Major-General Heusinger, the Chief of the Operations Department, had objected that this course of action would mean losing the initiative, and hence an unpredictable amount of time. And time was on the Soviet side. If they were to be forced to their knees at all, then the attempt would have to be made soon.

Halder had accepted this point. But in his opinion a renewed offensive should have been aimed at the heart of the Soviet Union—against Moscow.

But that was precisely what Hitler was stubbornly opposing. He seemed to have a positive horror of Moscow. Instead he had decided to try something entirely new after the unfortunate experiences on the Central Front in the previous year, and to seek the decision in the south by depriving Stalin of his Caucasian oil and by thrusting into Persia. Rommel's Africa Army played a part in this plan. The "desert fox," who was just then preparing his offensive from Cyrenaica against the British positions at Gazala and against Tobruk, the heart of the British defence of North Africa, was to advance right across Egypt and the Arabian Desert to the Persian Gulf. In this way Persia, the only point of contact between Britain and Russia, and after Murmansk the greatest supply base of US help for Russia, would be eliminated. Moreover, in addition to the Russian oilfields the very much richer

Arabian oilfields would fall into German hands. Mars had been appointed the god of economic warfare.

Halder's car stopped at the barrier at Gate 1 to Special Area I of the "Wolfsschanze," Hitler's Headquarters proper. The guard saluted. The barrier was raised. Along the narrow asphalted road the car drove on into Hitler's forest stronghold. The low concrete huts, with their camouflage paint and bush-planted flat roofs, lay perfectly hidden among the tall beeches. Even from the air they were impossible to make out. The whole area, over a wide radius, was hermetically sealed —protected by barbed-wire obstacles and belts of minefields. There were road-blocks on all roads. The small branch railway had been taken out of service and was now being used only by Goering's diesel coach plying to the Air Marshal's battle headquarters in the Johannisberg Forest near Lake Spirding, south of Rastenburg.

Colonel-General Jodl once remarked that the "Wolfsschanze" was a cross between a concentration camp and a monastery. It certainly was a Spartan military camp, differing from ordinary military establishments only in that Hitler turned night into day, working until two, three, or even four o'clock in the morning, and then sleeping until all hours. Whether they liked it or not, his closest collaborators also had to adapt themselves to this rhythm.

Halder drove past the information office of the Reich Press Chief. On the right were the radio and telephone exchange of the camp, and next to it Jodl's and Keitel's quarters. On the left of the road were the quarters of Bormann and the Reich Security Service. On the farthest edge of the forest was Hitler's hut, surrounded by one more high wire fence. Together with his Alsatian bitch Blondi, this wire fence was the last obstacle outside Hitler's Spartan hermitage in the Rastenburg forest.

For that conference on 28th March Hitler had invited only a small circle, only the most senior leaders of the armed forces—Keitel, Jodl, and Halder, and half a dozen other top-ranking officers of the three services. They were standing or sitting on wooden stools around the oak map-table. Hitler sat in the middle of one of its long sides; the Chief of the General Staff occupied one of its narrow sides.

Halder was given leave to speak and began to develop his plan. Its code name was "Case Blue." Originally it was to have been called "Case Siegfried," but Hitler no longer wanted to commit himself by choosing invincible mythological heroes as patrons for his military operations since the Emperor Frederick Barbarossa had let him down.

Hitler kept interrupting Halder with all kinds of questions. The conversation time and again went off at a tangent, but after three hours Hitler eventually gave his consent to the basic outline of the plan. This then was the scheme: Act 1: Two Army Groups to form a huge pair of pincers. The northern jaw of the pincers to advance from the Kursk–

446 Part Six: The Caucasus and the Oilfields

Kharkov area down the middle Don to the south-east, while the right jaw of the pincers drives rapidly eastward from the Taganrog area. West of Stalingrad the two jaws to meet, enclosing the bulk of the Soviet forces between Donets and Don, and annihilating them. Act 2: Advance into the Caucasus, that 700-mile-long range of high mountains between the Black Sea and the Caspian, followed by the conquest of the Caucasian oilfields.

It was noon when Halder left the "Wolfsschanze" to drive back to the Mauer Forest. He was weary and depressed, full of doubts and irritated by Hitler's know-all manner. But he nevertheless felt that he had won Hitler over for a plan that was at least practicable—a plan which used the German forces economically and which went for the

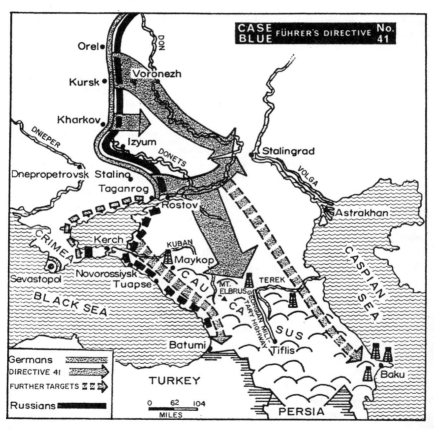

Map 25. By means of Operation Blue, the summer offensive of 1942, Hitler hoped to force the decision on the southern wing. West of Stalingrad the Soviets were to be surrounded by a gigantic pincer operation, and a thrust was then to be launched into the Caucasian oilfields.

objectives on the southern front a step at a time, with clearly defined strong points. If it came off, Stalin would lose the entire Caucasus, including Astrakhan and the Volga estuary—in other words, the overland as well as the shipping link with Persia. The southern objective of "Operation Barbarossa" would thus have been reached.

All that remained was to formulate the project into a clear directive for the separate branches of the armed forces.

Seven days later, on 4th April 1942, Colonel-General Jodl submitted his draft directive. The Armed Forces Operations Staff had solved the problem in the traditional German Staff manner: they had begun by briefly outlining the situation, by listing the objectives as separate "tasks," and, in this way, leaving considerable freedom to the Commander-in-Chief Army Group South, Field-Marshal von Bock, as to the actual execution of the vast operation. That had been a General Staff tradition for 130 years, from Scharnhorst to Schlieffen and Ludendorff.

But this High Command draft of "Operation Blue" was shot down almost at once. During the critical situations of the past winter Hitler had lost faith in the loyalty of his generals. Commanders-in-Chief and Corps commanders had often left no doubt that they were obeying his orders unwillingly. Following Brauchitsch's spectacular departure, Hitler had himself assumed supreme command of the Army, and he was not prepared now to have his authority diminished by "elastically framed tasks."

When he had read the draft he refused to give his consent. The plan, he argued, was leaving the Commander-in-Chief South far too much freedom of action. Hitler was not having any elastic directives.

He demanded detailed instructions. He wanted to see the execution of the operation laid down minutely to the last detail. When Jodl demurred Hitler took the papers out of his hand with the words: "I will deal with the matter myself." On the following day the result was available on ten pages of typescript—"Fuehrer Directive No. 41 of 5th April 1942." Alongside Plan Barbarossa, Directive No. 21, this new directive became one of the crucial papers of the Second World War, a blend of operational order, fundamental decisions, executive regulations, and security measures.

As this directive was not just the plan for another gigantic military campaign, but also the detailed time-table leading to Stalingrad—a document, in fact, which already contained in itself the turning-point of the war—its most important passages are worth quoting here.

Right in the preamble we find a bold claim: "The winter battle in Russia is drawing to its close. The enemy has suffered very heavy losses in men and material. In his anxiety to exploit what seemed like initial successes he has spent during this winter the bulk of his reserves earmarked for later operations."

448 Part Six: The Caucasus and the Oilfields

Proceeding from this thesis, the Order went on: "As soon as weather and ground conditions permit the German Command and the German forces, being superior to the enemy, must seize the initiative again in order to impose their will upon the enemy. The aim is to destroy what manpower the Soviets have left for resistance and to deprive them as far as possible of their vital military-economic potential."

This is how Hitler saw the execution of the plan: "While adhering to the original general outline of the campaign in the East, the task now is for the centre of the front to hold back temporarily... while all available forces are concentrated for the main operation in the southern sector, with the objective of annihilating the enemy on the Don and subsequently gaining the oilfields of the Caucasian region and the crossing of the Caucasus itself."

On the detailed execution of the campaign the directive stated: "It is the first task of the Army and Luftwaffe after the period of mud to create the prerequisites for the execution of the main operation. This entails the cleaning up and consolidation of the entire Eastern Front and the rearward military areas. The next tasks will be the clearing of the enemy from the Kerch Peninsula in the Crimea and the capture of Sevastopol."

A key problem of this extensive operation was the long flank along the Don. To avert the threat resulting from it Hitler took a fatal decision which did much to precipitate the disaster of Stalingrad. This was what he ordered: "As the Don front becomes increasingly longer in the course of this operation it will be manned primarily by formations of our Allies.... These are to be employed in their own sectors as far as possible, with the Hungarians being farthest north, then the Italians, and then, farthest to the south-east, the Rumanians."

So much for grand strategy and theory. As for the practical execution, this was to start with "Operation Bustard Hunt" in the Crimea. In his book *The Most Important Operations of the Great Fatherland War* the Soviet military historian Colonel P. A. Zhilin has the following to say about the situation in the Crimea in the spring of 1942: "The stubborn fighting of the Soviet troops and the Black Sea Fleet yielded us much strategic advantage and foiled the calculations of the enemy. The German Eleventh Army, tied down in the Crimea, could not be used for the attack against the Volga and the Caucasus."

That is entirely correct. And just because it was of such importance to the Soviets to keep Manstein's Eleventh Army immured in the Crimea, Stalin had mobilized a formidable force for this task.

Three Soviet Armies—the Forty-seventh, the Fifty-first, and the Forty-fourth—with seventeen rifle divisions, two cavalry divisions, three rifle brigades, and four armoured brigades, were blocking the

11-mile isthmus of Parpach, the passage from the Crimea to the Kerch Peninsula. Kerch in turn was the springboard to the eastern coast of the Black Sea, and hence into the foothills of the Caucasus.

Every mile of this vital neck of land was being defended by approximately 16,000 men—more than nine men to each yard.

The Soviet forces were established behind an anti-tank ditch 11 yards wide and 16 feet deep which ran across the entire width of the isthmus. Behind it extensive wire obstacles had been erected and thousands of mines laid. Massive girder-like structures, made of rails welded together like bristling hedgehogs, protected machine-gun posts, strongpoints, and gun emplacements. With water on both sides of this 11-mile front, all possibility of outflanking was ruled out.

"So that's where we are to drive through, Herr Generaloberst?" asked Manstein's driver and general factotum, Fritz Nagel, after looking through the trench telescope at the observation post of 114th Artillery Regiment, which offered a good view of the Soviet positions.

"Yes, that's where we've got to drive through, Nagel." Manstein nodded. He pushed his cap back and once more pressed his eyes to the telescope through which he had just let his sergeant have a look.

Fritz Nagel was always welcome at all headquarters. A native of Karlsruhe, he had been Manstein's driver since 1938. Whenever Manstein drove to the front Nagel was behind the wheel. He was calmness personified, and had more than once handled dangerous situations. Several times he had been wounded. But Manstein himself had never even been scratched: Nagel was a kind of talisman.

Manstein had driven out to the forward O.P. of 114th Artillery Regiment, in the sector of 46th Infantry Division, in the northern part of the front across the Parpach Isthmus, in order to have another look at the Soviet system of defences.

"Any other news?" he asked the commander of 46th Infantry Division. "Nothing special, Herr Generaloberst," Major-General Haccius replied.

"Well, good luck then, the day after to-morrow." Manstein nodded. "Come on, Nagel; we're driving home."

The day after to-morrow—8th May, D-Day for "Bustard Hunt," the code name for the break-through to Kerch.

If one is dealing with an enemy three times one's own strength and established, moreover, in a cleverly constructed defensive position, one can dislodge him only by courage and cunning. Manstein therefore based his plan on cunning.

The Soviet front in the isthmus had a curious shape: in its southern part it ran dead straight to the north, but in its northern part there was a big bulge to the west. This had originally been formed after the Soviets had dislodged the Rumanian 18th Division in the winter, and

P

German battalions had only just been able to seal off the Soviet penetration.

The obvious move would have been to strike at the flank of this bulge. But just because it was the obvious solution—and because the Russians expected it and had concentrated two Armies as well as nearly all their reserves in this sector—Manstein resisted the temptation. The fact that he chose a different plan again showed him to be one of the outstanding strategists of the Second World War.

Naturally, Manstein did everything to confirm enemy reconnaissance in the belief that he was going to strike in the north. Dummy artillery emplacements were built, troop movements were staged in the northern and central sectors of the front, radio signals intended for the enemy's monitoring service were sent out, and dummy reconnaissance actions were carried out.

But Manstein meanwhile was preparing to attack at the other end, in the southern sector of the line. The XXX Army Corps under Lieutenant-General Maximilian Fretter-Pico was to punch a hole with its three infantry divisions—the 50th, the 28th Light, and the 132nd—into the line of the Soviet Forty-fourth Army. After that the 22nd Panzer Division under Major-General Wilhelm von Apell, as well as a motorized brigade under Colonel von Groddeck, was to sweep through this breach deep into the Soviet hinterland in order subsequently to turn to the north, enveloping the Soviet forces, and then break through farther to the east.

It was a bold plan—five infantry divisions and one Panzer division against three Armies. Stuka formations of VIII Air Corps under Colonel-General Freiherr von Richthofen and units of Major-General Pickert's 9th Flak Division were available to support the infantry. Heavy Army artillery was brought over from Sevastopol for a concentrated bombardment.

To deal with the main obstacle, the anti-tank ditch, Manstein had thought up a particularly cunning move.

There was a great deal of strange activity on the beach east of Feodosiya during the night of 7th/8th May. Assault craft were being pushed into the water, and sappers and infantrymen of the Bavarian 132nd Infantry Division were getting in. But the engines remained silent. Boat after boat glided noiselessly away from the shore, propelled only by paddling. Soon the mysterious flotilla had been swallowed up by the night—four assault companies bobbing about on the Black Sea. Towards 0200 hours they were drifting along the coast to the east.

At 0315, like some primordial thunder-clap, the German artillery opened fire. Heavy mortars thundered, rocket batteries whined, AA guns hammered. Fire, smoke, and morning haze veiled the southern sector of the Parpach Isthmus. Stukas roared overhead and plummeted down. Their bombs tore into strongpoints and wire obstacles.

Prelude to Stalingrad 451

At 0325 hours pairs of white signals went up everywhere: the infantry was attacking. Right in front went the sappers. Theirs was the worst job—removing the mines and cutting the wire, always under enemy fire.

The Russians were putting up a barrage from all weapons. The Soviet machine-gunners behind the firing-slits of their pillboxes merely had to squeeze the trigger. They did not have to take aim. Their guns were emplaced for cross-fire, covering between them the entire forefield. All they had to do was shoot.

Soviet naval guns opened up. Mortars plopped. Shells, bombs, and bullets swept over the narrow neck of land across which the Germans must attack. Surely there was no other approach.

The moment the German artillery bombardment began the assault boats off the coast started up their engines. The Russians could not possibly hear the engine noise now.

Swift as arrows the boats streaked to the coast—towards the precise spot where the Soviet anti-tank ditch ran into the sea, wide as a barn-gate and filled with water.

The assault boats simply sailed into the ditch. The men leapt out and immediately started firing their machine-guns from their hips. The Soviets in their infantry dug-outs along the edge of the ditch were mown down before they even realized what was happening.

But there a built-in Russian flame-thrower opened up. The first German wave pressed themselves to the ground. They were pinned down.

A Messerschmidt fighter was coming in at low level from the sea. It roared along the trench, its guns blazing, and forced the Soviets to take cover.

The men of the German assault-boat commando leapt to their feet and burst into the trench. The first Russians raised their hands. There was utter confusion.

On the left of 132nd Infantry Division, along both sides of the Feodosiya–Kerch road, the 49th Jäger Regiment of the Silesian 28th Light Infantry Division was meanwhile working its way through the minefields. Captain Greve was leading the spearhead of 1st Battalion south of the road. He was racing through the enemy fire, along the narrow cleared lanes through the minefield.

The division had been assigned some self-propelled guns from Assault Gun Battalion 190. Lieutenant Buff, in charge of three of these steel fortresses, was advancing alongside the 1st Battalion, providing fire cover for Greve's men.

By 0430 hours the Jägers had reached the anti-tank ditch. Panting, the captain was lying by the edge. Sergeant Scheidt with his machine-gun was firing right and left. Sappers came running up with an assault ladder. Greve was the first to slide down into the ditch.

Major Kutzner, the commander of the 2nd Battalion, was severely wounded by the "Tartar Hill." There the Soviets had emplaced the anti-tank guns of an entire anti-tank regiment. The situation was saved by Second Lieutenant Fürnschuss with his self-propelled guns of Assault Gun Battalion 190. His 7·5-cm. long-barrel cannon shot up the Russian anti-tank guns.

First Lieutenant Reissner was charging at the head of his 7th Company. He ran through heavy enemy artillery-fire and flung himself down. He leapt to his feet again and ran on. There was the anti-tank ditch. Its edge had been smashed by gunfire. Reissner let himself roll down. A burst from a machine pistol cut him down. Though wounded, he continued to wave his Jägers on against the Soviet infantry dugout.

The 50th Infantry Division, on the left flank of the penetration area, was advancing through minefields and wire obstacles. Well-camouflaged machine-gun positions, which had survived the artillery barrage, caught them in enfilading and cross fire. The 1st Battalion, 123rd Infantry Regiment, suffered heavy casualties and was halted.

Lieutenant-Colonel von Viebahn, the regimental commander, was obliged to tackle the Soviet machine-gun posts by an attack at right angles to the front. By nightfall the 3rd Battalion eventually succeeded in penetrating as far as the anti-tank ditch.

Second Lieutenant Reimann with his 9th Company and parts of 10th Company, likewise put under his command, rolled up the Soviet positions along the ditch from the regiment's right wing as far as Lake Parpach; in furious hand-to-hand fighting he silenced all the machine-gun emplacements and strongpoints built into the anti-tank ditch and finally blasted the walls of the ditch to enable German armour to cross. The key obstacle of the Soviet defences had thus been taken along the whole front of the attack.

The companies of Colonel von Groddeck's Motorized Brigade—composed of Rumanian and German units, such as the Reconnaissance Detachment of 22nd Infantry Division—had managed, on the very afternoon of the first day of the attack, to reach the seashore in the sector of 132nd Infantry Division at the spot where earlier that day the assault boats had seized the anti-tank ditch, to clear the obstacle by quickly built crossings and to strike at the rear of the Soviet positions.

The spearheads of 22nd Panzer Division meanwhile were still waiting for their order to attack. But not until mid-morning of 9th May had the bridgeheads in the sector of the 28th Light and 50th Infantry Divisions been sufficiently enlarged to allow the rest of the units to be brought forward.

The Panzer companies and armoured cars deployed rapidly, burst into the Soviets' second and third lines of defence, broke all resistance,

Prelude to Stalingrad 453

reached the road bend to Arma-Eli, and crashed right into the assembly area of a Soviet armoured brigade.

As if the move had been rehearsed, six steel giants of Assault Gun Battalion 190 under Captain Peitz arrived on the scene at exactly the same moment. Before the Soviets could take up position they were smashed by the German tanks and assault guns.

As planned, 22nd Panzer Division now turned northward, behind the front of the two Soviet Armies which were still engaged with the Franconian–Sudeten 46th Infantry Division and the Rumanian brigades. Everything went according to Manstein's plan. But then abruptly the pattern changed. In the late afternoon of 9th May heavy spring rain began to fall. Within a few hours the tracks and the clay soil had turned into a bottomless morass. Jeeps and lorries were completely bogged down, and only tracked vehicles were able to make any progress. It was now Manstein's will against the forces of nature.

The armoured fighting vehicles of 22nd Panzer Division continued to struggle forward until late at night, and then took up position for all-round defence. Thus, when a clear day dawned on the following morning, 10th May, they were already deep in the flank and rear of the Soviet Fifty-first Army. A Soviet relief attack with strong armoured forces was repulsed. Wind sprang up and soon dried the soil. The division moved on towards the north. On 11th May it was at Ak-Monay by the sea, and thus in the rear of the Soviet Forty-seventh Army. Ten Russian divisions were in the bag. The remainder fled eastward. With this bold stroke the 22nd Panzer Division had wiped out a blot on its escutcheon—a blot dating back to 20th March 1942. On that day the newly organized Division, dispatched to the Crimea by the High Command of the Army without a single divisional exercise or even a test of co-operation between its formations, had been employed by Eleventh Army for a counter-attack on the Parpach front.

In the morning mist the formations had encountered Soviet forces preparing for an attack; they had got confused and had been shot up by the enemy. Field-Marshal von Manstein later admitted that it had been a mistake to send such an inexperienced division into a major operation. But what use was this admission by the Commander-in-Chief? Among the front-line formations of Eleventh Army the 22nd Panzer Division had been looked down upon since 20th March. Among the High Command of the Army it had likewise been in bad odour since that day. All the gallantry displayed by the division during the later part of the winter had been of no avail: the stigma of 20th March continued, unjustly, to stick to them.

Meanwhile Colonel von Groddeck and his fast brigade were boldly chasing eastward and preventing the Russians from establishing a line farther back. Wherever Soviet regiments tried to dig in von Groddeck struck. And then he raced on.

When the brigade had driven 30 miles deep into the hinterland and quite unexpectedly arrived at the "Tartar Ditch"—far behind the headquarters of Lieutenant-General D. T. Kozlov, the Commander-in-Chief Soviet Army Group Crimean Front—the Soviet command lost its nerve. Troops and headquarters disintegrated. Along the roads vast columns of fleeing formations were moving in the direction of Kerch, towards the eastern coasts of the peninsula. From there they hoped to save themselves across the straits to the mainland.

Desperately, Soviet tactical reserves attempted to halt the German spearheads, to enable as many formations as possible of the vast numbers accumulating on the beaches of the Kerch Peninsula to be ferried across to the mainland in motor-boats and light craft. They were hoping to repeat the feat accomplished by the British at Dunkirk almost exactly two years before.

But Manstein had no intention of having his victory diminished by any Soviet Dunkirk. He sent his armoured and motorized units, as well as Major-General Sander's North German 170th Infantry Division with 213th Infantry Regiment, to pursue and overtake the retreating Russians. But Colonel von Groddeck was no longer of the party. He had been severely wounded, and died of his wounds shortly afterwards. On 16th May Kerch was reached. The Soviet High Command did not succeed in pulling off a Dunkirk. Stalin was unable to save his Armies. The assault guns of XXX Army Corps, of Assault Gun Battalions 190, 197, and 249, soon put an end to the enemy's improvised naval transports.

The prize of the victory was 170,000 prisoners, 1133 guns, and 258 tanks. Three Soviet Armies had been defeated by half a dozen German divisions in eight days.

Early in the morning of 17th May Manstein and Colonel-General Freiherr von Richthofen were standing on a slight hill near Kerch. Before them lay the sea, the Kerch Strait, and beyond, barely 12 miles away, under a brilliant sun was the shore of the Taman Peninsula, the approaches to Asia, the gate to the Caucasus. With his victory Manstein had burst open the back-door to Stalin's fabulous oilfields.

At that very hour, when Manstein was gazing across to his great objective, 400 miles farther north, in the Kharkov area, the divisions of von Kleist's Army Group were mounting their attack which was to gain them the vital starting-line on the Donets for their summer offensive.

After many sleepless nights and anxious calculations, prompted by a Soviet surprise attack, Colonel-General von Kleist was now unleashing an offensive which has no parallel in terms of daring and strategic concept.

"Three o'clock," said Second Lieutenant Teuber, a Company Commander in 466th Infantry Regiment. No one made any reply. What was there to say? After all, it was a statement of fact. It meant that there were another five minutes to go.

In the east the sky was turning red. It was a cloudless sky. There was complete silence—so much so that the men's breathing could be heard. And also the ticking of the second lieutenant's large wrist-watch as he supported his hand against the edge of the trench. The seconds were ticking away—drops of time into the sea of eternity.

At last the moment had come. A roar of thunder filled the air. To the raw soldiers on the battlefield it was just an unnerving, deafening crash, but the old soldiers of the Eastern Front could make out the dull thuds of the howitzers, the sharp crack of the cannon, and the whine of the infantry pieces.

From the forest in front of them, where the Soviets had their positions, smoke was rising. Fountains of earth spouted into the air, tree-branches sailed up above the shell-bursts—the usual picture of a concentrated artillery bombardment preceding an offensive.

This then was the starting-line of the "Bear" Division from Berlin—but the picture was the same in the sectors held by the regiments of 101st Light Division, the Grenadiers of 16th Panzer Division, and the Jägers of 1st Mountain Division, the spearhead of attack of von Mackensen's III Panzer Corps. Along the entire front between Slavyansk and Lozovaya, south of Kharkov, the companies of von Kleist's Army Group were, on that morning of 17th May 1942, standing by to mount their attack under the thunderous roar of the artillery.

At last the barrage in front of the German assault formations performed a visible jump to the north. At the same moment Stukas of IV Air Corps roared over the German lines.

"Forward!" called Second Lieutenant Teuber. And, like him, some 500 lieutenants and second lieutenants were, at that very second, shouting out their command: "Forward!"

The question which had been worrying officers and men during the past few days and hours was forgotten—the great question of whether the German forces would succeed in striking at the root of the Russian offensive that had been moving westward for the past five days.

What was happening on that 17th May? What was the objective of the attack by Kleist's Army Group? To answer this question we must cast our eyes back a little.

For the purpose of gaining a proper starting-line for the great summer offensive of 1942 from the Kharkov area in the direction of the Caucasus and Stalingrad, Fuehrer Directive No. 41 had ordered that the Soviet bulge on both sides of Izyum, which represented a permanent threat to Kharkov, should be eliminated by a pincer operation. For this operation the C-in-C Army Group South, Field-Marshal von Bock, had made a simple plan: the Sixth Army under General Paulus was to attack from the north, and von Kleist's Group with units of

First Panzer Army and Seventeenth Army was to attack from the south. In this way Timoshenko's well-filled bulge was to have been pinched off and the Soviet Armies assembled in it annihilated in a battle of encirclement. The code word for this plan was "Fridericus."

But the Russians too had a plan. Marshal Timoshenko wanted to repeat his January offensive, and therefore he had prepared an attack with even stronger forces, an attack he hoped would decide the outcome of the war. With five Armies and a whole armada of armoured formations he intended to strike from the Izyum bulge and, north of it, from the Volchansk area, where his January offensive had ground to a halt, and burst through the German front with two wedges. In a big outflanking operation the town of Kharkov, the administrative centre of the Ukrainian heavy industry, was to have been retaken. This would have deprived the Germans of their vast supply base for the southern front, a base where enormous stores were situated.

Simultaneously Timoshenko wanted to repeat his earlier attempt of snatching Dnepropetrovsk from the Germans, as well as Zaporozhye, 60 miles farther on, with its huge hydro-electric power station which, in the forties, was a kind of eighth wonder of the world.

The realization of this plan would have been even more disastrous to the German Army Group South than the mere loss of their rearward base of Kharkov. Through Dnepropetrovsk and Zaporozhye ran the roads and railways to the lower reaches of the Dnieper; the river here was more like a string of lakes, and between those towns and the Black Sea there was no further certain crossing over it. All supplies for the German Armies on the southern wing, for the forces east of the Dnieper in the Donets area and in the Crimea, had to pass through these two traffic centres. Their loss would have precipitated disaster.

Thus in the spring of 1942 the attention of both sides was focused on the great bulge of Izyum, the fateful setting of future decisive battles for Bock as well as for Timoshenko. The question was merely: who would strike first, who would win the race against time—Timoshenko or Bock?

The German time-table envisaged 18th May as the day for the attack, but Timoshenko was quicker.

On 12th May he mounted his pincer operation against General Paulus's Sixth Army with surprisingly strong forces. The northern jaw of the pincers, striking from the Volchansk area, was represented by the Soviet Twenty-eighth Army with sixteen rifle and cavalry divisions, three armoured brigades, and two motorized brigades. That was an overwhelming force against two German Corps—General Hollidt's XVII Corps and General von Seydlitz-Kurzbach's LI Army Corps—with altogether six divisions.

Timoshenko's southern jaw struck with even more concentrated power from the Izyum bulge. Two Soviet Armies, the Sixth and the

Prelude to Stalingrad 457

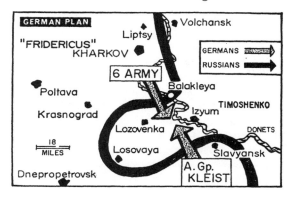

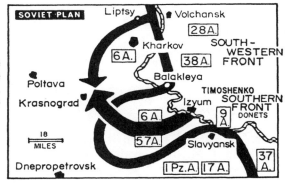

Map 26. The great battle south of Kharkov in the early summer of 1942, the curtain-raiser to Operation Blue.

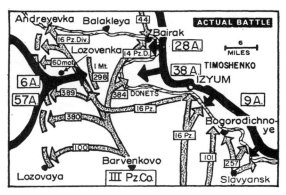

Fifty-seventh, pounced with twenty-six rifle and eighteen cavalry divisions, as well as fourteen armoured brigades, against the positions held by General of Artillery Heitz's VIII Corps and the Rumanian VI Corps. Half a dozen German and Rumanian infantry divisions, initially without a single tank, were finding themselves faced by a vastly superior enemy attacking with colossal armoured support.

There was no hope at all of intercepting the Russian thrust at its two focal points. The German lines were over-run. At the same time, just as during the winter battle, numerous German strongpoints held out in the rear of the advancing enemy.

General Paulus employed all available units of his Sixth Army against the Russian torrent bursting through his lines. Twelve miles before Kharkov he eventually succeeded, literally at the last moment, in halting Timoshenko's northern prong by striking at its flank with the hurriedly brought up 3rd and 23rd Panzer Divisions and 71st Infantry Division.

But Timoshenko's tremendously strong southern prong, striking from the Izyum bulge, was not to be stopped. Disaster seemed imminent. The Russians pursued their break-through far to the west, and on 16th May their cavalry formations were approaching Poltava, Field-Marshal von Bock's headquarters, more than 60 miles west of Kharkov. The situation was becoming dangerous. Bock was faced with a difficult decision.

In two days' time "Fridericus" was due to start. But the Soviet offensive had completely changed the situation. General Paulus's Sixth Army was pinned down and engaged in desperate defensive fighting. As an offensive striking force it had therefore to be written off. This meant that a pincer operation had become impossible.

Should he therefore drop the whole plan, or should he carry out "Operation Fridericus" with only one jaw? Bock's Chief of Staff, General of Infantry von Sodenstern, was urging him to adopt the "single-prong" solution. In view of the enemy's strength it would be a risky move—but an argument in its favour was the fact that with every mile he advanced farther westward Timoshenko's flank was getting more dangerously exposed.

That was Bock's chance. And in the end the Field-Marshal decided to take it. He decided to carry out "Operation Fridericus" with only one arm. To deny the Russians the possibility of screening their long flank he even advanced the date of attack by one day.

Thus von Kleist's Group—now called an Armeegruppe, or an Army-sized combat group—mounted its attack in the morning of 17th May from the area south of Izyum with units of First Panzer Army and Seventeenth Army. Eight infantry divisions, two Panzer divisions, and one motorized infantry division constituted Kleist's striking force. Rumanian divisions were covering its left wing.

At 0315 hours Second Lieutenant Teuber leapt out of his trench at the head of his company and with his men charged the Russian positions on the edge of the wood. Stukas screamed overhead, diving and dropping bombs on identified Soviet strongpoints, dugouts, and firing positions.

Some 2-cm. Army anti-aircraft guns on self-propelled carriages were driving along with Teuber's platoons, making up for their lack of tanks. Firing point-blank, these 2-cm. guns of Army Flak Battalion 616 slammed their shells into centres of Soviet resistance. The infantrymen were fond of this weapon and of their fearless crews who invariably rode with them into attack in the foremost line.

The first well-built Soviet positions collapsed under a hail of bombs and shells. Nevertheless those Russians who survived the artillery bombardment offered stubborn resistance. An assault battalion into whose position the 466th Infantry Regiment had driven held out to the last man. Four hundred and fifty killed Russians testified to the ferocity of the fighting.

Only slowly was the regiment able to gain ground through thick undergrowth, through minefields, and over obstacles made from tree-trunks. Second Lieutenant Teuber and his company found themselves up against the particularly stubbornly defended positions of the Mayaki Honey Farm, which was situated a short distance behind the main fighting-line. The Russians were using machine-guns, carbines, and mortars. The company made no headway at all.

"Demand artillery support," Teuber called over to the Artillery Liaison Officer. By walkie-talkie the Liaison Officer sent a message back: "Fire on square 14." A few minutes later a fantastic fireworks display broke out. Russian artillery, in turn, put down a barrage in front of the collective farm.

Teuber and his men charged. Over there was the Russian trench. The Soviets were still in it, cowering against its side. The charging German troops leapt in and likewise ducked close to the wall of the trench, seeking cover from the shells which were dropping in front, behind, and into the trench.

There they were crouching and lying shoulder to shoulder with the Russians. Neither side did anything to fight the other. Each man was clawing himself into the ground. For that moment they were just human beings trying to save themselves from the murderous screaming red-hot splinters of steel. It was as though enmity between man and man had been swept away by the insensate elemental force hammering down upon Russians and Germans.

Not till half an hour later, when the artillery-fire abruptly stopped, did Teuber's men leap to their feet and shout all along the trench: "*Ruki verkh*—hands up!" And the Russians dropped their machine pistols and rifles and raised their hands.

Teuber's platoons continued their advance. A little over a mile behind a bee-keeping farm they came across ten steaming Russian field kitchens which had just made tea and a millet porridge. The Russians were somewhat taken aback when German troopers suddenly lined up with their mess-tins. "Come on, Ivan, dish it out," they called. At first the Soviet cooks were apprehensive, but presently they grinned and piled the gruel into the Germans' mess-tins and filled their water-bottles with fragrant tea.

But breakfast ended on a different note. A Soviet biplane suddenly dipped low and machine-gunned the resting troops. The men of Teuber's Company opened up at the old-fashioned crate with their machine-guns and rifles. They scored several hits on the engine and ripped the wings to shreds. The machine reeled, went into a glide, and touched down barely 200 yards from where the troops had been taking things easy.

No. 1 Platoon charged the aircraft. But the pilot defended himself with his built-in machine-gun. When he had used up his ammunition he and his companion climbed out, both of them wearing leather flying-suits.

"*Ruki verkh!*" the Germans shouted. But the two Russians did not raise their arms. Instead they drew their pistols.

"Take cover!" the platoon commander shouted. But there was no need for that. The two airmen had no intention of resisting: they merely intended to escape captivity. First the officer accompanying him and then the pilot himself put bullets through their heads. When Teuber's men, still shaking their heads uncomprehendingly, went over to recover their bodies they discovered that the second officer was a girl holding the rank of second lieutenant.

By nightfall on 17th May the regiments of Colonel Puchler's 257th Infantry Division had reached the Donets along the entire width of their front. On 18th May they took their most northerly objective—Bogorodichnoye. Just as First Lieutenant Gust, commanding 3rd Battalion, 477th Infantry Regiment, reached the edge of the village with his foremost platoon, a river ferry crowded with thirty horses was making a last desperate effort to cast off from among the blazing barges. On catching sight of the Germans, however, the ferryman gave up the attempt. Burning boats were drifting down the river like meandering islands of fire.

Farther to the left the 101st Light Division also reached the Donets by the evening of 18th May. In a sweltering damp heat of 30 degrees Centigrade the battalions had to drive through a vast area of woodland, pick their way cautiously past well-camouflaged Soviet forest positions, moving in line abreast, and struggle laboriously through deep minefields. The sappers worked wonders. The Engineers Battalion 213, advancing with 101st Light Infantry Division, rendered harmless 1750 mines of all types on the first day.

For the first time since the previous summer's offensive mine dogs were encountered again—Alsatians and Doberman Pinschers with primed anti-tank mines on their backs. The dog-handlers, crouching in well-camouflaged positions, ordered the animals time and again to the advancing German formations. In a sickening kind of dog-hunt the animals were picked off and killed. But more and more of them came, entire packs of them, attempting, as they had been drilled, to get under the vehicles and gun-limbers. Wherever they succeeded and the projecting trigger-rod of the mine met with resistance the heavy charge exploded, together with the dog, and everything over an area of several yards was blown to pieces.

With the Donets line gained, 257th Infantry Division and 101st Light Division took over the eastern flank cover for the deep thrust by the armoured striking groups, a thrust aimed at the creation of a pocket. The 16th Panzer Division, acting as the spearhead of Lieutenant-General Hube's striking force, drove through the Russian positions with three combat groups under von Witzleben, Krumpen, and Sieckenius. They dislodged the enemy and repulsed strong counter-attacks. Then they drove on, straight through, into the suburbs of Izyum.

By 1230 hours on 18th May tanks and motor-cyclists of the Westphalian 16th Armoured Division were covering the only major east-west road crossing the Donets at Donetskiy. The combat group Sieckenius, the mainstay of which was 2nd Battalion, 2nd Panzer Regiment, left-turned and drove on westward, straight into the pocket.

The main blow of "Operation Fridericus," however, was to be dealt by General of Cavalry von Mackensen with his III Panzer Corps. He attacked with the 14th Panzer Division from Dresden in the centre, and with the Viennese 100th Light Division and the Bavarian 1st Mountain Division on the right and left respectively. The Russians were taken by surprise and routed by the swampy Sukhoy–Torets river. Barvenkovo was taken. A bridge was built. The 14th Panzer Division crossed over and pushed on towards the north. Eddying clouds of dust veiled the tanks. The fine black earth made the men look like chimney-sweeps.

In co-operation with the Panzer companies of Combat Group Sieckenius the Bereka river was crossed. Soviet armoured thrusts were successfully repulsed. In the afternoon of 22nd May, 14th Panzer Division reached Bayrak on the northern Donets bend.

That was the turning-point. For across the river, on the far bank, were the spearheads of Sixth Army—companies of the Viennese 44th Infantry Division, the "Hoch-und-Deutschmeister." With this link-up the Izyum bulge was pierced and Timoshenko's Armies, which had driven on far westward, were cut off. The pocket was closed.

Too late did Marshal Timoshenko realize his danger. He had not expected this kind of reply to his offensive. Now he had no other choice but to call off his promising advance to the west, turn his divisions about, and attempt to break out of the pocket in an easterly direction, with reversed fronts. Would the thin German sides of the pocket stand up to such an attempt? The decisive phase of the battle was beginning.

Colonel-General von Kleist was faced with the task of making his encircling front strong enough to resist both the Soviet break-out attempts from the west and their relief attempts mounted across the Donets from the east. Once more it was a race against time. With brilliant tactical skill General von Mackensen grouped all infantry and motorized divisions under his command like a fan around the axis of 14th Panzer Division. The 16th Panzer Division was first wheeled west and then moved north towards Andreyevka on the Donets. The 60th Motorized Infantry Division, the 389th Infantry Division, the 380th Infantry Division, and the 100th Light Division fanned out towards the west and formed the pocket front against Timoshenko's Armies as they flooded back east.

At the centre, like a spider in its web, was General Lanz's 1st Mountain Division; it had been detached from the front by von Mackensen to be available as a fire brigade.

This precaution finally decided the battle. For Timoshenko's Army commanders were driving their divisions against the German pocket front with ferocious determination. They concentrated their efforts in an attempt to punch a hole into the German front, regardless of the cost, in order to save themselves by reaching the Donets front only 25 miles away.

On Whit Monday the encircled Armies succeeded in steam-rolling their way through the barrier set up by 6th Motorized Infantry Division and 389th Infantry Division and in driving on to Lozovenka. It was clear that the Russians were trying to reach the main road to Izyum. It was then that Mackensen's precaution proved decisive. The Soviets encountered the 1st Mountain Division, which had taken up a switchline east of Lozovenka. The cover groups of 384th Infantry Division, supported by IV Air Corps, also flung themselves into the path of the Soviets. The action which followed was among the bloodiest of the whole war in Russia.

The following account is based on the report made by Major-General Lanz, then G.O.C. 1st Mountain Division, in his divisional records. By the light of thousands of white flares the Russian columns struck at the German lines. Officers and commissars were spurring on their battalions with shrill shouts of command. Arms linked, the Red Army men charged. The hoarse "*Urra*" rang eerily through the night.

"Fire," commanded the German corporals at the machine-guns and infantry guns. The first waves of attackers collapsed. The earth-brown columns wheeled to the north.

But there, too, they encountered the blocking positions of the Mountain Jägers. They ebbed back, and now, regardless of casualties, came pounding against the German front. They beat down and stabbed whatever opposed them, gained a few more hundred yards, and then sagged and collapsed in the enfilading fire of the German machine-guns. Whoever was not killed staggered, crawled, or stumbled back into the ravines of the Bereka river.

The following evening the same scene was repeated. But that time several T-34s accompanied the charging infantry. The Russian troops, their arms still linked, were under the influence of vodka. How else could the poor fellows find the courage to charge with shouts of "*Urra!*" into certain death?

Wherever a German strongpoint had been overwhelmed by the Soviets the bodies of its defenders were found, after a counter-attack had been launched, with their skulls cracked open, bayoneted, and trampled into unrecognizability. The fighting was marked by savage fury. It was an appalling highway of death.

On the third day, finally, the momentum of the Russians was broken. The two Commanders-in-Chief of the Soviet Sixth and Fifty-seventh Armies, Lieutenant-General Gorodnyanskiy and Lieutenant-General Podlas, as well as their staff officers, were lying dead on the battlefield. The great battle was over; Timoshenko was defeated. He had lost the bulk of twenty-two rifle divisions and seven cavalry divisions. Fourteen armoured and motorized brigades were completely routed. Some 239,000 Red Army men were wearily shuffling into captivity; 1250 tanks and 2026 guns had been destroyed or captured. That was the end of the battle south of Kharkov, the battle in which the Soviets had intended to surround the Germans and had been surrounded instead themselves. It was an unusual German victory—conjured up out of a defeat within a matter of days.

However, the victorious German divisions did not suspect that the success won by military skill and valour had merely opened the door for them to a sombre destiny: the men were now marching towards Stalingrad.

As yet the shadow of this city had not fallen upon the troops. Their minds and the High Command communiqués were still full of Kerch and Kharkov. After all, they had scored an astonishing success—two great battles of annihilation within three weeks. Six Soviet Armies had been smashed, 409,000 Soviet troops had been taken prisoner, 3159 guns and 1508 tanks had been destroyed or captured. The German Armies in the East had once more displayed their superiority. The fortunes of war were again following Hitler's colours. The terrible winter and the spectre of defeat had been forgotten.

And while the last few shots were still being exchanged in the pocket south of Kharkov, while small groups and handfuls of half-starved Russians were crawling out of their hide-outs, the machinery of a new battle was already turning—the battle for Sevastopol, the last Soviet strongpoint in the south-western corner of the Crimea, the strongest fortress in the world.

2. Sevastopol

A grave in the cemetery of Yalta–Between Belbek Valley and Rose Hill–324 shells per second–"Karl" and "Dora," the giant mortars–A fire-belching fortress–The "Maxim Gorky" battery is blown up–"There are twenty-two of us left.... Farewell"–Fighting for Rose Hill–Komsomols and commissars.

"WE'RE ready to cast off, Herr Generaloberst." The Italian naval lieutenant saluted. Manstein touched his cap, nodded with a smile, and turned to his entourage: "All right then, gentlemen, let's board our cruiser."

The cruiser was an Italian motor torpedo-boat, the only naval unit available to Manstein. Captain Joachim von Wedel, the Harbour Commandant of Yalta, had somehow got hold of it. Manstein wanted, on that 3rd June 1942, to sail along the southern coast of the Crimea to establish for himself whether the coastal road was under observation from the sea. It was along this road that all supplies for XXX Corps, which was holding the southern front of Sevastopol under General Fretter-Pico, had to be moved. Any threat to these supplies by Soviet naval units might have upset the programme for the battle of Sevastopol.

In brilliant sunshine the boat streaked along the Black Sea coast. The gardens of Yalta with their tall trees provided a beautiful setting for the white country houses and palaces. The boat held a westerly course until it was off Balaclava. The ancient fort on the bare, rocky hilltop towered into the blue sky with its two massive bastions.

The bay which cut into the shore at the foot of the rock was an iridescent blue. This was where in 1854–55, during the Crimean War, French, British, Turks, and Piedmontese, having landed at Yevpatoriya, fought their unending battle to bend the Tsar Nicholas to their will. The siege and battle of Sevastopol had gone on for nearly a year—347 days, to be exact—before the Russians surrendered. The number of casualties, including civilians, had been very high for those days. Estimates vary between 100,000 and 500,000.

Colonel-General von Manstein was acquainted with all these facts. He had read all the literature about the Crimean War. He also knew that under the ancient forts the Soviets had built entirely new modern defences—huge casemates, reinforced-concrete gun emplacements with armoured cupolas, and a labyrinth of underground supply stores. There was no doubt that in 1942 Stalin would defend this naval fortress every bit as stubbornly as the Tsar Nicholas I had done in 1854–55. Sevastopol with its favourable natural port was the main base of the Soviet Navy in the Black Sea. If it fell the Navy would have to withdraw to some hide-outs on the eastern coast.

Manstein and Captain von Wedel were engrossed in conversation when the boat was suddenly shaken by a crashing, splintering noise.

"Enemy aircraft," shouted Manstein's orderly, First Lieutenant Specht. The Italians racing to their anti-aircraft machine-gun were too late. Out of the sun two Soviet fighters from Sevastopol had pounced down on the boat and shot it up with their cannon.

The deck had been ripped open, and fire was beginning to spread. Captain von Wedel, who had been sitting next to Manstein, collapsed —dead. The Italian mate was slumped over the rail—also dead.

Fritz Nagel, Manstein's faithful companion in every battle since the first day of the war, had been flung against the after ventilation shaft with a severe thigh wound. His artery had been severed. Blood was welling out of his wound in quick spurts. The Italian commander tore off his shirt to tie up Nagel's artery.

Lieutenant Specht threw off his clothes, dived into the sea, and swam to the coast. Stark naked, he stopped a surprised lorry-driver and made him race to Yalta. There the lieutenant grabbed a motor-launch, raced back to the blazing motor torpedo-boat, and towed her back into Yalta harbour.

Manstein himself took Nagel to the military hospital. But it was too late. The sergeant was beyond help.

Two days later, when all round Sevastopol the squadrons of General Richthofen's VIII Air Corps were starting up their engines for the first act of the great battle, Manstein stood at the open grave of his driver in the cemetery of Yalta. The words spoken by the Colonel-General over the coffin of his sergeant are worthy of record in an otherwise so frightful chronicle of a frightful war: "Over the years during which we shared both the daily routine and major events we became friends. This bond of friendship cannot be severed even by the vicious bullet which struck you. My gratitude and loyal affection, the thoughts of all of us, accompany you beyond the grave into eternity. Rest in peace and farewell, my best comrade."

The rifle salvo rang out over the tree-tops. From the west came a sound like distant thunder: Richthofen's squadrons had gone into action against Sevastopol. The great twenty-seven-day battle against the world's strongest fortress had begun.

From the rocky hilltop there was a magnificent view of the entire area of Sevastopol. Sappers had blasted an observation post into the rock-face, reasonably secure against enemy artillery and aircraft. From there the entire town and fortified area could be surveyed by stereo-telescope as if from a viewing platform.

466 Part Six: The Caucasus and the Oilfields

At this observation post Manstein spent hour after hour with his chief of operations, Colonel Busse, and his orderly officer, "Pepo" Specht, observing the effect of the Luftwaffe's and artillery's bombardment. The date was 3rd June 1942.

At the spot where the ancient Greeks had established their first trading-post, where the Goths had built their hilltop castles during the great upheavals of the early centuries of the Christian era, where Genoese and Tartars had fought for the harbours and fertile valleys, and where in the nineteenth century rivers of British, French, and Russian blood had been shed, a German general was now sitting, pressed close to the rock, once more directing a battle for the harbours and bays of the Crimea, that idyllic peninsula in the Black Sea.

"Fantastic fireworks," Specht remarked. Busse nodded. But he was sceptical. "Even so I'm not sure that we'll punch sufficiently large holes into those fortifications for an infantry attack."

Manstein was standing by the trench telescope, gazing across to the Belbek Valley with its prominent peak which the troopers had named the Mount of Olives. Squadrons of Stukas were roaring overhead. They dived down on Sevastopol. They dropped their bombs and fired their cannon and machine-guns. Then they turned away again. Ground-attack aircraft skimmed over the plateau. Fighters tore across the sky at great height. Bombers droned along steadily. The Eleventh Army had unchallenged control of the air within a few hours of the beginning of the bombardment. The weak Soviet Air Force of the coastal Army had been smashed. It had gone into the battle with only fifty-three aircraft.

General von Richthofen's VIII Air Corps flew 1000, 1500, and even 2000 missions a day. "Continuous attack" was the Air Force experts' term for this conveyor-belt type of raid. And while a rain of bombs fell on Sevastopol from the sky, German guns of all calibres were pumping their shells into the enemy positions. The artillerymen sought out the enemy gun emplacements. They levelled trenches and wire obstacles. They sent shell after shell against the firing-slits and armoured cupolas of concrete gun positions. They fired their guns day and night—throughout five times twenty-four hours.

This was Manstein's idea of a really decisive overture to the attack—not, as usual, just one or two hours' concentrated artillery and Luftwaffe bombardment, followed by an infantry charge. Manstein knew that a conventional preliminary bombardment would be ineffective against Sevastopol's massive defences with their hundreds of concreted and armour-plated gun emplacements, the deep belt of pill-boxes, the powerful armoured batteries, the three defensive strips with their total of 220 miles of trenches, the extensive wire obstacles and minefields, and the rocket cannon and rocket mortars mounted in positions hewn out of the cliff-face.

That was why Manstein's plan provided for five days of massive annihilating fire by artillery, mortars, anti-aircraft guns, and assault guns. Altogether 1300 barrels hurled their missiles against the identified fortification works and field positions of the Soviets. To that must be added the bombs dropped by the squadrons of VIII Air Corps. The earth was covered with a hail of steel.

It was a murderous overture. Never during the Second World War, neither before Sevastopol nor after it, were such massed artillery forces employed by the Germans.

In North Africa at the end of October 1942 Montgomery opened the British offensive against Rommel's positions at El Alamein with his now historic 1000 barrels. Manstein employed 300 more at Sevastopol.

A particular rôle in the artillery bombardment was played by the mortars. This eerie weapon was here for the first time employed in heavy concentration. Two mortar regiments—the 1st Heavy Mortar Regiment and the 70th Mortar Regiment—as well as the 1st and 4th Mortar Battalions, had been concentrated in front of the fortress under the special command of Colonel Niemann—altogether twenty-one batteries with 576 barrels, including the batteries of 1st Heavy Mortar Regiment with their 11- and 12½-inch high-explosive and incendiary oil shells, which were particularly effective against fortifications.

Each second the bombardment lasted the barrels of this regiment alone spewed forth 324 mortar bombs against strictly limited sectors of the enemy's field fortifications. The effect on the Russians' morale was as destructive as the bombs' physical effects. The effect of thirty-six monstrous missiles with fiery tails whooshing from one single battery and slamming with a nerve-racking whine into an enemy position was unimaginable.

The fragmentation effect of a single mortar bomb was not as great as that of an artillery shell, but the blast of several of them exploding close to one another burst the troops' blood-vessels. Even the men lying a short distance away from the point of impact were demoralized by the deafening noise and paralysing pressure of the explosion. Terror and fear grew into panic. Only Stukas had been known to produce a similar effect on the usually so impassive Russians. It is only fair to say that, faced with concentrated fire from Russian rocket mortars, "Stalin's organ pipes," the German troops were often similarly gripped by fear and terror.

Among the conventional artillery battering against the doors of the fortress of Sevastopol there were three special giants which have gone down in military history—the Gamma mortar, the mortar "Karl" (also known as "Thor"), and the railway gun "Dora." These three miracles of modern engineering, the last word, as it were, in conventional artillery development, had been specially built for employment against fortresses. Before the war the only fortresses in existence, apart from those in Belgium and the French Maginot Line, were Brest Litovsk, Lomsha, Kronshtadt, and Sevastopol. Leningrad was no longer a fortress in the true meaning of the word, and the ancient French fortress towns along the Atlantic coast had long since ceased to count.

The Gamma mortar was a revival of "Big Bertha" of the First World War. Its 16·8-inch projectiles weighed 923 kg.—nearly a ton—and could be hurled at targets nearly 9 miles away. The length of its barrel was twenty-two feet. This unusual giant was serviced by 235 artillery men.

But "Gamma" was a pigmy compared with the 24·2-inch mortar known as "Karl" or "Thor"—one of the heaviest pieces of the Second World War and a special Army weapon against the most powerful concrete fortresses. Its 2200-kg. (2¼-ton) concrete-piercing bombs, which shattered the strongest concrete roofs, were hurled by a monster which barely resembled a conventional mortar at all. Its relatively short barrel of a little over sixteen feet and its colossal carriage and bogies made it look rather like some factory with an eerie stub of a smokestack.

But even "Karl" was not quite the last word in gunnery. That last word was stationed at Bakhchisaray, in the "Palace of Gardens" of the ancient residence of the Tartar Khans, and was called "Dora," or occasionally "Heavy Gustav." It was the heaviest gun of the last war. Its calibre was 31½ inches. Sixty railway carriages were needed to transport the parts of the monster. Its 107-foot barrel ejected high-explosive projectiles of 4800 kg.—*i.e.*, nearly five tons—over a distance of 29 miles. Or it could hurl even heavier armour-piercing missiles, weighing seven tons, at targets nearly 24 miles away. The missile together with its cartridge measured nearly twenty-six feet in length. Erect that would be about the height of a two-storey house.

"Dora" was able to fire three rounds in one hour. The giant gun stood on two double rails. Two flak battalions were permanently employed to guard it. Its operation, protection, and maintenance required 4120 men. The fire control and operation alone included one major-general, one colonel, and 1500 men.

These data are sufficient to show that here the conventional gun had been enlarged to gigantic, almost super-dimensional scale—indeed, to a point where one might question the economic return obtained from such a weapon. Yet one single round from "Dora" destroyed an ammunition dump in Severnaya Bay at Sevastopol although it was situated 100 feet below ground.

Manstein had been standing in his eyrie on the rock for the best part of three hours. He was closely watching the shell-bursts and comparing them with the exact data supplied to him by his Army's two chiefs of artillery, Lieutenant-General Zuckertort, the chief of artillery of LIV Corps, and Lieutenant-General Martinek, chief of artillery of XXX

Corps. With all his genius for strategy, Manstein was a man of detail. Indeed, this may have been the secret of his success.

"Wherever the 8·8 scores a direct hit there's no Ivan left looking out of the strongpoint," said Pepo Specht, who was just then looking through the telescope.

"Yes, the flak is quite indispensable against this kind of fortification," Manstein replied. As if to underline his words, the metallic crump of the 8·8-cm. guns rang out clearly through the hurricane of noise.

These anti-aircraft guns were indeed indispensable. It was in the siege of Sevastopol that the 18th Flak Regiment gained its fame. The flat-trajectory 'eight-eight' was the best weapon against fortifications projecting above ground-level. Employed in the foremost line, just like the mortars, the 8·8 guns, these fantastic miracle weapons of the Second World War, cracked pillboxes and gun emplacements at point-blank range. The 8·8-cm. batteries of 18th Flak Regiment alone fired 181,787 rounds in the course of the battle for Sevastopol.

From Manstein's observation post the three deeply echeloned defence systems protecting the core of the fortress were clearly visible.

The first of them was one to two miles deep with four sets of wire-protected trench positions in echelon, with timber strongpoints and concreted emplacements between them. Mines detonating by shell-bursts in front of and among the trenches indicated that the Russians had additionally laid thick belts of anti-tank mines. It was to be expected that many of these invisible obstacles had also been laid down against infantry assault.

The second belt of defences was about a mile deep and included, especially in its northern sector between the Belbek Valley and Severnaya Bay, a number of extremely heavy fortifications to which the German artillery observers had given easily remembered names—"Stalin," "Molotov," "Volga," "Siberia," "GPU," and—above all—"Maxim Gorky I," with its heavy 12-inch armoured batteries. The companion piece of this fort, "Maxim Gorky II," was south of Sevastopol and similarly equipped.

The eastern front of the fortress was particularly favoured by nature. Difficult country with deep, rocky ravines and fortified mountain-tops provided ideal ground for the defenders. "Eagle's Nest," "Sugar Loaf," "Northern Nose," and "Rose Hill" are the names that will always be remembered by those who fought on the eastern sector of the Sevastopol front.

A third belt of defences ran immediately round the town. This was a veritable labyrinth of trenches, machine-gun posts, mortar positions, and gun batteries.

According to Soviet sources, Sevastopol was defended by seven rifle divisions, one cavalry division on foot, two rifle brigades, three naval brigades, two regiments of marines, as well as an assortment of tank battalions and independent formations—a total of 101,238 men. Ten artillery regiments and two battalions of mortars, an anti-tank regiment and forty-five super-heavy coastal-defence naval gunnery units with altogether 600 guns and 2000 mortars were holding the defensive front. It was truly a fire-belching fortress that Manstein intended to take with his seven German and two Rumanian divisions.

The night of 6th/7th June was hot and sultry. Towards the morning a light sea-breeze sprang up. But it did not carry in any sea air—only dust from the churned-up approaches of Sevastopol. Clouds of this dust and smoke from the blazing ammunition dumps in the southern part of the town were drifting over the German lines.

At daybreak the German artillery bombardment once more stepped up its volume. Then the infantry leapt to their feet. Under this tremendous artillery umbrella assault parties of infantry and sappers charged against the enemy's main fighting-line at 0350 hours.

The main effort was made on the northern front. There the LIV Corps attacked with 22nd, 21st, 50th, and 132nd Infantry Divisions and with the reinforced 213th Infantry Regiment, which belonged to 73rd Infantry Division and formed the Corps reserve.

XXX Corps attacked from the west and south. But all that was not yet the main drive. The 72nd Infantry Division, the 28th Light Infantry Division, and the 170th Infantry Division, together with the Rumanian formations, were merely to gain a starting-line for the main attack scheduled for a few days later.

Up in the Belbek Valley and the Kamyshly Ravine the sappers were clearing lanes through the minefields, to enable the assault guns of Battalions 190 and 249 to be employed as quickly as possible in support of the infantry. Meanwhile the infantrymen were contending for the first enemy field positions. Although the artillery had smashed trenches and dugouts, the surviving Russians were resisting desperately. They had to be driven out of their well-camouflaged firing-pits with hand-grenades and smoke canisters.

Major-General Wolff's 22nd Infantry Division from Lower Saxony was once again assigned the difficult task of seizing Fort Stalin. During the previous winter the assault companies of 16th Infantry Regiment had scaled the outer ramparts of the fortress, but had then been compelled to withdraw again and had been taken back to the Belbek Valley.

Now they were to travel that costly road once more. Their first attempt, on 9th June, misfired. On 13th June the 16th, under the command of Colonel von Choltitz, again attacked the fort. Fort Stalin was a heap of shattered masonry, but fire was still coming from all directions. In the Andreyev wing the fortress commandant had employed only Komsomol and Communist Party members. In the operations report of the 22nd Infantry Division we read: "This was probably the toughest opponent we ever encountered."

To quote just one of many examples. Thirty men had been killed in one pillbox which had received a direct hit against its firing-slit. Nevertheless the ten survivors fought like demons. They had piled up their comrades' dead bodies like sandbags behind the shattered embrasures.

"Sappers forward!" the infantrymen shouted. Flame-throwers directed their jets of fire against this horrible barricade. Hand-grenades were flung. Some German troops were seen vomiting. But not until the afternoon did four Russians, trembling and utterly finished, reel out of the wreckage. They gave up after their political officer had shot himself.

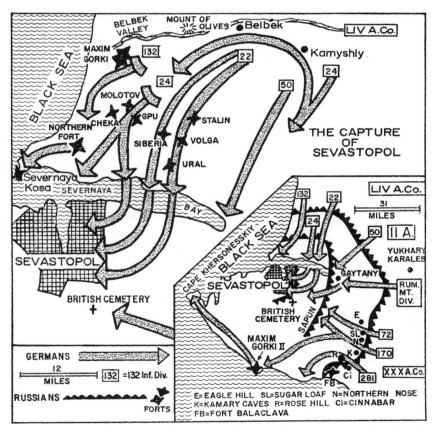

Map 27. The capture of Sevastopol. After five days of "annihilation bombardment" by artillery and Luftwaffe, Eleventh Army on 7th June 1942 launched its attack on the world's strongest fortress. The last fort fell on 3rd July 1942.

472 Part Six: The Caucasus and the Oilfields

The two attacking battalions of 16th Infantry Regiment suffered heavily in this savage fighting. Before long all their officers had fallen. A second lieutenant from the "Leaders' Reserve" took over command of the remnants of the rifle companies of the two battalions.

The fierce fighting for the second belt of defences raged in a sweltering heat until 17th June. An unbearable stench hung over the battlefield, which was covered with countless bodies over which flies buzzed in great clouds. The Bavarian 132nd Infantry Division on the right of the Lower Saxons had suffered such heavy casualties that it had to be temporarily withdrawn from the line. Its place was taken by the 24th Infantry Division, which—relieved in turn by the Rumanian 4th Mountain Division—was inserted between the 132nd and 22nd Infantry Divisions.

The situation was far from rosy for the German formations. Casualties were mounting ever higher, and an acute shortage of ammunition made it necessary from time to time to call a halt in the fighting. Already some commanding officers were suggesting that the attack should be suspended until new forces had been brought up. But Manstein knew very well that he could not count on any reinforcements.

On 17th June he gave orders for a renewed general attack along the entire northern front. The battered regiments once again went into action, firmly determined to take the main obstacles this time.

In the Belbek Valley, two and a half miles west of the "Mount of Olives," two 14-inch mortars were being moved into position. They belonged to the heavy motorized Army Artillery Battalion 641, and their instructions were to smash the armoured cupolas of "Maxim Gorky I." This Soviet fort's 12-inch shells were controlling the Belbek Valley and the way down to the coast.

It was a terrible job to get the two giants into firing position. After four hours of hard work by a construction party the battery commander, Lieutenant von Chadim, was at last able to give his first firing orders.

With a thunderous roar the monsters came to life. After the third salvo Sergeant Meyer, who was lying as the forward observer in the foremost line of 213th Infantry Regiment, reported that the hits scored by the concrete-piercing missiles had so far produced no effect on the armoured cupola.

"Special Röchling bombs," Chadim commanded. The nearly 12-feet-long projectiles, weighing one ton each, were handled into position with the aid of cranes. The "Röchlings" had already proved their value in the Western campaign, against the fortifications of Liège. They exploded not on impact, but only after they had penetrated some way into the resisting target.

Sergeant Friedel Förster and his fourteen comrades at No. 1 gun clapped their palms over their ears as the lieutenant raised his arm. "Fire!"

Twenty minutes later the operation was repeated. "Fire!"

Shortly afterwards came Sergeant Meyer's signal: "Armoured cupola blown off its hinges!"

"Maxim Gorky" had had its head cracked. The barrels of its 12-inch naval guns could be seen pointing at an angle at the sky. The battery was silent.

Now was the moment for Colonel Hitzfeld, the conqueror of the Tartar Ditch of Kerch. At the head of the battalions of his 213th Regiment he charged against the fort and occupied the armoured turrets and approaches.

"Maxim Gorky I" was no longer able to fire. But the Soviet defenders inside the vast chunk of concrete, which was over 300 yards long and 40 yards wide, did not surrender. Indeed, groups of them even staged lightning-like sorties through secret exits and ventilation-shafts.

The second company of Engineers Battalion 24 was ordered to put an end to the opposition. Demands for surrender were answered by the Soviets with machine-pistol fire. The first major blasting operation was staged with a mountain of dynamite, incendiary oil, and smoke canisters. When the fumes and smoke had dispersed the Soviets were still firing from embrasures and openings.

The second blasting eventually tore the concrete block wide open. Its vast interior was revealed to the sappers. "Maxim Gorky I" was three storeys deep—a veritable town.

The fort had its own water and power supply, a field hospital, canteen, engineering shops, ammunition lifts, arsenals, and deep battle stations. Every room and every corridor was protected by double steel doors. Each of these had to be blasted open individually.

The sappers flattened themselves against the walls. When the steel doors burst they flung their hand-grenades into the smoke and waited for the fumes to disperse. Then they went on to the next door.

The corridors were littered with Soviet dead. They looked like monsters, since they all wore gas-masks. The smoke and the stench had made this necessary.

In the next corridor the Germans suddenly came under machine-pistol fire. Hand-grenades were flung; pistol-shots rang out. Then a steel door slammed. The savage game started anew. Thus it went on hour after hour, as the fighting approached the nerve centre of the fortress—the command post.

The fighting in the fort "Maxim Gorky" was closely followed also in Sevastopol, in the battle headquarters of Vice-Admiral Oktyabrskiy, near the harbour. The wireless officer, Lieutenant Kuznetsov, was sitting at his receiver in the wireless-room listening. Every thirty minutes a report came through from "Maxim Gorky" on the situation. The

Admiral's order to all commanders and commissars had been: "Resistance to the last man."

There was the signal. Kuznetsov listened and took it down: "There are forty-six of us left. The Germans are hammering at our armoured doors and calling on us to surrender. We have opened the inspection hatch twice to fire at them. Now this is no longer possible."

Thirty minutes later came the last signal: "There are twenty-two of us left. We're getting ready to blow ourselves up. No more messages will be sent. Farewell."

They were as good as their word. The brain centre of the fort was blown up by its surviving defenders. The battle was over. Of the fort's complement of 1000 only fifty men were taken prisoners, and they were wounded. This figure speaks for itself.

On 17th June, while the battle for "Maxim Gorky I" was still raging, the Saxon battalions of 31st Infantry Regiment, 24th Infantry Division, took the forts "GPU," "Molotov," and "Cheka."

Major-General Wolff's 22nd Infantry Division from Bremen likewise made some headway towards the south, on the left of the Saxons, and on 17th June, with 65th Infantry Regiment, reinforced by 2nd Battery Assault Gun Battalion 190, took the fort "Siberia." The 16th Infantry Regiment cracked the forts "Volga" and "Ural." The 22nd were the first to reach Severnaya Bay on 19th June—the last barrier to the north of the southern part of Sevastopol.

Major-General Friedrich Schmidt's 50th Infantry Division from Brandenburg and Mecklenburg, together with General Laszar's Rumanian 4th Mountain Division, had the thankless task of making their way laboriously through the shrub-grown, rocky terrain from the north-east towards the high ground of Gaytany. They succeeded, and thus reached the eastern corner of Severnaya Bay.

On the western front XXX Corps under Lieutenant-General Fretter-Pico had mounted its attack on 11th June according to plan—first with the 72nd Infantry Division under Lieutenant-General Müllard-Gebhard and with Lieutenant-General Sinnhuber's 28th Light Division, and subsequently also with Major-General Sander's 170th Infantry Division. The divisions advanced along both sides of the main road leading from the coast to the town. Everything depended on gaining the commanding high ground of Sapun; these hills were the key to the southern part of the town, and battles were waged for mountain-tops and ravines. It was a miniature war against well-concealed strongpoints and fortified rock positions—"Northern Nose," "Chapel Mountain," and the Kamary caves were crucial points in the fighting. The men of 72nd Infantry Division will never forget those names.

The Jäger regiments of 28th Light Division fought their way over the steep rocks of the coastal mountains. Fort Balaclava had been taken by a surprise coup by 105th Infantry Regiment as long ago as the autumn of 1941—but even in June 1942 there was still plenty to do for the Jägers. It was a busy time for assault parties led by gallant men—men like Second Lieutenant Koslar, Sergeant Keding, and Sergeant Hindemith. "Tadpole Hill," "Cinnabar I, II, and III," "Rose Hill," and the notorious vineyard were key sectors in the savage fighting.

The reinforced 170th Infantry Division, until recently in reserve and now inserted between the two assault divisions, seized the vital Sapun Hills. The real inspiration of the attack was First Lieutenant Bittlingmeier with the 1st Battalion, 391st Infantry Regiment. In an hour and a half his battalion fought its way up to the ridge of the high ground. There, just as he had reached his target, Bittlingmeier was killed by a bullet, within sight of the town and harbour of Sevastopol.

On 18th June Major Baake with Reconnaissance Detachment 72 captured the "Eagle's Nest."

A macabre assignment was given to 420th Infantry Regiment, temporarily placed under 170th Infantry Division. Its task was to storm the old British cemetery where the dead of the Crimean War were buried. The Soviets had turned the cemetery into a heavy battery emplacement—a gruesome fortress.

On 20th June the reinforced 97th Infantry Regiment, 46th Infantry Division, took the Northern Fort and the notorious Konstantinovskiy Battery on the narrow Severnaya Kosa spit. It thus controlled the entrance to the harbour, and Sevastopol found itself in a German stranglehold. Manstein now held all the fortifications around Sevastopol. Nevertheless the Soviet High Command sent its 142nd Rifle Brigade into the town during the night of 26th June, by every possible craft they could lay their hands on. The reinforcements got into the fortress just in time to witness its fall.

The *coup de grâce* was administered by 22nd and 24th Infantry Divisions. The 22nd Artillery Regiment fired its 100,000th shell. It came down on the far shore of Severnaya Bay. In the dusty "Wolves' Glen" the regiments assembled for the final assault by moonlight.

On 27th June, shortly after midnight, the companies started crossing the bay in inflatable rubber dinghies and on rubber floats. The enemy spotted the move too late. The first assault parties had already reached and taken the power station.

Cautiously the battalions advanced to the edge of the town. At daybreak the Stukas came. They blasted a passage for the infantry. The last major anti-tank ditch was negotiated.

The Soviet defence collapsed in panic and chaos. Here and there a commissar, a commander, or a Komsomol member was found fighting to his last breath.

In a barricaded gallery, within the very cliffs of the bay, about 1000 women, children, and troops were sheltering. The commissar in command refused to open the doors. Sappers got ready to blow them in. At that moment the commissar blew up the entire gallery with everybody in it, himself included. A dozen German sappers were killed at the same time.

On 3rd July it was all over. Sevastopol, the strongest fortress in the world, had fallen. Two Soviet Armies had been smashed and 90,000 prisoners taken. On the devastated battlefield, among thousands of dead, were 467 pieces of artillery, 758 mortars, as well as 155 anti-tank and anti-aircraft guns.

The officers commanding the fortress, Admiral Oktyabrskiy and Major-General Petrov, were not found on the battlefield. They had been snatched out of the fortress by speedboat on 30th June.

Manstein's Eleventh Army was now available for the grand plan, for the offensive against Stalingrad and the Caucasus.

3. A Plan Betrayed to the Enemy

Venison and Crimean champagne–An interrupted feast–Major Reichel has disappeared–A disastrous flight–Two mysterious graves–The Russians know the plan for the offensive–The attack is mounted nevertheless–Birth of a tragedy.

THE commissar's villa was furnished in surprisingly good taste. Situated in a small garden of its own on the edge of the city of Kharkov, it was two-storeyed, with a properly constructed cellar. The comrade commissar had done himself proud. But then he had also been a man in a highly responsible job—in charge of the heavy industry of the Kharkov region. Now the villa had been taken over by General of Panzer Troops Stumme and the staff of his XL Panzer Corps.

Stumme was an excellent officer and a man who enjoyed life—short of stature, bursting with energy, and always on the go. He was never without the monocle which he had worn even while still a junior cavalry officer. His high blood-pressure gave his face a permanent flush. His physical and temperamental characteristics had earned him the nickname of "Fireball" among the officers and men of his headquarters. He knew his nickname, of course, but pretended not to, which saved him having to react every time he overheard some one using it.

A Plan Betrayed to the Enemy 477

Stumme was no scholarly General Staff officer, but a practical man with a genuine flair for spotting and grasping tactical opportunities. He was one of the best German tank commanders, clever in planning operations and resolute in executing them. He was a front-line officer, idolized by his soldiers, whose welfare was his constant concern. But he was also respected by his officers, who admired his energy and operational instinct.

His weakness, a pleasant weakness at that, was good food and drink. "War's bad enough—why eat badly as well? No, gentlemen, not me!" was a favourite saying of his. But the choice delicacies which the commander of the headquarters staff got hold of would invariably be shared with guests.

Just such a dinner-party was given by Stumme at his headquarters in the evening of 19th June 1942. The guests included the three divisional commanders of the Corps and the chief of corps artillery—Major-General von Boineburg-Lengsfeld, commanding 23rd Panzer Division, Major-General Breith, commanding 3rd Panzer Division, Major-General Fremerey, commanding 29th Motorized Infantry Division, and Major-General Angelo Müller, the artillery chief. Also present were Lieutenant-Colonel Franz, the chief of the Corps staff, Lieutenant-Colonel Hesse, the chief of operations, Second Lieutenant Seitz, the orderly officer, and Lieutenant-Colonel Harry Momm, the Corps adjutant and international show-jumper.

It was to be "the condemned man's last meal," as Stumme remarked jokingly. "Only a few more days of leisure, gentlemen—then we're off again. Let's hope we manage to force Stalin to his knees this time."

"Let's hope so," grunted General Breith, the robust Panzer leader from the Palatinate.

Two days previously the three divisional commanders had been informed verbally about the Corps' tasks during the first phase of "Operation Blue." Verbally only, because under Hitler's very strict security regulations a divisional commander was not allowed to know Corps orders for an offensive until that offensive had actually begun.

"Couldn't we have a few points in writing," one of the commanders had begged. It was against the strict security regulations, but Stumme consented.

"You can't lead a Panzer Corps on too tight a rein," he had said to his chief of staff and chief of operations, and had gone to dictate a brief outline of half a page of typescript: "For the eyes of divisional commanders only." And it had covered only the first phase of "Operation Blue." Lieutenant-Colonel Hesse had arranged for the top-secret document to go to the divisions by particularly reliable couriers.

This, in fact, was the usual practice with many Panzer Corps. After all, how could a divisional commander, in charge of a fast unit, take intelligent advantage of a sudden opportunity to break through if he did not know whether the further advance would aim north, south, or west?

Stumme's Corps had been assigned the task, under the first phase of "Operation Blue," of thrusting across the Oskol as part of Sixth Army and then wheeling north in order to encircle the enemy. If the division managed to get across the river quickly it was important for the commanders to know the general outlines of the plan, so that they should act correctly without losing time.

Stumme had always found his method of a brief written outline for his divisional commanders satisfactory. In this way he had never lost a chance, and nothing had ever gone wrong—at least not until that 19th June.

Stumme was enjoying his guests' surprise at the delicacies served. The main course was roast venison—a roebuck shot by Lieutenant-Colonel Franz on a reconnaissance outing. For an entrée there was caviare, washed down with Crimean champagne. Both these had been discovered in a Kharkov warehouse by a keen mess officer, and the visitors did not have to be asked twice to help themselves.

Nothing produces a merry atmosphere more quickly than sweet Crimean champagne—a fact confirmed at the Tsar's banqueting-tables in the old days and at many a Soviet festive gathering since. Around Stumme's dinner table, too, there was relaxed gaiety on that 19th June. The officers, who had all gone through the appalling winter, were beginning to see the future more optimistically.

Above all, the general commanding the Corps was full of energy and optimism. Earlier in the afternoon he had spoken to Army, where the mood had likewise been one of optimism. General von Mackensen with his reinforced III Panzer Corps had just opened a breach in the enemy lines for Sixth Army, in the Volchansk area north of Kharkov and east of the Donets, and thus enabled it to take up excellent starting positions for the great offensive along the Burluk, on the far side of the Donets.

In a bold encircling operation Mackensen with his four mobile and four infantry divisions had smashed greatly superior Soviet forces which had been firmly dug in along the commanding high ground on the Donets. The Corps had seized the high ground and taken 23,000 prisoners. In the impending large-scale offensive General Paulus's Sixth Army would not therefore now have to force a costly crossing of the Donets under enemy fire.

Lieutenant-Colonel Franz was using his knife, fork, dessert spoon, and brandy glass to illustrate Mackensen's interesting operation which had achieved such marked success at exceedingly low cost. His operation was seen as further evidence that the German Armies in the East had regained their old striking-power.

"And now Mackensen is about to repeat the same performance

south-east of us, in order to clear the enemy from the ground this side of the Donets and gain for us the Oskol as our starting-line for 'Operation Blue.' Splendid fellow Mackensen—he'll pull it off again, you'll see." Stumme raised his glass. Optimism and cheerfulness reigned unchallenged.

The time was five minutes to ten. No writing appeared on the wall as at Belshazzar's feast, nor did a bomb drop amid the gay revellers. All that happened was that Sergeant Odinga, the operations clerk, came in, bent down over Lieutenant-Colonel Hesse, and whispered something into his ear. The chief of operations rose from his chair and turned to Stumme. "If you'll excuse me, Herr General, I'm urgently wanted on the telephone."

Stumme laughed. "Don't come back with bad news!"

"I should hardly think so, Herr General," Hesse replied. "It's only the duty orderly officer of 23rd Panzer Division."

When they had closed the door behind them and were walking down the stairs to the map-room Sergeant Odinga remarked, "There seems to be quite a flap on at 23rd, Herr Oberstleutnant."

"Oh?"

"Yes—it seems Major Reichel, their chief of operations, has been missing since this afternoon."

"What?"

Hesse ran down the remaining stairs to the telephone. "Yes, what's up, Teichgräber?" He listened. Then he said, "No, he certainly isn't here." Hesse glanced at his wrist-watch. "He took off at 1400 hours, you say? But it's 2200 hours now. Tell me—what did he have with him?" Hesse listened intently. "His map-board? What? The file with the typed note too? But, for heaven's sake, that's not a thing to take with you on a reconnaissance flight?"

Hesse was stunned. He dropped the receiver on its rest and ran upstairs to the dining-room. The high spirits evaporated abruptly. They could tell from the chief of operations' expression that something had happened.

Briefly, turning alternately to Stumme and to von Boineburg-Lengsfeld, Lieutenant-Colonel Hesse reported what had occurred. Major Reichel, the chief of operations of 23rd Panzer Division, a brilliant and reliable officer, had taken off in a Fieseler Storch at 1400 hours, with Lieutenant Dechant as his pilot, to fly to XVII Army Corps Headquarters in order to have another look at the Division's deployment area as outlined in the typed note for the divisional commanders. Reichel must have flown beyond Corps Headquarters to the main fighting-line. He had not yet returned, nor had he landed anywhere within the division's area. He had had with him not only the typed note, but also his map with the Corps' divisions marked on it as well as the objectives of the first phase of "Operation Blue." Stumme had shot up from his chair. Boineburg-Lengsfeld tried to reassure the party: "He could have come down somewhere behind our divisions. There's no need to assume the worst straight away." He was fighting against the thought that was written in all their faces: the Russians have got him, complete with the directive and the objectives of "Operation Blue."

Stumme was living up to his nickname. All divisions along the front were instantly rung up: divisional commanders and regimental commanders were instructed to inquire from forward artillery observers and company commanders whether any incident had been observed at all.

Corps' headquarters was like a beehive. There was a continuous buzzing and ringing of telephones until, barely forty-five minutes later, the 336th Infantry Division came through. A forward artillery observer had seen a Fieseler Storch in the scorching hot afternoon haze, somewhere between 1500 and 1600 hours. The machine had banked and turned in the very low cloud, and finally, just as a heavy summer thunderstorm covered the whole sector, it had landed close to the Russian lines. "Strong assault party to be sent out at once," Stumme commanded.

Lieutenant-Colonel Hesse issued the detailed orders for the reconnaissance. The main interest, of course, was in the two men. If Reichel and his pilot could not be found, then a search must be made for a briefcase and a map-board. If the enemy had got to the spot first the ground must be searched for traces of fire or battle, or anything suggesting destruction of the papers.

In the grey dawn of 20th June the 336th Infantry Division sent out a reinforced company into the rather difficult ground. A second company provided flank cover and put up a show of activity to divert the Russians.

The aircraft was found in a small valley. It was empty. There was no briefcase and no map-board. The instruments had been removed from the dashboard—a favourite Russian practice whenever they captured a German machine. There were no traces of fire which might suggest destruction of the map or papers. Neither were there any traces of blood or any indication of a struggle. The aircraft's fuel-tank had a bullet-hole. The petrol had run out of the tank.

"Search the neighbourhood," the captain ordered. The men moved off in small groups. A moment later came the voice of a sergeant: "Over here!" He pointed to two mounds of earth, some 30 yards from the aircraft—two fresh graves. The company commander was satisfied. He recalled his parties and returned to base.

General Stumme shook his head when he received the report about the two graves. "Since when have the Russians shown such respect to our dead as to bury them? And alongside that aircraft, too!"

"Certainly looks odd to me," remarked Lieutenant-Colonel Franz. "I want to know more about this: it may be some piece of devilry," Stumme decided.

The 336th Infantry Division was ordered to send out a party again, to open up the graves and to find out whether they contained Reichel and Lieutenant Dechant.

The men of 685th Infantry Regiment set out again. With them went Major Reichel's batman, to identify him. The graves were opened. The lad thought he could recognize his major, although the body was in its underclothes and altogether was not a pleasant sight. In the second grave there were no items of uniform either.

Precisely what report XL Panzer Corps—at whose headquarters the entire investigation was concentrated—made to Army about the bodies found in the graves can no longer be reliably established. Certain staff officers do not even remember that any bodies were found at all. The intelligence officer of XL Panzer Corps, who was only a few miles away from the point where the aircraft came down, functioning as a kind of forward post of General Stumme's headquarters, and who was immediately enlisted in the search operation, considers that Major Reichel had vanished without trace. Lieutenant-Colonel Franz, as he then was, on the contrary believes that the bodies were identified beyond any doubt. In spite of such definite views expressed by staff officers of 336th Infantry Division there must remain a good deal of suspicion that the Russians may have staged an elaborate trick to deceive the Germans. Frau Reichel, admittedly, received a letter from Colonel Voelter, the chief of operations of Sixth Army, informing her that her husband had been "buried with full military honours in the German Army cemetery at Kharkov." She was even sent a photograph of the grave; but she did not receive the wedding-ring which her husband invariably wore. Consequently, a certain element of doubt continues to attach to the incident to this day.

For the German Command at the end of June 1942 it was, of course, of decisive importance to know whether Reichel was dead or whether he was alive in Russian captivity. If he was dead, then the Russians could know only what they had found on his map and in the typed note—the first phase of "Operation Blue." If they had caught the major alive, then there was the danger that GPU specialists would make him reveal all he knew. And Reichel, naturally, knew very nearly everything, in broad outline, about the grand plan for the offensive. He knew that it aimed at the Caucasus and at Stalingrad. The idea that the Soviet secret service had got Reichel and might make him talk did not bear thinking about. Yet there was every ground for suspecting that just that had happened.

It was no secret that Soviet front-line troops had strict orders to handle any officer with crimson stripes down his trousers—*i.e.*, every

Q

General Staff officer—like a piece of china and to take him at once to the next higher headquarters. Any German General Staff officers killed in action had to be brought in if at all possible, because in this way the Germans would be kept guessing uneasily whether they were alive or not. This uncertainty was deliberately fomented by skilful front-line propaganda.

Why should the Soviets suddenly make an exception? And even if they did, why the burial?

There is only one logical solution to the mystery. Reichel and his pilot had been taken prisoners by a Soviet patrol and subsequently killed. When the leader of the patrol brought the map and the briefcase to his commander the latter must have realized at once that this had been a senior German staff officer. To avoid unpleasantness and any possible questions about the bodies, he had sent the patrol back with orders to bury the two officers they had killed.

Needless to say, Stumme had to report the Reichel incident to Army at once. Lieutenant-Colonel Franz had already made a telephonic report during the night of 19th/20th June towards 0100 hours, to the chief of staff of Sixth Army, Colonel Arthur Schmidt, subsequently Lieutenant-General Schmidt. And General of Panzer Troops Paulus had no other choice but to report the matter, though with a heavy heart, via Army Group to the Fuehrer's Headquarters in Rastenburg.

Fortunately Hitler was just then in Berchtesgaden, and the report did not come to his ears straight away. Field-Marshal Keitel was in charge of the initial investigations. He was inclined to recommend to Hitler to take the severest measures against "the officers culpable as accomplices."

Keitel of course guessed Hitler's reaction. The Fuehrer's order had laid it down quite clearly that senior staffs must pass on operational plans only by word of mouth. In his Directive No. 41 Hitler had once again laid down strict security regulations for that vital operation, "Operation Blue." Hitler was constantly afraid of spies, and on every possible occasion had emphasized the principle that no person must know more than was absolutely necessary for the discharge of his task.

General Stumme, together with his chief of staff, Lieutenant-Colonel Franz, and General von Boineburg-Lengsfeld, the commander of 23rd Panzer Division, were relieved of their posts three days before the offensive, and Stumme and Franz were tried by a Special Senate of the Reich Military Court. Reich Marshal Goering presided. The indictment consisted of two charges—premature and excessive disclosure of orders.

In a twelve-hour hearing Stumme and Franz were able to prove that there could be no question of a "premature" issuing of orders. Moving the Panzer Corps into the Volchansk bridgehead over the only available Donets bridge alone required five of the short June nights.

That left the charge of "excessive disclosure of orders," and this became the core of the prosecution's case. It was pointed out that Corps had warned its Panzer divisions that, after crossing the Oskol and turning northward, they might encounter Hungarian formations in khaki uniforms similar to the Russian ones. This warning had been necessary since there was a danger that the German Panzer formations might otherwise mistake the Hungarians for Russians.

But the tribunal did not accept this excuse. The two defendants were sentenced to five and two years' fortress detention respectively. True, at the end of the hearing, Goering went and shook hands with both prisoners and said, "You argued your case honestly, courageously, and without subterfuge. I shall say so in my report to the Fuehrer."

Goering seems to have kept his promise. Field-Marshal von Bock likewise put in a good word for the two officers in a personal conversation with Hitler at the Fuehrer's Headquarters. Whose intervention it was that softened Hitler's heart it is impossible to establish to-day. But after four weeks Stumme and Franz were informed in identical letters that in view of their past services and their outstanding bravery the Fuehrer had remitted their sentences. Stumme went to Africa as Rommel's deputy, and Franz followed him as chief of staff of the Africa Corps. On 24th October General Stumme was killed in action at El Alamein. He lies buried there.

Following Stumme's recall, the XL Corps was taken over by General of Panzer Troops Leo Freiherr Geyr von Schweppenburg, the successful commander of XXIV Panzer Corps. He inherited a difficult task.

There was no doubt left: by 21st June, at the latest, the Soviet High Command knew the plan and order of battle for the first phase of the great German offensive. It was also known at the Kremlin that the Germans intended to make a direct west-east thrust from the Kursk area with extremely strong forces, and to gain Voronezh in an outflanking move by Sixth Army from the Kharkov area, in order thus to annihilate the Soviet forces before Voronezh in a pocket between Oskol and Don.

What the Soviets were not able to see from the map and piece of paper which the unfortunate Reichel had had with him was the fact that Weichs' Army Group was subsequently to drive south and southeast along the Don, and that the great strategic objectives were Stalingrad and the Caucasus. Unless, of course, Reichel had after all been taken alive by the Russians, and grilled, and the body in the grave by the aircraft had been some one else's.

Considering the cunning of the Soviet Intelligence Service, this possibility could not be entirely ruled out. The question, therefore, which the Fuehrer's Headquarters had to answer was: Should the plan of operation and the starting date be upset?

Both Field-Marshal von Bock and General Paulus opposed this

suggestion. The deadline for the offensive was imminent, which meant that it was too late for the Soviets to do very much about countering the German plans. Moreover, General Mackensen had mounted his second "trail-blazing operation" on 22nd June, and, with the objective of gaining a suitable starting-line for Sixth Army, had fought a successful minor battle of encirclement together with units of First Panzer Army in the Kupyansk area, resulting in the taking of 24,000 prisoners and the gaining of ground across the Donets to the Lower Oskol.

The launching platforms for "Operation Blue" had thus been gained. To interfere now with the complicated machinery of the great plan would mean to jeopardize everything. The machine, once started and so far running smoothly, must be allowed to run on. Hitler therefore decided to mount the offensive as envisaged: D-Day for Weichs' Army Group on the northern wing was 28th June, and for Sixth Army with XL Panzer Corps it was 30th June. The die was cast.

What followed is closely connected with the tragic affair of Major Reichel and contains the seed of the German disaster in Russia. It marks the beginning of a string of strategic mistakes which led inescapably to the disaster of Stalingrad, to the turning-point of the war in the East, and hence to Germany's defeat. To understand this turning-point, this change of fortunes which struck the German Armies in the East so suddenly, at the very peak of their success, it is necessary to look more closely at the involved strategic moves of "Operation Blue."

The basis of phase one of the German offensive in the summer of 1942 was the capture of Voronezh. This town, situated on two rivers, was an important economic and armaments centre, and controlled both the Don, with its numerous crossings, and the smaller Voronezh river. The town, moreover, was a traffic junction for all central Russian north-south communications, by road, rail, and river, from Moscow to the Black Sea and the Caspian. In "Operation Blue" Voronezh figured as the pivot point for movements to the south, and as a support base for flank cover.

On 28th June von Weichs' Army Group launched its drive against Voronezh with Second Army, the Hungarian Second Army, and the Fourth Panzer Army, Hoth's Fourth Panzer Army acting as the main striking force. Its core, in turn, its battering-ram as it were, was XLVIII Panzer Corps under General of Panzer Troops Kempf with 24th Panzer Division in the middle and 16th Motorized Infantry Division and the "Grossdeutschland" Division on the right and left respectively.

The 24th Panzer Division—formerly the East Prussian 1st Cavalry Division and the only cavalry division in the Wehrmacht to be re-equipped as a Panzer division during the winter of 1941–42—was assigned the task of taking Voronezh.

The division, under Major-General Ritter von Hauenschild, struck with all its might. Under cover of fire provided by VIII Air Corps the Soviet defences were over-run, the Tim river was reached, the bridge across it stormed, and the fuse, already lit for the demolition of the bridge, ripped out just in time. Then the divisional commander raced across in his armoured infantry carrier, ahead of the reinforced Panzer Regiment.

With the dash of cavalry the tanks raced down towards the Kshen river. Artillery and transport columns of the Soviet 160th and 6th Rifle Divisions were smashed. Another bridge was seized intact. It was a headlong chase. Divisional commander and headquarters group were right in front, regardless of exposed flanks, in accordance with Guderian's motto: "An armoured force is led from in front and is in the happy position of always having exposed flanks."

Whenever a refuelling halt had to be made the force was regrouped and quickly assembled combat groups raced on. By the evening of the first day of the attack motor-cyclists and units of 3rd Battalion, 24th Panzer Regiment, were charging the village of Yefrosinovka.

"Well, well, what have we got here?" Captain Eichhorn said to himself. On the edge of the village was a veritable forest of signposts, as well as radio vans, headquarters baggage trains, and lorries. This must be a senior command.

The motor-cyclists narrowly missed making a really big catch: the headquarters staff of the Soviet Fortieth Army, which had been stationed there, escaped at the last minute. But although they got away, their Army with the dispersal of its headquarters had lost its head.

In this fashion 24th Panzer Division, in the scorching summer of 1942, revived once more those classic armoured thrusts of the first few weeks of the war, and thereby demonstrated what a well-equipped, fresh, and vigorously led Panzer division was still capable of accomplishing against the Russians. Only a cloudburst stopped the confident formations for a while. They formed 'hedgehogs,' waited for the Grenadier regiments to follow up, and then the spearheads drove on under Colonel Riebel.

By 30th June the 24th Panzer Division had covered half the distance to Voronezh. It was facing a well-prepared Soviet position held by four rifle brigades. Behind them two armoured brigades were identified. Things were getting serious.

The Soviets employed three armoured Corps in their attempt to encircle the German formations which had broken through, and to cover Voronezh. Lieutenant-General Fedorenko, Deputy Defence Commissar and Commander-in-Chief of Armoured Troops, personally took charge of the operation. Clearly the Russians were aware of the significance of the German drive on Voronezh.

But Fedorenko was unlucky. His grandly conceived armoured thrust against the spearhead of Hoth's Fourth Panzer Army proved a failure. Superior German tactics, extensive reconnaissance, and a more elastic form of command ensured victory over the more powerful Soviet T-34 and KV tanks.

On 30th June, also, the day when 24th Panzer Division went into its first great tank battle, the German Sixth Army, 90 miles farther south, launched its drive towards the north-east, with Voronezh as its objective. The great pincers were being got ready to draw Stalin's first tooth. The operation was supported from the air by IV Air Corps, under Air Force General Pflugbeil.

The XL Panzer Corps burst forward from the Volchansk area, a powerful mailed fist of combat-tested units—the 3rd and 23rd Panzer Divisions, the 100th Jäger Division, and the 29th Motorized Infantry Division. Only the 23rd Panzer Division was still new in the East in

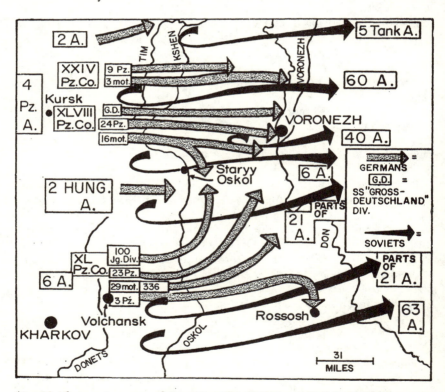

Map 28. The opening move of Operation Blue (28th June–4th July 1942). Voronezh was to be taken, and the first pocket to be formed in the Staryy Oskol area in co-operation between Fourth Panzer Army and Sixth Army. But for the first time the Soviet Armies refused battle and swiftly moved back across the Don.

1942. Its tactical sign was the Eiffel Tower, indicating where it came from; until recently it had been stationed in France as an occupation unit. The Soviets exploited this circumstance for their psychological warfare. Over the sector of the 23rd they dropped leaflets saying: "Men of 23rd Panzer Division, we welcome you to the Soviet Union. The gay Parisian life is now over. Your comrades will have told you what things are like here, but you will soon find out for yourselves." The ruse worked. The men of 23rd Panzer Division were shaken to know that the Russians were so well informed about their presence.

Freiherr von Geyr's first instruction was: On reaching the Oskol the troops will wheel north in order to form a pocket in the Staryy Oskol area in co-operation with Kempf's XLVIII Panzer Corps.

But something strange happened. The troops discovered that although enemy rearguards were fighting stubbornly in well-prepared defensive positions, the bulk of the Soviets was withdrawing eastward in good order. For the first time the Russians were refusing large-scale battle. They were pulling out of the incipient pocket. What did it mean? Were they so accurately informed about German intentions?

4. New Soviet Tactics

Fatal mistake at Voronezh–Timoshenko refuses battle–Hitler again changes his plan–Council of War at the Kremlin–The battle moves to the southern Don–Fighting for Rostov–Street fighting against NKVD units–The bridge of Bataysk.

WHEN the general commanding XL Panzer Corps was informed about the Soviet withdrawals he realized instantly that this move was jeopardizing the whole first phase of the German operation. In view of the changed situation he asked for authority to drive on eastward to the Don without further delay. But Sixth Army insisted on its plan for a pocket and ordered: "XL Panzer Corps will turn northward in order to link up with Fourth Panzer Army." Orders are orders. The pocket was sealed off. But inside there was nothing. The Russians had withdrawn even their heavy weapons. The mountain had laboured and a mere mouse had been born.

By then even the Fuehrer's Headquarters began to realize that things were not going according to plan. The Russians were rapidly withdrawing towards the Don. Would they be able to get away across the river while Fourth Panzer Army was still operating against Voronezh? In that case the entire first phase of "Operation Blue" would be a blow into thin air. The danger was considerable. There was no time to be lost.

Faced with this situation, Hitler on 3rd July arrived at the entirely correct opinion that clinging to the idea of taking Voronezh first might threaten the whole of "Operation Blue." On a lightning-like visit to von Bock's Headquarters he therefore informed the Field-Marshal: "I no longer insist on the capture of the town, Bock; nor, indeed, do I consider it necessary. You are free, if you wish, to drive southward at once." That was the moment of decision. The fortunes of war hung in the balance. Which way would the scales tip?

Geyr heaved a sigh of relief when, late at night on 3rd July, he received orders from Sixth Army to drive straight to the east towards the Don, in order to cut off the Russian retreat.

But by noon on the following day, 4th July, a new order arrived: he was not to drive east after all, but to the north, in the direction of Voronezh, in order to cover the southern flank of Fourth Panzer Army. What was up? What had happened at Voronezh? What was behind all this vacillation?

It is an odd fact that all Hitler's correct decisions during the first half of the war were made by him in a strange and otherwise quite untypical, diffident manner. That was true also of Voronezh.

He did not command Field-Marshal von Bock: you will bypass the town and pursue our schedule towards Stalingrad without losing any more time. No—he merely informed Bock that he no longer insisted on the capture of Voronezh. Thus the responsibility for the decision whether the force should be wheeled round without the previous seizure of this important traffic centre was left to the Commander-in-Chief Army Group South. It was a difficult decision for the Field-Marshal: should he take the town or should he bypass it? On careful consideration von Bock began to wonder whether it might not be better after all to take the cornerstone of Voronezh first—provided it could be taken quickly. Ought he not at least to try it? Bock hesitated and wavered.

At that point came the news that 24th Panzer Division with its reinforced 26th Rifle Regiment had gained a bridgehead over the Don at a crossing-point. Over a Soviet army bridge the German battalions were moving across the river, mixed up with retreating Russian columns. By nightfall reconnaissance units were within two miles of Voronezh.

On the left of the 24th the "Grossdeutschland" Motorized Infantry Division, which provided the northern flank cover for 24th Panzer Division, had likewise made rapid headway, and towards 1800 hours on 4th July had reached the Don. Farther south the 16th Motorized

Infantry Division had also reached the river with its reinforced motorcycle battalion.

At Semiluki the Soviets had left intact the bridge over the Don carrying the road to Voronezh. This circumstance proved that they were themselves hoping to get the bulk of their Armies across the river. By means of strong counter-attacks, supported by T-34s, they were trying to keep the Germans away from the bridge and to hold a wide bridgehead on the western bank.

Towards 2000 hours on 4th July Lieutenant Blumenthal with men of his 7th Company, 1st Motorized Infantry Regiment "Grossdeutschland," seized the road bridge over the Don to Voronezh and established a bridgehead on the eastern bank. The soviets tried to blow up the bridges at the last moment, but evidently had no electric detonation equipment in position. They therefore lit an ordinary fuse leading to stacks of dynamite under the piers. The small flame was already snaking along the cord.

Sergeant Hempel of Blumenthal's Company jumped into the river and, with the water reaching to his chin, waded underneath the bridge and wrenched away the burning cords within inches of the 120 lb. of explosives.

Meanwhile Russian columns were still moving over the bridge from the west, straight into the arms of a reception committee formed by Blumenthal's 7th Company on the eastern bank. "*Ruki verkh!*" ("Hands up!") The bridge had been taken. Could Voronezh be taken as easily?

Groups of 1st Infantry Regiment "Grossdeutschland," riding on assault guns, made a reconnaissance thrust in force against the town and got as far as the railway. Admittedly, they had to withdraw again in the face of furious counter-attacks by the strong defending forces—but nevertheless the Germans were virtually inside. It was this kind of support which induced Field-Marshal von Bock not to take up Hitler's suggestion to bypass Voronezh, but to attack it. He wanted to exploit the favourable opportunity as he saw it and to take the important town by a coup. He believed that his fast troops would still be able to get from Voronezh into the rear of Timoshenko's Armies in good time to cut off their retreat over the Don. That was the fundamental mistake from which, step by step, the tragedy of Stalingrad was to take shape.

At nightfall on 5th July, after a scorching day with the temperature at 40 degrees Centigrade, the fast formations of XLVIII and XXIV Panzer Corps, as well as the two motorized infantry regiments of the "Grossdeutschland" Division, the 24th Panzer Division, and the motorcyclists of the 3rd and 16th Motorized Infantry Divisions, were holding extensive bridgeheads east of the Don before Voronezh. In the north cover was provided by the approaching infantry divisions. But Army Group had made a mistake in assessing the enemy's strength. The town was crammed full with Soviet troops. At the last moment the Russians had reinforced Voronezh by a very special effort. Clearly, Timoshenko had drawn the right conclusion from the plans found on Major Reichel.

When Hitler was informed he was suddenly galvanized into action again. He now strictly vetoed any further attack on the town. The attack, he insisted, must be turned towards the south: that was where the objective lay.

But on 6th July units of 24th Panzer Division and "Grossdeutschland" Division were inside the town. The Russians appeared to be giving ground. Hitler, consequently, allowed this to influence him on the spur of the moment, and once again authorized the capture of Voronezh. However, he commanded that at least one Panzer Corps, the XL Corps, must continue the southward thrust launched on 4th July and drive on down the Don without further delay. Fourth Panzer Army was instructed to release further armoured formations as soon as possible in order to follow up the drive of XL Panzer Corps.

The second phase of "Operation Blue" therefore opened in a watered-down way. First the battles for the important town of Voronezh had been waged by armoured formations—not the most suitable for this kind of action—and now Bock was being progressively deprived of his most effective striking forces. To make matters worse, some of them presently ground to a standstill south of Voronezh for lack of fuel. As a result, Army Group South was no longer strong enough to force a decision in the battle for Voronezh itself, while for a drive to the south and a rapid cutting of the Don one Panzer Corps, even though reinforced by the subsequent assignment of further mobile formations, proved too weak.

On 7th July 3rd Motorized Infantry Division and 16th Motorized Infantry Division took the western part of Voronezh after heavy fighting. But the battalions were unable to get across the Voronezh river, which runs through the town from north to south. Time and again the Russians mounted counter-attacks, employing infantry and large packs of armour.

Timoshenko had concentrated at Voronezh the bulk of the Soviet Fortieth Army, with nine rifle divisions, four rifle brigades, seven armoured brigades, and two anti-tank brigades. This concentration left no doubt at all that Timoshenko was acquainted with Hitler's plan and was now making the correct counter-moves—tying down the bulk of the German forces on the northern wing outside Voronezh, in order to gain time to detach the bulk of his own Army Group from the Oskol and Donets and pull it back over the Don.

And in which direction was he withdrawing his force? Oddly enough, towards Stalingrad.

Although the German radio reported the capture of Voronezh on 7th July, fighting continued in the university quarter and in the woods north of the town until 13th July. Even after that date the Germans did not succeed in taking the eastern part of the town, or the bridge in its northern part, which would have enabled them to paralyse the north–south railway along the eastern bank of the river—a railway vital for Soviet supplies. The great supply road from Moscow to the south also remained in Russian hands.

The original plan had provided for the German motorized formations, after the rapid fall of Voronezh, to strike south down the Don in order to bar the way to Timoshenko's divisions withdrawing from the vast area between Donets and Don, and to intercept them on the Don. Instead, the precious motorized and Panzer divisions of XLVIII Corps and units of XXIV Panzer Corps were heavily involved in that accursed town, while 9th and 11th Panzer Divisions were still tied down in the northern blocking position of Fourth Panzer Army. Marshal Timoshenko personally conducted the operation. Voronezh was to be held as long as possible in order to delay the German drive to the south-east. Every day gained meant a clear advantage to Timoshenko.

In the evening of 6th July the spearheads of XL Panzer Corps were south of Voronezh, with the 1st Battalion, 3rd Rifle Regiment, of 3rd Panzer Division roughly 50 miles from Rossosh. But fuel was running low. Major Wellmann, with great faith in the supplies group, decided nevertheless to continue the drive with two armoured companies and one battery of 75th Artillery Regiment.

On a clear starlit night they drove through the steppe. In front was Busch's company, followed by Bremer's. The battalion commander has given the following account: "We knew that if the bridges over the Kalitva were to be captured intact we would have to reach Rossosh at dawn and would have to avoid all contact with the enemy, if only because of our shortage of ammunition and motor-fuel. Thus, keeping rigidly to our time-table, we drove on, past advancing Russian artillery and infantry units who, luckily, did not realize who we were."

Shortly before 0300 hours the first shabby houses of Rossosh lay ahead. The battalion's interpreter, Sergeant Krakowka, picked up a surprised Russian and grilled him. The terrified comrade revealed that in addition to the two bridges over the Kalitva marked on the map there was yet another—a tank bridge, completed only recently. Bremer and Busch, the company commanders, made their plan of operations with the battalion commander.

In the grey light of dawn Wellmann's columns drove through Rossosh, still sound asleep and unsuspecting. On a sports ground stood a number of Kurier aircraft. There was an occasional tank. In front of a massive three-storeyed building stood a few sentries, but they did not associate the approaching cloud of dust with anything hostile.

Major Wellmann's command carrier was driving a short way behind the armoured carriers of 1st Company. The company crossed the bridge. Wellmann reached the Soviet bridge guard on the northern bank. The sentry realized what had happened and snatched his rifle from his shoulder. Wellmann's radio operator, Private Tenning, leapt from the vehicle like lightning, rammed his machine pistol into the Russian's stomach, knocked his rifle out of his hands, and hauled him back to the command carrier—their first prisoner, and an important one. The Russian stated that Rossosh contained a very high-ranking headquarters and that its defending forces included at least eight tanks.

At that moment the first shots came from the far river-bank. They were followed by nearly five hours of ferocious fighting against the town's surprised but tough defenders.

Firing came from all directions. T-34s roamed all over the place. Soviet infantry reformed. But Wellmann's men held the bridges. Their salvation was the field howitzer battery which they had brought along with them; its pieces had been positioned so cleverly by their experienced crews that they dominated the wide road along the river.

The fighting was fierce and confused. But the greater dash and stronger nerves of the Germans gave them victory. The Soviet tanks were mostly immobilized in hand-to-hand fighting. Sergeant Naumann made a particular catch: he cleared out the map section of Timoshenko's Army Group headquarters and captured twenty-two senior staff officers, mostly of the rank of colonel. Timoshenko himself had still been in Rossosh during the night, but must have got away at the last moment.

But in spite of all their gallantry the engagement would probably have ended badly for Wellmann's group if the bulk of 3rd Panzer Division had not got to Rossosh in time. Soviet resistance was broken. Major-General Breith's Berlin Division had reached one more important milestone along its road to the Don.

Nevertheless the upheaval which the German time-table had suffered as a result of the fighting for Voronezh was being felt everywhere. In the area south of Rossosh, around Millerovo, fairly strong enemy forces were suspected to be present; these were now to be destroyed by direct attack before any further advance was made. This was yet another deviation from the original plan, another sin against the spirit of a fast operation towards Stalingrad.

It was amid this rather confused situation that the third phase of "Operation Blue" began, the phase which, according to Directive No. 41, was to usher in the decisive stage of the great summer offensive of 1942—the attack of the southern prong with General Ruoff's Seventeenth Army and Colonel-General von Kleist's First Panzer Army on 9th July. The objective was a link-up in the area—note: the area, not the city—of Stalingrad, with a view to encircling and destroying the Russian forces between Donets and Don.

But just as he had done in the north, so Timoshenko resisted in the south at a few chosen points only, while the bulk of his armies were pulling out towards the east and the south.

As a result, the attack of the southern prong achieved nothing beyond pushing the retreating Russians in front of it into the great Don loop. But there no German line had yet been established which might have cut off the path of the withdrawing Russian formations.

When Hitler realized that an encircling operation on the middle Don was no longer possible because of the Russians' quick withdrawal and the delay suffered at Voronezh, he wanted at least to intercept, encircle, and destroy the enemy forces which he believed to be still grouped along the lower Don. In order to achieve this objective he dropped on 13th July the key feature of his great plan—the rapid drive to Stalingrad with all forces in order to bar the lower Volga.

Hitler would have been well able to carry out this operation—indeed, in the circumstances, it would have been the only correct thing to do. For if an enemy refuses to be encircled and withdraws instead, then he must be pursued. He must not be allowed time to establish a new line of defence. The German objective now was the elimination of enemy forces in the Stalingrad area, and that objective could have been achieved by an energetic pursuit of the Russians.

After all, Hitler had two Panzer Armies at his disposal, and some important crossings over the Don had already been gained. He could have reached Stalingrad in a very short period of time. But Hitler was suffering from a great delusion: he believed the Soviets to be at the end of their strength. He regarded the Soviet retreat as nothing more than flight, as organizational and moral collapse, whereas in actual fact it was a planned withdrawal.

Such incidents of panic as occurred in many places were due to the incompetence of the lower ranks of the Russian command. Strategically, Timoshenko had the withdrawal well under control. He had set it swiftly into motion. His objective was to save the bulk of the Soviet forces for determined resistance far back in the interior of the country.

Hitler did not see that danger, or else did not want to see it. He believed he could manage Stalingrad "with one hand" and simultaneously fight a large-scale battle of encirclement on the lower Don with Rostov at its centre. For that purpose he cut short Fourth Panzer Army's advance along the Don towards Stalingrad, halting it in front of the great Don loop, and, in complete divergence from phase three of the grand plan, turned it straight down to the south. Just as he had halted the advance on Moscow in the early autumn of 1941 and switched round Guderian's fast troops to fight their battle of encirclement at Kiev, he now wanted to defeat the Russians at Rostov by another improvised surprise operation. It was to be the greatest battle of encirclement of the war.

The Sixth Army meanwhile continued its lonely advance towards Stalingrad, now deprived of its spearhead, the fast units of XL Panzer Corps, which had also been switched to Rostov.

On the same day that this fateful decision was taken Field-Marshal von Bock was relieved of his post. He had opposed Hitler's strategic plans and had wanted to keep the Army Group together as an integrated fighting force under his command.

However, the Fuehrer's Headquarters had already issued orders for Army Group South to be divided up. On 7th July Field-Marshal von Bock noted in his diary: "Orders received that Field-Marshal List will assume command of Eleventh and Seventeenth Armies and of First Panzer Army. This means that the battle is being chopped in two."

That was precisely what was happening: the battle was being chopped in two. Hitler was changing not only the time-table of his great summer offensive, but the entire structure of the southern front.

Field-Marshal List's Army Group A, to which Fourth Panzer Army was later temporarily attached, was informally known as the Caucasus Front. Army Group B, consisting of Sixth Army, the Hungarian Second Army, and Second Army, and since Bock's recall under the command of Colonel-General von Weichs, retained its original assignment—Stalingrad.

This regrouping makes it clear that on 13th July Hitler believed that he could simultaneously achieve both great strategic objectives of the 1942 summer offensive, originally scheduled one after the other, by the simple expedient of dividing his forces. He was hopelessly blinded by his mistaken belief that the Russians were 'finished.'

But the Russians were anything but 'finished.' On the very day that Hitler ordered the disastrous turn to the south, split up his forces, and sacked von Bock, a council of war was held at the Kremlin under Stalin's chairmanship.

Present were Foreign Minister Molotov, Marshal Voroshilov, Chief of the General Staff Shaposhnikov, as well as an American, a British, and a Chinese liaison officer. The Soviet General Staff had made it clear to Stalin that he could not afford any more battles like Kiev or Vyazma—in other words, that holding on at all costs was out. Stalin had accepted their view. He endorsed the decision of the Great General Staff which was expounded by Shaposhnikov at the meeting on 13th July. The Soviet troops would withdraw to the Volga and into the Caucasus; there they would offer resistance, forcing the Germans to spend the coming winter in inhospitable territory. All key industries would be evacuated to the Urals and to Siberia.

From the middle of July the German General Staff had known from an agent's report about this important meeting, but Hitler had regarded it as a canard.

If there was anyone left who doubted that Timoshenko was in fact withdrawing his Army Group, down to the last man and gun, from the area between Donets and Don, then he was soon convinced at Millerovo. The XL Panzer Corps, acting as the outer eastern prong of the pincers, thrust straight into this Russian withdrawal after having wheeled south from Rossosh, with all its three divisions moving in the foremost line.

All along the railway and the road south of Millerovo the Soviet masses were pouring to the south-east. The divisions of the German Panzer Corps were not strong enough to halt these enemy columns. Nor, in view of the resistance offered them around Millerovo, were they able to establish an intercepting line farther south on the lower Don.

The battle was moving south. It was in the south that Hitler was seeking out the enemy. Indeed, he was so confident of victory in the south that he deleted Manstein's Eleventh Army—which was standing ready in the Crimea to strike across the Kerch Strait—from his plan of operations. Instead the Eleventh Army was entrained for the north. It was to take Leningrad.

After fierce fighting Geyr's XL Panzer Corps reached the lower Don on 20th July and established bridgeheads at Konstantinovka and Nikolayevskaya.

In the meantime the First Panzer Army, forming the inner prong of the new pincer operation, had likewise fought its way towards the south, crossed the Donets, and begun, jointly with Seventeenth Army advancing from the Stalino area, to drive upon Rostov, which was being defended with particular determination by the Soviets as a key bridgehead on the Don.

West of Rostov the Seventeenth Army had broken through enemy positions on 19th July and was now advancing towards the Don between Rostov and Bataysk with LVII Panzer Corps on the left and V Corps on the right. General Kirchner, again supported by his well-tried Colonel Wenck, mounted a bold thrust against Rostov with LVII Panzer Corps, with a view to taking this important city on the Don estuary by surprise and capturing the great Don bridge between Rostov and Bataysk intact. To his Corps belonged the 13th Panzer Division, the SS Panzer Grenadier Division "Viking," the 125th Infantry Division, and the Slovak Fast Division.

From the north, leading the First Panzer Army, General von Mackensen's III Panzer Corps was advancing on Rostov with 14th and 22nd Panzer Divisions. Once again, as in November 1941, von Mackensen's formations were engaged in fighting for this city. On 22nd July Colonel Rodt's 22nd Panzer Division was involved in heavy fighting north-east of Rostov. The 204th Panzer Regiment was driving to the south. The 14th Panzer Division was wheeling against Novocherkassk.

All through the day and night furious fighting raged in the strongly fortified approaches to the city.

On the same day the 13th Panzer Division under Major-General Herr and the SS Panzer Grenadier Division "Viking" under General of Waffen SS Steiner attacked from the west and north-west. Rostov itself had been reinforced into a strongly defended town since the beginning of the year and, in addition to strong defences in the approaches, was surrounded by three defensive rings with deep minefields, anti-tank ditches, and anti-tank obstacles. Nevertheless assault parties of LVII Panzer Corps succeeded in breaching the covering-lines on the edge of the city by surprise. The non-armoured group of 13th Panzer Division attacked from the west with 93rd Rifle Regiment, while the armoured group of the reinforced 4th Panzer Regiment advanced along the Stalino–Rostov road and penetrated into the northern part of the city. On its right the Armoured Group Gille of the SS Panzer Grenadier Division "Viking" struck right through numerous strongpoints and anti-tank ditches of the outer defensive ring and seized the airfield of Rostov with Sturmbannführer Mühlenkamp's SS Panzer Battalion.

On 23rd July the 22nd Panzer Division slowly gained ground from the north towards the edge of the city. In the sector of LVII Panzer Corps the 13th Panzer Division continued its attack into the city with tanks, rifle companies, and motor-cyclists. The SS Panzer Grenadier Division "Viking" got stuck initially in heavy street fighting, and 125th Infantry Division closed up behind it. At first light First Lieutenant von Gaza burst through the enemy positions with his 2nd Company, 66th Rifle Regiment, forced a small river, and seized the road bridge.

The Motor-cycle Battalion 43 charged into the city, mounted. The 13th Panzer Division cleared numerous road-blocks and barricades and slowly gained ground towards the Don. But while its spearheads were moving forward, enemy resistance flared up again behind them from side-streets, from strongly reinforced blocks of buildings, and, in particular, from open squares along its flanks.

To begin with, the tanks of the "Viking" got stuck in the street fighting. Then Sturmbannführer Dieckmann succeeded with his battalion in dislodging the enemy and resuming the attack in a south-westerly direction.

By afternoon the Motor-cycle Battalion, 13th Panzer Division, had reached the northern bank of the Don, but in the maze of industrial and port installations they had reached the river rather too far east of the main road bridge. Before the motor-cyclists were able to reach the bridge over the Don, leading to Bataysk, one of its spans was blown up and crashed into the water. While 13th Panzer Division was clearing up the area around the bridge the sappers worked feverishly until the following day, making the bridge serviceable again, though at first only for pedestrians and light vehicles. By nightfall the district north of the bridge was in German hands. The 1st Battalion, 66th Rifle Regiment, took the district around the General Post Office and the NKVD Headquarters, where the enemy resisted stubbornly and skilfully. By nightfall the infantry had taken up positions covering the tanks from all directions. There were fires in many parts of the town. In the early hours of the night units of 22nd Panzer Division, coming from the north, accomplished the first link-up between the spearheads of III and LVII Panzer Corps in the centre of Rostov.

Early in the morning of 24th July the fighting for the city was resumed. In the area of the post-office the enemy was overwhelmed fairly quickly, but the NKVD Headquarters was being skilfully defended by a crack force. Not until noon did riflemen of 13th Panzer Division, supported by tanks of 22nd Panzer Division, succeed in breaking enemy resistance and taking the building.

Other units of 13th Panzer Division and "Viking" had meanwhile succeeded in mopping up much of the city centre and pushing the stubbornly resisting enemy out eastward or westward. While 13th Panzer Division was holding the district north of the bridge to Bataysk, the Panzer Battalion "Viking" under Sturmbannführer Mühlenkamp thrust along the northern bank of the Don and by a surprise coup took a ford six miles west of the city—a ford used by the enemy for his withdrawal—thereby enabling the foremost units of XLIX Mountain Corps and the vanguards of 73rd and 298th Infantry Divisions to cross the Don there during the night of 24th/25th July.

In the centre of Rostov meanwhile fierce street fighting continued, and did not, in fact, cease for several days. The operation is described in an account by General Alfred Reinhardt, who, then a colonel, commanded 421st Infantry Regiment, 125th Infantry Division, in July 1942. He describes the savage street fighting, house by house, right across a barricaded metropolis—an operation which has probably never been equalled. It was the kind of fighting which the German troops would have encountered in Moscow or Leningrad.

By evening of 23rd July, a scorching hot day, the battalions of the Swabian 421st Infantry Regiment had gained the northern part of Rostov. Panzer companies and riflemen of 13th and 22nd Panzer Divisions as well as of the "Viking" SS Panzer Grenadier Division had reached the Don to both sides of the city. They were also fighting hard in the city centre, but were unable to pierce the heavily fortified built-up area, especially as they lacked the necessary infantry for that kind of operation. But the city had to be penetrated if the great Don bridge was to be gained for a drive to the south, towards the Caucasus.

NKVD troops and sappers had barricaded Rostov and were now defending it to the last bullet. That speaks for itself. This force, the political guard of the Bolshevik regime, Stalin's SS, the backbone of the State police and the secret service, was in its own way a crack force—fanatical, brilliantly trained, tough to the point of cruelty, familiar with all ruses of war, and unconditionally loyal. Above all, the NKVD troops were past masters at street fighting. After all, as the guard of the regime against possible rebellion, that was their main field of action.

What these street fighting experts had done to Rostov defies imagination. The streets had been torn open, and paving-stones had been piled up to make barricades several feet thick. Side-streets were blocked off by massive brick strongpoints. Steel girders planted in the ground and buried mines made any sudden rushing of defence posts impossible. The entrances of buildings had been bricked up; windows had been sandbagged to make firing positions; balconies had been turned into machine-gun nests. On the roofs were well-camouflaged hideouts for NKVD snipers. And in the basements lay tens of thousands of Molotov cocktails, those primitive but highly effective weapons against tanks, simply bottles filled with petrol and touched off with phosphorus or other chemicals which burst into flame upon contact with air.

Wherever a front door had not been bricked up one could be certain that a hidden booby-trap would go off the moment the door-handle was depressed. Or else a fine trip-wire stretched over the threshold would touch off a load of dynamite.

This was no ground for armoured formations, and one that offered little prospect of quick victory. True, the Panzer troops had made the first, decisive breach. But the city centre of Rostov was the battlefield of the assault parties. Laboriously they had to clear house after house, street after street, pillbox after pillbox.

Reinhardt's Swabian troops got to work against this skilfully fortified area. But the colonel tackled his cunning opponents with his own medicine—with equal cunning, with precision, and with fierce resolution.

The 1st Battalion, 421st Infantry Regiment, under Major Ortlieb and the 3rd Battalion under Captain Winzen were divided into three assault companies each. Each company was given one heavy machine-gun, one anti-tank gun, one infantry gun, and one light field howitzer for the main streets.

The direction of the drive was north to south. The city plan was divided into precise operation sectors. Each assault company was allowed to advance only as far as a fixed line across the north-south road allotted to it, a line drawn for all companies right across the town plan from west to east—the A, B, C, and D lines.

Next, the whole district had to be mopped up and contact made with the assault groups on both sides. Each unit had to wait along these lines until its neighbours had come abreast of it, and until orders for the resumption of the attack came down from regiment. In this way the six assault companies always fought in line abreast, and if any company should find itself making faster headway it could not be attacked from the flank by the enemy provided it stuck to the rules. In this way the operation in the thick maze of buildings and streets remained firmly under control from the top.

As soon as the assault companies of 1st and 3rd Battalions had cleared their allotted district Reinhardt immediately sent in six more assault wedges of 2nd Battalion. Their task was a "second picking over"—to search every building from roof-top to cellar. All civilians, including women and children, were taken from the fighting area to special collection points.

No one who might throw a hand-grenade or fire a machine pistol was left in the buildings behind the assault troops. The companies fighting their way forward had to be safe from their rear.

The plan worked with precision. It was probably only thanks to it that Rostov was cleared so quickly of the stubbornly resisting enemy forces—in a mere fifty hours of savage and relentless fighting.

In his account of the operation General Reinhardt reports: "The fighting for the city centre of Rostov was a merciless struggle. The defenders would not allow themselves to be taken alive: they fought to their last breath; and when they had been bypassed unnoticed, or wounded, they would still fire from behind cover until they were themselves killed. Our own wounded had to be placed in armoured troop carriers and guarded—otherwise we would find them beaten or stabbed to death."

Fighting was fiercest in the Taganrog road, which led straight to the Don bridge. There the German attack was held up repeatedly because it was impossible to pin-point the well-camouflaged NKVD men behind their machine-guns.

Dust, smoke, and showers of sparks from the blazing buildings enveloped the street. Keeping close to the walls of the houses, Major Ortlieb ran along the pavement to the big barricade in front. From there he waved the light field howitzer forward. "To start with, we'll shave all those balconies off."

The anti-tank gun was hauled up the road, pulled by men at the double, and was likewise brought into position along the barricade. Finally an infantry piece was also brought forward.

Then began the bombardment of "suspicious points"—chimneys, basements, and sandbagged balconies. Reinhardt himself came running up to the front line. He stood behind the foremost barricade in the main street, his binoculars at his eyes. Time and again bursts from a heavy Maksim machine-gun swept over the pavement.

"Büsing," Reinhardt called out. First Lieutenant Büsing, commanding the 13th Company, came crawling over to the general, pressed flat to the ground. Reinhardt pointed at a balcony on the second floor of a building. "Over there, Büsing—that balcony with the orange boxes.

You can see a wisp of dust there now. That's where the Russians are. Let's have that balcony off!"

Büsing hurried back to his heavy infantry gun.

"Fire!"

The second round brought the balcony down. Among the confusion of masonry they could see the Russians and their machine-gun hurled down to the street. Eventually Reinhardt brought up a few tanks of 13th Panzer Division to support his infantry. They zigzagged along the street, from one side to the other. Under their cover several small assault parties worked their way forward.

Things were worst in the old town and in the harbour district. There the streets, until then more or less straight and regular, degenerated into a maze of crooked lanes. It was no place for infantry guns: even machine-guns were no use there.

It was a case of hand-to-hand fighting. The men had to crawl right up to basement windows, doors, and the corners of houses. They could hear the enemy breathe. They could hear them slam their bolts home. They could even hear them whisper among themselves. They took a firm grip on their machine pistols—they leapt to their feet—a rapid burst of fire—and down they flopped behind cover again.

On the other side of the street a flame-thrower roared. Hand-grenades crashed. The cry of a wounded man rang eerily through the ghostly street—the long-drawn-out cry of pain: "Stretcher, stretcher...."

The wooden houses were consumed by fire. The pungent smoke made fighting more difficult, even though the wind was favourable and drove the smoke towards the Don. By the time the D line was reached it was dark. Only a few hundred yards divided the companies of 421st Infantry Regiment from the combat groups of the Panzer formations of LVII Panzer Corps on the northern bank of the Don to both sides of the road bridge to Bataysk. Night fell. The men were lying among wooden huts, tool-sheds, and heaps of rubble. The night was riddled with machine-gun fire. Flares lit up the spectral scene as bright as day for seconds at a time.

Sergeant Rittmann with his platoon of 11th Company lay in a shed in the harbour. The Russians were firing from a weighing-shed.

"Now," Rittmann commanded. With three men he overwhelmed the Russian machine-gun in the weighing-shed. Then they raced on, flinging hand-grenades to the right and left. Towards 2300 hours Rittmann and his men reached the bank of the Don and dug in.

On 25th July before daybreak the assault companies of 125th Infantry Division resumed the attack. But suddenly progress was easy. The last enemy units on the river-bank had withdrawn across the Don during the night. At 0530 hours all the assault companies of the regiment had reached the Don. Rostov was fully in German hands. But

Rostov was important as the gateway to the Caucasus only if the gate itself was held by the Germans—the bridge over the Don and the four miles of causeway across the swampy ground which were the continuation of the bridge and which presently became the great bridge into Bataysk. Beyond Bataysk was the plain—a clear road for the drive to the south, towards the Caucasus.

That gateway was finally opened by the "Brandenburg" Regiment, that mysterious, much-maligned, but incredibly brave special formation of daredevil volunteers, in co-operation with units of 13th Panzer Division.

On 24th July the Motor-cycle Battalion 43 was the first German formation to cross the Don. Second Lieutenant Eberlein, commanding 1st Company, had been ferried across the river with twenty-eight volunteers by sappers of 13th Panzer Division. Simultaneously, though at a different point, half a company of the "Brandenburg" also crossed the Don. Their intention was to capture the important bridges outside Bataysk—above all, the long causeway on the southern bank of the Don, a causeway consisting of a multitude of lesser bridges and carrying the only road to the south.

During the night of 24th/25th July 1942 First Lieutenant Grabert with his half-company charged along the causeway towards Bataysk. The handful of men of Motor-cycle Battalion 43 under Second Lieutenant Eberlein were already in position in front of the big bridge, in order to keep down the Soviet bridge guard.

But the motor-cyclists were scarcely able to raise their heads from the mud: the moment they moved they came under fire from the piers of the railway bridge about 200 yards on their left, where the Soviets had a machine-gun in position. There was also mortar-fire. Lynx-eyed, the men were watching out for the Russian muzzle flashes in order to aim their own mortars.

At 0230 hours Lieutenant Grabert raced on to the bridge with his leading section. The men by the machine-guns were under cover, their fingers taut on their triggers. But nothing stirred on the Russian side. Like phantoms Grabert and his section flitted across the bridge, on both sides of the roadway, followed at short intervals by the other two platoons. Now the Russians had noticed something. Their machine-guns opened up; mortars plopped. The German covering party immediately put up all the fire they could. Everything depended on whether Grabert would get through.

He did get through, overwhelmed the strong Soviet bridge guard, and established a small bridgehead. Throughout twenty-four hours he held it against all enemy counter-attacks.

The companies and their commanders literally sacrificed themselves for the sake of the bridge. Lieutenant Grabert and Second Lieutenant Hiller of the "Brandenburg" were both killed in action. NCOs and men were mown down in large numbers by the infernal fire of the Soviets.

The Stukas arrived in the nick of time. Then the first reinforcements came up over the causeway and the bridge. By its last pier lay Siegfried Grabert—dead. Some 200 yards farther on, in a swampy hole, lay Second Lieutenant Hiller. Next to him, his hand still clutching his first-aid kit, lay a Medical Corps NCO, with a bullet-hole in his head. But on 27th July the Panzer and rifle companies of LVII Panzer Corps were moving over the bridges towards the south, towards the Caucasus.

5. Action among the High Mountains

A blockhouse near Vinnitsa—Fuehrer Directive No. 45—By assault boat to Asia—Manychstroy and Martinovka—The approaches to the Caucasus—Chase through the Kuban—Mackensen takes Maykop—In the land of the Circassians.

IN July 1942 the Fuehrer's Headquarters were deep in Russia, near Vinnitsa in the Ukraine. The staffs of the High Command of the Army, with the Chief of the General Staff, had taken up quarters on the edge of the town of Vinnitsa. For Hitler and his operations staff the Todt Organization had built a number of well-camouflaged blockhouses under the tall pines of an extensive forest. Hitler had moved in on 16th July. The weather was scorching hot, and the shade of the fragrant pines brought no relief. Even at night the heat was sweltering and heavy. The climate did not agree with Hitler, who was for the most part bad-tempered, aggressive, and deeply mistrustful of every one. Generals, officers, and political liaison personnel among Hitler's entourage all agree that the period of his stay in the Ukraine was full of tensions and conflicts. The code name for the Fuehrer's Headquarters near Vinnitsa was "Werewolf." And, indeed, Hitler raged like a werewolf in his small blockhouse.

On 23rd July Colonel-General Halder was summoned to report. Hitler was suffering terribly from the heat, and the news from the front made his temper even worse. Victory succeeded victory, the Russians were in flight—but, oddly enough, the expected large-scale annihilation of enemy forces had not come off between Donets and Don, either at Staryy Oskol or at Millerovo. Nor, for that matter, did it seem to be taking shape at Rostov. What was the reason? What was happening?

"The Russians are systematically avoiding contact, my Fuehrer," Halder argued.

"Nonsense." Hitler cut him short. "The Russians are in full flight, they're finished, they are reeling from the blows we have dealt them during the past few months."

Halder remained calm, pointed at the map which lay on the big table, and contradicted: "We have not caught the bulk of Timoshenko's forces, my Fuehrer. Our encircling operations at Staryy Oskol and Millerovo were punches at nothing. Timoshenko has pulled back the bulk of his Army Group, as well as a good part of his heavy equipment, across the Don to the east, into the Stalingrad area, or else southward, into the Caucasus. We've no idea what reserves he has left there."

"You and your reserves! I'm telling you we didn't catch Timoshenko's fleeing masses in the Staryy Oskol area, or later at Millerovo, because Bock spent far too much time at Voronezh. We were too late to intercept the southern group north of Rostov, as it was flooding back in panic, simply because we wheeled our fast troops south too late and because Seventeenth Army began frontally pushing them back east too soon. But I'm not having that happen to me again. We've now got to unravel our fast troops in the Rostov area and employ our Seventeenth Army, as well as First Panzer Army and also Fourth Panzer Army, with the object of coming to grips with the Russians south of Rostov, in the approaches to the Caucasus, encircling them and destroying them. Simultaneously, Sixth Army must administer the final blow to the remnants of the Russian forces which have fled to the Volga, to the area of Stalingrad. At neither of these two vital fronts must we allow the reeling enemy any respite at all. But the main weight must be with Army Group A, with the attack against the Caucasus."

Colonel-General Halder, Chief of the Army General Staff, tried in vain in his conversation with Hitler on 23rd July 1942 to disprove the Fuehrer's thesis. He implored Hitler not to split his forces and not to strike at the Caucasus until after Stalingrad had been taken and the German flank and rear on the Don, as well as between Don and Volga, sufficiently protected.

Hitler brushed aside these misgivings of the General Staff. He was confident of victory and completely obsessed by the belief that the Red Army had already been decisively defeated. This is borne out by several more positively dumbfounding decisions. He transferred the bulk of Field-Marshal von Manstein's Eleventh Army with five divisions from the Crimea, where it was standing by to operate against the

Caucasus, all the way up to Leningrad in order to take this irritating fortress at long last.

But that was not all. Hitler also pulled out the magnificently equipped SS Panzer Grenadier Division "Leibstandarte" from the Eastern Front and transferred it to France for rest and reorganization into a Panzer Division. Yet another crack formation from the southern front, the Motorized Infantry Division "Grossdeutschland," was similarly withdrawn from the fighting-line shortly afterwards. Hitler commanded that as soon as the Manych Dam was reached this division was to be pulled out of the front and transferred to France to remain at the disposal of the High Command. This decision was partly due to the shortage of motor-fuel on the southern front. The principal reason, however, which Hitler advanced for these decisions was that, according to his information, the invasion of Western Europe was imminent. It was an incomprehensible and fatal mistake. These seven divisions, quite needlessly, withdrawn from the southern front would certainly have been enough to avert the catastrophe of Stalingrad.

Halder was bitter as he returned to his headquarters on the edge of Vinnitsa from his interview on 23rd July. He wrote in his diary: "His persistent underestimation of the enemy's potential is gradually taking on grotesque forms and is beginning to be dangerous."

But Hitler stuck to his mistaken assessment of the situation, and summed up his ideas in a fundamental "Fuehrer Directive No. 45" which he dictated on the same day, 23rd July, following his argument with Halder.

The Directive was delivered to the Army Groups on 25th July. In its introduction it declared—contrary to the actual facts and to the experience of the previous three weeks' fighting—that only weak enemy forces of Timoshenko's Armies had succeeded in escaping German encirclement and reaching the southern bank of the Don.

Contrary to Directive No. 41 for "Operation Blue," which envisaged that Stalingrad should first be reached before the offensive was launched into the Caucasus for the seizure of the Russian oil, the new directive laid down the various objectives as follows:

> (1) The first task of Army Group A is the encirclement and annihilation in the area south and south-east of Rostov of the enemy forces now escaping across the Don. For this purpose powerful fast formations must be employed from the bridgeheads, to be formed in the Konstantinovskaya–Tsimlyanskaya area, in a general south-westerly direction, roughly towards Tikhoretsk across the Don; these formations to consist of infantry, Jäger, and mountain divisions. The cutting of the Tikhoretsk–Stalingrad railway-line with advanced units is to be effected simultaneously. . . .
>
> (2) Following the annihilation of the enemy force south of the

Don the main task of Army Group A is the seizure of the entire eastern coast of the Black Sea, with a view to eliminating the enemy's Black Sea ports and his Black Sea Fleet....

Another force, to be formed by the concentration of all remaining mountain and Jäger divisions, will force a crossing of the Kuban and seize the high ground of Maykop and Armavir....

(3) Simultaneously another force, to be composed of fast formations, will take the area around Groznyy, with some of its units cutting the Ossetian and Georgian Military Highways, if possible in the passes. Subsequently this force will drive along the Caspian to take possession of the area of Baku....

The Italian Alpini Corps will be assigned to the Army Group at a later date.

This operation of Army Group A will be known under the code name "Edelweiss."

(4) Army Group B—as previously instructed—will, in addition to organizing the defence of the Don line, advance against Stalingrad, smash the enemy grouping which is being built up there, occupy the city itself, and block the strip of land between Don and Volga.

As soon as this is accomplished fast formations will be employed along the Volga with the object of driving ahead as far as Astrakhan in order to cut the main arm of the Volga there too.

These operations of Army Group B will be known by the code name of "Heron."

This was followed by directives for the Luftwaffe and the Navy.

Field-Marshal List, a native of Oberkirch in Bavaria, a man with an old Bavarian General Staff training and with distinguished service in the campaigns in Poland and France, was the General Officer Commanding Group A. He was a clever, cool, and sound strategist—not an impulsive charger at closed doors, but a man who believed in sound planning and leadership and detested all military gambles.

When a special courier handed him Directive No. 45 at Stalino on 25th July, List shook his head. Subsequently, in captivity, he once remarked to a small circle of friends that only his conviction that the Supreme Command must have exceptional and reliable information about the enemy's situation had made the new plan of operations seem at all comprehensible to him and his chief of staff, General von Greiffenberg.

Always form strongpoints—that had been the main lesson of Clausewitz's military teaching. But here this lesson was being spectacularly disregarded. To quote just one instance: moving up behind Sixth Army, which was advancing towards Stalingrad and the Volga valley, were formations of the reinforced Italian Alpini Corps with its excellent mountain divisions. List's Army Group A, on the other hand, which was now faced with the first real alpine operation of the war in the East—the conquest of the Caucasus—had at its disposal only three mountain divisions, two of them German and one Rumanian. The Jäger divisions of Ruoff's Army-sized combat group (the reinforced Seventeenth Army) were neither trained for alpine warfare nor did they possess the requisite clothing and equipment. Four German mountain divisions—hand-picked men from the Alpine areas and thoroughly trained in mountain warfare—were employed piecemeal all over the place. A few weeks later, when it was too late, the Fuehrer's Headquarters was to be painfully reminded of that fact when General Konrad's Mountain Jäger battalion found themselves pinned down along the ridges of the Caucasus, within sight of their objective.

Allowing for the forces at his disposal, Field-Marshal List turned Directive No. 45 into a passable plan of operations. Ruoff's group, the reinforced Seventeenth Army, was to strike south frontally from the Rostov area towards Krasnodar. The fast troops of Kleist's First Panzer Army—followed on their left wing by Hoth's Fourth Panzer Army—were given the task of bursting out of their Don bridgeheads and driving on Maykop as the outer prong of a pincer movement. In this way, by the collaboration between Ruoff's slower infantry divisions and Kleist's fast troops, the enemy forces presumed to be south of Rostov were to be encircled and destroyed.

Colonel-General Hoth's Fourth Panzer Army on the eastern wing was to provide flank cover for this operation. Its first objective was Voroshilovsk.[1]

This then was the plan for the attack towards the south, for an operation which followed a highly dramatic course and proved decisive for the whole outcome of the war in the East.

While Ruoff's group was still fighting for Rostov some of the units of the First and Fourth Panzer Armies had advanced as far as the Don. By 20th July the Motor-cycle Battalion of 23rd Panzer Division had succeeded in crossing the river at Nikolayevskaya and establishing a bridgehead on the southern bank of the Don. Three days later a combat group of 3rd Panzer Division thrust south and crossed the Sal at Orlovka. From there the XL Panzer Corps drove against the Manych sector with 3rd and 23rd Panzer Divisions.

The Soviet Command was clearly determined not to allow its forces to be encircled again. The Soviet General Staff and the commanders in the field stuck strictly to their new strategy—essentially the old strategy that had defeated Napoleon—of enticing the enemy into the wide-open spaces of their vast country in order to make him fritter away his strength until they could pounce upon him on a broad front at the right moment.

[1] Now Stavropol.

Action among the High Mountains 507

The German formations encountered entirely novel combat conditions south of the Don. Ahead of them lay 300 miles of steppe, and beyond it one of the mightiest mountain ranges in the world, extending from the Black Sea to the Caspian, right across the path of the attacking German armies.

The steppe north of the Caucasus provided the enemy with excellent opportunities for elastic resistance. The countless water-courses, big and small, running from the watershed of the Caucasus towards the Caspian as well as towards the Black Sea, were obstacles which could be held by defenders with relatively slight forces.

As in the desert, the route of advance through the steppe was dictated to the attacker by watering-points. The war was moving into a strange and unfamiliar world. The more than 400-mile-long Manych, eventually, formed the boundary between Europe and Asia: to cross it meant to leave Europe.

The Westphalian 16th Motorized Infantry Division of the III Panzer

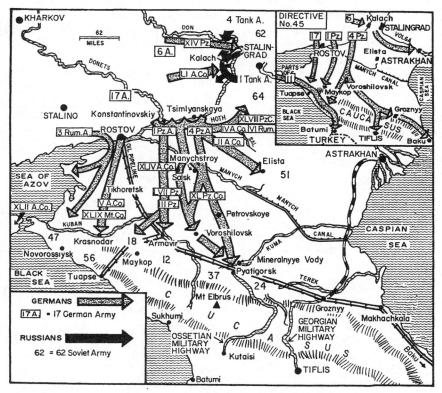

Map 29. The situation on the southern front between 25th July and early August 1942. The inset map shows the position envisaged in Directive No. 45.

Corps and the Berlin–Brandenburg 3rd Panzer Division of XL Panzer Corps were the first German formations to cross into the Asian continent.

General Breith's 3rd Panzer Division, the spearhead of XL Panzer Corps, had pursued the retreating Russians from the Don over the Sal as far as Proletarskaya on the Karycheplak, a tributary of the Manych. Breith's Panzer troops had thus reached the bank of the wide Manych river. Strictly speaking, this river was a string of reservoirs backed up by dams; these lakes were often nearly a mile wide. The reservoirs and massive dams formed a hydroelectric power system known as Manych-Stroy.

On the far bank, well dug in, were the Soviet rearguards. The Manych was an ideal line of defence for the Soviets, a solid barrier across the approaches to the Caucasus.

"How are we going to get across there?" General Breith anxiously inquired of his chief of operations, Major Pomtow, and of the commander of 3rd Rifle Regiment, Lieutenant-Colonel Zimmermann.

"Where the river is narrowest the Russian defences are strongest," replied Pomtow, pointing to a file of aerial reconnaissance reports.

"According to prisoners' evidence, the far bank is held by NKVD troops," Zimmermann added.

"And well dug in, too, by the look of these aerial pictures," Breith nodded.

"Why not outwit them by choosing the widest spot—near the big dam, where the river is nearly two miles across? They won't expect an attack there," Pomtow suggested.

It was a good idea, and it was adopted. Fortunately, the Panzer Engineers Battalion 39 was still dragging twenty-one assault boats with them. They were brought up. The scorching heat had so desiccated them that, when they were tried out, two of the boats sank like stones at once. The other nineteen were also leaky, but would be all right provided the men baled vigorously.

Second Lieutenant Moewis and a dozen fearless "Brandenburgers" reconnoitred two suitable crossing-points, almost exactly at the widest point of the river. Both crossings were upstream from the small town of Manych-Stroy, which was situated directly at the far end of the dam. The dam itself appeared to have been blocked and mined in a few places only. The small town would have to be taken by surprise, so as to prevent any Soviet demolition parties from wrecking the dam completely.

For this action a combat group was formed from units of 3rd Panzer Grenadier Brigade. The 2nd Battalion, 3rd Panzer Grenadier Regiment, attacked on the left and the 1st Battalion on the right. A strong assault company was formed from units of 2nd Battalion, 3rd Panzer Grenadier Regiment, and put under the command of First Lieutenant

Action among the High Mountains

Tank, the well-tried commander of 6th Company. Its orders were: "Under cover of darkness a bridgehead will be formed on the far bank of the reservoir. Following the crossing by all parts of the combat group, the enemy's picket line will be breached and the locality of Manych-Stroy taken by storm."

To ensure effective artillery support from the north-eastern bank an artillery observer was attached to the combat group. The bold attack by XL Panzer Corps across the Manych was successful. At the focal point of the action 3rd Panzer Division feinted an attack from the north-west with one battalion of 6th Panzer Regiment, while the 1st Battalion, 3rd Panzer Grenadier Regiment, actually struck across the river. The action was prepared by a sudden concentrated bombardment by divisional artillery between 2400 and 0100 hours.

Tank's men were lying on the bank. The sappers had pushed their craft into the water. The shells were whining overhead, crashing on the far bank and enveloping it in smoke and dust.

"Now!" ordered Tank. They leapt into the boats and pushed off. They had to bale feverishly with empty food-tins to prevent the boats being flooded. The noise of the motors was drowned by the artillery bombardment. Not a shot was fired by the Russians.

The river was crossed without casualties. The keels of the nineteen boats scraped over the gravel on the far bank. Tank was the first to leap ashore. He was standing in Asia.

"White Very light," Tank called out to the squad commander. The white flare soared skyward from his pistol. Abruptly the German artillery lengthened range. The sappers got into their boats again to bring over the next wave.

Tank's men raced over the flat bank. The Soviets in the first trench were completely taken by surprise and fled. Before they could raise the alarm in the next trench Tank's machine-guns were already mowing down the enemy's outposts and sentries.

But by then the Russians to the right and left of the landing-point had been alerted. When the assault craft came over with their second load they were caught in the crossfire of Soviet machine-guns. Two boats sank. The remaining seventeen got across, with 120 men and supplies of ammunition, as well as the 2nd Battalion headquarters.

But that was the end of the ferrying. Major Boehm, the battalion commander, succeeded in extending the bridgehead on the southern bank of the Manych. Then he was severely wounded. Lieutenant Tank, the senior company commander in 2nd Battalion, assumed command in the bridgehead. The Russians were covering the entire bank with enfilading fire. Soviet artillery of all calibres pounded the area. In any case, with increasing daylight all further ferrying operations had to come to an end.

Lieutenant Tank and his men were still lying on the flat ground by the river, in the captured Soviet trenches and in hurriedly dug firing-pits. The Russians were mortaring and machine-gunning them, and also launched two counter-attacks which got within a few yards of Tank's position.

The worst of it was that they were running out of ammunition. The machine-gun on the right wing had only two belts left. Things were not much better with the others. The mortars had used up all their ammunition.

"Why isn't the Luftwaffe showing up?" Tank's men were asking as they looked at the hazy, overcast sky. Towards 0600 hours, almost as if the commodore commanding the bomber Geschwader had heard their prayers, German fighter-bombers roared in, just as the sun was breaking through, the sun which a short while earlier had dispersed the fog from their airstrips. They shot up the Soviet artillery emplacements and machine-gun nests. Under cover of the hail of bombs and machine-gun fire a third wave of infantry was eventually ferried across the river.

Lieutenant Tank made good use of their waiting time. He skipped from one platoon commander to another, instructing them in detail. Then the attack was launched, platoon by platoon—against Manych-Stroy.

The Soviets were completely taken by surprise. They had not expected the strongly defended township to be attacked from the rear and from the side. Their entire attention had been focused forward, towards the dam. Tank's men quickly rolled up the rearward positions of the Russians.

By the time the Soviet commander had reorganized his defence and drawn up his men with their backs to the dam the first German tanks and armoured infantry carriers of Lorried Infantry Battalion Wellmann were already roaring over the narrow roadway along the top of the dam.

Manych-Stroy fell. Wellmann's battalion got across the dam unscathed. The Manych had been conquered: the last major obstacle on the road to the south, towards the Caucasus and the oilfields, had been overcome.

In the morning of 2nd August the 3rd Panzer Division thrust through as far as Ikituktun[1] with the Combat Group von Liebenstein, while the Group Pape established a bridgehead at Pregatnoye. Geyr von Schweppenburg's XL Panzer Corps and, on its right, General von Mackensen's III Panzer Corps were now fighting in Asia.

The bold crossing of the Manych and the opening of the door to the Caucasus were supplemented by an equally bold and successful operation by the 23rd Panzer Division from Baden-Württemberg. They wiped out a strong and cunningly positioned Soviet ambush which

[1] Now Pushkinskoye.

had seriously threatened the German flank without anyone even suspecting the danger.

At the Sal crossing near Martynovka, Timoshenko had placed an entire motorized Corps in an excellently camouflaged ambush, complete with many tanks.

Major-General Mack was advancing behind the 3rd Panzer Division, driving with his reinforced Motor-cycle Battalion 23 towards Martynovka, which had been reported by German aerial reconnaissance to be "only lightly held."

Mack's attack was launched at the very moment when the Russian Corps was moving into position. Mack instantly realized the danger. He tied down the enemy by frontal attack, surrounded him in a bold operation with the reinforced 201st Panzer Regiment of the Combat Group Burmeister, and in the early hours of 28th July struck at the rear of the Russians, who were completely taken by surprise.

In confused tank duels, often at such extremely short range as 20 or 30 yards, the Russians' T-34s were knocked out and their anti-tank guns smashed. The 9th Company, 201st Panzer Regiment, alone, the first German unit to penetrate into Martynovka, destroyed twelve T-34s and six T-70s, as well as several anti-tank and infantry guns. Captain Fritz Fechner immobilized several T-34s by means of 'sticky bombs.'

The tank battle of Martynovka was the first operation for a long time in which superior tactical leadership and skilful operation of tank against tank succeeded in pinning down a major Soviet formation and annihilating it. Altogether seventy-seven enemy tanks were destroyed and numerous guns captured.

While the grenadiers and tanks of the 3rd Panzer Division were pursuing the retreating Soviets along the Manych river through the Kalmyk Steppe in a scorching heat of 50 degrees Centigrade, past huge herds of cattle, past curious camels and dromedaries, Hitler was sitting in his stifling-hot block-house at his Ukrainian headquarters near Vinnitsa, looking at his large situation map. General Jodl was making a report.

The subject under discussion, however, was not the successful operation on the Manych, reported in the High Command communiqués, but the nasty situation in which the Sixth Army was finding itself in the great Don bend.

General Paulus, admittedly, had reached the Don with his northern and southern attacking force, but the bridgehead at Kalach, which controlled access to the narrow strip of land between Don and Volga, was not only being held by the Soviets, but in fact being turned into a springboard for a counter-offensive.

Lieutenant-General Gordov, the Soviet Commander-in-Chief of the "Stalingrad Front," had already lined up four Soviet Armies—the

Twenty-first, the Sixty-second, the Sixty-third, and the Sixty-fourth—as well as two Tank Armies in the process of formation—the First and the Fourth—in front of the German Sixth Army.

The Soviet Fourth Tank Army had begun to encircle Paulus's XIV Panzer Corps. General von Seydlitz-Kurzbach's LI Army Corps on the southern wing was already in serious trouble. The entire Sixth Army was beginning to be paralysed by a shortage of ammunition and a total lack of fuel.

Hitler's decision to push ahead simultaneously with the operations against the Caucasus and against Stalingrad had meant that supplies too had had to be divided. And since the greater distances had to be tackled in the south, the Quartermaster-General of the Army General Staff, General Wagner, had given the Caucasus front priority in fuel-supplies. Many motorized long-distance supply columns originally destined for Sixth Army were redirected to the south.

By 31st July Hitler was at last compelled to realize that his optimism had been unfounded. He could no longer shut his eyes to the fact that the strength of Sixth Army, impaired as it was by serious supply shortages, was no longer sufficient to take Stalingrad against the strong Soviet opposition.

On that day he therefore decreed yet another change of plan. The Fourth Panzer Army—though stripped of XL Panzer Corps—was detached from the Caucasus front, put under Army Group B, and moved south of the Don towards the north-east, in order to strike at the flank of the Soviet front at Kalach, before Stalingrad.

It was a good plan, but it came too late. The dispatch of Fourth Panzer Army changed nothing as far as the dissipation of forces was concerned. The units which Hitler was taking away from Army Group A merely weakened that Army Group's offensive striking power against the Caucasus; as a reinforcement for Army Group B these units were too few and arrived too late to ensure an early capture of Stalingrad. Two equally strong Army groups were advancing in divergent directions, at right angles to each other, towards objectives which were a long way from each other. The most acute problem, that of supplies, became completely insoluble because the overall operation continued to lack a clear centre of gravity.

The German High Command had manœuvred itself into a hopeless situation and had allowed itself to become dependent on the opponent's decisions. In the Stalingrad area the Soviets were already dictating the time and place of the battle.

The Fuehrer's Directive of 31st July demanded that on the Caucasus front the second phase of Operation Edelweiss was now to begin—the capture of the Black Sea coast. Army Group A was to employ its fast formations, now grouped under the command of First Panzer Army, in the direction of Armavir and Maykop. Other units of the

Army Group, the Army-sized combat group Ruoff with General Kirchner's LVII Panzer Corps, were to drive via Novorossiysk and Tuapse along the coast to Batumi. The German and Rumanian mountain divisions of General Konrad's XLIX Mountain Corps were to be employed on the left wing across the high Caucasian passes to outflank Tuapse and Sukhumi.

To begin with, everything went according to plan—with absolutely breathtaking precision. On the day when the new Fuehrer Directive was issued the III and LVII Panzer Corps also made a big leap forward in the direction of the Caucasus. General von Mackensen with the 13th Panzer Division, newly placed under his command, took Salsk on the same evening. Advancing across several anti-tank ditches, the division on 6th August gained Kurgannaya on the Laba, and the 16th Motorized Infantry Division took Labinsk.

In the evening of 9th August Major-General Herr's 13th Panzer Division stormed the oil town of Maykop, the administrative seat of a vast region of oilfields. Fifty aircraft were captured intact. The oil storage-tanks, however, had been destroyed and the plant itself paralysed by the removal of all key equipment.

Progress was also made by XLIX Mountain Corps and by V Army Corps, which had forced a crossing of the Don east of Rostov. By 13th August the divisions took Krasnodar and forced a crossing of the Kuban.

The advance of LVII Panzer Corps proceeded equally successfully. After a rapid southward thrust through the Kuban steppe the Panzer Combat Group Gille of the "Viking" SS Panzer Grenadier Division, and Combat Groups "Nordland" and "Germania" behind it, were deployed along the northern bank of the Kuban. The Panzer Group Gille crossed the river; the group under von Scholtz put across at Kropotkin and swiftly established a bridgehead, thereby clearing the road to the southern bank of the Kuban for the Army-sized Combat Group Ruoff.

The "Viking" Division was then turned south-west, in the direction of Tuapse, at the head of LVII Panzer Corps. Under the command of General of Waffen SS Felix Steiner, the Scandinavian, Baltic, and German volunteers who were grouped together in the "Viking" Division penetrated into the north-westerly and south-westerly part of the Maykop oilfields.

During the first few days of August 1942 the fast formations of Army Group A were thus sweeping along their entire front through the Kuban and Kalmyk steppes in order to engage the elastically resisting and slowly withdrawing Russian divisions before they reached the Caucasus, and in order to prevent them from escaping into the mountains and there establishing a new defensive line.

R

Signaller Otto Tenning, who was then driving in the command car of the point battalion of 3rd Panzer Division, has given the author the following report: "The next place we came to was Salsk. For our further advance through the Kalmyk steppe orders were that we must not fire at enemy aircraft. In this way the Russians were to be prevented from making out the position of our most advanced units, since through the raised clouds of dust it was probably impossible from the air to tell friend from foe. I was detailed with my scout car to 1st Company and was doing a reconnaissance with Sergeant Goldberg. We were cautiously approaching a small village when the recce leader suddenly spotted something suspicious and sent a radio signal: 'Enemy tanks lined up along the edge of the village.' To our surprise we discovered a little later that these supposed tanks were in fact camels. There was much laughter. From then onwards dromedaries and camels were no longer anything unusual. Indeed, our supply formations made a lot of use of these reliable animals."

The advanced formations of 3rd Panzer Division reached the town of Voroshilovsk on 3rd August. The Russian troops in the town were taken by surprise, and the town itself was captured after a brief skirmish towards 1600 hours. A Russian counter-attack with tanks and cavalry was repulsed.

The advance continued. Men of the "Brandenburg" Regiment went along with the forward units, always ready for special assignments. Rumanian mountain troops too were included in the formations under 3rd Panzer Division. The native Caucasian population was friendly and welcomed the Germans as liberators.

There simply is no denying the fact that entire tribes and villages readily, and, indeed, against the wishes of the German High Command, volunteered to fight against the Red Army. These freedom-loving people believed that the hour of national independence had come for them. Stalin's wrath, when it struck them later, was terrible: all these tribes were expelled from their beautiful homeland and banished to Siberia.

The faster the advance towards the Caucasus was gaining ground the clearer it became that the Russians were still withdrawing without any great losses in lives or material. The German formations were gaining territory—more and more territory—but they did not succeed in savaging the enemy, let alone annihilating him. A few upset peasant carts and a few dead horses were all the booty lining the route of the German advance.

In order to cover the increasingly long eastern flank of the drive to the Caucasus, General Ott's LII Army Corps with 111th and 370th Infantry Divisions were wheeled eastward on a broad front and deployed towards the Caspian. Elista, the only major town in the Kalmyk steppe, fell on 12th August.

Action among the High Mountains 515

Meanwhile the 3rd and 23rd Panzer Divisions continued to move southward. The Kalmyk steppe lay parched under a scorching sun. The thermometer stood at 55 degrees Centigrade. A long way off, in the brilliantly blue summer sky, the troops saw a white cumulus cloud. But the cloud did not move. The following day, and the day after, it was still in the same spot. It was no cloud. It was Mount Elbrus, 18,480 feet high, with its glistening glaciers and its eternal snow—the greatest mountain massif of the central Caucasus.

"How many miles have we done to-day?" Colonel Reinhardt, the commander of 421st Infantry Regiment, asked his adjutant. First Lieutenant Boll looked at his map, on which the routes of advance of 125th and, next to it, 198th Infantry Divisions—forming V Corps—were marked. He measured off the distance. "Forty miles, Herr Oberst."

Forty miles. That was the distance the infantry had marched that day. Under a searing sky, through the treeless Kuban steppe. On the march the columns were enveloped in thick clouds of greyish-brown dust. Only the heads of the horsemen showed above it. The farther south they advanced the looser the connection became between individual regiments. Only the distant trails of dust indicated that somewhere far to the right and far to the left there were other marching columns similarly advancing to the south.

In the shade of his command car Reinhardt studied the map.

"Absolutely terrifying, those distances," the adjutant observed.

Reinhardt nodded. His finger moved across the map to the Kalmyk steppe. "Kleist's tanks are no better off either."

Indeed, they were no better off. The XL Panzer Corps—since 2nd August subordinated to First Panzer Army—had taken Pyatigorsk with 3rd Panzer Division on 10th August and Mineralnyye Vody with 23rd Panzer Division, and had thus reached the foot of the Caucasus. The last great obstacle still ahead was the Terek river. Would they be able to cross it and then gain the high mountain-passes by way of the Ossetian and Georgian Military Highways?

The III and LVII Panzer Corps at the centre of the front were meanwhile driving on through heat and dust, from the Don into the Maykop oilfield region, attempting to overtake the retreating enemy. Colonel Reinhardt stabbed the map at Krasnodar. "That's our objective."

Then he pointed to Maykop. "And that's where Kleist has to get. Then we'll see what we've collected in the pocket formed by our Seventeenth Army and Kleist's First Panzer Army between these two cornerstones."

The adjutant nodded. "It's a good plan, Herr Oberst, but I have a feeling that the Russians are not going to oblige us by waiting for the bag to close."

Reinhardt passed the map back to Boll. "We'll see," he grunted. "Got any water left?"

"Not a drop, Herr Oberst. My tongue's been sticking to the roof of my mouth like a piece of fly-paper for this past hour."

They climbed back into their car. "Let's go; we've got to drive another six miles to-day."

That was what the advance was like for 421st Infantry Regiment, 125th Infantry Division, and it was much the same for all the infantry, Jägers, and mountain units of Ruoff's group in early August 1942. For a while the war on the southern front assumed the character of desert warfare. The pursuit of the Soviets through the Kuban steppe turned into a race from watering-point to watering-point. There were few stops for food. Admittedly, emergency supplies of drinking-water were carried for the troops in large water-cisterns, but these could not, of course, carry enough for the horses as well. As a result, new watering-places had to be captured every day.

On the right wing of Army Group A the Russians withdrew elastically in the face of pressure by the German Seventeenth Army, in the same way as they had done successfully on the middle Don. The Soviets would systematically hold on with strong rearguards to the few villages and numerous river-beds: at first they would defend them stubbornly, and presently they would abandon them so quickly that they lost scarcely any prisoners. In this way they implemented Marshal Timoshenko's new instruction: the enemy's advance was to be delayed, but at the decisive moment the units were to be withdrawn to avoid encirclement at all costs.

That was the new elastic strategy of the Russians. The Soviet General Staff had dropped Stalin's old method of contesting every inch of ground, a method which had time and again led to encirclement and gigantic losses.

The lower Soviet commands very soon learned these tactics of delaying actions and elastic resistance, a technique which had been struck from the German training manuals since 1936. By making skilful use of the numerous river-courses running across the line of the German attack, the Russians time and again delayed the German advance and pulled back their own infantry.

In these circumstances the German divisions of the Army-sized Combat Group Ruoff and of First Panzer Army did not succeed in implementing the key task of Directive No. 45: "The enemy forces escaping over the Don are to be encircled and annihilated in the area south and south-east of Rostov." Once again Hitler's plan had misfired.

They moved on interminably—pursuing, marching, and driving. The advance continued, farther and farther. The troops moved from river to river: the Kagalnik was overcome, the Yeya was crossed. There

Action among the High Mountains 517

were eight more rivers to cross before the Württemberg V Corps reached the Kuban. This Corps was employed against Novorossiysk between the Rumanian Third Army on its right and the LVII Panzer Corps and, behind it, the XLIX Mountain Corps on its left. The XLIV Jäger Corps followed behind General Kirchner's fast divisions.

At Tikhoretsk the oil pipeline from Baku to Rostov crossed the railway and road. The Russians were stubbornly defending this key point with strong artillery, anti-tank guns, and three armoured trains.

The 8·8 flak combat parties, put under 125th Infantry Division, had a difficult task. But at last the advanced units of 125th and 198th Infantry Divisions linked up. Tikhoretsk fell. The Russians gave way. But they did not flee in panic.

Striking suddenly from vast fields of man-high sunflowers, the Russians frequently caught the German troops in their fire. But as soon as the Germans tried to come to grips with them they were gone. At night individual vehicles were ambushed. It was no longer possible to send out motor-cycle dispatch-riders.

In this manner V Corps with 125th, 198th, 73rd, and 9th Infantry Divisions reached the Krasnodar area by 10th August 1942. In a mere sixteen days the infantrymen had covered the roughly 200 miles from Rostov to the capital of the Kuban Cossacks, fighting and marching through the scorched earth of the sun-seared Kuban steppe, but also through magnificently fertile river-valleys.

Around them stretched unending fields of sunflowers, huge areas of wheat, millet, hemp, and tobacco. Enormous herds of cattle moved across the limitless steppe. The gardens of the Cossack villages were veritable oases. Apricots, mirabels, apples, pears, melons, grapes, and tomatoes grew in luxuriant profusion. Eggs were as plentiful as sand on the seashore, and there were gigantic herds of pigs. It was a splendid time for the cooks and paymasters.

Krasnodar, the centre of the Kuban district, on the northern bank of the Kuban river, then had about 200,000 inhabitants. It was a town of large oil refineries.

General Wetzel employed his V Corps for a concentric attack on the town—the 73rd Infantry Division from Franconia from the north-west, the Hessian Regiments of the 9th from the north, and the men from Württemberg of the 125th and 198th Infantry Divisions from the north-east and east. The Russians offered stubborn and furious resistance in the orchards and the suburbs. They wanted to hold open the town centre, with its bridge over the Kuban, in order to take across as many human beings as possible and, even more important, whatever material they could move. Anything that could not be taken across to the southern bank was set on fire, including the vast oil-storage tanks.

By noon on 11th August Major Ortlieb with the 1st Battalion, 421st

Infantry Regiment, had worked his way forward to within charging distance of the bridge—a mere 50 yards. Packed closely together, Russian columns were moving over the bridge.

The 2nd Company was given the order to attack. Captain Sätzler sprang to his feet, his pistol in his raised hand. He took only three steps before he collapsed, shot through the head.

The company charged on. Its forward sections were within 20 yards of the ramp. At that moment the watchful Soviet bridge officer detonated the charges.

At half a dozen separate points the bridge went up with a roar like thunder, complete with the Russian columns on it. Among the smoke and dust, men and horses, wheels and weapons, could be seen sailing through the air. Horse-drawn vehicles, the horses bolting, raced over the collapsing balustrades, hurtling into the river and disappearing under the water.

This action proved that the Russians had learnt a lot in recent months. The demolition of the bridge cost the Germans two days. Not until the night of 13th/14th August did the 125th Infantry Division succeed in crossing the river by assault boats and rafts.

During the preceding day Major Ortlieb had reconnoitred the crossing-point right under the watchful eyes of the Soviets entrenched on the far side. Disguised as a peasant woman, a hoe over his shoulder and a basket over his arm, he had calmly walked along the river.

Under the concentrated fire of German artillery and the 3·7-cm. flak battery the troops accomplished the crossing of the Kuban and succeeded in building a pontoon bridge. The V Corps was marching into the land of the Circassians. The Muslim population had hoisted the crescent, the flag of Islam, over their houses and was welcoming the Germans as their liberators from the atheist Communist yoke.

6. Between Novorossiysk and the Klukhor Pass

"The sea, the sea!"–The mountain passes of the Caucasus –Fighting for the old Military Highways–Expedition to the summit of Mount Elbrus–Only 20 miles to the Black Sea coast–For the lack of the last battalion.

WITH the crossing of the Kuban the last great river obstacle across the path of Ruoff's Armeegruppe—the new-style Army-sized combat group—had been overcome. The divisions were now able to attack their real strategic objective—the ports of Novorossiysk, Tuapse, Sochi, Sukhumi, and Batumi. These were objectives of exceptional importance. Not only would the Soviet Black Sea fleet be denied its last bases, and thus the conditions created for the German Caucasus front to be supplied by the sea-route, but an even greater prize might be won. Once the last Soviet coastal strip on the Black Sea was occupied by German troops, Turkey would very likely change over into the German camp. This might have far-reaching consequences upon the Allied conduct of the war. The British–Soviet positions in Northern Persia would collapse, and the southern supply route for American military aid to Stalin—by way of the Persian Gulf, the Caspian, and up the Volga—would be severed.

Even the bold plan of directing Rommel's Africa Corps via Egypt into Mesopotamia would enter the realm of the possible. At that time the men of the German–Italian Panzer Army in Africa were standing at El Alamein, at the gates of Cairo, after their brilliantly fought pursuit during the late summer of 1942. The sappers were already calculating the number of bridging columns they would need for crossing the Nile, and whenever a trooper was asked, "Where's our next stop?" he would frivolously reply, "Ibn Saud's palace."

These fantastic long-range objectives were exceedingly popular among the men of Ruoff's combat group. As soon as the formations of XLIX Mountain Corps heard that they were moving into the Caucasus they too coined their slogans. In his book *Mountain Jägers on All Fronts* Alex Buchner reports the answer of a Jäger to the question about the purpose and objective of the long trek through the steppe: "Down the Caucasus, round the corner, slice the British through the rear, and say to Rommel, 'Hello, General, here we are!'"

Towards the end of August 1942 the divisions of V Corps began their attack against Novorossiysk, the first major naval fortress on the

520 Part Six: The Caucasus and the Oilfields

eastern coast of the Black Sea. Novorossiysk, which had 95,000 inhabitants at that time, was an important harbour and industrial town with extensive cold-storage plants and shipbuilding yards, with a large fish-processing industry and cement-mills.

The 125th and 73rd Infantry Divisions fought their way forward through the foothills of the Caucasus to the approaches of the town. Quite suddenly before them they saw the sea. Colonel Friebe, commanding 419th Infantry Regiment, on catching sight of the sea-coast from some high ground, ordered the old Greek tag to be radioed to his neighbour, Colonel Reinhardt of 421st Infantry Regiment: "Thalassa, thalassa—the sea, the sea!" With these words, according to the ancient Greek historian Xenophon, the Greek vanguards, 2400 years before, had hailed the sea when they first caught sight of it after their arduous retreat through the waterless deserts and mountains of Asia Minor, when they reached the coast near Trebizond, exactly opposite Novorossiysk.

But a great deal of hard and costly fighting was needed before the regiments of 125th and 73rd Infantry Divisions gained control of Novorossiysk, which was being stubbornly defended by units of the Soviet Forty-seventh Army.

On 6th September 1942 the 1st Battalion, 186th Infantry Regiment, under Lieutenant Ziegler launched its assault against the port and harbour at the head of 73rd Infantry Division.

By 10th September the town and its surroundings were firmly in German hands. The first objective of Ruoff's Army-sized combat group had been reached. The next objective was Tuapse—a keypoint in the narrow coastal plain. Tuapse became a turning-point in the destinies of List's Army Group.

In addition to V Infantry Corps, XLIV Jäger Corps, and LVII Panzer Corps, Seventeenth Army also included XLIX Mountain Corps, with its 1st and 4th Mountain Divisions, as well as the Rumanian 2nd Mountain Division. There was a special purpose behind this combination of infantry, Jägers, and mountain troops. While General Wetzel's infantry divisions were taking Novorossiysk by frontal attack across the wooded foothills of the north-western Caucasus, the 97th and 101st Jäger Divisions, advancing behind LVII Panzer Corps via Maykop, were already fighting their way across the "Wooded Caucasus" towards the port of Tuapse. These Jäger divisions were experts in operating in hilly country. General Konrad's mountain Jägers, on the other hand, were to drive across the 10,000 to 14,000 feet high passes of the Central Caucasus towards the Black Sea coast, bursting in, as it were, through the back door. Their objective was Sukhumi, the palm-lined town on the sub-tropical coast and capital of the Abkhaz Autonomous Soviet Socialist Republic. From there it was about 100 miles to the Turkish frontier at Batumi.

Behind advanced motorized combat groups of the SS Panzer Grenadier Division "Viking" and the Slovak Fast Division, General Konrad's mountain Jägers on 13th August mounted their attack from the steppe against the high passes of the Caucasus—the 4th Mountain Division on the right, to gain the passes in the headwaters region of the Laba river, the 1st Mountain Division on the left, to charge over the mountain passes along the glaciers of Mount Elbrus, where the Kuban river has its source. The most important crossing was the Klukhor Pass, 9230 feet high, the starting-point of the old Sukhumi Military Highway.

In the sector of 1st Mountain Division Major von Hirschfeld made a rapid dash with the 2nd Battalion, 98th Mountain Jäger Regiment, as far as the entrance to the pass, which was barricaded and defended by strong Russian forces. The position could not be taken by frontal assault. But von Hirschfeld gave the Russians a demonstration of German mountain warfare. While cleverly deceiving the enemy by engaging him frontally, he outflanked the pass by negotiating the sheer sides of the mountains, and presently rolled up the Soviet positions from the rear. The highest point of the Sukhumi Military Highway was in German hands by the evening of 17th August.

Quick as lightning Major von Hirschfeld continued his dash into the Klydzh Valley, took the village of Klydzh at the foot of the mountains, and thus found himself in the middle of the luxuriant forests of the Black Sea coast. One last leap, and the coastal plain would be gained.

But a surprise advance into the plain was not to be accomplished with the weakened forces. The Russians were furiously and stubbornly defending the exit from the mountains. Sukhumi, the great objective, was a mere 25 miles away. But Major von Hirschfeld, far ahead of the bulk of his forces, with a mere handful of men entirely self-dependent, was in a dangerous position. On his left flank was a big void; Kleist's Panzer Army was still in the steppe, north of Mount Elbrus.

Faced with this situation, General Konrad decided upon a bold operation in order to cover the Corps' left flank. Captain Groth, with a high alpine company composed of mountain guides and climbers, was given the task of getting into the Mount Elbrus passes, which were over 13,000 feet high, and of cutting off the Baksan Valley from where the Russians were threatening the German flank.

This was probably the most spectacular battlefield of the war. Deeply creviced, the sheer faces of rust-red porphyry dropped precipitously over several thousand feet from the rocky mass of Mount Elbrus. The distant ice-fields of the great Asau Glacier glistened in the sun—ice-falls, cleft rocks, and vast expanses of scree.

Over the savage mountain fighting for the former Tsarist hunting-lodge of Krugozor, situated at an altitude of nearly 10,000 feet over the deep cleft of the Baksan Valley, towered Mount Ushba, 15,411 feet high, and one of the most beautiful mountains in the world. It was topped only by Mount Kazbek, farther east on the Georgian Military Highway, and by the twin peaks of Mount Elbrus.

Naturally enough, the men of 1st Mountain Division, in whose line of advance Mount Elbrus was situated, wanted to conquer the giant mountain. Such an operation, of course, was of no military value, but the world might prick up its ears if German troops planted the swastika on the highest mountain in the Caucasus.

General Konrad therefore authorized the proposed climb. He made it a condition, however, that the ascent should be made jointly by men of 1st and 4th Divisions. It was a wise decision: it avoided wounding the mountaineering pride of 4th Division.

The expedition was led by Captain Groth. The participants from 4th Mountain Division were under the command of Captain Gämmerler. The climbers had a curious surprise very early in the proceedings. First Lieutenant Schneider had set out from the base camp with his signals party ahead of the bulk of the climbers, because the heavy signals equipment they were carrying would later be bound to slow them down. A long way ahead, on the far side of the huge glacier, the men saw the fantastic Intourist House which the Soviets had erected at an altitude of 13,800 feet—a massive oval shape of concrete, without any kind of ledge or projection, entirely clad with aluminium sheeting. It looked like a gigantic airship gondola. There were forty rooms with sleeping accommodation for a hundred people in that amazing glacier hotel. Above it was a meteorological station, and below the main structure was the kitchen building.

Schneider and his party made fast progress over the snow of the glacier, which had not yet been softened by the day's sunshine. Suddenly, through his binoculars, he spotted a Soviet soldier in front of the house. "Careful," Schneider called out to his men. He made them turn off the direct route and bypass the hotel. Among the rocks above the building they took up battle positions.

Just then Captain Groth came trudging along, all alone. Before it was possible to warn him the Russians had him covered. The Soviet garrison consisted of only three officers and eight men. They had come up only that morning.

Groth instantly grasped the situation and kept his head. One of the Russian officers spoke German: to him Groth explained the hopelessness of their situation. He pointed to the German rope parties approaching in the distance and to the signals platoon which had taken up positions among the rocks. In this way he eventually persuaded the Soviets to withdraw voluntarily. Four of the Red Army men, however, preferred to stay with Groth and await the arrival of the bulk of the German climbing party in order to offer their services as porters.

The following day, 18th August, was declared a day of rest. The mountain Jägers were to get acclimatized to the altitude. On 19th August the assault on the summit was to start. But the plan was foiled by a sudden blizzard. On 20th August heavy thunderstorms with gusts of hail kept the men again at the Mount Elbrus house.

On 21st August at last a brilliantly sunny morning promised a fine day. They had set out at 0300 hours—Captain Groth with sixteen men and Captain Gämmerler with five.

By 0600 hours the fine weather was at an end. A Föhn came up from the Black Sea. Fog, and later a snowstorm, defended the peak of the giant mountain. In a small refuge Groth and Gämmerler stopped with their men for a break. Should they return again? No—the mountain Jägers wanted to go on.

On they went. The climb in the rarefied air and the biting cold became an eerie race. The men's eyes were caked with snow. A gale was howling over the icy flank of the ridge. Visibility was barely 10 yards.

By 1100 they had conquered the ice-slope. Captain Gämmerler stood at the highest point of the ridge. In front of him the ridge began to drop again. Clearly he was on the summit.

Sergeant Kümmerle of the 1st Mountain Division rammed the shaft of the Reich War Flag deep into the soft snow. Then the standards of 1st and 4th Mountain Divisions with the edelweiss and the gentian were thrust into the ground. A brief handshake and the party quickly climbed into the eastern face where the force of the westerly gale was somewhat diminished. Presently an amazed world was told that the swastika was flying from the highest peak in the Caucasus.

The conquest of Mount Elbrus by German mountain Jägers, a successful climb of a mountain entirely unknown to them, and in appalling weather, was an outstanding mountaineering feat. It is not made any less remarkable by the fact that a few days later, when the weather had cleared, Dr Rümmler, a special correspondent with the Corps, discovered that the flags had been planted not at the trigonometric point of the highest peak, but on an eminence of the summit ridge about 130 feet below the main summit. In the fog and icy blizzard of 21st August the mountain Jägers had mistaken that point for the actual summit.

To return to the fighting in the mountain passes. While the battalions of 1st Mountain Division were forcing their way through the Klukhor Pass and along the old and dilapidated Sukhumi Military Highway, always within sight of the 18,480-foot peak of Mount Elbrus, Major-General Eglseer took his 4th Mountain Division from Austria and Bavaria through the high-level passes of the main range. Colonel von

Stettner with the 1st and 3rd Battalions, 91st Mountain Jäger Regiment, gained the Sancharo and Alustrakhu Passes at altitudes between 8500 and 10,000 feet. The main range of the mountains had thus been crossed, and the further advance now was downhill towards the passes of the foothills and into the sub-tropical forests of the Sukhumi area.

Major Schulze with the 3rd Battalion, 91st Mountain Jäger Regiment, stormed through the Bgalar Pass, and thus found himself immediately above the wooded slopes dropping steeply towards the coastal plain. The coast, the great objective, was a mere 12 miles away.

The Jägers had covered more than 120 miles of mountains and glaciers. With exceedingly weak forces they had fought engagements at altitudes of nearly 10,000 feet, overwhelming the enemy, charging vertiginous rocky ridges and windswept icy slopes and dangerous glaciers, clearing the enemy from positions considered impregnable. Now the men were within sight of their target. But they were unable to reach it.

Von Stettner's combat group had only two guns with twenty-five rounds each at its disposal for the decisive drive against the coast. "Send ammunition," he radioed. "Are there no aircraft? Aren't the Alpini coming with their mules?"

No—there were no aircraft. As for the Alpini Corps, they were marching to the Don, towards Stalingrad.

Colonel von Stettner, the commander of the gallant 91st Mountain Jäger Regiment, was in the Bzyb Valley, 12 miles from Sukhumi.

Major von Hirschfeld was in the Klydzh Valley, 25 miles from the coast.

Major-General Rupp's 97th Jäger Division had fought its way to within 30 miles of Tuapse. Included in this division were also the Walloon volunteers of the "Wallonie" Brigade under Lieutenant-Colonel Lucien Lippert.

But nowhere were the troops strong enough for the last decisive leap. Soviet resistance was too strong. The attacking formations of Army Group A had been weakened by weeks of heavy fighting, and supply lines had been stretched far in excess of any reasonable scale. The Luftwaffe had to divide its forces between the Don and the Caucasus. Suddenly the Soviet Air Force controlled the skies. Soviet artillery was enjoying superiority. The German forces lacked a few dozen fighter aircraft, half a dozen battalions, and a few hundred mules. Now that the decision was within arm's reach these vital elements were missing.

It was the same as on all other fronts: there were shortages everywhere. Wherever operations had reached culmination point and vital objectives all but attained, the German Armies suffered from the same fatal shortages. Before El Alamein, 60 miles from the Nile, Rommel was crying out for a few dozen aircraft to oppose British air power, and for a few hundred tanks with a few thousand tons of fuel. In the villages west of Stalingrad the assault companies of Sixth Army were begging for a few assault guns, for two or three fresh regiments with some anti-tank guns, assault engineers, and tanks. On the outskirts of Leningrad and in the approaches of Murmansk—everywhere the troops were crying out for that famous last battalion which had always decided the outcome of every battle.

But Hitler was unable to let any of the fronts have this last battalion. The war had grown too big for the Wehrmacht. Everywhere excessive demands were being made on the troops, and everywhere the fronts were dangerously over-extended.

The battlefields everywhere, from the Atlantic to the Volga and the Caucasus, were haunted by the spectre of impending disaster. Where would it strike first?

7. Long-range Reconnaissance to Astrakhan

By armoured scout-car through 80 miles of enemy country–The unknown oil railway–Second Lieutenant Schliep telephones the station-master of Astrakhan–Captain Zagorodnyy's Cossacks.

IN the area of First Panzer Army, which formed the eastern part of Army Group A, the 16th Motorized Infantry Division was covering the exposed left flank by means of a chain of strongpoints.

The date was 13th September 1942, and the place was east of Elista in the Kalmyk steppe.

"Hurry up and get ready, George—we're off in an hour!"

"Slushayu, gospodin Oberleutnant—yes, sir," shouted George the Cossack, and raced off. First Lieutenant Gottlieb was delighted with his eagerness.

George came from Krasnodar. He had learnt his German at the Teacher Training College there. In the previous autumn, while acting as a messenger, he had run straight into the arms of the motor-cyclists of 16th Motorized Infantry Division. Since then he had been doing all kinds of services for 2nd Company—first as assistant cook and later, after volunteering for the job, as interpreter. George had numerous good reasons for disliking Stalin's Bolshevism, and there was not a man in the company who did not trust him. In particularly critical situations George had even helped out as a machine-gunner.

Lieutenant Gottlieb had just returned from a conference with the Commander of Motor-cycle Battalion 165—the unit which later became Armoured Reconnaissance Battalion 116. There the last details had been discussed for a reconnaissance operation through the Kalmyk steppe to the Caspian. Lieutenant-General Henrici, commanding 16th Motorized Infantry Division, who had recently relieved LII Corps at Elista, wanted to know what was going on in the vast wilderness along the flank of the Caucasus front. Between the area south of Stalingrad and the Terek river, which 3rd Panzer Division had reached near Mozdok on 30th August with 394th Panzer Grenadier Regiment under Major Pape, there was a gap nearly 200 miles wide. Like a huge funnel this unknown territory extended between Volga and Terek, the base of the triangle being the coast of the Caspian. Any kind of surprise might come from there. That was why the area needed watching.

The task of guarding this huge no-man's-land had been assigned at the end of August to virtually a single German division—16th Motorized Infantry Division. It was based on Elista in the Kalmyk steppe. The actual surveillance and reconnaissance as far as the Caspian Sea and the Volga Delta was done, to begin with, by long-range reconnaissance formations. Reinforcements were not to be expected until the end of September, when Air Force General Felmy would bring up units under his Special Command "F."

It was then that the 16th Motorized Infantry Division earned its name of "Greyhound Division"—a name which the subsequent 16th Panzer Grenadier Division and, later still, 116th Panzer Division continued to bear with pride.

Apart from a few indispensable experts, the operation was mounted by volunteers alone. The first major expedition along both sides of the Elista–Astrakhan road was staged in mid-September. Four reconnaissance squads were employed. These were their tasks:

(1) Reconnoitre whether any enemy forces were present in the gap between Terek and Volga, and if so where; whether the enemy was attempting to ferry troops across the Volga; which were his bases; and whether any troop movements were taking place along the riverside road between Stalingrad and Astrakhan.

(2) Supply detailed information on road conditions, the character of the coast of the Caspian Sea and the western bank of the Volga, as well as about the new, and as yet unknown, railway-line between Kizlyar and Astrakhan.

The force started out on Sunday, 13th September, at 0430 hours. A cutting wind was blowing from the steppe: it was going to be exceedingly cold until the sun broke through.

Long-range Reconnaissance to Astrakhan

For their adventurous drive 90 miles deep into unknown, inhospitable enemy country the reconnaissance squads were appropriately equipped. Each squad had two eight-wheeled armoured scout cars with 2-cm. anti-aircraft guns, a motor-cycle platoon of twenty-four men, two or three 5-cm. anti-tank guns—either self-propelled or mounted on armoured infantry carriers—and one engineer section with equipment. There were, moreover, five lorries—two each carrying fuel and water and one with food-supplies—as well as a repair and maintenance squad in jeeps. Finally, there was one medical vehicle with a doctor, and signallers, dispatch-riders, and interpreters.

Second Lieutenant Schroeder's reconnaissance squad had bad luck from the start. Shortly after setting out, just beyond Utta, the squad made contact with an enemy patrol. Second Lieutenant Schroeder was killed; Maresch, the interpreter, and Sergeant Weissmeier were wounded. The squad returned to base and set out again on the following day under the command of Second Lieutenant Euler.

Lieutenant Gottlieb, Second Lieutenant Schliep, and Second Lieutenant Hilger had meanwhile advanced with their own long-range reconnaissance squads to the north, to the south, and immediately along the great road from Elista to Astrakhan. Lieutenant Gottlieb, having advanced first along the road and then turned away north-east, into the steppe in the direction of Sadovskaya, had reached a point 25 miles from Astrakhan on 14th September. On 15th September he was within 15 miles of the Volga. From the high sand-dunes there was an open view all the way to the river. Sand and salt swamps made the ground almost impassable—but armoured reconnaissance squads invariably found a way.

The maps which Gottlieb had taken with him were not much good. At every well, therefore, George the Cossack had to engage in lengthy palavers with nomadic Kalmyks to find out about roads and tracks. These Kalmyks acted in a friendly way towards the Germans.

"The great railway? Yes—there are several trains each day between Kizlyar and Astrakhan."

"And the Russkis?"

"Yes—they ride about a lot. Only yesterday a large number of them spent the night by the next well, about an hour's march to the east from here. They came from Sadovskaya. There must be a lot of them there."

"Really?" George nodded and gave the friendly nomad a few cigarettes.

Their laughter was abruptly cut short by a shout. One of them pointed to the north. Two horsemen were approaching at a gallop—Soviets.

The Kalmyks melted away. The two German armoured scout cars were behind a dune and not visible to the Russians. Lieutenant

Gottlieb called out to George: "Come back!" But the Cossack did not reply. He stuffed his forage-cap under his voluminous cloak, sat down on the well-head, and lit a cigarette.

Cautiously the two Russians approached—an officer and his groom. George called out something to them. The officer dismounted and walked up to him.

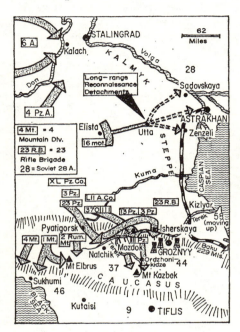

Map 30. The gap between the Caucasus front and Stalingrad was nearly 200 miles wide. Reconnaissance patrols of 16th Motorized Infantry Division got as far as the approaches of Astrakhan.

Lieutenant Gottlieb and his men could see the two talking and laughing together. They were standing next to each other. "The dirty dog," the wireless operator said. But just then they saw George quickly whip out his pistol. Clearly he was saying *"Ruki verkh"* because the Soviet officer put up his hands and was so much taken by surprise that he called out to his groom to surrender as well.

Gottlieb's reconnaissance squad returned to Khalkhuta with two valuable prisoners.

Second Lieutenant Euler's special task was to find out exactly what the defences of Sadovskaya were like and whether any enemy troops were being ferried across the Volga in this area north of Astrakhan.

The distance from Utta to Sadovskaya, as the crow flew, was about 90 miles. Euler almost at once turned off the great road towards the north. After driving some six miles Euler suddenly caught his breath: a huge cloud of dust was making straight for his party at considerable speed. "Disperse vehicles!" he commanded. He raised his binoculars to his eyes. The cloud was approaching rapidly. Suddenly Euler started laughing. The force that was charging them was not Soviet armour, but a huge herd of antelopes, the sayga antelope which inhabits the steppe of Southern Russia. At last, getting scent of the human beings, they wheeled abruptly and galloped away to the east. Their hooves swept over the parched steppe grass, raising a cloud of dust as big as if an entire Panzer regiment were advancing across the endless plain.

Euler next reconnoitred towards the north-east and found the villages of Yusta and Khazyk strongly held by the enemy. He bypassed them, and then turned towards Sadovskaya, his principal target.

On 16th September Euler with his two armoured scout cars was within 3 miles of Sadovskaya, and thus only just over 4 miles from the lower Volga. The distance to Astrakhan was another 20 miles. Euler's reconnaissance squad probably got farthest east of any German unit in the course of Operation Barbarossa, and hence nearer than any other to Astrakhan, on the finishing-line of the war in the East.

What the reconnaissance squad established was of prime importance: the Russians had dug an anti-tank ditch around Sadovskaya and established a line of pillboxes in deep echelon. This suggested a well-prepared bridgehead position, evidently designed to cover a planned Soviet crossing of the Lower Volga.

When the Russian sentries recognized the German scout cars wild panic broke out in their positions. The men, until then entirely carefree, raced to their pillboxes and firing-pits and put up a furious defensive fire with anti-tank rifles and heavy machine-guns. Two Russians, who raced across the approaches in the general confusion, were intercepted by Euler with his scout car. He put a burst in front of their feet. *"Ruki verkh!"*

Terrified, the two Red Army men surrendered—a staff officer of Machine-gun Battalion 36 and his runner. It was a rare catch.

Second Lieutenant Jürgen Schliep, the commander of the Armoured Scout Company of 16th Motorized Infantry Division, had likewise set out with his party on 13th September. His route was south of the main road. His chief task was to find out whether—as the interrogation of prisoners seemed to suggest—there was really a usable railway-line from Kizlyar to Astrakhan, though no such line was marked on any map. Information about this oil railway was most important, since it could be used also for troop transports.

Schliep found the railway. He recalls: "In the early morning of our second day out we saw the distant salt lakes glistening in the sun. The motor-cycles had great difficulty in negotiating the deep sand, and our two-man maintenance team with their maintenance vehicle was kept busy with minor repairs."

When Schliep eventually spotted the railway-track through his binoculars he left the bulk of his combat group behind, and with his two scout cars and the engineer party drove on towards a linesman's hut. In fact, it was the station of Zenzeli.

Schliep's account continues: "From afar we saw fifty or sixty civilians working on the permanent way. It was a single-track line, protected along both sides by banks of sand. The men in charge of the party skedaddled the moment they saw us, but the rest of the civilian workers welcomed us with cheers. They were Ukrainian families—old men, women, and children—who had been forcibly evacuated from their homes and had been employed there on this work for the past few months. Many of the Ukrainians spoke German and hailed us as their liberators."

While the troops were talking to the Ukrainians a wisp of smoke suddenly appeared in the south. "A train," the workers shouted.

Schliep brought his scout car into position behind a sand-dune. An enormously long goods train, composed of oil and petrol tankers, was approaching with much puffing. It was hauled by two engines. Six rounds from the 2-cm. guns—and the locomotives went up. Steam hissed from their boilers and red-hot coal whirled through the air. The train was halted. And now one tanker wagon after another was set on fire.

"A damned shame—all that lovely fuel!" the gunners were grumbling. But the Ukrainians clapped their hands delightedly each time another tanker went up in flames. Finally the German engineers blew up the rails and the permanent way.

Just as they were getting ready to blow up the station shack the telephone rang. The engineers nearly jumped out of their skins. "Phew—that gave me quite a turn," said Sergeant Engh of the maintenance squad. But he quickly collected himself and shouted across to Schliep: "Herr Leutnant—telephone!"

Schliep instantly grasped the situation, grabbed his interpreter, and rushed back into the shack. "Zenzeli station, station-master speaking," the interpreter said in Russian, with a broad grin. "*Da, da—da, tovarishch,*" he kept saying.

At the other end of the line was Astrakhan goods station. Astrakhan —the southern terminal of the A–A Line, the Astrakhan–Archangel line, the finishing-post of the war. The spearheads of the German forces were talking to Astrakhan over the telephone.

The traffic controller in Astrakhan wanted to know whether the oil train from Baku had passed through; a train in the opposite direction was waiting at the bypass point of Bassy.

A train in the opposite direction! The interpreter tried to persuade the comrade in Astrakhan to send it off at once. But this advice aroused the suspicions of the comrade in Astrakhan. He asked a few questions to trap his interlocutor. And the inexpert replies clearly justified his suspicions.

He started shouting and cursing terribly. At that the interpreter stopped play-acting and said, "Just you wait, Papushka—we'll soon be in Astrakhan."

With the most obscene of Russian oaths the comrade in Astrakhan flung down the receiver. He did not therefore hear the bang two minutes later when the wooden station building of Zenzeli was blown up with a couple of high-explosive charges.

Second Lieutenant Schliep, who had lost all radio contact with Division ever since his first day out, now tried to reconnoitre towards Bassy. But evidently the station official in Astrakhan had raised the alarm. Soviet artillery and heavy machine-guns had taken up positions outside the village. Schliep's long-range reconnaissance squad turned away, and on 17th September returned safe and sound to Utta. Still on the same day, Schliep made his report to Division, as well as to Colonel-General von Weichs, the C-in-C Army Group B, who happened to be present at Lieutenant-General Henrici's headquarters. The 16th Motorized Infantry Division was now part of Army Group B.

Everybody breathed a sigh of relief. So far there was no danger yet from the steppe or the Lower Volga—*i.e.*, the Caucasian flank. That was the main information brought back by the reconnaissance squads. It was important because, ever since the end of August, Army Group A had been trying to get the offensive on its left wing going again. Kleist's Panzer Army was to make an all-out effort to burst open the gate to Baku, in order to capture the Soviet oilfields and thus reach at least one of the key objectives of the summer offensive.

The last obstacle on the way to this objective was the Terek river, in front of which the armoured spearheads of Kleist's Army had come to a halt. Kleist once more tried his luck, and, indeed, this time the fortunes of war seemed to hold out the prospect of victory to the German Wehrmacht.

After consultation with the General Officer Commanding XL Panzer Corps, Colonel-General von Kleist pulled back the 3rd Panzer Division from the stubbornly defended Baksan valley in a skilful transversal manœuvre, and moved it behind the lines of 23rd Panzer Division eastward along the Terek. After fierce street fighting the division took Mozdok on 25th August. It next got a second combat group to mount a surprise river crossing at Ishcherskaya. This vital leap across the Terek was performed by the 394th Panzer Grenadier Regiment from Hamburg—formed in 1940—41 from the Harburg 69th Motorized Infantry Regiment.

The date was 30th August 1942 and the time nearly 0300 hours. The assault-boats, the engineers, and the Panzer grenadiers were ready.

They were merely waiting for the artillery barrage which was to cover their crossing operation.

At the appointed time it came: a distant rumbling in the rear, a howling whine overhead, and the crash of bursting shells on the enemy bank. For fully ten minutes this hail of fire from eighty-eight barrels beat down upon the Russian positions. That was ample time for the engineers and grenadiers. They sprang from behind cover and got the assault-boats into the water.

The first groups of 1st Battalion were being ferried across. But now the Russians were waking up. Their field guns, those excellent 'crash-boom' pieces, time and again put their shells on the German crossing-points. These field guns were among the most effective and most dangerous Soviet weapons.

The Terek, about 275 yards wide at the crossing-point, was a treacherous mountain river with a powerful current and swirling eddies. Fountains of white spume rose high all round the boats—near misses by enemy mortars.

Among the turbulent waves the small assault-boats were bobbing about, their bows high above the water, the grenadiers crouched low in the stern. Somehow they got through the inferno.

At the very beginning of the attack, while still on the German bank, the commander of 1st Battalion, Captain Freiherr von der Heyden-Rynsch, and his adjutant, Second Lieutenant Ziegler, were killed. Second Lieutenant Wurm also heeled over, mortally wounded. First Lieutenant Dürrholz, commanding 2nd Company, was wounded during the crossing and fell overboard into the river. He was listed missing believed killed.

An assault-boat swished round the bend. "Up—at the double!" Quick as lightning the next group jumped into the boat. "Three, four, five—one more," the engineer by the tiller counted them in. With a whine the motor sprang to life. They were off.

Shell-bursts to the right and left of them. The water seethed with spume and spray. The engineer by the tiller stood upright, unperturbed, steering the boat safely across the river. And there was the far bank. A slight curve, and the men leapt ashore.

Under the protective barrage of their own artillery the riflemen fought their way forward, yard by yard. The beginnings of a bridgehead had been gained—but they were no more than the beginnings. Very soon the enemy turned out to be stronger than had been assumed. The Soviets were well dug in along the edge of the village of Mundar-Yurt, and offering stubborn resistance. From well-prepared field positions, as well as an anti-tank ditch, they were keeping the German grenadiers, who were lying in open ground, under constant heavy fire.

In the afternoon Major Günther Pape, the young regimental

commander, himself crossed the Terek with the operations staff of 394th Panzer Grenadier Regiment to see for himself what the situation was like. The main fighting-line was so arranged and the troops so organized that the bridgehead that had been gained could also be defended with what few forces were available.

Throughout five days the men of 394th Panzer Grenadier Regiment held out on the far bank of the Terek. They were fighting south of the 44th parallel. The only units farther south were the forward units of 1st Mountain Division in the Klydzh Valley, and they were nearly on the 43rd parallel—to be exact, at latitude 43 degree 20 minutes, the most southerly point reached by the German forces on Soviet territory in the course of Operation Barbarossa.

In unfavourable country, without heavy equipment, Pape's men were facing a far superior and stubbornly fighting opponent. The regiment was tying down three Soviet divisions. This compelled the Soviets to withdraw troops from elsewhere. The bridgehead established by 3rd Panzer Division at considerable cost provided the prerequisite for the attack by the newly brought-up LII Army Corps. As a result, the 111th and 370th Infantry Divisions succeeded likewise in crossing the Terek at Mozdok and in establishing a further bridgehead. The 394th were thus able to abandon their unfavourable position.

But at Mozdok, as elsewhere, the forces lacked the strength to continue the offensive. The Russians were simply too strong, and the German troops were too few and too battle-weary. The last chance to conquer the Baku oilfields was allowed to slip by unused.

Just as in the western foothills of the Caucasus, by the Black Sea, so operations also ground to a standstill on the Terek. The front froze. Within a short distance of the target-line of the whole campaign the offensive vigour of Operation Barbarossa was spent. The Terek became the ultimate boundary of German military conquest.

In the defensive positions along the Terek, right among the battalions of 3rd Panzer Division at Ishcherskaya, a strange formation was fighting side by side with the German grenadiers—a Cossack unit. The manner in which Captain Zagorodnyy's Cossack squadron came to be fighting on the German side was typical of the war in the East.

When General Freiherr von Geyr's XL Panzer Corps had taken 18,000 prisoners at Millerovo in the summer, the greatest problem was: Who was going to take the Soviet prisoners to the rear? The shrunken units of the German divisions were unable to spare any men for such duties. It was then that Captain Kandutsch, the Corps Intelligence Officer, conceived the idea of separating the rather pro-German Kuban and Don Cossacks from the rest of the prisoners, mounting them on the countless stray horses that were wandering all over the place, and using them as an escort for the Red Army prisoners. The Cossacks, who had never been enamoured of Bolshevism, were delighted. In no time Captain Zagorodnyy had organized a squadron and moved off with the 18,000 Soviet prisoners. No one at Corps headquarters thought he would ever see Zagorodnyy or his Cossacks again.

But during the first week of September there was a knock at the Intelligence Officer's door at XL Panzer Corps headquarters in Russkiy on the Terek, and in stepped a colourfully attired Cossack officer and reported in broken German: "Captain Zagorodnyy with his squadron reporting for duty." Kandutsch was speechless. There he was landed with those Cossacks again.

What was to be done about those Cossacks? Kandutsch telephoned the Chief of Staff, Lieutenant-Colonel Carl Wagener. There was a lot of argument. Eventually it was decided that Zagorodnyy's men would be regrouped as Cossack Squadron 1/82, given four weeks' training, and then employed at the front.

It worked out very well. At the front, in the Ishcherskaya position, Captain Zagorodnyy enforced the strictest order and discipline. Only once, during the first night, did he find two trench sentries asleep. His pistol barked twice. Never again did a Cossack sleep on sentry duty, nor was there a single deserter.

The Captain's most reliable helper was the commander of his 1st Troop, Lieutenant Koban, a broad-shouldered Cossack who remained faithful to his squadron—as Zagorodnyy himself—to the very last. Whenever Koban was sick his wife would take the troop's parade. This attractive, brave Cossack woman had ridden in her husband's troop from the start. Like any other Cossack, she would ride out on patrol. In the end she died with the squadron.

The squadron's death occurred in grim and tragic circumstances, thousands of miles away from their homeland, in the liberation of which they were hoping in 1942 to play their part; it occurred after many a hard and gallant action on the Eastern front.

Captain Kandutsch reports: "At the end of May 1944, when XL Panzer Corps was crossing the Rumanian frontier in a westerly direction, the Cossack squadron was ordered to be transferred to France. Deputizing for General von Knobelsdorff, the Corps Commander, Major Patow, the Corps Adjutant, said good-bye to the Cossacks. Captain Zagorodnyy at last received the Iron Cross 1st Class which he had coveted so passionately. He had thoroughly earned it. After that the Cossacks once more formed up in column—probably for the last time—for a march-past at the gallop. It was an unforgettable sight."

Six weeks later the squadron was caught in a heavy fighter-bomber raid near Saint-Lô in France, during the invasion, and was completely routed.

Only a few men escaped with their lives. They brought the news of the fate of the Cossacks to Germany. Among those killed were all the officers, as well as Lieutenant Koban's wife. But to this day the men of XL Panzer Corps have not forgotten their comrades-in-arms of many tough engagements.

8. The Terek marks the Limit of the German Advance

Hitler's clash with Jodl–The Chief of the General Staff and Field-Marshal List are dismissed–An obsession with oil–Panzer Grenadiers on the Ossetian Military Highway –The Caucasus front freezes.

ON 7th September 1942 the heat of the late summer lay heavily on the Ukrainian forests. In the airless blockhouses of the Fuehrer's Headquarters, known as "Werewolf," the thermometer rose to 30 degrees Centigrade. Hitler was suffering more than usually from the heat. It served to increase his anger at the situation between Kuban and Terek. All the reports from the "oilfield front" indicated that the troops had reached the limit of their strength.

Army Group A was stuck in the Caucasus and on the Terek. The valleys leading down to the Black Sea coast, above all to Tuapse, were blocked by the Soviets, and the Terek proved to be a strongly reinforced obstacle, the last obstacle before the old military highways to Tiflis, Kutaisi, and Baku.

We can't make it, the divisions reported. "We can't make it, we can't make it—how I hate these words!" Hitler fumed. He refused to believe that further progress was impossible on the Terek or on the mountain front merely because the forces were inadequate. He was putting the blame on the commanders in the field and on what he called their mistakes in mounting the operations.

For that reason Hitler had sent the Chief of the Wehrmacht Operations Staff, General of Artillery Jodl, to Stalino in the morning of 7th September, to see Field-Marshal List and to find out for himself why no progress was being made along the road to Tuapse. Jodl was to lend emphasis to Hitler's orders.

Late in the evening Jodl returned. His report triggered off the worst crisis in the German High Command since the beginning of the war. Jodl defended Field-Marshal List and supported his view that the forces were too weak for the objectives assigned to them. Like List, he demanded a radical regrouping of the front.

Hitler refused. He suspected Jodl of having allowed himself to be bamboozled by List. The general, irritable from the heat and his exhausting day, blew up. Furiously and in a loud voice he quoted to Hitler his own orders and directives of the past few weeks, which List had observed meticulously, and which had led to the very difficulties in which Army Group A was now finding itself.

Hitler was flabbergasted at Jodl's accusations. His most intimate general was not only in revolt against him, but was clearly questioning his strategic skill and blaming him for the crisis in the Caucasus and for the emerging bogey of defeat on the southern front.

"You're lying," Hitler screamed. "I never issued such orders—never!" Then he left Jodl standing and stormed out of his blockhouse into the darkness of the Ukrainian woods. It was hours before he came back—pale, shrunken, with feverish eyes.

The extent to which Hitler had been upset by his encounter was shown by the fact that from that date onward he no longer took his meals with his generals. From then until his death he would sullenly take his meals in the Spartan solitude of his headquarters, with Blondi, the Alsatian bitch, as his only company.

But that was not the only reaction to Jodl's accusations. There were much more far-reaching consequences: Colonel-General Halder, the Chief of the General Staff, and Field-Marshal List were relieved of their posts. Hitler even decided to dismiss his devoted Generals Keitel and Jodl, and envisaged their replacement by Field-Marshals Kesselring and Paulus—a plan which unfortunately was not put into effect, since the appointment of generals with front-line experience would at least have avoided the disaster of Stalingrad.

In the end, however, Hitler did not part with his military aides Keitel and Jodl, who had served him for so many years. He merely ordered that in the future every one of his words, as well as every remark by any general at military conferences, was to be taken down in shorthand. Otherwise he clung stubbornly to his order that the attack on the Caucasus front must be continued. On no account would he renounce the main objective of his summer offensive. The oil of the Caucasus, Groznyy, Tiflis, and Baku, as well as the transhipment ports on the Black Sea, must be captured at all costs. The autumn of 1942 was to bring the German forces to the objectives of the Russian campaign, at least in the south.

Hitler's attitude was one of many indications of his increasing stubbornness also in the military field. This side of his character was presently to bring about the doom of the fighting front. In other ways Hitler's obsessions had been patent for some time.

In the economic field Hitler's obsession was oil. Oil to him was the element of progress, the driving force of the machine age. He had read everything that had ever been written about oil. He was acquainted with the history of the Arabian and American oilfields, and knew about oil extraction and refining. Anyone turning the conversation to oil could be sure of Hitler's attention. Goering was put in charge of the economic four-year plan because he was playing Hitler's favourite card—oil.

Typical of Hitler's attitude is an attested remark he made about an efficient civil servant in the Trade Policy Department of the German Foreign Office: "I can't bear the man—but he does understand about oil." Hitler's Balkan policy was based entirely on Rumania's oil. He had built into the Barbarossa directive a special campaign against the Crimea, merely because he was worried about the Rumanian oilfields, which he believed could be threatened by the Soviet Air Force from the Crimea.

Above all, Hitler's obsession with oil led him to neglect the most revolutionary scientific development of the twentieth century—atomic physics. There was no room left in his mind for understanding the decisive military significance of nuclear fission, discovered in Germany and first developed by German physicists. Here again it was evident that Hitler was essentially a man of the nineteenth not the twentieth century.

Every one of Hitler's *idées fixes* played its fatal part in the war against Russia—but most decisive of all was his obsession with oil. It dominated the campaign in the East from the start, and in the summer of 1942 it was this obsession that led Hitler to take decisions and make demands on the Southern Front which eventually decided the campaign of 1942, and hence the course of the war. One last glance at the oilfield front in 1942 will support this contention.

Army Group A was stuck on the northern and western edge of the Caucasus. But Hitler refused to acknowledge that German strength was at an end. He wanted to drive to Tiflis and to Baku over the ancient Caucasian military highways. He therefore ordered that the offensive must be resumed across the Terek.

Orders were orders. In weeks of heavy fighting the First Panzer Army attempted to extend its bridgehead over the Terek towards the south and the west, a step at a time. All forces were concentrated: LII Army Corps was reinforced with units of XL Panzer Corps and also received 13th Panzer Division from III Panzer Corps. It was this division which on 20th September succeeded in crossing the Terek south-west of Mozdok. On 25th September General von Mackensen launched an attack with the whole of III Panzer Corps against Ordzhonikidze, on the road to Tiflis. While 23rd Panzer Division was slowly advancing with units of 111th Infantry Division, the SS Panzer Grenadier Division "Viking," brought up from the Western Caucasus.

pushed through farther south against the Georgian Military Highway. The ancient road to Tiflis was reached.

The combat group of the SS Panzer Grenadier Regiment "Nordland" arrived on the battlefield from the lower, wooded part of the Caucasus, and with it the "Viking" Division was able to force its way into the northern part of the Groznyy oilfields and to block the Georgian Military Highway at two points. The keypoint known as Hill 711 was stormed, at heavy cost, by a battalion of Finnish volunteers fighting within the "Viking" Division, and was held against all enemy counter-attacks. But would the troops have any strength left for the final push, for the last 60 miles?

Four weeks passed before III Panzer Corps had accumulated the necessary reserves in manpower, fuel, and replacements to launch a new—and as they hoped the final—attack.

On 25th and 26th October the Corps moved forward from its bridgehead on the western bank of the Terek in order to break through towards the south-east. The battalions fought stubbornly. An enemy force of four divisions was smashed and about 7000 prisoners were taken. Rumanian mountain troops blocked the valleys leading to the south. The 13th and 23rd Panzer Divisions drove on to the south-east, and by a vigorous attack on 1st November took Alagir and cut the Ossetian Military Highway on both sides of the town. Major-General Herr's 13th Panzer Division, following up this bold armoured thrust, reached a point three miles west of Ordzhonikidze on 5th November.

By then the last remnants of strength were spent. Soviet counter-attacks from the north cut off the divisions from their rearward communications. To begin with, the First Panzer Army was unable to help, and in the teeth of opposition from the Fuehrer's Headquarters ordered the severed groups to break through backward. The most forward combat group of SS Panzer Grenadier Division "Viking" arrived in the nick of time to meet their old comrades-in-arms of 13th Panzer Division half-way, to get them out of the enemy trap and absorb them.

During the night of 11th/12th November the 13th Panzer Division linked up again with Corps. In bitter fighting the 13th and the "Viking" beat off attacks by pursuing enemy units.

About the middle of November a break in the weather put an end to all attempts at restarting the operation.

On the right wing, in the sector of Seventeenth Army, the mountain troops had already abandoned the snowbound passes of the High Caucasus because supplies were no longer getting through. The infantry and Jäger regiments dug in. The attacks on the Black Sea ports, on the oilfields, and on Baku, Tiflis, and Batumi suffered shipwreck within sight of the objectives. The entire front was at a standstill.

Why?

The Terek Marks the Limit of the German Advance

Because the new Soviet tactics of evasion had foiled the boldly conceived pincer operations between Don and Donets. Because the Soviet commanders-in-chief had succeeded at the last moment in regaining control over their formations withdrawing from the Lower Don into the Caucasus. Because, finally, American supplies were reaching the battered Soviet Armies from Iran via the Caspian. The battle-weary German formations were too weak to break this last resistance. Here as elsewhere the German forces lacked the last battalion.

PART SEVEN: *Stalingrad*

1. Between Don and Volga

Kalach, the bridge of destiny over the Don–Tank battle in the sands of the steppe–General Hube's armoured thrust to the Volga–"On the right the towers of Stalingrad"–Heavy anti-aircraft guns manned by women–The first engagement outside Stalin's city.

ANYONE studying the battle of Stalingrad is struck first of all by the strange circumstance that this city did not rank as a principal objective in the plans for the great summer offensive. In "Operation Blue" the city was only a marginal consideration. It was to be "brought under military control"—in other words, it was to be eliminated as an armaments centre and as a port on the Volga. That was a task for aircraft and long-range artillery, but not an assignment for an entire Army. The purpose could have been achieved equally well with bombs and shells; as a city Stalingrad was of no strategic importance. The operations of the Sixth Army were therefore designed, under the general strategic plan, to cover the flank of the Caucasus front and its important military-economic targets. For this task the capture of Stalingrad might be useful, but it was by no means indispensable. That this flank-cover assignment of Sixth Army should eventually lead to the turning point of the war and to a battle which decided the fate of the entire campaign was one of the tragic aspects of the disaster of Stalingrad. It shows how much the outcome of a war can be determined by accidents and mistakes.

In September 1942, when the main operation of the summer offensive, the battle in the Caucasus and on the Terek, was grinding to a standstill, encouraging news was arriving at the Fuehrer's Headquarters from the Stalingrad front. In a sector where the capture of the Don and Volga bends at Stalingrad was envisaged merely for the sake of flank and rear cover for the battle for the oilfields, progress was suddenly being made after weeks of crisis. On 13th September a report came in from Sixth Army that the 71st Infantry Division, belonging to LI Corps under General of Artillery von Seydlitz-Kurzbach, had penetrated the deeply echeloned fortified approaches of Stalingrad and stormed the high ground just outside the city centre.

Part Seven: Stalingrad

On the following day, 14th September 1942, Lieutenant-General von Hartmann with parts of his Lower Saxon 71st Infantry Division broke through to the Volga after some costly street fighting past the northern of the two railway stations. Hartmann's assault squads, admittedly, represented only a thin wedge, but nevertheless the city had been pierced and the swastika flag was flying over the city centre. It was a gratifying success, and encouraged the hope that at least the Don–Volga operation would be victoriously completed before the onset of winter, so that, with flank cover ensured, the offensive could then be resumed in the Caucasus.

How did this gratifying success of 14th September 1942 come about? To answer this question we must cast our minds back to the summer, to the operation between Donets and Don, when during the second half of July the Sixth Army was advancing solitarily down the Don towards Stalingrad, while the bulk of Army Group South—the First and Fourth Panzer Armies—were wheeled southward by Hitler to fight the battle of encirclement of Rostov.

At the head of Sixth Army moved General von Wietersheim's XIV Panzer Corps. This was the only Panzer Corps under the Army, and consisted of 16th Panzer Division and 3rd and 60th Motorized Infantry Divisions. In the face of this mailed fist the Russians withdrew over the Don, towards the north and the east, in the direction of Stalingrad.

This retreat, undoubtedly ordered by the Soviet Command and envisaged by it as a strategic withdrawal, nevertheless turned into wild flight in the sectors of many Soviet divisions, largely because the order for the withdrawal came unexpectedly and was not clearly formulated. The retreat was poorly organized. Officers and troops were not yet experienced in these new tactics. The result was that the middle and lower commands lost control of their units. In many places there was panic. It is important to realize these circumstances in order to understand why this withdrawal was interpreted on the German side as a Soviet collapse.

Undoubtedly there were symptoms of collapse in many places, but the higher Soviet command remained untouched by this. The higher command had a clear programme: Stalingrad, the city on the Volga bend which bore Stalin's name, the ancient Tsaritsyn, was earmarked by the Soviet General Staff as the final centre of resistance. Stalin had reluctantly allowed his generals to withdraw from the Donets and the Don. But he now drew a line at the Volga.

"I order the formation of an Army Group Stalingrad. The city itself will be defended by Sixty-second Army to the last man," Stalin had said to Marshal Timoshenko on 12th July 1942. In a strategically favourable area Stalin intended to bring about a turn of the tide, just as he had done once before—during the Revolution in 1920, against the White Cossack General Denikin. All he needed was time—time to bring up reserves, time to build defensive positions along the northern approaches to the city on the strip of land between Don and Volga, as well as along the favourable line of high ground stretching south of Stalingrad as far as the Kalmyk steppe.

But would the Germans allow the Red Army enough time to mobilize all its strength and re-form in the Stalingrad area?

Major-General Kolpakchi was then Commander-in-Chief Sixty-second Army. His staff officers stood at the Don crossings in the Kalach area, their machine pistols at the ready, trying to bring some sort of order into the flood of retreating Soviet regiments.

But the Germans did not come. "No more enemy contact," the Russian rearguards reported. Kolpakchi shook his head. He reported to Army Group: "The Germans are not following up."

"What does it mean?" Marshal Timoshenko asked his chief of staff. "Have the Germans changed their plans?"

The excellent Soviet espionage organizations knew nothing of a change of plan. Neither Richard Sorge from the German Embassy in Tokyo nor Lieutenant Schulze-Boysen from the Air Ministry in Berlin had reported anything about changes in the plans for the German offensive. Nor was there anything from the top-level agents Alexander Rado in Switzerland or Gilbert in Paris. Surely one of them would have uncovered something. For there was no doubt that there was still a leak in the German High Command. Indeed, the reports from Rössler, one of the Soviet agents in Switzerland, which quoted "Werther" as their source and which came from a well-informed official in the German High Command, proved that these channels of information were just then working very smoothly. There was therefore no indication that the Germans had changed their plans regarding the operation at Stalingrad.

But it was quite definite: the much-feared armoured spearheads of General Paulus's forces were not coming on. Soviet aerial reconnaissance reported that the German advanced formations had halted in the area north of Millerovo. The Soviets could not understand it. They never suspected the real reason for this halt: XIV Panzer Corps had run out of fuel.

Following the decision taken at the Fuehrer's Headquarters on 3rd July—the decision to push ahead with the Caucasus operation without waiting for Stalingrad to be eliminated—the major part of the fuel supplies originally earmarked for Sixth Army were switched round to the Caucasus front, since it was there that Hitler wanted to concentrate his main effort. A considerable proportion of the fast troops and supply formations of Sixth Army were abruptly paralysed as a result.

In this way the bulk of Sixth Army, in particular XIV Panzer Corps, remained immobilized for eighteen days. Eighteen days was a long time.

The Russians made good use of the time thus gained. "If the Germans are not following up there is time to organize the defence on the western bank of the Don," Timoshenko decided. Major-General Kolpakchi assembled the bulk of his Sixty-second Army in the great Don bend and established a bridgehead around Kalach. In this manner the vital crossing of the Don was blocked 45 miles west of Stalingrad. The fortified loop of the Don projected towards the west like a balcony, flanking the river to the north and south.

About 20th July, when the Sixth Army was once more ready to resume its advance, General Paulus found himself faced with the task of first having to burst open the Soviet barrier around Kalach in order to continue his thrust across the Don towards Stalingrad. Thus began the battle for Kalach, an interesting operation and one of considerable importance for the further course of events—in fact, the first act of the battle of Stalingrad.

General Paulus mounted his attack on the Kalach bridgehead as a classical battle of encirclement. He made his own XIV Panzer Corps reach out in a wide arc on the left, and the XXIV Panzer Corps, assigned to him from Hoth's Panzer Army, similarly on the right wing, the two to link up at Kalach. The VIII Infantry Corps covered the Army's deep flank in the north, while Seydlitz's LI Corps was making a frontal attack on Kalach between the two Panzer Corps.

The main brunt of the heavy fighting in the great Don bend was borne, above all, by the two Panzer divisions—the 16th Panzer Division of XIV Panzer Corps and the 24th Panzer Division of XXIV Panzer Corps. The motorized divisions covered their flanks.

The East Prussian 24th Panzer Division under Major-General von Hauenschild received orders to cross the Chir and to wheel northward along the Don towards Kalach. It was opposed by strong forces of the Soviet Sixty-fourth Army, then still under the command of Lieutenant-General Chuykov.

The first attack with two Panzer companies and units of Panzer grenadier regiments did not, to begin with, get through the minefields behind which the Russians were well dug in. But on 25th July, towards 0330 hours, the 24th renewed their attack, and this time succeeded in dislodging the enemy from his well-established positions and in capturing the vital high ground west of the Solenaya stream.

The 21st Panzer Grenadier Regiment under Colonel von Lengerke repulsed dangerous Soviet attacks against the northern flank. In the afternoon there was a heavy cloudburst which made the attack increasingly difficult on the rain-softened ground. The weather, together with the Soviet 229th and 214th Rifle Divisions, which resisted stubbornly and furiously in their positions, made a surprise drive to the Don impossible.

On 26th July, at last, progress was made. The 26th Panzer Grenadier

Regiment punched a hole in the enemy lines on the Solenaya stream. Riding on top of light armour, the grenadiers drove on to the east. The break-through was accomplished.

The Panzer Grenadier Regiment and one Panzer battalion raced towards the Chir crossing at Nizhne-Chirskaya. At 1400 hours the spearheads reached the river and wheeled south towards the bridge. In street fighting that night the large village was occupied, and shortly before midnight the ford and the bridge over the Chir east of it were captured.

While the Panzer grenadiers were establishing a bridgehead on the eastern bank, tanks and armoured infantry carriers advanced through the enemy-held forest as far as the bridge over the Don. By dawn they had reached the huge river—the river of destiny for Operation Barbarossa.

Enemy attempts to blow up the bridge were fortunately unsuccessful. Only a small section was demolished, and that was soon repaired. Once again the 24th Panzer Division had seized an important bridge almost undamaged.

However, the drive over the river on to the narrow neck of land between Don and Volga, in the direction of Stalingrad, could not yet be attempted. First of all, the strong Russian force west of the river had to be destroyed, especially since the Russians had meanwhile concentrated two Armies east of the Don, against whom the weak armoured spearheads of the German Sixth Army could not possibly achieve any success single-handed.

On 6th August the last round opened in the battle for Kalach. An armoured assault group of 24th Panzer Division under Colonel Riebel, the commander of 24th Panzer Regiment, advanced from the Chir bridgehead and drove through the covering units of 297th Infantry Division northward, in the direction of Kalach. The objective was another 22 miles away.

The Russians resisted desperately. They realized what was at stake: if the Germans got through, then all their forces west of the river would be cut off and the door to Stalingrad would be burst open.

However, the "mailed fist" of the 24th battered a way through the Soviet defensive positions and minefields, repulsed numerous counter-attacks by enemy armour, and escorted the unarmoured units of the division through the Soviet defensive lines, which were still intact in many places.

Then, in many columns abreast, the 24th Panzer Division roared forward in a wild hunt through the steppe, and at nightfall had reached the commanding Hill 184, just before Kalach, in the rear of the enemy.

Along the left prong of the pincers, in the sector of XIV Panzer Corps, the operation had likewise been going according to schedule.

S

Lieutenant-General Hube's 16th Panzer Division from Westphalia launched its attack on 23rd July with four combat groups attacking from the upper Chir. A volunteer division of the Soviet Sixty-second Army offered the first furious resistance on the hills of Roshka. Mues's battalion drove right up to the enemy pillboxes and field positions with its armoured infantry carriers, the Panzer grenadiers on top of them. The enemy was kept down by machine-gun fire. The grenadiers leapt from their vehicles and flushed the Russians out of their dugouts with hand-grenades and pistols.

By the afternoon a wide gap had been punched into the enemy lines. The combat group under von Witzleben was able to break through towards the south-east, mounted on armour, and on the following day, 24th July, it reached the Liska sector, north-west of Kalach. It was only another 12 miles to their objective.

The Panzer battalion under Count Strachwitz—the 1st Battalion 2nd Panzer Regiment, reinforced by artillery, motor-cycle units, and grenadiers mounted on armour—raced eastward under the command of Colonel Lattmann's combat group, and at dawn had reached the last enemy barrier north of Kalach. After heavy fighting the Soviets were dislodged from their positions. Count Strachwitz wheeled south and rolled up the entire Soviet defences. Only 6 more miles to go.

Meanwhile units of the 60th and 3rd Motorized Infantry Divisions, coming from the north-west, had moved between 16th Panzer Division and the Don, facing south. There they were engaged in exceedingly tough defensive fighting with enemy armoured brigades and rifle divisions brought up from beyond the river over the bridges at Kalach and Rychov. In consequence, units of both German attacking groups were already fighting in the rear of the Soviet bridgehead forces. The pocket behind General Kolpakchi's divisions was beginning to take shape.

The Soviets realized the danger and flung all available forces against the northern prong. It was a life-and-death struggle, a battle fought by the Soviets not only with furious determination but also with surprisingly strong armour.

The official history of the 16th Panzer Division provides a dramatic picture of the tank battles at the time. Strong mobile armoured forces were facing each other. They stalked one another, each side trying to surround and cut off the other. There was no front line proper.

Like destroyers and cruisers at sea, the tank units manœuvred in the sandy ocean of the steppe, fighting for favourable firing positions, cornering the enemy, clinging to villages for a few hours or days, bursting out again, turning back, and again pursuing the enemy. And while these armoured forces were getting their teeth into each other in the grass-grown steppe, the cloudless sky above the Don became the scene of fierce fighting between the opposing air forces, with each side trying to strike at the enemy in the numerous gorges which crossed the territory, to blow up his ammunition columns and to set fire to his fuel supplies.

In the sector of Reinisch's combat group alone the Russians employed 200 tanks. Sixty-seven of them were shot up. The remainder turned tail.

Colonel Krumpen's group was surrounded by the Soviets. The division switched all available forces to the danger-point. There were no rearward communications left: the fighting units had to be supplied with fuel from the air. The crisis was averted only by a supreme effort.

On 8th August the spearheads of 16th and 24th Panzer Divisions linked up at Kalach. The pocket was firmly closed. The ring itself was formed by XIV and XXIV Panzer Corps, as well as XI and LI Infantry Corps. Inside the pocket were nine Soviet rifle divisions, two motorized and seven armoured brigades of the Soviet First Tank Army and the Sixty-second Army. One thousand tanks and armoured vehicles as well as 750 guns were captured or destroyed.

At long last another successful battle of encirclement had been fought—the first since the early summer, since the battle of Kharkov. It was also to be the last of Operation Barbarossa. It was fought 40 miles from the Volga, and it is worth noting that here, outside the gates of Stalingrad, the officers and men of the Sixth Army once again demonstrated their marked superiority in mobile operations against a numerically far superior enemy. Once more it was made patent that, provided their material strength was anything like adequate to the fighting conditions, the German formations could deal with any Soviet opposition.

Mopping-up operations in the Kalach area and the capture of bridges and bridgeheads across the Don for the advance on Stalingrad took another fortnight in view of the tough opposition offered by the Soviets. Meanwhile the 24th Panzer Division and 297th Infantry Division were returned from Sixth Army to Hoth's Fourth Panzer Army.

All the courage of their desperation was of no avail to the Russians. On 16th August the great bridge of Kalach was taken by Second Lieutenant Kleinjohann with units of 3rd Company Engineers Battalion 16 by a daring coup which involved putting out a fire on the bridge. The damage done to its roadway and sub-structure was quickly repaired. And now developments followed one another in rapid succession.

On 21st August infantry units of von Seydlitz's Corps—the 76th and 295th Infantry Divisions—crossed the river Don at two points, where it was about 100 yards wide and flowed between steep banks, and established bridgeheads at Luchinskoy and Vertyachiy. Paulus's plan was clear: he intended to drive a corridor from the Don to the Volga, to block off Stalingrad in the north and then take the city from the south.

Lieutenant-General Hube, originally an infantryman but now a brilliant tank commander, was crouching by the pontoon bridge of Vertyachiy, in the garden of a peasant cottage, together with Lieutenant-Colonel Siekenius, commanding 2nd Panzer Regiment. Spread out on a little hummock of grass in front of them was a map.

Hube moved his right hand over the sheet. The left sleeve of his tunic was empty, its end tucked into a pocket. Hube had lost an arm in World War I. The commander of 16th Panzer Division was the only one-armed tank general in the German Wehrmacht.

"We have here the narrowest point of the neck of land between Don and Volga, just about 40 miles," he was saying. "The ridge of high ground marked as Hill 137, which Army orders have assigned to us as our route of attack, is ideal ground for armour. There are no streams or ravines crossing our line of advance. Here's our opportunity to drive a corridor right through the enemy to the Volga in one fell swoop."

Siekenius nodded. "The Russians are bound to try to defend this neck of land with everything they've got, Herr General. Indeed, it's an

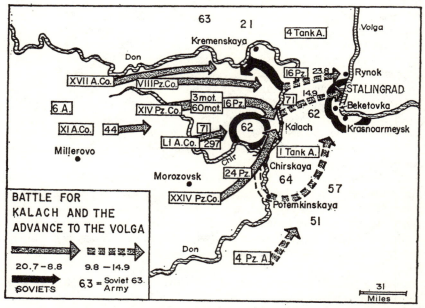

Map 31. The battle of Stalingrad was opened at Kalach on the Don. The Soviet forces west of the Don were surrounded and the path cleared to the strip of land between Don and Volga.

ancient defensive position of theirs. The Tartar Ditch running across from the Don to the Volga was an ancient defensive rampart against incursions from the north which aimed at the Volga estuary."

Hube traced the Tartar Ditch with his forefinger. He said, "No doubt the Russians will have developed it into an anti-tank ditch. But we've taken anti-tank ditches before. The main thing is that it's got to be done fast—quick as lightning, in the usual way."

A dispatch-rider came roaring up on his motor-cycle. He was bringing last-minute orders from Corps for the thrust to the Volga.

Hube glanced at the sheet of paper. Then he rose and said, "The balloon goes up at 0430 hours to-morrow, Sieckenius."

The lieutenant-colonel saluted. Every detail of the attack, with the exception of the time of attack, had been laid down by Army order ever since 17th August. Now they also knew H hour—0430 on 23rd August.

The 16th Panzer Division was to drive through to the east as far as the Volga in one continuous movement, close to the northern edge of Stalingrad. The flanks of this bold armoured thrust were to be covered on the right by the 60th Motorized Infantry Division from Danzig and on the left by the 3rd Motorized Infantry Division from Brandenburg. It was an operation entirely to Hube's taste, entirely in the manner of the armoured thrusts of the early months of the war.

To-morrow they would reach Stalingrad. They would stand on the Volga. Hube and Sieckenius both realized that Stalingrad and the Volga were the ultimate targets, the easternmost points to be reached. There the offensive war would end; there Operation Barbarossa would come to its final stop, culminating in victory.

"Till to-morrow, Sieckenius."

"Till to-morrow, Herr General."

Hube's right hand touched the peak of his cap. Then he turned once more and added, "To-morrow night in Stalingrad."

During the night the 16th Panzer Division moved in a huge column into the bridgehead which 295th Infantry Division had established at Luchinskoy. Ceaselessly Russian bombers attacked the vital bridge, guided to their target by blazing vehicles. But the Russians were unlucky. The bridge remained intact. About midnight the formations were in position close behind the main fighting-line, in ground providing no cover. The grenadiers immediately dug foxholes for themselves, and for additional safety the armoured vehicles were driven on top of them. Throughout the night Soviet artillery and "Stalin's organ-pipes" smothered the bridgehead, about three miles long and one and a half miles deep, with carpet fire. It was not an enjoyable night.

In the morning of 23rd August 1942 the spearheads of 16th Panzer Division crossed the pontoon bridge of Vertyachiy. On the far side the formations fanned out to form a broad wedge. In front was the Combat

Group Sieckenius; behind it, in echelon, the Combat Groups Krumpen and von Arenstorff.

Undeterred by the presence of enemy forces to the right and left of the ridge of high ground, as well as in the small river-courses and ravines, the tanks, armoured infantry carriers, and towing vehicles of 16th Panzer Division and the armoured units of 3rd and 60th Motorized Infantry Divisions rolled eastward. Above them droned the armoured ground-attack and Stuka formations of VIII Air Corps on their way to Stalingrad. On their return flight the machines dipped low over the tanks, exuberantly sounding their sirens.

The Soviets tried to halt the German armoured thrust along the Tartar Ditch. It was in vain. The Russian opposition was overcome and the ancient ditch with its high dykes over-run. Clearly the Soviets were taken by surprise at the vigorous attack, and—as nearly always in such a situation—lost their heads and were unable to improvise an effective defence against the Germans.

Frequently the penetrations were no more than 150 to 200 yards wide. General Hube was leading the attack from the command vehicle of the Signals Company, in the foremost line. In this way he was kept fully informed about the situation at any one moment. And full information was the secret of successful armoured attack.

It was a field day for the signallers—Sergeant Schmidt and Corporals Quenteux and Luckner. Altogether, they had an important share in the success of the offensive. The signals section of the division dealt with 456 coded radio signals on the first day of the fighting alone.

A particular problem were the Soviet nests of resistance which, commanded by resolute officers and commissars, continued to fight on along the narrow penetrations. They had to be overcome by a new technique. Reconnaissance aircraft reported their positions by radio or smoke markers, and individual combat groups would then hive off the main attacking wedge to deal with them.

In the early afternoon the commander in the lead tank called out to his men over his throat microphone: "Over on the right the skyline of Stalingrad." The tank commanders were all up in their turrets, looking at the long-drawn-out silhouette of the ancient Tsaritsyn, now a modern industrial city extending some 25 miles along the Volga. Pit-head gear, factory smoke-stacks, tall blocks of buildings, and, farther south in the old city, the onion-topped spires of the cathedrals were towering into the sky. Clouds of smoke were hanging over those parts of the city where Stukas were bombing road intersections and barracks.

The tanks' tracks crunched through the scorched grass of the steppe. Trails of dust rose up behind the fighting vehicles. The leading tanks of Strachwitz's battalion were making for the northern suburbs of Spartakovka, Rynok, and Latashinka. Suddenly, as if by some secret command, an artillery salvo came from the outskirts of the city—Soviet heavy flak inaugurating the defensive battle of Stalingrad.

Strachwitz's battalion fought down gun after gun—thirty-seven emplacements in all. One direct hit after another was scored against the emplacements, and the guns together with their crews were shattered.

Strangely enough, the battalion suffered hardly any losses itself. The reason why was soon to become plain. As the Panzer crews penetrated into the smashed gun emplacements they found to their amazement and horror that the crews of the heavy anti-aircraft guns consisted of women—workers from the "Red Barricade" ordnance factory. No doubt they had had some rudimentary training in anti-aircraft defence, but clearly they had no idea of how to use their guns against ground targets.

As 23rd August drew to its end the first German tank reached the high western bank of the Volga close to the suburb of Rynok. Nearly 300 feet high, the steep bank towered over the river, which was well over a mile wide at that point. The water looked dark from the top. Convoys of tugs and steamers were moving up and down stream. On the far bank glistened the Asian steppe, losing itself into infinity.

Near the river the division formed a hedgehog for the night, hard by the northern edge of the city. Right in the middle of the hedgehog were divisional headquarters. Wireless-sets hummed; runners came and went. All through the night work continued: positions were being built, mines laid, tanks and equipment serviced, refuelled, and restocked with ammunition for the next day's fighting, for the battle for the industrial suburbs of Stalingrad North.

The men of 16th Panzer Division were confident of victory and proud of the day's successes; no one as yet suspected that these suburbs and their industrial enterprises would never be entirely conquered. No one suspected that just there, where the first shot had been fired in the battle of Stalingrad, the last one would be fired also.

The division no longer had any contact with the units following behind; the regiments of 3rd and 60th Motorized Infantry Divisions had not yet come up. That was hardly surprising, since Hube's armoured thrust to the Volga had covered over 40 miles in a single day. The objective—the Volga—had been reached; all communications across the 40-mile-wide neck of land between Don and Volga had been cut. The Soviets had clearly been taken entirely by surprise by these developments. The division's positions came only under random artillery fire during the night. Maybe Stalingrad would fall the next day, dropping into Hube's lap like a ripe plum.

2. Battle in the Approaches

*The Tartar Ditch–T-34s straight off the assembly line–
Counter-attack by the Soviet 35th Division–Seydlitz's
Corps moves up–Insuperable Beketovka–Bold manœuvre
by Hoth–Stalingrad's defences are torn open.*

ON 24th August, at 0440 hours, the Combat Group Krumpen launched its attack against Spartakovka, Stalingrad's most northerly industrial suburb, with tanks, grenadiers, artillery, engineers, and mortars, preceded by Stukas.

But the enemy they encountered was neither confused nor irresolute. On the contrary: the tanks and grenadiers were met by a tremendous fireworks. The suburb was heavily fortified, and every building barricaded. A dominating hill, known to the troops as "the big mushroom," was studded with pillboxes, machine-gun nests, and mortar emplacements. Rifle battalions and workers' militia from the Stalingrad factories, as well as units of the Soviet Sixty-second Army, were manning the defences. The Soviet defenders fought stubbornly for every inch of ground. The order which pinned them to their positions had said clearly: "Not a step back!"

The two men who saw that this order was ruthlessly implemented were Colonel-General Andrey Ivanovich Yeremenko, Commander-in-Chief Stalingrad and South-east Front, and his Political Commissar and Member of the Military Council, Nikita Sergeyevich Khrushchev. It was then, over twenty years ago, that the officers of the 16th Panzer Division heard this name for the first time from Soviet prisoners.

With the forces available Spartakovka could clearly not be taken. The Soviet positions were impregnable. The determination displayed by the Soviets in holding their positions was further illustrated by the fact that they launched an attack against the northern flank of Hube's "hedgehog" in order to relieve the pressure on Spartakovka. The Combat Groups Dörnemann and von Arenstorff were hard pressed to resist the increasingly vigorous Soviet attacks.

Brand-new T-34s, some of them still without paint and without gunsights, attacked time and again. They were driven off the assembly line at the Dzerzhinskiy Tractor Works straight on to the battlefield, frequently crewed by factory workers. Some of these T-34s penetrated as far as the battle headquarters of 64th Panzer Grenadier Regiment and had to be knocked out at close quarters.

The only successful surprise coup was that by the engineers, artillery men, and Panzer Jägers of the Combat Group Strehlke in taking the landing-stage of the big railway ferry on the Volga and thereby cutting the connection from Kazakhstan via the Volga to Stalingrad and Moscow.

Strehlke's men dug in among the vineyards on the Volga bank. Large walnut-trees and Spanish chestnuts concealed their guns which they had brought into position against river traffic and against attempted landings from the far bank.

But in spite of all their successes the position of 16th Panzer Division was highly precarious. The Soviets were holding the approaches to the northern part of the city, and simultaneously, with fresh forces brought up from the Voronezh area, put pressure on the "hedgehog" formed by the division. Everything depended on securing the German corridor across the neck of land, and the 16th were therefore anxiously awaiting the arrival of 3rd Motorized Infantry Division.

The advanced units of that division had left the Don bridgeheads side by side with 16th Panzer Division on 23rd August and moved off towards the east. At noon, however, their ways had parted. Whereas the 16th had continued towards the northern part of Stalingrad, Major-General Schlömer's regiments had fanned out towards the north in order to take up covering positions along the Tartar Ditch in the Kuzmichi area.

The general was moving ahead with the point battalion. Through his binoculars he could see goods trains being feverishly unloaded at Kilometre 564, west of Kuzmichi.

"Attack!"

The motor-cyclists and armoured fighting vehicles of Panzer Battalion 103 raced off. Gunners of Army Flak Battalion 312 sent over a few shells. The Russian columns dispersed.

The goods wagons contained a lot of useful things from America. These had been shipped across the Atlantic and Indian Oceans, through the Persian Gulf, across the Caspian Sea, and up the Volga as far as Stalingrad, and thence by railway to the front, to the halt at Kilometre 564. Now these supplies were being gratefully received by Schlömer's 3rd Motorized Infantry Division—magnificent brand-new Ford lorries, crawler tractors, jeeps, workshop equipment, mines, and supplies for engineering troops.

The tanks of the advanced battalion had continued on their way when suddenly five T-34s appeared, evidently in order to recapture the precious gifts from the USA. Their 7·62-cm. shells quite literally dropped into the pea soup which was just being dished up for the division's operations section. The general and the chief of operations dropped their mess-tins and took cover. Fortunately two tanks of the point battalion had got stuck near the goods train with damaged tracks. They knocked out two of the T-34s and saved the situation. The remainder turned tail.

While Schlömer's formations were still following behind 16th Panzer Division more disaster loomed up; a Soviet Rifle Division, the 35th, reinforced with tanks, was driving down the neck of land from the north in forced marches. Its aim—as revealed by the papers found on a captured courier—was to seal off the German bridgeheads over the Don and keep open the neck of land for the substantial forces which were to follow.

The Soviet 35th Division moved southward in the rear of the German 3rd Motorized Infantry Division; it over-ran the rearward sections of the two foremost divisions of von Wietersheim's Panzer Corps, forced its way between the bridgehead formed by the German VIII Infantry Corps and the German forces along the Tartar Ditch, and thereby prevented the German infantry, which was just then moving across the Don into the corridor, from closing up on the forces ahead of them.

As a result, the rearward communications of the two German lead divisions were cut off, and those divisions had to depend upon themselves. True, the 3rd Motorized Infantry Division and the 16th Panzer Division succeeded in linking up, but these two divisions now had to form a "hedgehog" 18 miles wide, extending from the Volga to the Tartar Ditch, in order to stand up to the Soviet attacks from all sides. Supplies had to be brought up by the Luftwaffe, or else escorted through the Soviet lines by strong Panzer convoys.

This unsatisfactory and critical situation persisted until 30th August. Then, at long last, the infantry formations of LI Corps under General of Artillery von Seydlitz moved up with two divisions on the right flank. The 60th Motorized Infantry Division likewise succeeded in insinuating itself into the corridor front after heavy fighting.

As a result, by the end of August, the neck of land between Don and Volga was sealed off to the north. The prerequisites had been created for a frontal attack on Stalingrad, and the outflanking drive by Hoth's Panzer Army from the south was now covered against any surprises from the northern flank.

General von Seydlitz-Kurzbach had been wearing the Oak Leaves to the Knights Cross of the Iron Cross since the spring of 1942. It was then that this outstanding commander of the Mecklenburg 12th Infantry Division had punched and gnawed his way through to the Demyansk pocket with his Corps group and freed Count Brockdorff-Ahlefeldt's six divisions from a deadly Soviet stranglehold.

That was why Hitler was again placing great hopes for the battle of Stalingrad on the personal bravery and tactical skill of this general, born in Hamburg-Eppendorf and bearing the name of an illustrious Prussian military family.

At the end of August Seydlitz launched his frontal attack against

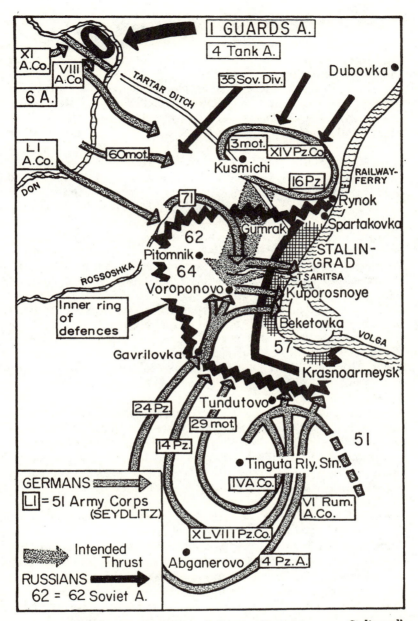

Map 32. On 30th August the Fourth Panzer Army tore open Stalingrad's inner belt of defences. Paulus's units were to have driven down from the north at the same time. But XIV Panzer Corps was tied down by enemy attacks. When Hoth's divisions linked up with 71st Infantry Division they were two days too late: the Russians had fallen back to the city outskirts at the last minute.

the centre of Stalingrad with two divisions striking across the neck of land from the middle of Sixth Army. His first objective was Gumrak, the airport of Stalingrad.

The infantry had a difficult time. The Soviet Sixty-second Army had established a strong and deep defensive belt along the steep valley of the Rossoshka river. These defences formed part of Stalingrad's inner belt of fortifications, which circled the city at a distance of 20 to 30 miles.

Until 2nd September Seydlitz was halted in front of this barrier. Then, suddenly, on 3rd September the Soviets withdrew. Seydlitz followed up, pierced the last Russian positions before the city, and on 7th September was east of Gumrak, only five miles from the edge of Stalingrad.

What had happened? What had induced the Russians to give up their inner and last belt of defences around Stalingrad and to surrender the approaches to the city? Had their troops suddenly caved in? Was the command no longer in control? Those were exciting possibilities.

There can be no doubt that this particular development in the battle of Stalingrad was of vital importance for the further course of operations. The events in this sector have not yet received adequate attention in German publications about Stalingrad—but the battle for the Volga metropolis certainly hung in the balance during these forty-eight hours of 2nd and 3rd September. The fate of the city appeared to be sealed.

Marshal Chuykov, then still a lieutenant-general and Deputy Commander-in-Chief Sixty-fourth Army, casts some light in his memoirs on the mystery of the sudden collapse of Russian opposition in the strong inner belt of fortifications along the Rossoshka stream. The solution is to be found in the actions and decisions of the two outstanding contestants in this mobile battle of Stalingrad—Hoth and Yeremenko.

Yeremenko, the bold and dashing, yet also strategically gifted Commander-in-Chief of the "Stalingrad Front," has revealed some interesting details of this great battle in his most recent publications. Chuykov's memoirs fill in many gaps and cast additional light on various aspects.

Colonel-General Hoth, Commander-in-Chief of the Fourth Panzer Army, now living in Goslar, where, before the war, he had served with the Goslar Rifle Regiment, just as Guderian and Rommel, has made available to the present author his personal notes about the planning and execution of the offensive which brought about the collapse of the Soviet front.

At the end of July, Hoth's Fourth Panzer Army had wheeled away from the general direction of attack against the Caucasus and had been re-directed from the south through the Kalmyk steppe against the Volga bend south of Stalingrad. Its thrust was intended to relieve Paulus's Sixth Army, which even then was being hard pressed in the Don bend.

But once again the German High Command had contented itself with a half-measure. Hoth was approaching with only half his strength: one of his two Panzer Corps, the XL, had had to be left behind on the Caucasus front. His effective strength, in consequence, consisted only of Kempf's XLVIII Panzer Corps, with one Panzer and one motorized division, as well as von Schwedler's IV Corps, with three infantry divisions. Later, Hoth also received the 24th Panzer Division. The Rumanian VI Corps under Lieutenant-General Dragalina with four infantry divisions was subordinated to Hoth to protect his flank.

The Soviets instantly realized that Hoth's attack spelled the chief danger to Stalingrad. After all, his tanks were already across the Don, whereas Paulus's Sixth Army was still being pinned down west of the river by the Soviet defenders.

If Hoth, coming from the Kalmyk steppe, were to succeed in gaining the Volga bend with the commanding high ground of Krasnoarmeysk and Beketovka, Stalingrad's doom would be sealed and the Volga would be severed as the main supply artery for American deliveries through the Persian Gulf.

On 19th August Hoth reached the southernmost line of defence of the Soviet Sixty-fourth Army, and at the first attempt achieved a penetration at Abganerovo. Kempf's Panzer Corps pushed through with 24th and 14th Panzer Divisions as well as with 29th Motorized Infantry Division, followed on the left by Schwedler's infantrymen.

Twenty-four hours later Hoth's tanks and grenadiers were attacking the high ground of Tundutovo, the southern cornerstone of Stalingrad's inner ring of defences.

Colonel-General Yeremenko had concentrated all his available forces in this favourable and vital position. Armoured units of the Soviet First Tank Army, regiments of the Soviet Sixty-fourth Army, militia, and workers' formations were holding the line of hills with their wire obstacles, blockhouses, and earthworks established in deep echelon. Krasnoarmeysk in the Volga bend was only 9 miles away.

The companies of 24th Panzer Division attacked again and again, swept forward by their experienced commanders and combat-group leaders. But success continued to be denied to them. Colonel Riedel, commanding 24th Panzer Regiment, and for many years Guderian's ADC, was killed in action. Colonel von Lengerke, commanding 21st Panzer Grenadier Regiment, was mortally wounded in an attack against the railway to Krasnoarmeysk. Battalion commanders, company commanders, and the old and experienced NCOs were killed in the infernal defensive fire of the Soviets.

At that stage Hoth called a halt. He was a cool strategist, not a gambler. He realized that his attacking strength was inadequate.

At his battle headquarters at Plotovitoye Hoth sat bent over his maps. His chief of staff, Colonel Fangohr, was entering the latest situation reports. Only two hours before, Hoth had visited General Kempf at his Corps Headquarters and had driven with him to General Ritter von Hauenschild to hear about the situation at 24th Panzer Division. He had also called on Major-General Heim at the railway station of Tinguta. In a balka, one of those typical deep ravines of Southern Russia, Heim had explained the difficult situation in which 14th Panzer Division found itself. Here, too, further advance seemed impossible.

"We've got to tackle this thing differently, Fangohr," Hoth was thinking aloud. "We are merely bleeding ourselves white in front of these damned hills: that's no ground for armour. We must regroup and mount our attack somewhere else, somewhere a long way from here. Now, listen carefully...."

The colonel-general was developing his idea. Fangohr was busily drawing on his map, checking reconnaissance reports and measuring distances. "That should be possible," he would mumble to himself now and again. But he was not entirely happy about Hoth's plan, mainly because time would again be lost with regrouping. Besides, a lot of fuel would be needed for all this driving around. And fuel was very short. And ultimately those "damned hills" in front of Krasnoarmeysk and Beketovka would have to be tackled one way or another, for they dominated the entire southern part of the city and its approaches. Exactly the same arguments against regrouping were advanced also by General Kempf. But in the end both Fangohr and Kempf let themselves be persuaded by their commander-in-chief.

Hoth rang up Army Group. He had a half-hour conversation with Weichs. Weichs agreed and promised to come round in person to discuss the operational problems, and especially fuel supplies.

Everything sprang into action: orderlies raced off with orders; telephone wires buzzed ceaselessly. The entire headquarters personnel were moving in top gear. A regrouping operation was being carried out.

Unnoticed by the enemy, Hoth pulled out his Panzer and motorized formations from the front during the night and replaced them by infantry of the Saxon 94th Division. In a bold move, rather like a castling in chess, he moved his mobile formations past the rear of IV Corps in the course of two nights and reassembled them 30 miles behind the front in the Abganerovo area to form them into a broad wedge of attack.

On 29th August this armada struck northward at the flank of the Soviet Sixty-fourth Army, to the complete surprise of the enemy. Instead of fighting his way frontally towards the Volga bend, across the heavily fortified hills of Beketovka and Krasnoarmeysk, which were studded with tanks and artillery, Hoth intended to bypass these positions and enemy forces hard to the west of Stalingrad, in order then to wheel round and attack the entire high ground south of the town with an outflanking attack which would simultaneously trap the left wing of the Soviet Sixty-fourth Army.

The operation started astonishingly well. Jointly with the assault infantry of IV Corps the fast formations on 30th August burst through Stalingrad's inner belt of fortifications at Gavrilovka and over-ran the rearward Soviet artillery positions. By the evening of 31st August Hauenschild with his 24th Panzer Division had reached the Stalingrad–Karpovka railway-line—an unexpected penetration 20 miles deep.

The entire picture, as a result, was changed. A great opportunity was offering itself. The prize was no longer merely the capture of the high ground of Beketovka and Krasnoarmeysk, but the encirclement of the two Soviet Armies west of Stalingrad, the Sixty-second and Sixty-fourth. This prize was suddenly within arm's reach, provided only Sixth Army could now drive southward with its fast formations, towards Hoth's units, in order to close the trap. Hoth's bold operation had created an opportunity for annihilating the two enemy Armies covering Stalingrad.

Army Group headquarters instantly realized this opportunity. In an order to General Paulus, transmitted by radio at noon on 30th August, it was stated:

> In view of the fact that Fourth Panzer Army gained a bridgehead at Gavrilovka at 1000 hours to-day, everything now depends on Sixth Army concentrating the strongest possible forces, in spite of its exceedingly tense defensive situation... on its launching an attack in a general southerly direction... in order to destroy the enemy forces west of Stalingrad in co-operation with Fourth Panzer Army. This decision requires the ruthless denuding of secondary fronts.

When Army Group, moreover, received information on 31st August of the deep penetration made by 24th Panzer Division west of Voroponovo, Weichs sent another order to Paulus on 1st September, couched in considerable detail and no doubt intended as a reminder. Under Figure 1 it said: "The decisive success scored by Fourth Panzer Army on 31.8 offers an opportunity for inflicting a decisive defeat on the enemy south and west of the Stalingrad–Voroponovo–Gumrak line. It is important that a link-up should be established quickly between the two Armies, to be followed by a penetration into the city centre."

The Fourth Panzer Army reacted swiftly. On the same day, 1st September, General Kempf led the 14th Panzer Division and the 29th Motorized Infantry Division in the direction of Pitomnik, having quite ruthlessly denuded the sectors hitherto held by 24th Panzer Division.

But the Sixth Army did not come. General Paulus found himself unable just then to release his fast forces for a drive to the south, in view of the strong Soviet attacks being made against his northern front. He considered it impossible to hold the northern barrier successfully with his Panzer Jägers and a few tanks and assault guns, even if supported by ground-attack aircraft of VIII Air Corps, while hiving off an armoured group to be formed from the five Panzer battalions of XIV Panzer Corps for a drive to the south. He was afraid that, if he did so, his northern front would collapse.

Perhaps he was right. Perhaps any other decision would have been a gamble. In any event, a great opportunity was missed. Twenty-four hours later, in the morning of 2nd September, operational reconnaissance by 24th Panzer Division established that there was no enemy left in front of the German lines. The Russians had pulled out of the southern defensive position, just as on the same day they had abandoned a defensive position facing Seydlitz's Corps in the western sector. What had induced the Russians to take this surprising step?

General Chuykov, Deputy Commander-in-Chief of the Sixty-fourth Army, had realized the dangerous situation which had arisen as a result of Hoth's advance. He gave the alarm to Colonel-General Yeremenko. Yeremenko not only saw the danger, but also acted in a flash, in complete contrast to the former ponderous way in which the Soviet Commands used to react to such situations. Yeremenko took the difficult and dangerous decision—but the only correct one—of abandoning the well-prepared inner belt of defences. He sacrificed strongpoints, wire obstacles, anti-tank barriers, and infantry trenches in order to save his divisions from the threatening encirclement, and retreated with his two Armies to a new, improvised defensive line close by the edge of the city.

This operation showed once more how consistently the Soviets were implementing the new tactics adopted by the Soviet High Command early in the summer. In no circumstances again were major formations to allow themselves to be encircled. For the sake of this new principle they were prepared to risk the loss of the city of Stalingrad.

In the afternoon of 2nd September General Paulus decided after all to dispatch fast units of his XIV Panzer Corps to the south, and on 3rd September the infantrymen of Seydlitz's Corps linked up with Hoth's armoured spearheads. Thus the pocket envisaged by Army Group on 30th August was, in fact, formed and closed, but no enemy was trapped inside it. The manœuvre had been accomplished forty-eight hours too late. This delay was to cost Stalingrad. But as yet nobody suspected this.

Army Group thereupon issued orders to Paulus and Hoth to exploit the situation and to penetrate into the city as fast as possible.

3. The Drive into the City

General Lopatin wants to abandon Stalingrad–General Chuykov is sworn in by Khrushchev–The regiments of 71st Infantry Division storm Stalingrad Centre–Grenadiers of 24th Panzer Division at the main railway station–Chuykov's last brigade–Ten crucial hours–Rodimtsev's Guards.

RIGHT through the middle of Stalingrad runs the river Tsaritsa. Its deep gorge divides the city into a northern and a southern half. The Tsaritsa kept its name when Tsaritsyn became Stalingrad, and it still bears its name to-day now Stalingrad has become Volgograd. In 1942 the famous or notorious gorge of the Tsaritsa formed the junction between Hoth's and Paulus's Armies. Along it the inner wings of the two Armies were to advance swiftly through the city as far as the Volga. Everything seemed to suggest that the enemy was fighting rearguard actions only and was about to abandon the city.

Marshal Chuykov's memoirs reveal the disastrous situation in which the two Soviet Armies in Stalingrad found themselves following the surrender of the approaches to the city. Even experienced commanders did not rate Stalingrad's chances high. General Lopatin, the Commander-in-Chief of the Sixty-second Army, was of the opinion that the city could not be held. He therefore decided to abandon it. But when he tried to give effect to his decision the Chief of Staff, General Krylenko, refused his consent and sent an urgent message to Khrushchev and Yeremenko. Lopatin was relieved of his command, although he was anything but a coward.

Lopatin's decision is not hard to understand if one reads Chuykov's account of the situation before Stalingrad. Chuykov writes: "It was bitter to have to surrender these last few kilometres and metres outside Stalingrad, and to have to watch the enemy's superiority in numbers, in military skill, and in initiative."

The Marshal describes how the machine operators of the State farms, where the various headquarters of Sixty-fourth Army had been set up, were sneaking off to safety. "The roads to Stalingrad and to the Volga were jammed. The families of collective farmers and State farm workers were on the move, complete with their livestock. They were all making for the Volga crossings, driving their animals in front of them and carrying their chattels on their backs. Stalingrad was in flames. Rumours that the Germans were in the city already added to the panic."

That then was the situation. But Stalin was not prepared to surrender his city without bitter struggle. He had sent one of his most reliable supporters, an ardent Bolshevik, to the front as the Political Member of the Military Council, with orders to inspire the Armies and the civilian population to fight to the end—Nikita Sergeyevich Khrushchev. For the sake of Stalin's city he made self-sacrifice a point of honour for every Communist.

Lieutenant-General Platanov's three-volume documentary history of the Second World War contains several figures illustrating the situation: 50,000 civilian volunteers were incorporated in the "People's Guard"; 75,000 inhabitants were assigned to Sixty-second Army; 3000 young girls were mobilized for service as nurses and as telephone and radio operators; 7000 Komsomol members between thirteen and sixteen were armed and absorbed in the fighting formations. Everybody became a soldier. Workers were ordered to the battlefield with the weapons they had just produced in their factories. The guns made in the Red Barricade ordnance factory went into position in the factory grounds straight from the assembly line and opened fire on the enemy. The guns were manned by the factory workers.

On 12th September Yeremenko and Khrushchev entrusted General Chuykov with the command of Sixty-second Army, which had been led since Lopatin's dismissal by the Chief of Staff, Krylenko, and charged him with the defence of the fortress on the Volga. It was an excellent choice. Chuykov was the best man available—hard, ambitious, strategically gifted, personally brave, and incredibly tough. He had no share in the Red Army's disasters of 1941 because at that time he had still been in the Far East. He was an unspent force, and he was not haunted by past calamities, as were so many of his comrades.

On 12th September, at 1000 hours sharp, Chuykov reported to Khrushchev and Yeremenko at Army Group headquarters in Yamy, a small village on the far, left bank of the Volga. It is interesting that the talking was done by Khrushchev and not by Yeremenko, the military boss.

According to Chuykov's memoirs, Khrushchev said: "General Lopatin, the former C-in-C Sixty-second Army, believes that his Army cannot hold Stalingrad. But there can be no more retreat. That is why he has been relieved of his post. In agreement with the Supreme Commander-in-Chief, the Military Council of the Front invites you, Comrade Chuykov, to assume command of Sixty-second Army. How do you see your task?"

"The question took me by surprise," Chuykov records, "but I had no time to consider my reply. So I said, 'The surrender of Stalingrad would wreck the morale of our people. I swear not to abandon the city. We shall hold Stalingrad or die there.' N. S. Khrushchev and

A. I. Yeremenko looked at me and said that I had correctly understood my task."

Ten hours later Seydlitz's Corps launched its attack against Stalingrad Centre. Chuykov's Army headquarters on Hill 102 was shattered by bombs, and the general had to withdraw to a dug-out in the Tsaritsa gorge, close by the Volga, complete with his staff, cook, and serving-girl.

On the following day, 14th September, General von Hartmann's men of 71st Infantry Division were already inside the city. In a surprise drive they pushed through to the city centre and even gained a narrow corridor to the Volga bank.

At the same hour the Panzer Grenadiers of 24th Panzer Division stormed south of the Tsaritsa gorge through the streets of the old city, of ancient Tsaritsyn, captured the main railway station, and on 16th September likewise reached the Volga with von Heyden's battalion. Between Beketovka and Stalingrad, in the suburb of Kuporoznoye, units of 14th Panzer Division and 29th Motorized Infantry Division had been established since 10th September, cutting off the city and the river from the south. Only in the northern part of the city was Chuykov still able to hold on. "We've got to gain time," he said to his commanders. "Time to bring up reserves, and time to wear out the Germans."

"Time is blood," he remarked, thus modifying the American motto "Time is money." Indeed, time was blood. The battle of Stalingrad was epitomized in this phrase.

Glinka, Chuykov's cook, heaved a sigh of relief when he got to his kitchen in the new dug-out. Above him was thirty feet of undisturbed earth. Happily he observed to the general's serving-girl, "Tasya, my little dove, we shan't get any shell-splinters dropping into our soup down here. There's no shell that will go through this ceiling."

"But there is," replied Tasya, who knew Glinka's fears. "A one-ton bomb would crash through all right—the general said so himself."

"A one-ton bomb—are there many of those?" the cook asked anxiously.

Tasya reassured him. "It would be a chance in a million; it would have to drop directly on our dug-out. That's what the general said."

The noise of the front came only as a distant rumble into the deep galleries. Ceiling and walls were neatly clad with boards. There were dozens of underground rooms for the personnel of Army headquarters. Right in the centre was a big room for the general and his chief of staff. One exit of this so-called Tsaritsyn dug-out, built during the preceding summer as a headquarters for the Army Group, led into the Tsaritsa gorge, closed by the steep bank of the Volga, while the other opened into Pushkin Street.

564   *Part Seven: Stalingrad*

Nailed to the boarded wall of Chuykov's office was a hand-drawn city plan of Stalingrad, nearly ten feet high and six feet wide—the General Staff map of the battle. These were no longer fronts in the hinterland; the scale of the battle maps was no longer in kilometres but in metres. It was a matter of street corners, blocks, and individual buildings.

General Krylov, Chuykov's chief of staff, was entering the latest situation reports—the German attacks in blue, the Russian defensive positions in red. The blue arrows were getting closer and closer to the battle headquarters.

"Battalions of the 71st and 295th Infantry Divisions are furiously attacking Mamayev Kurgan and the main railway station. They are being supported by 204th Panzer Regiment, which belongs to 22nd Panzer Division. The 24th Panzer Division is fighting outside the southern railway station," Krylov reported.

Chuykov stared at the plan of the city. "What has become of our counter-attacks?"

"They petered out. German aircraft have been over the city again since daybreak. They are pinning down our forces everywhere."

A runner came in with a situation sketch from the commander of 42nd Rifle Brigade, Colonel Batrakov. Krylov picked up his pencil and drew a semicircle around the battle headquarters. "The front is still half a mile away from us, Comrade Commander-in-Chief," he announced in a deliberately official manner.

Only half a mile. The time was 1200 hours on 14th September. Chuykov knew what Krylov was implying. They had one tank brigade left as a last reserve, with nineteen T-34s. Should it be sent into action?

"What's the situation like on the left wing of the southern city?" Chuykov asked.

Krylov extended the blue arrow indicating the German 29th Motorized Infantry Division beyond Kuporoznoye. The suburb had fallen. General Fremerey's Thuringians were pushing on, in the direction of the grain elevator. The sawmills and the food cannery were already inside the German lines. Only from the southern landing-stage of the ferry to the tall grain elevator was there still a Soviet defensive line. Chuykov picked up the telephone and rang up Army Group. He described the situation to Yeremenko. Yeremenko implored him: "You must hold the central river port and the landing-stage at all costs. The High Command is sending you the 13th Guards Rifle Division. This is 10,000 men strong and a crack unit. Hold the bridgehead open for another twenty-four hours, and try also to defend the landing-stage of the ferry in the southern part of the city."

Beads of sweat stood on Chuykov's forehead. The air in the gallery was suffocating. "All right then, Krylov, scrape together anything you can find; turn the staff officers into combat-group commanders. We've got to hold the crossing open for Rodimtsev's Guards."

The last brigade with its nineteen tanks was thrown into the fighting—one battalion in front of the Army headquarters, from where the central railway station and main river port could also be covered, and another into the line between the grain elevator and the southern landing-stage.

At 1400 hours Major-General Rodimtsev, a legendary commander in the field and Hero of the Soviet Union, turned up at headquarters, bleeding and covered in dirt. He had been chased by German fighter-bombers. He reported that his division was standing by on the far bank and would cross the Volga at night. Frowning, he gazed at the blue and red lines on the town plan.

At 1600 hours Chuykov again spoke to Yeremenko on the telephone. There were five hours to go before nightfall. In his memoirs Chuykov describes his feelings during those five hours: "Would our shattered and battered units and fragments of units in the central sector be able to hold out for another ten or twelve hours? That was my main worry at the time. If the men and their officers were to prove unequal to this well-nigh superhuman task the 13th Guards Rifle Division would not be able to cross over, and would merely witness a bitter tragedy."

Shortly before dusk Major Khopka, commander of the last reserves employed in the river port area, appeared at headquarters. He reported: "One single T-34 is still capable of firing, but no longer of moving. The brigade is down to 100 men." Chuykov regarded him coldly: "Rally your men around the tank and hold the approaches to the port. If you don't hold out I'll have you shot."

Khopka was killed in action; so were half his men. But the remainder held out.

Night came at last. All the staff officers were in the port. As the companies of Rodimtsev's Guards Division came across the Volga they immediately went into action at the main points of defence in order to check the advance of 71st Infantry Division and to hold the 295th Infantry Division on the Mamayev Kurgan, the dominating Hill 102. Those were crucial hours. Rodimtsev's Guards prevented Stalingrad Centre from being taken by the Germans on 15th September. Their sacrifice saved Stalingrad. Twenty-four hours later the 13th Guards Rifle Division had been smashed, bombed to smithereens by Stukas and mown down by shells and machine-gun fire.

In the southern city, too, a Guards division was fighting—the 35th under Colonel Dubyanskiy. Its reserve battalions were brought across by ferry from the left bank of the Volga to the southern landing-stage and immediately employed against spearheads of 29th Motorized Infantry Division in order to hold the line between the landing-stage and the grain elevator.

But the Stukas of Lieutenant-General Fiebig's VIII Air Corps pounded the battalions with bombs, and the remainder were crushed between the jaws of 94th Infantry Division and 29th Motorized Infantry Division. Only in the grain elevator, which was full of wheat, did fierce fighting continue for some time: it was a huge concrete block, as solid as a fortress, and every floor was furiously contested. It was there that assault parties and engineers of 71st Infantry Regiment were in action against the remnants of the Soviet 35th Guards Rifle Division amid the smoke and stench of the smouldering grain.

In the morning of 16th September the situation again looked bad on Chuykov's city plan. The 24th Panzer Division had conquered the southern railway station, wheeled westward, and shattered the defences along the edge of the city and the hill with the barracks. Costly fighting continued on Mamayev Kurgan and at the main railway station.

Chuykov telephoned Nikita Sergeyevich Khrushchev, Member of the Army Group's Military Council: "A few more days of this kind of fighting and the Army will be finished. We have again run out of reserves. There is absolute need for two or three fresh divisions."

Khrushchev got on to Stalin. Stalin released two fully equipped crack formations from his personal reserve—a brigade of marines, all of them tough sailors from the northern coasts, and an armoured brigade. The armoured brigade was employed in the city centre around the main river port in order to hold open the supply pipeline of the front. The marines were employed in the southern city. These two formations prevented the front from collapsing on 17th September.

On the same day the German High Command transferred to Sixth Army the complete authority over all German formations engaged on the Stalingrad front. Thus the XLVIII Panzer Corps was detached from Hoth's Fourth Panzer Army and placed under General Paulus's command. Hitler was getting impatient: "This job's got to be finished: the city must finally be taken."

Why the city was not taken, even though the German tankmen, grenadiers, engineers, Panzer jägers, and flak troops fought stubbornly for each building, is explained by the following figures. Thanks to Khrushchev's determined fight for the Red Army's last reserves, Chuykov received one division after another between 15th September and 3rd October—altogether six fresh and fully equipped infantry divisions. They were fully rested formations, two of them Guards divisions. All these forces were employed in the ruins of Stalingrad Centre and in the factories, workshops, and industrial settlements of Stalingrad North, which had all been turned into fortresses.

The German attack on the city was conducted, during the initial stage, with seven divisions—battle-weary formations, weakened by weeks of fighting between Don and Volga. At no stage were more than ten German divisions employed simultaneously in the fighting for the city.

True enough, even the once so vigorous Siberian Second Army was no longer quite so strong during the first phase of the battle. Its physical and moral strength had been diminished by costly operations and withdrawals. At the beginning of September it still consisted, on paper, of five divisions and five armoured and four rifle brigades—a total of roughly nine divisions. That sounds a lot, but the 38th Mechanized Brigade, for instance, was down to a mere 600 men and the 244th Rifle Division to a mere 1500—in other words, the effective strength of a regiment.

No wonder General Lopatin did not think it possible to defend Stalingrad with this Army, and instead proposed to surrender the city and withdraw behind the Volga. But determination counts for a great deal, and the fortunes of war, which so often follow the vigorous commander, have tipped the scales in more than one battle in the past.

By 1st October Chuykov, Lopatin's successor, already had eleven divisions and nine brigades—*i.e.*, roughly fifteen and a half divisions—not counting Workers' Guards and militia formations.

On the other hand, the Germans enjoyed superiority in the air. General Fiebig's well-tried VIII Air Corps flew an average of 1000 missions a day. Chuykov in his memoirs time and again emphasizes the disastrous effect which the German Stukas and fighter-bombers had upon the defenders. Concentrations for counter-attack were smashed, road-blocks shattered, communication lines severed, and headquarters levelled to the ground.

But what use were the successes of the "flying artillery" if the infantry was too weak to break the last resistance? Admittedly the Sixth Army was able, following the settling down of the situation on the Don, to pull out the 305th Infantry Division and to use it later for relieving one of the worn-out divisions of LI Army Corps. But General Paulus did not receive a single fresh division. With the exception of five engineer battalions, flown out from Germany, all the replacements he received for his bled-white regiments had to come from within the Army's zone of operations. In the autumn of 1942 the German High Command had no reserves whatever left on the entire Eastern Front. Serious crises had begun to loom up in the areas of all the Army Groups from Leningrad down to the Caucasus.

In the north Field-Marshal von Manstein was compelled to use the bulk of his former Crimean divisions for counter-attacks against Soviet forces which had penetrated deeply into the German front. After fierce defensive fighting on the Volkhov, continuing until 2nd October, Army Group North was forced to gain a little breathing-space for itself in the first battle of Lake Ladoga.

In the Sychevka–Rzhev area Colonel-General Model had to employ all his skill and all his forces in order to ward off Russian breakthrough attempts. He had to stand up to three Soviet Armies.

In the centre and on the southern wing of the Central Front, Field-Marshal von Kluge likewise had to employ all his forces to prevent a break-through towards Smolensk.

In the passes of the Caucasus and on the Terek, finally, the Armies of Army Group A were engaged in a desperate race against the threatening winter, trying once more, with a supreme effort, to get through to the Black Sea coast and the oilfields of Baku.

In France, Belgium, and Holland, on the other hand, there were a good many divisions. They spent their time playing cards. Hitler, who persistently under-rated the Russians, made the opposite mistake of over-rating the Western Allies. Already, in the autumn of 1942, he feared an Allied invasion. The American, British, and Soviet secret services nurtured his fear by clever rumours about a second front. The skilfully launched spectre of an invasion which was not to materialize for another twenty months was already tying down twenty-nine German divisions, including the magnificently equipped "Leibstandarte" and the 6th and 7th Panzer Divisions. Twenty-nine divisions! A quarter of them might have turned the tide on the Stalingrad–Caucasus front.

4. Last Front Line along the Cliff

Chuykov's escape from the underground passage near the Tsaritsa–The southern city in German hands–The secret of Stalingrad: the steep river-bank–The grain elevator–The bread factory–The "tennis racket"–Nine-tenths of the city in German hands.

DURING the night of 17th/18th September Chuykov had to clear out of his bomb-proof shelter near the Tsaritsa. It was virtually a flight, for grenadiers of the Lower Saxon 71st Infantry Division, the division with the clover-leaf for its tactical sign, suddenly appeared at the Pushkin Street entrance to the dug-out towards noon. Chuykov's staff officers had to grab their machine pistols. The underground gallery was rapidly filling with wounded and with men who had become separated from their units. Drivers, runners, and officers smuggled their way into the safety of the dug-out under all kinds of pretexts, "in order to discuss urgent matters." As the underground passages had no ventilation system they were soon filled with smoke, heat, and stench. There was only one thing to do—get out.

The headquarters guard covered the retreat by way of the second exit, into the Tsaritsa gorge. But even there German assault parties of Major Fredebold's 191st Infantry Regiment were already in evidence. Carrying only his most important papers and the situation map, Chuykov surreptitiously made his way through the German lines to the Volga bank, through the night and the fog, and together with Krylov crossed to the eastern side by boat.

Chuykov at once boarded an armoured cutter and recrossed the Volga to the upper landing-stage in the northern city. There he established his battle headquarters in the steep cliff towering above the river, behind the "Red Barricade" ordnance factory—a few caves blasted into the 650-foot-high bluff in the blind angle of the German artillery. The various dug-outs were linked by well-camouflaged communication trenches in the steep scarp.

Glinka's kitchen was accommodated in the inspection shaft of the effluent tunnel of the "Red Barricade" works. Tasya, the serving-girl, had to perform real acrobatics dragging her pots and pans up the steel ladder of the shaft into the daylight and then balancing them down a cat-walk along the cliff-face into the Commander-in-Chief's dug-out.

Admittedly, the number of mouths to be fed at headquarters had greatly diminished. Various senior officers, including Chuykov's deputies for artillery and engineering troops, for armour and for mechanized troops, had slipped away quietly during the move of headquarters and had stayed behind on the left bank of the Volga. "We shed no tears over them," Chuykov records. "The air was cleaner without them."

The move which the C-in-C Stalingrad had to perform was symbolical: the focus of the fighting was shifting to the north. The southern and central parts of the city could no longer be held.

On 22nd September the curtain went up over the last act in the southern city. Assault parties of 29th Motorized Infantry Division, together with grenadiers of 94th Infantry Division and the 14th Panzer Division, stormed the smoke-blackened grain elevator. When engineers blasted open the entrances a handful of Soviet marines of a machine-gun platoon under Sergeant Andrey Khozyaynov came reeling out into captivity, half insane with thirst. They were the last survivors.

Men of the 2nd Battalion of the Soviet 35th Guards Division were lying about the ruins of the concrete block—suffocated, burnt to death, torn to pieces. The doors had been bricked up: in this way the commander and commissar had made all retreat or escape impossible.

The southern landing-stage of the Volga ferry was likewise occupied. Grenadiers of the Saxon 94th Infantry Division under Lieutenant-General Pfeiffer, the division whose tactical sign was the crossed swords found on Meissen porcelain, took over cover duty along the Volga bank on the southern edge of the city.

In Stalingrad Centre, in the heart of the city, Soviet opposition also crumbled. Only a few fanatical nests of resistance, manned by remnants of the Soviet 34th and 42nd Rifle Regiments, were holding out among the debris of the main railway station and along the landing-stage of the big steam ferry in the central river-port.

By 27th September—applying the customary criteria of street fighting—Stalingrad could be said to have been conquered. The 71st Infantry Division, for example, had reached the Volga over the division's entire width—211th Infantry Regiment south of the Minina gorge, 191st Infantry Regiment between the Minina and Tsaritsa gorges, and 194th Infantry Regiment north of the Tsaritsa.

The fighting now centred on the northern part of the city with its workers' settlements and industrial enterprises. The names have gone down not only in the history of this war, but in world history generally —the "Red Barricade" ordnance factory, the "Red October" metallurgical works, the "Dzerzhinskiy" tractor works, the "Lazur" chemical works with its notorious "tennis racket," as the factory's railway sidings were called because of their shape. These were the "forts" of the industrial city of Stalingrad.

The fighting for Stalingrad North was the fiercest and the most costly of the whole war. For determination, concentration of fire, and high density of troops within a very small area these operations are comparable only to the great battles of material of World War I, in particular the battle of Verdun, where more than half a million German and French troops were killed during six months in 1916. The battle in Stalingrad North was hand-to-hand fighting. The Russians, who were better at defensive fighting than the Germans anyway, benefited from their superior possibilities of camouflage and from the skilful use of their home ground. Besides, they were more experienced and better trained in street fighting and barricade fighting than the German troops. Finally, Chuykov was operating right under Khrushchev's eyes, and therefore he whipped up Soviet resistance to red heat. As each company crossed the Volga into Stalingrad three slogans were impressed on it:

Every man a fortress!
There's no ground left behind the Volga!
Fight or die!

This was total war. This was the implementation of the slogan "Time is blood." The chronicler of 14th Panzer Division, Rolf Grams, then a major commanding Motor-cycle Battalion 64, quotes a very illuminating account of an engagement: "It was an uncanny, enervating battle above and below ground, in the ruins, the cellars, and

Last Front Line along the Cliff

the sewers of the great city and industrial enterprises—a battle of man against man. Tanks clambering over mountains of debris and scrap, crunching through chaotically destroyed workshops, firing at point-blank range into rubble-filled streets and narrow factory courtyards. . . . But all that would have been bearable. What was worse were the deep ravines of weathered sandstone dropping sheer down to the Volga, from where the Soviets would throw ever new forces into the fighting.

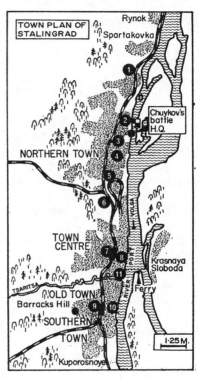

Map 33. 1 Tractor works; 2 "Red Barricade" ordnance factory; 3 bread factory; 4 "Red October" metallurgical works; 5 "Lazur" chemical works with "tennis racket" sidings; 6 Mamayev Kurgan hill; 7 central railway station; 8 Red Square with department store; 9 southern railway station; 10 grain elevator; 11 Chuykov's dugout in the Tsaritsa gorge.

Across the river, in the thick forests of the lower, eastern bank of the river, the enemy lurked invisible, his batteries and his infantry hidden from sight. But he was there nevertheless, firing, and night after night, in hundreds of boats across the river, sending reinforcements into the ruins of the city."

These Soviet supplies flowing in steadily across the river to buttress the defenders, this fresh blood continually pumped into the city through the vital artery that was the Volga—these constituted the key problem of the battle. The key to it all was in the weathered sandstone gorges of the Volga bank. The steep bluff, out of reach of the German artillery, contained the Soviet headquarters, the field hospitals, the ammunition dumps. Here were ideal assembly points for the shipments of men and material across the river at night. Here were the starting-lines for counter-attacks. Here the tunnels carrying the industrial sewage emerged on the surface—now empty underground galleries leading into the rear of the German front. Soviet assault parties would creep through them. Cautiously they would lift a manhole cover and get a machine-gun into position. Suddenly their bursts of fire would sweep the rear of the advancing German formations, mowing down cookhouse parties and supply columns. A moment later the manhole covers would drop back into place and the Soviet assault parties would have vanished.

German assault troops assigned to deal with such ambushes were helpless. The steep western bank of the Volga was worth as much as a deeply echeloned bomb-proof belt of fortifications. Frequently only a few hundred yards divided the German regiments in their operations sectors from the Volga bank.

General Doerr in his essay on the fighting in Stalingrad observes quite correctly: "It was the last hundred yards before the Volga which held the decision both for attacker and defender."

The way to this vital bank in Stalingrad North led through the fortified workers' settlements and industrial buildings. They formed a barrier in front of the vital steep bank. It would take an entire chapter to describe these operations. A few typical examples will testify to the heroism displayed on both sides.

At the end of September General Paulus tried to storm the last bulwarks of Stalingrad, one after another, by a concentrated assault. But his forces were insufficient for an all-embracing large-scale attack on the entire industrial area.

The well-tried 24th Panzer Division from East Prussia, advancing from the south across the airfield, stormed the "Red October" and "Red Barricade" housing estates. The Panzer Regiment and units of 389th Infantry Division also captured the housing estate of the "Dzerzhinskiy" tractor works, and on 18th October fought their way into the brickworks. The East Prussians had thus reached the steep Volga bank. In this sector, at least, the objective had been attained. The division then moved south again into the area of the "Lazur" chemical works and the "tennis racket" railway sidings.

The 24th had tackled their task—but at what cost! Each of the grenadier regiments was just about large enough to make a battalion, and the remnants of the Panzer Regiment were no more than a reinforced company of armoured fighting vehicles. Those crews without tanks were employed as rifle companies.

The huge "Dzerzhinskiy" tractor works, one of the biggest tank-manufacturing enterprises in the Soviet Union, was stormed on 14th October by General Jaenecke's 389th Infantry Division from Hesse and by the regiments of the Saxon 14th Panzer Division. Across the debris of the vast factory grounds the tanks and grenadiers of the 14th drove through to the Volga bank, wheeled south, penetrated into the "Red Barricade" ordnance factory, and thus found themselves immediately in front of the steep bank close to Chuykov's battle headquarters.

The ruins of the gigantic assembly buildings of the tractor plant, where Soviet resistance kept flaring up time and again, were gradually being captured by battalions of the 305th Infantry Division from Baden-Württemberg, the Lake Constance Division, which had been brought from the Don front on 15th October to be employed against Stalingrad's tractor plant. The men from Lake Constance were engaged in protracted fighting with companies of the Soviet 308th Rifle Division under Colonel Gurtyev. The operation was a perfect illustration of General Chuykov's remark in his diary: "The General Staff map is now replaced by the street plan of part of the city, by a sketch plan of the labyrinth of masonry that used to be a factory."

On 24th October the 14th Panzer Division reached its objective—the bread factory at the southern corner of the "Red Barricade." The Motor-cycle Battalion 64 was heading the attack. On the first day of the fighting Captain Sauvant supported the assault on the first building with units of his 36th Panzer Regiment.

On 25th October the attack on the second building collapsed in the fierce defensive fire of the Russians. Sergeant Esser was crouching behind a wrecked armoured car. Across the road, at the corner of the building, lay the company commander—dead. Ten paces behind him the platoon commander—also dead. By his side a section leader was groaning softly—delirious with a bullet through his head.

Quite suddenly Esser went berserk. He leapt to his feet. "Forward!" he screamed. And the platoon followed him. It was some 60-odd yards to the building—60 yards of flat courtyard without any cover. But they made it. Panting, they flung themselves down alongside the wall, they blasted a hole into it with an explosive charge, they crawled through, and they were inside. At the windows across the room crouched the Russians, firing into the courtyard. They never realized what was happening to them when the machine pistols barked out behind them: they just slumped over.

Now the next floor. Cautiously the men crept up the stone staircase. Each door-frame was covered by one man. "*Ruki verkh!*" Aghast, the Russians raised their hands. In this way Esser captured the building with a mere twelve men, taking eighty prisoners and capturing an anti-tank gun and sixteen heavy machine-guns. Hundreds of Soviet dead were left behind on the macabre battlefield of the second block of a bread factory.

Across the road, in the line of buildings forming the administrative block, Captain Domaschk was meanwhile fighting with the remnants of the 103rd Rifle Regiment. All the company commanders had been killed.

The brigade sent Second Lieutenant Stempel from its headquarters personnel, so that at least one officer should be available as company commander. A sergeant put him in the picture about the situation.

A moment later Stempel moved off with his motor-cycle troops to attack between a railway track and a shattered wall. In front of him Stukas were pasting all nests of resistance. In short bounds the men followed the bombs, seizing the ruins of the administrative block and approaching the steep Volga bank.

But there were only two dozen men left. And from the gorges of the steep bank ever new masses of Soviet troops were welling up. Wounded men with bandages, commanded by staff officers, drivers from transport units, even the sailors from the ferries. They were mown down, and dropped to the ground like dry leaves in the autumn. But they kept on coming.

Stempel sent a runner: "I cannot hold out without reinforcements!"

Shortly afterwards came seventy men, thrown into the fighting by a forward command. They were led by a lieutenant. Two days later all seventy were dead or wounded. Stempel and the men of 103rd Rifle Regiment had to withdraw and give up the river-bank.

Nevertheless some four-fifths of Stalingrad were in German hands during those days. Towards the end of October, when the Westphalian 16th Panzer Division and the infantrymen of 94th Infantry Division had at last captured the hotly contested suburb of Spartakovka, which had been fought over ever since August, and smashed the Soviet 124th and 149th Rifle Brigades, as much as nine-tenths of the city were in German hands.

Outside Chuykov's headquarters in the steep cliff the Soviet 45th Rifle Division was holding only a short strip of bank, approximately 200 yards across. South of it, in the "Red October" metallurgical works, only the ruins of its eastern block, the sorting department, the steel foundry, and the tube mill remained in Russian hands. Here units of the 39th Guards Rifle Division under Major-General Guryev were fighting stubbornly for every piece of projecting masonry. Every corner, every scrapheap, had to be paid for dearly with the blood of the assault parties of 94th and 79th Infantry Divisions. Contact towards the north, with 14th Panzer Division, was maintained by the companies of 100th Jäger Division, which at the end of September had been switched from the Don bend to Stalingrad—a further illustration of how the long Don front was being everywhere denuded of German troops for the sake of capturing that accursed city of Stalingrad. South of the "Red October" metallurgical works only the "Lazur" chemical works with its "tennis racket" sidings, as well as a minute bridgehead around the steam ferry landing-stage in the central river port, were still being held by the Soviets.

By the beginning of November Chuykov was altogether holding only one-tenth of Stalingrad—a few factory buildings and a few miles of river-bank.

5. Disaster on the Don

Danger signals along the flank of Sixth Army–Tanks knocked out by mice–November, a month of disaster–Renewed assault on the Volga bank–The Rumanian-held front collapses–Battle in the rear of Sixth Army–Breakthrough also south of Stalingrad–The 29th Motorized Infantry Division strikes–The Russians at Kalach–Paulus flies into the pocket.

STALINGRAD is on the same parallel as Vienna, Paris, or Vancouver. At that latitude the temperature in early November is still fairly mild. That was why General Strecker, commanding XI Corps in the great Don bend, was still wearing his light-weight overcoat as he drove to the headquarters of the Austrian 44th Infantry Division, the Hoch-und-Deutschmeister Division.

In the fields the soldiers were busy lifting potatoes and fodder beet, and raking maize straw and hay—supplies for the winter.

General Strecker's XI Corps was to have covered the left flank of Stalingrad along the big Don bend. But this loop of the river was 60 miles long—and 60 miles cannot be held by three divisions. As a result, the general was compelled to adopt a position along a chord of the arc; in this way he saved about 30 miles, but it meant surrendering to the Soviets the river-bend at Kremenskaya.

Lieutenant-General Batov, commanding the Soviet Sixty-fifth Army, immediately seized his opportunity, crossed the Don, and was now established in a relatively deep bridgehead on the southern bank. Batov's regiments made daily attacks on the positions of Strecker's divisions in an attempt to bring about the collapse of the German flank on the Don.

But Strecker's divisions were established in good positions. Colonel Boje, for instance, when he welcomed the Corps Commander at the headquarters of 134th Infantry Regiment, was able to point to such a clever system of positions on the high ground behind the river that he confidently assured him: "There will be no Russian getting through here, Herr General."

Strecker asked for very detailed reports, especially about everything that had been noticed from the division's observation post at the edge of a small wood south-west of Sirotinskaya since the end of October.

From the edge of that wood there was an excellent view far across the Don. Through the trench telescope it was even possible to make out the German positions of VIII Corps all the way across to the Volga. But above all the enemy's hinterland lay revealed to the eye, like a relief map. And, indeed, a great many significant moves had been spotted: the Russians were bringing up troops and materials to the Don in continuous day-and-night transports, both against Strecker's front and against that of the Rumanian Third Army adjoining it on the left.

Anxiously Corps headquarters recorded these reports every evening. They were fully confirmed by aerial reconnaissance of Fourth Air Fleet. Every morning Strecker passed the reports on to Golubinskaya, where General Paulus had his headquarters. And Paulus, in turn, had been passing the reports on to Army Group since the end of October.

Army Group's reports to the Fuehrer's Headquarters stated: The Russians are deploying in the deep flank of Sixth Army.

On this flank along the Don there stood, next to Strecker's Corps, the Rumanian Third Army along a front of about 90 miles. Next to it was the Italian Eighth Army, and next to that the Hungarian Second Army.

"Why is such a broad sector held only by Rumanians, Herr General?" the staff officers would ask their GOC. They had nothing against the Rumanians—they were brave soldiers—but it was common knowledge that their equipment was pitiful, even more pitiful than that of the Italians. Their weapons were antiquated, they lacked adequate anti-tank equipment, and their supplies were insufficient. Everybody knew that.

But Marshal Antonescu, the Rumanian Head of State, had insisted— as had also Italy's Mussolini—that the forces he was making available for the Eastern Front must be employed as complete units only, and under their own officers. Hitler had reluctantly agreed, although he would have preferred to follow his generals' advice to adopt the "boned corset" method—*i.e.*, to employ alternate foreign and German formations, the latter acting as stiffening units. This idea, however, was ruled out by the national susceptibilities of Germany's allies. As a result, the flank cover of the main German forces at Stalingrad, with their thirteen infantry divisions, three motorized divisions and three Panzer divisions, was entrusted to foreign Armies whose operational effectiveness was inadequate.

Disaster on the Don 577

Naturally, Hitler too read the reports about Soviet troop concentrations opposite the Rumanian front. At his situation conferences he heard the Rumanian Colonel-General Dumitrescu warn of the danger and ask that the Rumanian Third Army should be given anti-tank and Panzer formations to support it, or else should be allowed to shorten its front. To shorten a front was a proposal which invariably aroused Hitler's indignation. To yield ground was not part of his tactics. He wanted to hold everything, forgetting Frederick the Great's old adage: "He who would defend everything defends nothing at all."

In judging the situation on the Don front in the autumn of 1942 Hitler was confirmed in his optimistic assessment by a paper prepared by the Army General Staff, a document so far not widely known. This suggested that an analysis of the General Staff section for "Foreign Armies East" of 9th September 1942 showed that the Russians had no operational reserves of any importance left on the Eastern Front. This Hitler was only too ready to believe. Why then yield ground?

As for the Rumanians' request for anti-tank and Panzer support, Hitler proved reasonable. But the only major formation that could be made available and directed behind the Rumanian Third Army—apart from a few formations of flak, Panzers, Jäger battalions, and Army artillery—was Lieutenant-General Heim's XLVIII Panzer Corps with one German and one Rumanian Panzer division, as well as units of 14th Panzer Division. This Corps was temporarily detached from Fourth Panzer Army and transferred to the area south of Serafimovich.

Normally a German Panzer Corps represented a very considerable fighting force, and more than adequate support for an infantry Army. It would have been quite sufficient to protect the threatened front of the Rumanian Third Army. But Heim's Corps was anything but a Corps. Its centre-piece was the German 22nd Panzer Division. This division had been lying behind the Italian Eighth Army since September in order to be rested and replenished. Contrary to the plans of the Army High Command, it had been only partially re-equipped with German tanks, to take the place of the Czech-manufactured ones, and as yet had few Mark IIIs and Mark IVs. Moreover, the division had parted with its 140th Panzer Grenadier Regiment under Colonel Michalik a few months before to send it to Second Army in the Voronezh area. There the "Brigade Michalik" was made into 27th Panzer Division. The division's Panzer Engineers Battalion finally had been engaged in street fighting in Stalingrad for several weeks.

It is important to remember these facts to understand with what kind of shadow unit the German High Command was hoping to meet a very palpable threat to the Rumanian front on the Don.

Was Hitler aware of all this? Was he informed of the fact that 22nd Panzer Division had not yet been re-equipped? There are many indications that this had been kept from him.

T

On 10th November Corps headquarters and 22nd Panzer Division received orders for the division to move into the sector of the Rumanian Third Army. The division's last units left for the south on 16th November, making for the big Don loop. It was a 150-mile journey through frost and snow.

But neither the frost nor the snow was the main problem. There seemed to be a jinx on this Panzer Corps: one nasty surprise was followed by another.

While stationed on a "quiet front" the 22nd Panzer Division had received practically no fuel for training or testing runs. Its 204th Panzer Regiment, consequently, had been lying scattered behind the Italian Don front, camouflaged under reeds and entirely immobile. The tanks had been well hidden in pits dug into the ground and protected against the frost with straw. The Panzer men had been unable to convince their superior commands that a motorized unit must keep its vehicles moving even during rest periods, and for that purpose required fuel. But no fuel was assigned to it, and engines therefore could not be tested. That then was how Colonel von Oppeln-Bronikowski found the 204th Panzer Regiment shortly before it was moved. When departure was suddenly decided upon and the tanks were to be brought out hurriedly from their pits, only 39 out of 104 could be started up, and that only with difficulty. A further 34 dropped out in the course of the move: the engines simply conked out and the turrets of many tanks refused to turn. In short, the electrical equipment broke down.

What had happened? The answer is staggeringly simple. Mice, nesting in the straw with which the tank-pits were covered, had entered the tanks in search of food and had nibbled the rubber insulation of the wiring. As a result, faults developed in the electrical equipment, and ignition, battery-feeds, turret-sights, and tank guns were out of action. Indeed, several tanks caught fire from short circuits and sparking. And since disasters never come singly, there was a severe drop in temperature just as the unit set out on its march—but the Panzer Regiment had no track-sleeves for winter operations. Somewhere these had been lost on the long journey to the Don.

The result was that the tanks slithered from one side of the icy roads to the other and made only very slow headway. The Tank Workshop Company 204 had not been taken along on this move because of fuel shortage, which meant that no major repairs could be carried out en route.

Instead of the 104 tanks listed in the Army Group records as constituting the strength of 22nd Panzer Division, the Division in fact reached the assembly area of XLVIII Panzer Corps with 31 armoured fighting vehicles. Another 11 followed later. On 19th November, therefore, the Division could boast 42 armoured vehicles—just about enough to amalgamate the tanks, armoured carriers, and motor-cycles, as well as a motorized battery, under the name of Panzer Combat Group Oppeln.

The second major formation of the Corps—the Rumanian 1st Panzer Division—had 108 tanks at its disposal on 19th November. But of that total 98 were Czech 38-T types—perfectly good armoured fighting vehicles, but inferior in armour and fire-power even to the Soviet medium tanks. The "corset boning" designed to stiffen the Rumanian Third Army on the middle Don about mid-November was therefore no real stiffening at all. Yet it was here that the Russian Armies were massing.

November 1942 was a month of disasters. On 4th November Rommel's Africa Army was badly mauled by Montgomery at El Alamein and had to save itself by withdrawing from Egypt into Tripoli. Four days later Eisenhower's invasion army landed in the rear of the retreating German forces, on the west coast of Africa, and started advancing on Tunis.

The long-range effects of the shocks in Africa were felt on all German fronts. Hitler now found himself compelled to secure also that part of Southern France which had hitherto been unoccupied. As a result, four magnificently equipped major mobile formations, which might otherwise have been available for the Eastern Front, were tied down in France—the 7th Panzer Division, and the "Leibstandarte," "Reich," and "Death's Head" Waffen SS Divisions. Against the fire-power and effective combat strength of these four divisions Chuykov with his troops on the Volga bank would not have stood up for forty-eight hours.

On 9th November Hitler returned to Berchtesgaden from a visit to the Munich Löwenbräu Cellar, where he had assured his old comrades of the 1923 putsch: "No power on earth will force us out of Stalingrad again!"

Jodl now handed him the latest reports. They indicated that the Russians were deploying not only north-west of Stalingrad, on the middle Don, opposite the Rumanian Third Army, but also south of the hotly contested city, where two Corps of the Rumanian Fourth Army were covering the flank of Hoth's Fourth Panzer Army. These Soviet moves, reported from various sources, indicated an early attack.

Scowling, Hitler read the reports and bent down over his map. One glance was enough to show him what was at stake. The Soviet deployment along both wings of the Stalingrad front suggested an intended pincer operation against Sixth Army.

Although he was still inclined to under-rate Soviet reserves, Hitler nevertheless realized the danger threatening along the extensive Rumanian sectors of the front. "If only this front were held by German formations I wouldn't lose a moment's sleep over it," he observed. "But this is different. The Sixth Army really must make an end of this business and take the remaining parts of Stalingrad quickly."

Quick action was what Hitler wanted. He was anxious to put an end to the strategically useless tying down of so many divisions in one city; he wanted to regain his freedom of operation. "The difficulties of the fighting at Stalingrad and the reduced combat strength of the units are well known to me," the Fuehrer said in a radio message to General Paulus on 16th November. "But the difficulties on the Russian side must be even greater just now with the ice drifting down the Volga. If we make good use of this period of time we shall save a lot of blood later on. I therefore expect that the commanders will once again display their oft-proved energy, and that the troops will once again fight with their usual dash in order to break through to the Volga, at least at the ordnance factory and the metallurgical works, and to take these parts of the city."

Hitler was right about Russian difficulties due to the ice on the river. This is confirmed by Lieutenant-General Chuykov's notes. In connection with the situation reports of the Soviet Sixty-second Army and its supply difficulties, Chuykov observes in his diary:

"14th November. The troops are short of ammunition and food. The drifting ice has cut communications with the left bank.

"27th November. Supplies of ammunition and evacuation of wounded have had to be suspended."

The Soviet Command thereupon got Po-2 aircraft to carry ammunition and foodstuffs across the Volga. But these machines were not much help since they had to drop their cargoes over a strip only about 100 yards wide. The slightest error, and the supplies dropped either into the river or into enemy hands.

Paulus had Hitler's message urging him to make a quick end at Stalingrad read out to all commanding officers on 17th November. On 18th November the assault parties of the Stalingrad divisions renewed their attack. They hoped that this would be the final charge.

And so they stormed against the Russian positions—the emaciated men of Engineers Battalions 50, 162, 294, and 336. The grenadiers of 305th Infantry Division leapt out of their dug-outs, bent double, weapons at the ready, knapsacks bulging with hand-grenades. Panting, they dragged machine-guns and mortars across the pitted ground and through the maze of ruined factory buildings. Bunched around self-propelled AA guns, behind tanks or assault guns, they attacked—amid the screaming roar of Stukas and the rattle of enemy machine-guns. Soaked to the skin by drizzling rain and driving snow, filthy, their uniforms in tatters. But they stormed—at the landing-stage of the ferry, at the bread factory, at the grain elevator, among the sidings of the "tennis racket." On their first day they "conquered" 30, 50, or even 100 yards. They were gaining ground—slowly but surely. Another twenty-four hours, or perhaps forty-eight hours, and the job would be done.

However, on the following morning, 19th November, at first light, just as the assault parties were resuming their step-by-step advance through the labyrinth of masonry among the factory buildings, storming barricades made of old Russian gun-barrels, flinging explosive charges down manholes into effluent tunnels, slowly inching their way to the Volga bank, the Russians launched their attack against the Rumanian Third Army on the Don, 90 miles away to the north-west.

Colonel-General von Richthofen, commanding Fourth Air Fleet, notes in his diary: "Once again the Russians have made masterly use of the bad weather. Rain, snow, and freezing fog are making all Luftwaffe operations on the Don impossible."

The Soviet Fifth Tank Army was striking from the Serafimovich area—the exact spot where there should have been a strong German Panzer Corps, but where in fact there was only the shadow of a Panzer Corps, Heim's Corps. The Soviets came in strength of two armoured Corps, one cavalry Corps, and six rifle divisions. On the left of the Fifth Tank Army the Soviet Twenty-first Army simultaneously struck southward from the Kletskaya area with one armoured Corps, one Guards cavalry Corps, and six rifle divisions.

This multitude of Soviet Corps sounds rather frightening. But a Soviet Army as a rule had only the fighting strength of a German Corps at full establishment, a Soviet Corps more or less equalled a German division, and a Soviet division was roughly the strength of a German brigade. Colonel-General Hoth very rightly observes: "We over-rated the Russians on the front, but we invariably under-rated their reserves."

The Soviet attack was prepared by eighty minutes of concentrated artillery-fire. Then the first waves came on through the thick fog. The Rumanian battalions resisted bravely. Above all, the 1st Cavalry Division and the regiments of the Rumanian 6th Infantry Division, belonging to General Mihail Lascar, fought stubbornly and held their positions.

But the Rumanians soon found themselves faced with a situation they were not up to. They fell victim to what Guderian has called "tank fright," the panic which seizes units inexperienced in operations against armour. Enemy tanks, which had broken through the line, suddenly appeared from behind, attacking. A cry went up: "Enemy tanks in the rear!" Panic followed. The front reeled. Unfortunately the Rumanian artillery was more or less paralysed by the fog, and fire at pin-point targets was impossible.

By mid-day on 19th November the catastrophe was taking shape.

582 *Part Seven: Stalingrad*

Entire divisions of the Rumanian front, in particular the 13th, 14th, and 9th Infantry Divisions, disintegrated and streamed back in panic.

The Soviets thrust behind them, westward towards the Chir, south-westward, and towards the south. Presently, however, their main forces wheeled towards the south-east. It was becoming obvious that they were making for the rear of Sixth Army.

Now it was up to XLVIII Panzer Corps. But everything suddenly seemed to go wrong with General Heim's formations. Army Group directed the Corps to counter-attack in a north-easterly direction towards Kletskaya—*i.e.*, against the infantry of the Soviet Twenty-first Army, which had 100 tanks at their disposal. But no sooner had the Corps been set in motion than an order came from the Fuehrer's Headquarters at 1130 hours, countermanding the original order: the attack was to be directed towards the north-west, against what was realized to be the much more dangerous breakthrough of the fast formations of the Soviet Fifth Tank Army in the Blinov–Peschanyy area. Everything about turn! To support its operations the Corps was assigned the three divisions of the Rumanian II Corps—badly mauled and disintegrating units with little fight left in them.

By nightfall on 19th November the Soviet armoured spearheads had penetrated some 30 miles through the gap at Blinov.

The German Corps, in particular the armoured group of 22nd Panzer Division under Colonel von Oppeln-Bronikowski, performed an exemplary wheeling manœuvre through an angle of 180 degrees and flung itself into the path of the enemy armoured forces at Peschanyy. But the full damage done by the mice now began to show: the forced march through icy gorges, without track-sleeves to stop the tanks slithering about, resulted in further losses. As a result, the gallant but unlucky division arrived at the battlefield of Peschanyy with only twenty tanks, to face a vastly superior opponent. Fortunately the Panzer Jäger Battalion was near by and, in some dashing actions and hotly fought duels of anti-tank gun against tank, succeeded in battering the Soviet armoured spearhead.

Twenty-six T-34s lay blazing in front of the hurriedly established defensive lines. If there had been just one Panzer regiment on the right and left of them, one single Panzer regiment, the Red storm might have been broken here, at its most dangerous point. But there was nothing at all to the right or left—nothing except fleeing Rumanians. The Soviets simply streamed past.

The 22nd Panzer Division, which apart from the Armoured Group Oppeln had nothing left except its Panzer jägers, one Panzer grenadier battalion, and a few batteries, was threatened with encirclement. It was forced to take evasive action.

As a result, the Rumanian 1st Armoured Division, engaged in gallant fighting under General Radu farther to the east, now became separated from 22nd Panzer Division. The Corps was split up and its fighting power gone. Army Group realized the danger and hurriedly sent an order by radio to the Rumanian 1st Armoured Division to wheel to the south-west to regain contact with Oppeln's group. But things continued to go wrong with Heim's Corps—almost as if there was a curse on it. The German signals unit with the Rumanian 1st Armoured Division had been knocked out and so did not receive the order. As a result, instead of facing south-west, the gallant division continued to fight with its front towards the north. Meanwhile the Russians were driving south-east unopposed.

The intentions of the Soviets now emerged clearly. They were aiming at Kalach. There was nothing left to oppose them with. The bulk of the Rumanian Third Army was in a state of dissolution and panic. Within four days it lost 75,000 men, 34,000 horses, and the entire heavy equipment of five divisions.

The Soviet offensive was well conceived and followed the pattern of the German battles of encirclement of 1941. While its two-edged northern prong was cutting through the shattered Rumanian Third

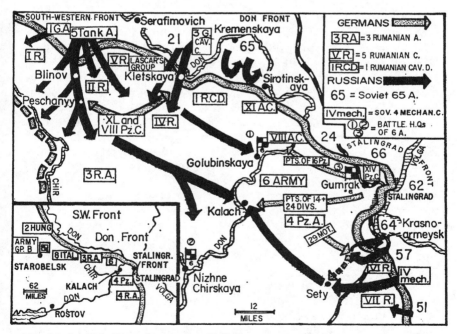

Map 34. On 19th November, just as Sixth Army was mounting one more attack to storm the last Soviet positions, four Soviet Armies and one Armoured Corps burst through the Rumanian-held sectors on the northern and southern flanks of Sixth Army and raced towards Kalach. The inset map shows the front line of Army Group B before the Soviet breakthrough.

Army, the southern prong launched its attack on 20th November against the southern flank of the Stalingrad front, from the Beketovka–Krasnoarmeysk area and from two other concentration points farther south.

Here too the Soviets had chosen for their offensive an area held by Rumanian units. It was the sectors of the Rumanian VI and VII Corps. With two fully motorized Corps, so-called mechanized Corps, as well as a cavalry Corps and six rifle divisions, the Soviet Fifty-seventh and Fifty-first Armies of Yeremenko's Army Group launched their attack. Between these two Armies lurked the IV Mechanized Corps with a hundred tanks. As soon as a breakthrough was achieved this Corps was to race off for a wide outflanking attack on Kalach.

The bulk of the Soviet Fifty-seventh Army, with its tanks and motorized battalions, encountered the Rumanian 20th Division west of Krasnoarmeysk and smashed it with the first blow.

A dangerous situation developed, since that blow was aimed directly, and by the shortest route, at the rear of the German Sixth Army.

But now it was seen what a single experienced and well-equipped German division was able to accomplish; it was also seen that the Soviet offensive armies were by no means outstanding fighting units.

When the disaster struck, the experienced 29th Motorized Infantry Division from Thuringia and Hesse was stationed in the steppe some 30 miles south-west of Stalingrad, as an Army Group reserve. It had been pulled out of the Stalingrad front at the end of September, reinforced to full fighting strength, and earmarked by the Fuehrer's Headquarters for the drive to Astrakhan. At the beginning of November, in view of the difficult situation on the Caucasus front, it received orders through Hoth's Panzer Army to prepare to leave for the Caucasus at the end of November. Once there, the 29th were to prepare for the spring offensive. Such was the optimism in the German High Command at the beginning of November—notwithstanding the situation at Stalingrad. Shortly afterwards a special leave train took some thousand men of the division back to Germany.

Then, on 19th November, this division in full combat strength, under the command of Major-General Leyser, was a real godsend. Since Colonel-General Hoth was unable to get through to Army Group on the telephone, he acted independently, and at 1030 hours on 20th November dispatched Leyser's Division straight from a training exercise to engage the units of the Soviet Fifty-seventh Army which had broken through south of Stalingrad.

The 29th set off hell for leather. The Panzer Battalion 129 roared ahead, in a broad wedge of fifty-five Mark III and Mark IV tanks. Along the flanks moved the Panzer jägers. Behind came the grenadiers on their armoured carriers. And behind them was the artillery. In spite of the fog they drove forward, towards the sound of the guns.

The commanders were propped up in the open turrets. Visibility was barely 100 yards. Suddenly the fog cleared.

At the same moment the tank commanders jerked into action. Immediately ahead, barely 400 yards away, the Soviet tank armada of the XIII Mechanized Corps was approaching. Tank-hatches were slammed shut. The familiar words of command rang out: "Turret 12 o'clock—armour-piercing—400—numerous enemy tanks—fire in your own time!"

Everywhere there were flashes of lightning and the crash of the 7·5-cm. tank cannon. Hits were scored and vehicles set on fire. The Soviets were confused. This kind of surprise engagement was not their strong suit. They were milling around among one another, falling back, getting stuck, and being knocked out.

Presently a new target was revealed. A short distance away, on a railway-line, stood one goods train behind another, disgorging masses of Soviet infantry. The Russians were being shipped to the battlefield by rail.

The artillery battalions of the 29th Motorized Infantry Division spotted the promising target and started pounding it. The breakthrough of the Soviet Fifty-seventh Army was smashed up.

But no sooner was this breach successfully sealed than the alarming news came that 18 miles farther south, in the area of the Rumanian VI Corps, the Soviet Fifty-first Army had broken through at the centre along the southern wing, and was now driving towards Sety with its fast IV Corps. A crucial moment in the battle had come. The 29th Motorized Infantry Division was still in full swing. If this unit could keep up its offensive defence by driving south-west into the flank of the Soviet mechanized Corps, which had about ninety tanks, it seemed very likely that this penetration too would be sealed off. Colonel-General Hoth was therefore getting ready to deliver this second blow at the flank of Major-General Volskiy's Corps.

But just then, on 21st November, an order came down from Army Group: Break off attack; adopt defensive position to protect southern flank of Sixth Army. The 29th Division was detached from Hoth's Fourth Panzer Army and, together with General Jaenecke's IV Corps, subordinated to Sixth Army. But it was not till the morning of 22nd November that General Paulus was informed that the 29th Motorized Infantry Division was now under his command.

In this way a magnificent fighting unit with considerable striking power was held back and employed defensively in a covering line as though it were an infantry division, although in fact there was nothing to defend. Admittedly, orthodox military principles demanded that the flank of an Army threatened by enemy penetrations should be protected—but in this particular instance Army Group should have realized that the southern prong of the Soviet drive was not for the moment directed at Stalingrad at all, but at Kalach, with a view to linking up with the northern prong on the Don and closing the big trap behind Sixth Army.

Weichs's Army Group has been accused, and not without justification, of having pursued a strategy of piecemeal solutions, a strategy of "first things first." Naturally, it is easy to be wise after the event. In all probability, the Army Group did not at the time realize the aim of the Russian attacks. Nevertheless a properly functioning reconnaissance should have revealed what was happening within the next few hours. Major-General Volskiy's IV Mechanized Corps had meanwhile got as far as Sety. Even before nightfall the Russians took up rest positions. They halted their advance. What was the reason? The answer is of some interest.

The surprising appearance of the 29th Motorized Infantry Division on the battlefield had caused the Soviet Corps commander, Major-General Volskiy, who had just then been informed by radio of the disaster that had overtaken the Soviet Fifty-first Army, to lose his nerve. He was afraid of being attacked along his extended unprotected flank. In fact, he was afraid of the very thing that Hoth intended to do. He therefore halted his force even though his Army commander furiously demanded that he should continue to advance. But not until the 22nd, when no German attack came and when he had received another brusque order from Yeremenko, did he resume his advance, wheel towards the north-west, and reach Kalach on the Don twenty-four hours later.

This course of events shows that a well-aimed thrust by 29th Motorized Infantry Division and units of Jaenecke's Corps could have changed the situation and prevented the encirclement of Sixth Army from the south. But when are reliable reconnaissance reports ever available during major breakthroughs? To make matters worse, Paulus and his chief of staff had spent most of their time on the move during these decisive days and hours.

On 21st November Paulus had transferred his Army headquarters from Golubinskaya on the Don to Gumrak, close to the Stalingrad front. Meanwhile, accompanied by Arthur Schmidt, his chief of staff, and by his chief of administration, he had flown to Nizhne-Chirskaya because there, at the point where the Chir ran into the Don, a well-equipped headquarters had been built for Army, with direct lines to Army Group, to the Army High Command, and to the Fuehrer's Headquarters. Nizhne-Chirskaya had been intended for the winter headquarters of Sixth Army—for the period after the capture of Stalingrad.

Paulus and his chief of staff had intended to make use of the good communications facilities at Nizhne-Chirskaya in order to acquaint themselves thoroughly and comprehensively with the situation before moving on to Gumrak. There never was the slightest shadow of suspicion—nor is there to this day—that Paulus intended to remain outside the pocket, away from his headquarters. But Hitler clearly misunderstood the motives and intentions of the Commander-in-Chief Sixth Army. Paulus had barely arrived at Nizhne-Chirskaya when Hitler peremptorily ordered him to return into the pocket.

Colonel-General Hoth had also gone to Nizhne-Chirskaya in the morning of 22nd November, on orders from Army Group, in order to discuss the situation with Paulus. He found him irritable and profoundly upset by the humiliating order he had received from Hitler. The features of this military intellectual bore a pained expression and reflected his deep anxiety over the confused situation. Major-General Schmidt, the chief of staff, on the other hand, was calmness itself. He was constantly on the telephone to the various commanders in the field, collecting information, compiling a picture of the enemy's intentions, and discussing defensive measures. He was the typical detached, calm, professional General Staff officer. He was to prove his strength of character during twelve years of Soviet captivity.

The details which Schmidt entered on his map, which lay spread out before him by the telephone, were anything but encouraging. The situation looked bad in the rear of Sixth Army, west of the Don. And it was not much better along its south-western flank.

6. Sixth Army in the Pocket

"Get the hell out of here"–"My Fuehrer, I request freedom of action"–Goering and supplies by air–The Army High Command sends a representative into the pocket–General von Seydlitz calls for disobedience–Manstein takes over–Wenck saves the situation on the Chir.

THE sky was covered with lowering clouds and a blizzard was blowing from the steppe, blinding the eyes of ground and aerial reconnaissance and rendering impossible the employment of ground-attack aircraft and Stukas. Once again the weather was on Stalin's side. In desperate operations the Luftwaffe, hardly ever able to operate with more than two machines at a time, pounced upon the enemy's spearheads at the penetration points. Hurriedly rounded-up units of supply formations of Sixth Army, rearward services, Army railway companies, flak units, and Luftwaffe ground personnel were strenuously building a first line of defence along the Chir in order, at least, to prevent an extension of the Russian breakthrough into the empty space towards the south-west, in the direction of Rostov.

Particularly grim was the news that the forward air strip at Kalach had been over-run and the short-range reconnaissance planes of VIII Air Corps wrecked. North of Kalach the 44th Infantry Division was still established in good positions west of the Don. Admittedly, it was cut off from its supply units and had to depend on itself, but it acted as a vital crystallizing point west of the river. That in itself was hopeful. It was not to last.

In Stalingrad, General Paulus had suspended all offensive operations in the evening of 19th November, on orders from Army Group. Only a few hundred yards from the objectives a halt had to be called. Units of the three Panzer divisions—the 14th, 16th, and 24th—were formed into combat groups, pulled out of the front, and dispatched towards the Don against the enemy advancing from the north-west.

But in view of the headlong development of the situation in the breakthrough area these weak forces were unable to achieve anything decisive.

On 22nd November at 1400 hours Paulus and Schmidt flew back over the enemy lines to Gumrak, inside the pocket. The new Army headquarters were a little over a mile west of the small railway station.

At nightfall on 22nd November the northern wedge of the Soviets had reached the high ground by the Don and taken the bridge of Kalach by a coup. The southern attacking group was likewise outside the town. On 23rd November Kalach fell. The trap was closed behind Sixth Army.

What was to be done now?

This is the question that has been asked time and again in the voluminous literature that has since appeared about Stalingrad, and that has been answered by a number of conflicting theories. It is a well-known fact, after all, that once a battle has been lost every young subaltern knows how it might have been won. What interests the military historian is what led to the mistakes and errors of judgment. After all, most battles are lost through mistakes and errors of judgment. And the mistakes and errors which led the Sixth Army into the pocket of Stalingrad did not just date from the beginning of November. They cannot be put at Paulus's door, but sprang from the directives issued by the most senior German commands in the late summer.

It is probably true that the period between 19th and 22nd November represented the last chance of rectifying these mistakes and errors. The German High Command ought perhaps to have realized on 19th November the extent of the danger threatening the Army: by ordering it to disengage from the Volga and abandon Stalingrad it might have saved the situation. But this was not a decision that Sixth Army could take on its own authority. General Paulus could not have a sufficiently clear picture of the overall situation for taking such a far-reaching decision on his own authority, a decision which might have threatened the entire southern front, such as withdrawing Sixth Army from its position and starting a precipitate retreat. Besides, a sober assessment of the situation compels one to admit that on 19th, 20th, and even on 22nd November disaster did not yet appear to be inevitable. This is borne out by a careful examination of the state of affairs.

At the General Staff College of Military District I in Königsberg in East Prussia, Arthur Schmidt and Wolfgang Pickert had both been pupils of the late General Osswald, an expert on tactics. "The Southern Cross" his students nicknamed him. His particular trick was to give a brief outline of a situation and then say to his class, "Gentlemen, you have ten minutes—then I want your decision with brief statement of reasons." It was a phrase that none of Osswald's students ever forgot.

When General Pickert, commanding 9th Flak Division, arrived at Nizhne-Chirskaya on the morning of 22nd November he was greeted by his old friend Arthur Schmidt with Osswald's stock phrase: "Pickert —decision with brief statement of reasons."

Pickert's reply came at once: "Get the hell out of here."

Schmidt nodded: "That's what we'd like to do, too, but—" And then Paulus's chief of staff explained to his old friend the official view of the Army: there was no cause for panicky measures; there was nothing yet in the tactical situation to justify independent local decisions in disregard of the overall situation. The most important thing was to cover the Army's rear. Any precipitate withdrawal from the safe positions in Stalingrad might have disastrous consequences. That these considerations were in fact justified was shown only a few days later.

But on 22nd November, when he had that conversation, Schmidt could not know that Hitler had already decided to pin down the Army in Stalingrad. At the time of his discussion with Pickert at Nizhne-Chirskaya, therefore, there were only two things to be done: secure the threatened rear of the Army—i.e., establish a solid front to the west and south—and then prepare for a break-out towards the southwest. What was needed for this, more than anything else, was fuel, and this would have to be flown in by the Luftwaffe. Fuel for the tanks and fuel for the gun-tractors.

This view was in line with the ideas of Weichs's Army Group, which had issued orders in the evening of 21st November to hold Stalingrad and the Volga front "in all circumstances" and to prepare for a breakout. But Pickert doubted that the Luftwaffe would be able to supply the Army even for a short period, and again urged an early breakout. Schmidt pointed out that one could not leave behind the units of XIV and XI Corps which were still on the western bank of the Don or the 10,000 wounded. "That would be like a Napoleonic retreat," he said.

The fact that Paulus and Schmidt were also firmly resolved to break out eventually, after appropriate preparation, is proved by what happened during the next few hours. During the afternoon of 22nd November Paulus received an order by radio from Army High Command via Army Group: "Hold on and await further orders." Quite clearly this was intended as a bar to any overhasty disengagements. Paulus meanwhile had formed an accurate picture of the situation on his south-western flank, where Soviet forces were operating with about a hundred tanks, and sent a signal to Army Group B at 1900 hours, containing the following passage:

> South front still open east of the Don. Don frozen over and crossable. Fuel almost used up. Tanks and heavy weapons then immobilized. Ammunition short. Food sufficient for six days. Army intends to hold remaining area of Stalingrad down to both sides of Don and has set into motion appropriate measures. Indispensable for this is successful sealing off of southern front and continuous plentiful supplies by air. Request freedom of action for the event that hedgehog formation in the south does not come off. Situation might then compel abandonment of Stalingrad and northern front in order to defeat enemy with full force on southern front between Don and Volga and regain contact with Rumanian Fourth Army...

The signal made it perfectly clear what Paulus had in mind. He had made careful plans for all eventualities. He intended to form a hedgehog, but he also demanded freedom of action—*i.e.*, the freedom to disengage rapidly, if the situation should make this necessary.

At 2200 hours a personal signal arrived from Hitler. It refused freedom of action and ordered the Army to stay put. "Sixth Army must know," it said in the signal, "that I am doing everything to help and to relieve it. I shall issue my orders in good time."

Thus the break-out from the pocket was explicitly and firmly forbidden. Paulus reacted instantly. At 1145 hours on 23rd November he radioed to Army Group: "I consider a break-out towards the south-west, east of the Don, by pulling XI and XIV Army Corps over the Don, still possible at present moment, even though material will have to be sacrificed."

Weichs supported this demand in a teleprinted message to Army High Command, emphasizing: "Adequate supply by air is not possible."

At 2345 hours on 23rd November Paulus, after careful reflection and further conversation with the GOCs in his Army, sent another radio message direct to Hitler, urgently requesting permission to break out. All the Corps commanders, he pointed out, shared his view. "My Fuehrer," Paulus radioed,

> since the arrival of your signal of the evening of 22.11 there has been a rapid aggravation of the situation. It has not been possible to seal off the pocket in the south-west and west. Enemy break-

throughs are clearly imminent there. Ammunition and fuel are nearly used up. Numerous batteries and anti-tank weapons have run out of ammunition. Timely and adequate supplies are out of the question. The Army is facing annihilation in the immediate future unless the enemy attacking from the south and west is decisively defeated by the concentration of all available forces. This demands the immediate withdrawal of all divisions from Stalingrad and of strong forces from the northern front. The inescapable consequence must then be a breakthrough towards the south-west, since the eastern and northern fronts, thus depleted, can no longer be held. Admittedly, a great deal of material will be lost, but the bulk of valuable fighting men and at least part of the material will be saved. I continue to accept full responsibility for this far-reaching appraisal, although I wish to record that Generals Heitz, von Seydlitz, Strecker, Hube, and Jaenecke share my assessment of the situation. In view of the circumstances I once more request freedom of action.

Hitler's reply came at 0838 hours on 24th November by a radio signal headed "Fuehrer Decree"—the highest and strictest category of command. Hitler issued very precise orders for the establishment of the pocket fronts and the withdrawal across the Don into the pocket of all Army units still west of the river. The order concluded: "Present Volga front and present northern front to be held at all costs. Supplies coming by air."

Now the Sixth Army was definitely pinned down in Stalingrad by supreme order, even though Army Group, Army, and the local Luftwaffe commander questioned the practicability of aerial supplies. How could such a thing have happened?

It has been generally accepted that Goering had personally guaranteed to supply the Army from the air and had thus been responsible for Hitler's disastrous decision. But historical fact does not entirely bear out this theory.

Contrary to all legend, the decisive conversation with Hitler at the Berghof in Berchtesgaden was conducted not by Goering, but by his chief of staff, Jeschonnek, a sound and sensible man. He reported Goering's affirmative answer to the question of supplying Sixth Army by air, but tied it to a number of conditions such as the indispensable holding on to airfields near the front and passable flying weather.

To represent this qualified undertaking to supply the Army by air as the sole reason for Hitler's mistaken decision would be an unjustified shifting of responsibility from Hitler to Goering—*i.e.*, on to the Luftwaffe. Hitler was only too ready to snatch at Goering's straw, for he did not want to surrender Stalingrad. He was still hoping to strike the Russians mortally by the conquest of territory.

No retreat whatsoever! He implored his generals to remember the winter of 1941 before Moscow, when his rigid orders to hold on had saved Army Group Centre from annihilation. He forgot that what was the correct decision at Moscow in the winter of 1941 need not necessarily apply on the Volga in the winter of 1942. Holding out inflexibly was no panacea.

Besides, there was no strategic necessity for holding on to Stalingrad at the risk of endangering an entire Army. Surely the real task of Sixth Army was to protect the flank and rear of the Caucasus operation. That, at least, was how it had been clearly laid down in the time-table for "Operation Blue." And this task could be implemented even without the capture of Stalingrad—for instance, along the Don.

In a lecture to officers of the German Bundeswehr, Colonel-General Hoth formulated this important aspect of the problem of Stalingrad in the following way:

"From Directive No. 41 it emerges that the main target of the campaign in the summer of 1942 was not the capture of Stalingrad, but the seizure of the Caucasus with its oilfields. This area was indeed vital to the Soviet conduct of the war, and it was also of outstanding economic and political importance to the German Command. At the end of July 1942, when the spearheads of the two German Army Groups were approaching the lower Don earlier than expected, and while the Armies of the Russian South-West Front were falling back in disorder across the middle Don, Hitler on 23rd July ordered the continuation of the operation towards the south, into the Caucasus, by Army Group A, which was assigned four Armies for this purpose. Only the Sixth Army continued to be deployed against Stalingrad. The Chief of the Army General Staff, who had from the outset opposed the far-reaching objective of an operation across the Caucasus, considered it necessary to seek out the enemy grouping at Stalingrad and to defeat it before the Caucasus was crossed. He therefore urged that Sixth Army should be reinforced by two Panzer divisions, which were therefore detached from Fourth Panzer Army. Shortly afterwards Army Group A, although with the focus of the campaign in its sector, was deprived of the Fourth Panzer Army and the Rumanian Third Army, which were both dispatched up the Don to Army Group B. The focus of the campaign had thus been switched to the capture of Stalingrad. Army Group A, thus weakened, ground to a halt north of the Caucasus."

At that moment the Sixth Army's operations at Stalingrad lost their strategic meaning. According to the laws of strategy, the Army should now have been pulled back from its positions jutting out far to the east, in order to dodge the enemy counter-blow that was to be expected and in order to gain reserves. Paulus himself flew to the Fuehrer's Headquarters on 12th September and tried to win Hitler over for such a decision. It was in vain. Hitler remained stubborn. He was unfortunately confirmed in his attitude by the disastrous reports from the

Eastern Department of the General Staff, to the effect that the Russians had no appreciable reserves left along the Eastern Front.

Hitler stuck to his orders that Stalingrad must be taken, and the weakened Sixth Army got its teeth into the city. The longer the fighting continued the more did the capture of the last few workshops and the last few hundred yards of river-bank become a matter of prestige for Hitler, especially as he believed that, after the reverses in Africa and in the Caucasus, he must not give ground at Stalingrad. Prestige, and not strategic considerations, demanded the struggle for the last ruin.

This view was also shared by Weichs's Army Group, whose shrewd chief of staff, General von Sodenstern, has said: "Stalingrad had been taken and eliminated as an armaments centre; shipping on the Volga had been cut. The few technical bridgeheads which the enemy had in the city were no objective that justified the pinning down and using up of the bulk of available German forces. Army Group command was, instead, vitally interested in getting the troops into adequate winter positions as soon as possible, reinforcing the fast formations, and making them mobile for the winter; in addition, it was anxious to form the urgently necessary tactical reserves behind the expected key points of the defensive fighting, and in particular behind the three Armies of Germany's allies on the Don. Such reserves could be drawn only from Sixth Army. That was why about the end of September or the beginning of October, as soon as it was found that Stalingrad could not be taken at the first assault, the command of Army Group B proposed that the offensive against Stalingrad should be suspended altogether. It had also asked for permission to evacuate the front bulge of Stalingrad and, instead of holding the arc, to adopt a position along a chord covering the area between Volga and Don; the left wing of Fourth Panzer Army was to have been bent back south-west of Stalingrad and the new line to have run north-west towards the Don. The Chief of the Army General Staff agreed. But he did not succeed in getting the proposal approved by Hitler."

This then was the background to Hitler's disastrous order to Paulus on 24th November, with its two key demands: hold out and await supplies by air. Goering's promise merely supported Hitler in his attitude against his generals—but it was not the decisive motive for his order. It sprang not from the grandiloquence of one of Hitler's paladins, but from Hitler's own intentions. Stalingrad was the brain child of his strategy, the product of his war which had been a gigantic gamble from the outset, based on victory or ruin.

One often hears the view to-day that because Hitler's hold-on order with its reference to airborne supplies was an unmistakable death sentence on the Army Paulus should not have obeyed it.

But how could Paulus and his closest collaborators in Gumrak judge the strategic motives behind the decision of the Supreme Command?

Besides, had not 100,000 men been encircled in the Demyansk pocket for some two and a half months the previous winter, supplied only by air, and had they not eventually been got out? And had not Model's Ninth Army held out in the Rzhev pocket in accordance with orders? And what about Kholm? Or Sukhinichi?

At the operations centre of the surrounded Sixth Army there was a witness from 25th November onward whose observations about Stalingrad have not up to the present received the attention they deserve—Coelestin von Zitzewitz, now a businessman in Hanover, but then a General Staff major in the Army High Command. On 23rd November he was dispatched to Stalingrad with a signals section by General Zeitzler, the Chief of the General Staff, as his personal observer with instructions to send a daily report to Army High Command about the situation of the Sixth Army. Zitzewitz was summoned to Zeitzler at 0830 hours on 23rd November and informed of his assignment.

The way in which Zitzewitz received his orders from the Chief of the General Staff throws an interesting light on how the situation was assessed at Army High Command. This is Zitzewitz's account:

"Without any preamble the general stepped up to the map spread out on the table: 'Sixth Army has been encircled since this morning. You will fly out to Stalingrad to-day with a signals section of the Operations Communications Regiment. I want you to report to me direct, as fully as possible and as quickly as possible. You will have no operational duties. We are not worried: General Paulus is managing very nicely. Any questions?' 'No, sir.' 'Tell General Paulus that everything is being done to restore contact. Thank you.' With that I was dismissed."

On 24th November Major von Zitzewitz with his signals section—one NCO and six men—flew from Lötzen[1] via Kharkov and Morozovsk into the pocket. What opinions did he find there?

Zitzewitz reports: "General Paulus's first question, naturally, was how did the Army High Command see the relief of Sixth Army. That I could not answer. He said that his principal worry was the supply problem. To supply an entire Army from the air was a task never accomplished before. He had informed Army Group and Army High Command that his requirements would be at first 300 tons a day and later 500 tons if the Army was to survive and remain capable of fighting. These quantities had been promised him.

"The Commander-in-Chief's view seemed to me entirely reasonable: the Army could hold out only if it received the supplies it needed, above all fuel, ammunition, and food, and if relief from without could be expected within a foreseeable time. It was up to the Supreme Command to do the necessary staff work and plan these supplies and the Army's relief, and then to issue appropriate orders.

[1] Now Gizycko.

"Paulus himself took the view that a withdrawal of Sixth Army would be useful within the general picture. He kept emphasizing that Sixth Army could be employed much more usefully along the breached front between Voronezh and Rostov than here in the Stalingrad area. Moreover, the railways, the Luftwaffe, and the entire supply machinery would then be freed for tasks serving the general situation.

"However, this was not a decision he could take on his own authority. Nor could he foresee that his demands concerning relief and supplies would not be fulfilled; for that he lacked the necessary information. The Commander-in-Chief had communicated all these considerations to his generals—all of whom were in favour of breaking out, like himself—and had then given them his orders for their defensive operations."

What else could Paulus have done—Paulus, a typical product of German General Staff training? A Reichenau, a Guderian, or a Hoepner might have acted differently. But Paulus was no rebel; he was a pure strategist.

There was one general in Stalingrad whose views differed fundamentally from Paulus's and who was unwilling to accept the situation created by the Fuehrer's order—General of Artillery Walter von Seydlitz-Kurzbach, commanding LI Corps. He urged Paulus to disregard the Fuehrer's order, and demanded a break-out from the pocket on his own responsibility.

In a memorandum of 25th November he set out for the Commander-in-Chief Sixth Army the views he had already passionately expressed at a meeting of all GOCs on 23rd November, but had then failed to carry his point. His point had been: immediate break-out.

The memorandum began as follows: "The Army is faced with a clear alternative: breakthrough to the south-west in the general direction of Kotelnikovo or annihilation within a few days."

The main arguments of the memorandum about the necessity of a break-out did not differ from the views of the other GOCs in Sixth Army, or from the views held by Paulus himself. The accurate assessment of the situation, worked out by Colonel Clausius, the brilliant chief of staff of LI Army Corps, voiced the opinions of all General Staff officers at all the headquarters in the pocket.

Seydlitz proposed that striking forces should be built up by means of denuding the northern and the Volga fronts, that these forces should attack along the southern front, that Stalingrad should be abandoned and a breakthrough made in the direction of the weakest resistance—*i.e.*, towards Kotelnikovo.

The memorandum said:

> This decision involves the abandonment of considerable quantities of material, but on the other hand it holds out the prospect of smashing the southern prong of the enemy's encirclement, of saving

a large part of the Army and its equipment from disaster and preserving it for the continuation of operations. In this way part of the enemy's forces will continue to be tied down, whereas if the Army is annihilated in its hedgehog position all tying down of enemy forces ceases. Outwardly such an action could be represented in a way avoiding serious damage to morale: following the complete destruction of the enemy's armaments centre of Stalingrad the Army has again detached itself from the Volga, smashing an enemy grouping in doing so. The prospects of a successful breakthrough are the better since past engagements have shown the enemy's infantry to have little power of resistance in open ground.

All this was correct, convincing and logical. Any General Staff officer could subscribe to it. The problem lay in the final passage of the memorandum. This is what it said:

Unless Army High Command immediately rescinds its order to hold out in a hedgehog position it becomes our inescapable duty before our own conscience, our duty to the Army and to the German people, to seize that freedom of action that we are being denied by the present order, and to take the opportunity which still exists at this moment to avert catastrophe by making the attack ourselves. The complete annihilation of 200,000 fighting men and their entire equipment is at stake. There is no other choice.

This highly emotional appeal for disobedience carried no conviction with Paulus, the cool General Staff type. Nor did it convince the other Corps commanders. Besides, a few polemically coloured and factually untenable statements left Paulus unimpressed. "The Army's annihilation within a few days" was a wild exaggeration, and Seydlitz's argument on the issue of supplies was unfortunately also incorrect. Seydlitz had said: "Even if 500 aircraft could land every day they could bring in no more than 1000 tons of supplies, a quantity insufficient for the needs of an Army of roughly 200,000 men now facing large-scale operations without any stocks in hand."

If the Army had in fact received 1000 tons a day it would probably have been able to get away.

Nevertheless Paulus passed on the memorandum to Army Group. He added that the assessment of the military situation conformed with his own views, and therefore asked once more for a free hand to break out if it became necessary. However, he rejected the idea of a breakout against the orders of Army Group and the Fuehrer's Headquarters. Colonel-General Freiherr von Weichs passed on the memorandum to General Zeitzler, the Chief of the General Staff.

Paulus did not receive permission to break out. Was Seydlitz therefore right in demanding disobedience? Setting aside for the moment the moral or philosophical aspect of the matter, the question remains of whether this proposed disobedience was in fact practicable.

How had Khrushchev acted when General Lopatin wanted to withdraw his Sixty-second Army from Stalingrad at the beginning of October because, recalling its frightful losses, he could foresee only its utter destruction? Khrushchev had deposed Lopatin before he could even set the withdrawal in motion.

Paulus, similarly, would not have got far with open insubordination to Hitler. It was a delusion to think that in the age of radio and teleprinter, of ultra-shortwave transmitter and courier aircraft, a general could act like a fortress commander under Frederick the Great, taking decisions against the will of his Supreme Commander and watching while his sovereign could not do anything about it. Paulus would not have remained in command for another hour once his intention had been realized. He would have been relieved of his post and his orders would have been countermanded.

Indeed, an incident affecting Seydlitz personally shows how reliable and quick communications were between Stalingrad and the Fuehrer's Headquarters at the Wolfsschanze, thousands of miles away. The incident, moreover, illustrates the dangers inherent in a precipitate retreat from the safe positions along the Volga.

During the night of 23rd/24th November—*i.e.*, before handing in his memorandum—General Seydlitz had pulled back the left wing of his Corps on the Volga front of the pocket, contrary to explicit orders. This move was intended by Seydlitz as a kind of signal for a break-out, as a priming of the fuse for a general withdrawal from Stalingrad. It was designed to force Paulus's hand.

The 94th Infantry Division, which was established in well-built positions and had not yet lost touch with its supply organization, detached itself from its front in accordance with Seydlitz's orders. All awkward or heavy material was burnt or destroyed—papers, diaries, summer clothing, were all flung on bonfires. The men then abandoned their bunkers and dug-outs and withdrew towards the northern edge of the city. Foxholes in the snow and icy ravines took the place of the warm quarters the troops had left behind: that was how they now found themselves, this vanguard of a break-out. But far from triggering off a great adventure, the division suddenly found itself engaged by rapidly pursuing Soviet regiments. It was over-run and shot up. The entire 94th Infantry Division was wiped out.

That then was the outcome of a spontaneous withdrawal aiming at a break-out. What was far more significant was that even before Sixth Army headquarters got to know of these developments along its own left flank Hitler was already informed. A Luftwaffe signals section in the disaster area had sent a report to the Luftwaffe Liaison Officer at the Fuehrer's Headquarters. A few hours later Hitler sent a radio signal to Army Group: "Demand immediate report why front north of Stalingrad was pulled back."

Paulus made inquiries, established what had happened, and—left the query from the Fuehrer's Headquarters unanswered. Seydlitz was not denounced to Hitler. In this way Hitler was not informed about the whole background and did not know that Seydlitz was responsible for the disaster. By his silence Paulus accepted responsibility. How many Commanders-in-Chief would have reacted in this way to a patent infringement of military discipline? Hitler's reaction, however, was a shattering blow to Paulus. Hitler had held Seydlitz in high regard ever since the operations of the Demyansk pocket, and now regarded him as the toughest man in the Stalingrad pocket; he was convinced that Paulus was responsible for the shortening of the front. He therefore decreed, by radio signal of 24th November at 2124 hours, that the northern part of the fortress area of Stalingrad should be "subordinated to a single military leader" who would be personally responsible to him for an unconditional holding out.

And whom did Hitler appoint? He appointed General von Seydlitz-Kurzbach. In accordance with the principle "divide and rule" Hitler decided to set up a second man in authority by the side of Paulus, as a kind of supervisor to ensure energetic action. When Paulus took the Fuehrer's signal to Seydlitz in person and asked him, "And what are you going to do now?" Seydlitz replied, "I suppose there is nothing I can do but obey."

During his captivity and after his release General Paulus referred to this conversation with Seydlitz time and again. General Roske, the commander of Stalingrad Centre, recalls that General Paulus told him even before he was taken prisoner that he had said to Seydlitz, "If I were now to lay down the command of Sixth Army there is no doubt that you, being *persona grata* with the Fuehrer, would be appointed in my place. I am asking you: would you then break out against the Fuehrer's orders?" After some reflection Seydlitz is reported to have replied, "No, I would defend."

This sounds strange in view of Seydlitz's memorandum, but his answer is attested. And officers well acquainted with Seydlitz do not consider it improbable.

"I would defend." That is precisely what Paulus did.

Like Chuykov on the other side of the line, Paulus and his staff also lived below the ground. In the steppe, four miles west of Stalingrad, close to the railway station of Gumrak, Army Headquarters were established in twelve earth bunkers. The bunker where Colonel-General Paulus lived was twelve foot square. With some six feet of solidly frozen soil as their ceilings these dugouts provided adequate cover against bombardment by medium artillery. Internally they were finished with wooden planks and any material that was to hand. Homemade clay stoves provided heat whenever there was enough fuel available; such fuel had to be brought from Stalingrad Centre. Blankets were fitted across the entrances, as protection against the wind and to prevent too rapid loss of heat. The vehicles were parked some distance away from the bunkers, so that practically no change in the steppe landscape was observable from the air. Only here and there a thin wisp of smoke could be seen coming from a snowy hummock.

On that eventful 24th November, shortly after 1900 hours, Second Lieutenant Schätz, the signals officer, entered General Schmidt's bunker with a decoded signal from Army Group. It was headed: "Top secret—Commander-in-Chief only"—*i.e.*, the highest classification. It ran: "Assuming command of Army Group Don 26.11. We will do everything to get you out. Meanwhile Army must hold on to Volga and northern fronts in accordance with Fuehrer's order and make strong forces available soonest possible to blast open supply route to south-west at least temporarily." The signal was signed "Manstein." Paulus and Schmidt heaved a sigh of relief.

It was not an easy task that confronted the Field-Marshal. He was bringing with him no fresh forces, but was taking over the encircled Sixth Army, the shattered Rumanian Third Army, the Army-sized Combat Group Hollidt consisting of scraped together forces on the Chir, and the newly formed Army-sized Combat Group Hoth.

The headquarters of the new Army Group Don, under which Paulus now came, were in Novocherkassk. Manstein arrived in the morning of 27th November and assumed command at once.

In spite of all difficulties Manstein's plan looked promising and bold. He intended to make a frontal attack from the west, from the Chir front, with General Hollidt's combat group direct against Kalach, while Hoth's combat group was to burst open the Soviet ring from the south-west, from the Kotelnikovo area.

To understand the general picture we must cast back our minds to the situation on the Chir and at Kotelnikovo, the two cornerstones of the starting-line of the German relief attack.

The situation between Don and Chir had stabilized beyond all expectation. That was very largely due to the work of a man we have come across before—Colonel Wenck, on 19th November still chief of staff of LVII Panzer Corps, which was engaged in heavy fighting for Tuapse on the Caucasus front. On 21st November he was ordered by Army High Command to fly immediately to Morozovskaya by a special aircraft made available by the Luftwaffe in order to take up the post of chief of staff with the Rumanian Third Army.

That same evening Wenck arrived at this badly mauled Rumanian Third Army. He gives the following account: "I reported to Colonel-General Dumitrescu. Through his interpreter, Lieutenant Iwansen, I was acquainted with the situation. It looked pretty desperate. On the following morning I took off in a Fieseler Storch to fly out to the front in the Chir bend. Of the Rumanian formations there was not much left. Somewhere west of Kletskaya, on the Don, units of Lascar's brave group were still holding out. The remainder of our allies were in headlong flight. With the means at our disposal we were unable to stop this retreat. I therefore had to rely on the remnants of XLVIII Panzer Corps, on *ad hoc* units of the Luftwaffe, on such rearward units of the encircled Sixth Army as were being formed into combat groups by energetic officers, and on men from Sixth Army and Fourth Panzer Army gradually returning from leave. To begin with, the forces along the Don–Chir arc, over a sector of several hundred miles, consisted merely of the groups under Lieutenant-General Spang, Colonel Stahel, Captain Sauerbruch, and Colonel Adam, of *ad hoc* formations from rearward services and Sixth Army workshop personnel, as well as of tank crews and Panzer companies without tanks, and of a few engineer and flak units. To these was later added the bulk of XLVIII Panzer Corps which fought its way through to the south-west on 26th November. But I was not able to make contact with Heim's Panzer Corps until Lieutenant-General Heim had fought his way through to the southern bank of the Chir with 22nd Panzer Division. The Army Group responsible for us, at first, was Army Group B under Colonel-General Freiherr von Weichs. However, I frequently received my orders and directives direct from General Zeitzler, the Chief of the Army General Staff, since Weichs's Army Group was more than busy with its own affairs and probably could not form a detailed picture of my sector anyway.

"My main task, to start with, was to set up blocking units under energetic officers, which would hold the long front along the Don and Chir along both sides of the already existing Combat Groups Adam, Stahel, and Spang, in co-operation with Luftwaffe formations of VIII Air Corps—at least on a reconnaissance basis. As for my own staff, I literally picked them up on the road. The same was true of motorcycles, staff cars, and communications equipment—in short, all those things which are necessary for running even the smallest headquarters. The old NCOs with experience of the Eastern Front were quite invaluable in all this: they adapted themselves quickly and could be used for any task.

"I had no communication lines of my own. Fortunately, I was able to make use of the communications in the supply area of Sixth Army, as well as of the Luftwaffe network. Only after countless conversations over those connections was I able gradually to form a picture of the situation in our sector, where the German blocking formations were engaged and where some Rumanian units were still to be found. I myself set out every day with a few companions to form a personal impression and to make what decisions were needed on the spot—such as where elastic resistance was permissible or where a line had to be held absolutely.

"The only reserves which we could count on in our penetration area was the stream of men returning from leave. These were equipped from Army Group stores, from workshops, or quite simply with 'found' material.

"In order to collect the groups of stragglers who had lost their units and their leaders after the Russian breakthrough, and to weld these men from three Armies into new units, we had to resort sometimes to the most out-of-the-way and drastic measures.

"I remember, for instance, persuading the commander of a Wehrmacht propaganda company in Morozovskaya to organize film shows at traffic junctions. The men attracted by these events were then rounded up, reorganized, and re-equipped. Mostly they did well in action.

"On one occasion a Field Security sergeant came to me and reported his discovery of an almost abandoned 'fuel-dump belonging to no one' by the side of a main road. We did not need any juice ourselves, but we urgently needed vehicles for transporting our newly formed units. I therefore ordered signposts to be put up everywhere along the roads in the rearward area, lettered 'To the fuel-issuing point.' These brought us any number of fuel-starved drivers with their lorries, staff cars, and all kinds of vehicles. At the dump we had special squads waiting under energetic officers. The vehicles which arrived were given the fuel they wanted, but they were very thoroughly screened as to their own functions. As a result of this screening we secured so many vehicles complete with crews—men who were merely driving about the countryside trying to get away from the front—that our worst transport problems were solved.

"With such makeshift contrivances new formations were created. Although they were officially known as *ad hoc* units, they did in fact represent the core of the new Sixth Army raised later. Under the leadership of experienced officers and NCOs these formations acquitted themselves superbly during those critical months. It was the courage and steadfastness of these motley units that saved the situation on the Chir, halted the Soviet breakthroughs, and barred the road to Rostov."

That was the account of Colonel—subsequently General of Armoured Troops—Wenck.

A firm rock in the battle along Don and Chir was the armoured group of 22nd Panzer Division. By its lightning-like counter-attacks during those difficult weeks it gained an almost legendary reputation among the infantry. Admittedly, after a few days this group was down to about six tanks, twelve armoured infantry carriers, and one 8·8 flak gun. Its commander, Colonel von Oppeln-Bronikowski, sat in a Mark III Skoda tank, leading his unit from the very front, cavalry-style. This armoured group acted as a veritable fire-brigade on the Chir. It was flung into action by Wenck wherever a dangerous situation arose.

When Field-Marshal von Manstein took over command of the new

Army Group Don on 27th November, Wenck reported to him at Novo-cherkassk. Manstein knew the colonel. His laconic order to him was therefore simply: "Wenck, you'll answer to me with your head that the Russians won't break through to Rostov in the sector of your Army. The Don–Chir front must hold. Otherwise not only the Sixth Army in Stalingrad but the whole of Army Group A in the Caucasus will be lost." And Army Group A meant one million men. It was hardly surprising that in such a situation the commanders in the field would frequently resort to desperate means.

Above all, there was a severe shortage of fast armoured tactical reserves to deal with the enemy tanks which popped up all over the place, spreading terror in the rearward areas of the Army Group. Wenck's staff thereupon raised an armoured unit from damaged tanks and immobilized assault guns and armoured troop carriers; this unit was used very effectively at the focal points of the defensive battle between Don and Chir.

Naturally, this unit had to be replenished. And so Wenck's officers conceived the idea of "securing occasional tanks from the tank transports passing through their area on their way to Army Group A or Fourth Panzer Army, manning them with experienced tank crews, and incorporating them in their Panzer companies." Thus, gradually, Wenck collected his "own Panzer Battalion." But one day, when his chief of operations, Lieutenant-Colonel Hörst, in his evening situation report was careless enough to refer to the clearing up of a dangerous penetration on the Chir by "our Panzer Battalion" the Field-Marshal and his staff sat up. Wenck was summoned to Army Group headquarters.

"With what Panzer Battalion did your Army clear up the situation?" Manstein asked. "According to our records it has no such battalion." There was nothing for it—Wenck had to confess. He made a factual report, adding, "We had no other choice if we were to cope with all those critical situations. If necessary, I request that my action be examined by court martial."

Field-Marshal von Manstein merely shook his head, aghast. Then a suspicion of a smile flickered round his lips. He decided to overlook the whole thing, but forbade all further "tank-swiping" in future. "We passed on some of our tanks to 6th and 23rd Panzer Divisions, and from then onward employed our own armoured units in no more than company strength so that they should not attract attention from higher commands."

In this manner the wide breach which the Soviet offensive had torn in the German front in the rear of Sixth Army was sealed again. It was a tremendous triumph of leadership. For weeks a front about 120 miles long was held by formations consisting largely of Reich railway employees, Labour Servicemen, construction teams of the Todt

Organization, and of volunteers from the Caucasian and Ukrainian Cossack tribes.

It should also be recorded that numerous Rumanian units which had lost contact with their Armies placed themselves under German command. There, under German leadership and, above all, with German equipment, they often acquitted themselves excellently, and many of them remained with these German formations for a long time at their own request.

The first major regular formation to reach the Chir front arrived at the end of November, when XVII Army Corps under General of Infantry Hollidt fought its way into the area of the Rumanian Third Army. Everybody heaved a sigh of relief.

At Wenck's suggestion Army Group now subordinated to General Hollidt the entire Don–Chir sector with all formations which had been fighting there; these were formed into the "Armeeabteilung Hollidt." Thus the motley collection of units known to the troops as "Wenck's Army" ceased to exist. It had accomplished a task with few parallels in military history.

Its achievement, moreover, provided the foundation for the second act of the operations on the Chir—the recapture of the high ground on the river's south-western bank, indispensable for any counter-attack. This task was accomplished at the beginning of December by the 336th Infantry Division brought up for the purpose and by the 11th Panzer Division following behind it.

The high ground was taken in fierce fighting and held against all Soviet attacks. These positions on the Chir were of vital importance for the relief of Stalingrad as planned by Manstein, an offensive for which the Field-Marshal employed Hoth's Army-sized combat group from the Kotelnikovo area east of the Don. The Chir front provided flank and rear cover for this rescue operation for Sixth Army. More than that—as soon as the situation permitted XLVIII Panzer Corps, now under the command of General of Panzer Troops von Knobelsdorff, was to support Hoth's operation by attacking in a north-easterly direction with 11th Panzer Division, 336th Infantry Division, and a Luftwaffe field division. The springboard for this auxiliary operation was to be Sixth Army's last Don bridgehead at Verkhne-Chirskaya, at the exact spot where the Chir ran into the Don. There Colonel Adam, General Paulus's Adjutant, was holding this keypoint with hurriedly collected *ad hoc* units of Sixth Army in truly heroic 'hedgehog' fighting.

Thus all steps were being taken and everything humanly possible was being done at this eleventh hour by dint of courage and military skill to rectify Hitler's great mistake and to rescue the Sixth Army.

7. Hoth launches a Relief Attack

"Winter Storm" and "Thunder-clap"—The 19th December—Another 30 miles—Argument about "Thunder-clap"—Rokossovskiy offers honourable capitulation.

ON 12th December Hoth launched his attack. The task facing this experienced, resourceful, and bold tank commander was difficult but not hopeless.

Hoth's right flank had been secured, like the line on the Chir, by drastic means. Colonel Doerr, who was Chief of the German Liaison Staff with the shattered Rumanian Fourth Army, had built up a thin covering line with *ad hoc* units and scraped-together parts of German mobile formations, in much the same way as Colonel Wenck had done in the north. The combat groups under Major Sauvant with units of 14th Panzer Division, and under Colonel von Pannwitz with his Cossacks, flak units, and *ad hoc* formations, restored some measure of order among the retreating Rumanian troops and the German rearward services to whom the panic had spread. The 16th Motorized Infantry Division retreated from the Kalmyk steppe to prepared positions. In this way it also proved possible on the southern wing to foil Russian attempts to strike from the east at the rear of the Army Group Caucasus and to cut it off.

One would have thought that Hitler would now have made available whatever forces he could for Hoth's relief attack, to enable him to strike his liberating blow across 60 miles of enemy territory with the greatest possible vigour and speed. But Hitler was again stingy with his formations. With the exception of 23rd Panzer Division, which was coming up under its own steam, he did not release any of the forces in the Caucasus. The only fully effective formation allocated to Hoth was General Raus's 6th Panzer Division with 160 tanks, which had to be brought from France. It arrived on 12th December with 136 tanks. The 23rd Panzer Division arrived with 96 tanks.

Sixty miles was the distance Hoth had to cover—60 miles of strongly held enemy territory. But things started well. Almost effortlessly the 11th Panzer Regiment, 6th Panzer Division, under its commander Colonel von Hünersdorff on the very first day dislodged the Soviets, who fell back to the east. The Russians abandoned the southern bank of the Aksay, and Lieutenant-Colonel von Heydebreck established a bridgehead across the river with units of 23rd Panzer Division.

The Soviets were taken by surprise. Colonel-General Yeremenko telephoned Stalin and reported anxiously: "There is a danger that Hoth may strike at the rear of our Fifty-seventh Army which is sealing off the south-western edge of the Stalingrad pocket. If, at the same time, Paulus attacks from inside the pocket, towards the south-west, it will be difficult to prevent him from breaking out."

Stalin was angry. "You will hold out—we are getting reserves down to you," he commanded menacingly. "I'm sending you the Second Guards Army—the best unit I've left."

But until the Guards arrived Yeremenko had to manage alone. From his ring around Stalingrad he pulled out the XIII Armoured Corps and flung it across the path of Hoth's 6th Panzer Division. He also ruthlessly denuded his Army Group of its last reserves and threw in the 235th Tank Brigade and the 87th Rifle Division against the spearheads of Hoth's attack. Fighting for the high ground north of the Aksay went on for five days. Fortunately for Hoth, the 17th Panzer Division, which Hitler had at last made available, arrived just in time. Consequently the enemy was dislodged on 19th December.

After a memorable all-night march the armoured group of 6th Panzer Division reached the Mishkova sector at Vasilyevka in the early morning of 20th December. But Stalin's Second Guards Army was there already. Nevertheless General Raus's formations succeeded in establishing a bridgehead two miles deep. Only 30 to 35 miles as the crow flew divided Hoth's spearheads from the outposts of the Stalingrad front.

What meanwhile was the situation like inside the pocket? The supply position of the roughly 230,000 German and German-allied troops was pitiful. It soon turned out that the German Luftwaffe was in no position to keep a whole Army in the depth of Russia supplied from temporary air-strips in mid-winter. There were not enough transport aircraft. Bombers had to do service as transport machines. But these could not carry more than 1½ tons of cargo. Moreover, their withdrawal from operational flying had unpleasant consequences in all sectors of the front. Once again the crucial problem of the campaign was clearly revealed: Germany's material strength was insufficient for this war.

General von Seydlitz had put the daily supply requirements at 1000 tons. That was certainly too high. Sixth Army regarded 600 tons as desirable and 300 tons as the minimum figure for keeping the Army in some sort of fighting condition. Bread requirements alone for the defenders in the pocket amounted to forty tons a day.

Fourth Air Fleet tried to fly in these 300 tons a day. Lieutenant-General Fiebig, the experienced Commander of VIII Air Corps, was assigned this difficult task—and at first it looked as though it could be accomplished. Soon, however, frost and bad weather proved insuperable enemies, more dangerous than Soviet fighters or the Soviet heavy flak. Icing up, poor visibility, and the resulting accidents caused more casualties than enemy action. Nevertheless the air crews displayed a dash and gallantry as in no previous operation. Never before in the history of flying had men set out with such disdain of death and such firm resolution as for the supply air-lift to Stalingrad. Some 550 aircraft were totally lost. That means that one-third of the aircraft employed were lost together with their crews—the victims of bad weather, fighters, and flak. One machine in every three was lost—a terrifying rate which no air power in the world could have kept up.

Only on two occasions was the minimum cargo of 300 tons delivered—or very nearly so. On 7th December, according to the diary of the Chief Quartermaster of Sixth Army, 188 aircraft landed at the airfield of Pitomnik and delivered 282 tons. On 20th December the figure was 291 tons. According to Major-General Herhudt von Rohden's excellent essay, based on the records of the Luftwaffe, the peak day of the air-lift was 19th December, when 154 aircraft delivered 289 tons of supplies to Pitomnik and evacuated 1000 wounded.

On an average, however, the daily deliveries between 25th November and 11th January totalled 104·7 tons. During that period a total of 24,910 wounded were evacuated. At this rate of supplies the men in the pocket had to go hungry and were seriously short of ammunition.

Nevertheless the divisions held out. To this day the Soviets have not published any definite figures of German deserters. But according to all available German sources their number, until mid-January, must have been negligible. Indeed, as soon as the news spread among the troops that Hoth's divisions had launched their relief attack a real fighting spirit spread among them. There was hardly a trooper or an officer who was not firmly convinced that Manstein would get them out. And even the most battle-weary battalion felt strong enough to strike at the ring of encirclement to meet their liberators half-way. That such a plan existed was generally known inside the pocket. After all, units of two motorized divisions and one Panzer division were standing by on the southern front of the pocket, ready to strike in the direction of Hoth's divisions the moment these were close enough and the order for "Winter Storm" was given.

The afternoon of 19th December was cold but clear—magnificent flying weather. Over Pitomnik there was a continuous roar of transport aircraft. They touched down and unloaded their cargoes, were packed full with wounded, and took off again. Petrol-drums were piled high, packing-cases were stacked on top of one another. Shells were trundled away. If only they had this kind of weather every day!

Twenty-four hours earlier an emissary from Manstein had arrived in the pocket in order to acquaint Army with the Field-Marshal's ideas about the break-out. Major Eismann, the Intelligence Officer of Army Group Don, had meanwhile flown back again. No one suspected as yet that his visit was to become an irritating episode in the Stalingrad

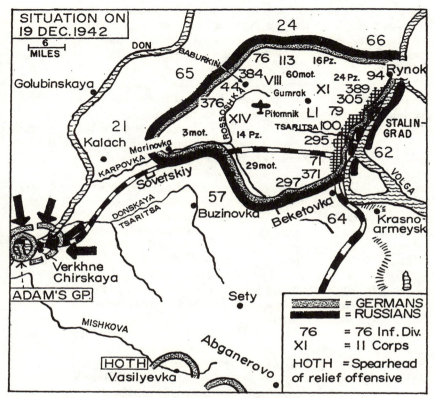

Map 35. The Stalingrad pocket before the Soviet full-scale attack.

tragedy—simply because no written record is extant of the conversations, and the account written by the major from memory ten years later has given rise to many conflicting theses. To this day it has not been definitely established what Paulus, Schmidt, and Eismann really said and what they meant. Did Eismann convey clearly and accurately Manstein's view that the present situation offered only the brutal alternative of early break-out or annihilation? Did he convey clearly that Hollidt's group on the Chir was so busy defending itself against Soviet counter-attacks that there could be no question of its launching an attack in support of Hoth? Did he report that ever-stronger Soviet formations were being deployed against Hoth? Above all, did he state unambiguously that the Field-Marshal was entirely clear about one thing—that the break-out demanded the surrender of Stalingrad in several stages, no matter what label was given to the operation, in order not to arouse Hitler's suspicions too soon? And what did Paulus and Schmidt say in reply? Questions and more questions—and none of them capable of being answered to-day. Eismann's mission is likely to engage the attention of military historians for a long time yet.

The 19th December might be called the day of decision, the day when the drama of Stalingrad reached its culmination.

Paulus and his chief of staff, Major-General Schmidt, were standing in the dugout of the Army chief of operations, in front of a teleprinter which had been connected to a decimetre-wave instrument, a radio circuit which could not be monitored by the Soviets. In this way Sixth Army had an invaluable, even though technically somewhat cumbersome, direct line to Army Group Don in Novocherkassk.

Paulus was waiting for the arranged contact with Manstein. Now the time had come. The machine started ticking. It wrote: "Are the gentlemen present?"

Paulus ordered the reply to be sent: "Yes."

"Will you please comment briefly on Eismann's report," came the message from Manstein.

Paulus formulated his comment concisely.

Alternative 1: Break-out from pocket in order to link up with Hoth is possible only with tanks. Infantry strength lacking. For this alternative all armoured reserves hitherto used for clearing up enemy penetrations must leave the fortress.

Alternative 2: Break-out without link-up with Hoth is possible only in extreme emergency. This would result in heavy losses of material. Prerequisite is preliminary flying-in of sufficient food and fuel to improve condition of troops. If Hoth could establish temporary link-up and bring in towing vehicles this alternative would be easier to carry out. Infantry divisions are almost immobilized at moment and are getting more so every day as horses are slaughtered to feed men.

Alternative 3: Further holding out in present situation depends on aerial supplies on sufficient scale. Present scale utterly inadequate.

And then Paulus dictated into the teleprinter: "Further holding out on present basis not possible much longer." The teleprinter tapped three crosses.

A moment later Manstein's text came ticking through: "When at the earliest could you start Alternative 2?"

Paulus answered: "Time needed for preparation three to four days."

Manstein asked: "How much fuel and food required?"

Paulus replied: "Reduced rations for ten days for 270,000 men."

The conversation was interrupted. A quarter of an hour later, at 1830 hours, it was resumed, and Manstein and Paulus once more talked to each other through the keyboards of their teleprinters. In a strangely anonymous way the words appeared, clicking, on the paper:

"Colonel-General Paulus here, Herr Feldmarschall."

"Good evening, Paulus."

Manstein reported that Hoth's relief attack with General Kirchner's LVII Panzer Corps had got as far as the Mishkova river.

Paulus in turn reported that the enemy had attacked his forces concentrated for a possible break-out at the south-western corner of the pocket.

Manstein said: "Stand by to receive an order."

A few minutes later the order came clicking over the teleprinter. This is what it said:

Order!

To Sixth Army.

(1) Fourth Panzer Army has defeated the enemy in the Verkhne-Kumskiy area with LVII Panzer Corps and reached the Mishkova sector. An attack has been initiated against a strong enemy group in the Kamenka area and north of it. Heavy fighting is to be expected there. The situation on the Chir front does not permit the advance of forces west of the Don towards Stalingrad. The Don bridge at Chirskaya is in enemy hands.

(2) Sixth Army will launch attack "Winter Storm" as soon as possible. Measures must be taken to establish link-up with LVII Panzer Corps if necessary across the Donskaya Tsaritsa in order to get a convoy through.

(3) Development of the situation may make it imperative to extend instruction for Army to break through to LVII Panzer Corps as far as the Mishkova. Code name: "Thunder-clap." In that case the main task will again be the quickest possible establishment of contact, by means of tanks, with LVII Panzer Corps with a view to getting convoy through. The Army, its flanks having been covered along the lower Karpovka and the Chervlenaya, must then be moved forward towards the Mishkova while the fortress area is evacuated section by section.

Operation "Thunder-clap" may have to follow directly on attack "Winter Storm." Aerial supplies will, on the whole, have to be brought in currently, without major build-up of stores. Airfield of Pitomnik must be held as long as possible.

All arms and artillery that can be moved at all to be taken along, especially the guns needed for the operation, and to be ammunitioned, but also such weapons and equipment as are difficult to replace. These must be concentrated in the south-western part of the pocket in good time.

(4) Preparations to be made for (3). Putting into effect only upon express order "Thunder-clap."

(5) Report day and time of attack (2).

It was a historic document. The great moment had come. The Army was to assemble for its march into freedom. For the moment, however, only "Winter Storm" was in force—*i.e.*, a corridor was to be cleared to Hoth's divisions, but Stalingrad was not to be evacuated for the time being.

During the afternoon Manstein had again tried to obtain Hitler's consent to an immediate total break-out by Sixth Army, to Operation "Thunder-clap." But Hitler only approved "Winter Storm," while refusing his consent to the major solution. Nevertheless Manstein, as this document reveals, issued orders to Sixth Army to prepare for "Thunder-clap," and explicitly stated under (3): "Development of the situation may make it imperative to extend instruction for Army to break through." To extend it, that is, into a break-out.

The drama had reached its climax. The fate of a quarter million troops depended on two code names—"Winter Storm" and "Thunder-clap."

At 2030 hours the two chiefs of staff were again sitting in front of their teleprinters. General Schmidt reported that enemy attacks were engaging the bulk of Sixth Army's tanks and part of its infantry strength. Schmidt added: "Only when these forces have ceased to be tied down in defensive fighting can a break-out be launched. Earliest date 22nd December." That was three days ahead.

It was an icy night. In the bunkers at Gumrak there was feverish activity. On the following morning at 0700 hours Paulus was already on his way to the crisis points of the pocket. Throughout the day there was local fighting in many sectors. In the afternoon, when the two chiefs of staff, Schultz and Schmidt, had another conversation over the teleprinters, Schmidt reported: "As a result of losses during the past few days manpower situation on the western front and in Stalingrad exceedingly tight. Penetrations can be cleared up only by drawing upon the forces earmarked for 'Winter Storm.' In the event of major penetrations, let alone breakthroughs, our Army reserves, in particular the tanks, have to be employed if the fortress is to be held at all. The situation could be viewed somewhat differently," Schmidt added, "if it were certain that 'Winter Storm' will be followed immediately by 'Thunder-clap.' In that event local penetrations on the remaining fronts could be accepted provided they did not jeopardize the withdrawal of the Army as a whole. In that event we could be considerably stronger for a break-out towards the south, as we could then concentrate in the south numerous local reserves from all fronts."

It was a vicious circle, a problem that could be solved only if permission for "Thunder-clap" was obtained.

General Schultz replied, unfortunately through the medium of the teleprinter so that the imploring note in his voice was lost, as he dictated to his clerk: "Dear Schmidt, the Field-Marshal believes that Sixth Army must launch 'Winter Storm' as soon as possible. You cannot wait until Hoth has got to Buzinovka. We fully realize that your attacking strength for 'Winter Storm' will be limited. That is why the Field-

Hoth Launches a Relief Attack 611

Marshal is trying to get approval for 'Thunder-clap.' The struggle for this approval has not yet been decided at Army High Command in spite of our continuous urging. But regardless of the decision on 'Thunder-clap' the Field-Marshal emphatically points out that 'Winter Storm' must be started as soon as possible. As for fuel supplies, foodstuffs, and ammunition, over 3000 tons of stores loaded on columns are already standing behind Hoth's Army and will be ferried through to you the moment the link-up has been established. Together with this cargo column numerous towing vehicles will be sent to you in order to make your artillery mobile. Moreover, thirty buses are standing by to evacuate your wounded."

Thirty buses! Nothing, evidently, had been forgotten. And all that stood between Sixth Army and salvation was 30 miles as the crow flew, or 40 to 45 miles by road.

At that moment, right in the middle of these considerations and calculations, planning and preparations, a new disaster befell the German front in the East: three Soviet Armies had launched an attack against the Italian Eighth Army on the middle Don on 16th December. Once again the Russians had chosen a sector held by the weak troops of one of Germany's allies.

After short savage fighting the Soviets broke through. The Italians fled. The Russians raced on to the south. One Tank Army and two Guards Armies flung themselves against the laboriously established weak German line along the Chir. If the Russians succeeded in overrunning the German front on the Chir there would be nothing to halt them all the way to Rostov. And if the Russians took Rostov, then Manstein's Army Group Don would be cut off and von Kleist's Army Group in the Caucasus would be severed from its rearward communications. It would be a super-Stalingrad. What would be at stake then was no longer the fate of 200,000 to 300,000 men, but a million and a half.

On 23rd December, while the men of Sixth Army were still hopefully awaiting their liberators, enemy armoured spearheads were already striking down from the north towards the airfield of Morozovsk, 95 miles west of Stalingrad, on which the surrounded Army's entire supplies depended. The disastrous situation was thus plain. Hollidt's group on the Chir no longer had any flank cover.

In this situation Manstein had no other choice than to order Hoth to switch one of his three Panzer divisions immediately to the left, to the lower Chir, in order to forestall a further breakthrough by the Russians. Hoth did not hesitate, and made his strongest unit available for this vital task.

The 6th Panzer Division was in the middle of its attack in the direction of Stalingrad when the order to turn away reached it. Left with only two battle-weary divisions, Hoth now found it impossible to continue his attack towards Stalingrad. Indeed, under pressure from the Soviet Second Guards Army, he even had to withdraw behind the Aksay on Christmas Eve.

Field-Marshal von Manstein was a very worried man. He sent an urgent teleprinter signal to the Fuehrer's Headquarters, imploring him:

> The turn taken by the situation on the left wing of the Army Group requires the immediate switching of forces to that spot. This measure means dropping for an indefinite period the relief of Sixth Army, which in turn means that this Army would now have to be adequately supplied on a long-term basis. In Richthofen's opinion no more than a daily average of 200 tons can be counted on. Unless adequate aerial supplies can be ensured for Sixth Army the only remaining alternative is the earliest possible breakout of Sixth Army at the cost of a considerable risk along the left wing of Army Group. The risks involved in this operation, in view of that Army's condition, are sufficiently known.

In military officialese this message said all there was to be said: Sixth Army must now break out or else it is lost.

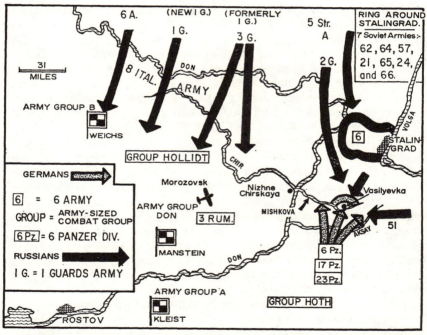

Map 36. On 22nd December 1942 the armoured spearheads of Hoth's Army-size combat group were within 30 miles of the Soviet ring around Stalingrad. However, the Russian breakthrough in the sector held by the Italian Eighth Army prevented the continuation of the relief offensive. As the Russian thrust was aimed at Rostov, threatening both Manstein's and Kleist's Army Groups with encirclement, Sixth Army had to be sacrificed in order to avert this greater danger.

Hoth Launches a Relief Attack 613

Tensely the reply was awaited at Novocherkassk. Zeitzler sent it by teleprinter: The Fuehrer authorizes the withdrawal of forces from Army Group Hoth to the Chir, but he orders that Hoth should hold his starting-lines in order to resume his relief attack as soon as possible.

It was beyond comprehension. Admittedly, Hitler had a cogent argument against authorizing "Thunder-clap": Paulus, he argued, did not have enough fuel to get through to Hoth. This view was based on a report by Sixth Army to the effect that the tanks had enough fuel left only for a fighting distance of 12 miles. This report has since been frequently questioned, but General Schmidt has recalled his strict controls designed to establish stocks of 'black' petrol, and Paulus himself has pointed out, and justly so, that no Army could base a life-and-death operation on the suspected existence of 'black' petrol supplies.

Faced with this situation, Manstein once more had himself put through to Paulus by teleprinter in the afternoon of 23rd December, and asked him to examine whether "Thunder-clap" could not after all be carried out if no other choice was left.

Paulus asked: "Does this conversation mean that I have authority to initiate 'Thunder-clap'?"

Manstein: "I cannot give you this authority to-day, but am hoping for a decision to-morrow." The Field-Marshal added: "The point at issue is whether you trust your Army to fight its way through to Hoth if long-term supplies cannot be laid on for you."

Paulus: "In that case we have no other alternative."

Manstein: "How much fuel do you need?"

Paulus: "One thousand cubic metres."

But a thousand cubic metres meant about a quarter of a million gallons or a thousand tons.

Why, one might ask, did Paulus not mount his operation at that moment in spite of all risks and all his misgivings? Why did he not comply with the order to launch "Winter Storm"—regardless of fuel supplies and foodstuffs, considering that in any case the survival of the Army was at stake?

In his memoirs Field-Marshal von Manstein outlines the responsibility which that order placed on Paulus. The three divisions in the south-western corner of the pocket, where the break-out was to be made, were extensively involved in defensive fighting. Could Paulus run the risk of launching his attack with only parts of these divisions, in the hope of bursting through the powerful ring of encirclement? Besides, would the Soviet attacks even give him a chance of doing so? And would he be able to hold the remaining fronts until Army Group issued the command "Thunder-clap," thus authorizing him to launch the full-scale break-out? And would the tanks have enough fuel to get back again into the pocket in the event of "Winter Storm" being a failure? And what would become of the 6000 wounded and sick?

Paulus and Schmidt could see only the possibility of launching "Winter Storm" and "Thunder-clap" simultaneously. And even that would be practicable only after sufficient quantities of fuel had been flown in.

Army Group, on the other hand, wanted to initiate the full-scale break-out by "Winter Storm" alone, taking the view that the Soviet ring must first be breached along the south-western front before the separate sectors of the pocket front could be dismembered one by one —in other words, before "Thunder-clap" could be set in motion.

Quite apart from military considerations, Manstein's schedule was based on the conviction that only such a phased evacuation would lead Hitler to accept the inevitability of the abandonment of Stalingrad; only then would he be unable to countermand it. Field-Marshal von Manstein knew very well that if Army Group were to order Sixth Army from the outset to launch its full-scale break-out and abandon Stalingrad this order would undoubtedly be countermanded by Hitler without delay.

Paulus, however, tied down in his pocket and fully engaged with improvisations against Soviet attacks, was unable at the time to see the overall picture.

Clearly there is nothing to be gained from seeking the causes of the Stalingrad tragedy at the level of Sixth Army or of Army Group Don, or by trying to pin responsibility on any individual in the sector.

Hitler's strategic mistakes, based as they were on underrating the enemy and overrating his own forces, had brought about a situation which could no longer be remedied by makeshift expedients, ruses of war, or hold-on orders. Only the timely withdrawal of Sixth Army in October could have averted the catastrophe which befell a quarter-million troops on the Volga. Admittedly, even that would no longer have changed the outcome of the war.

To-day, moreover, it is clear from what we know about Russian strength, as revealed by Soviet military writers, that even "Winter Storm" and "Thunder-clap" would no longer have saved a combat-worthy Army. But there might possibly have been a hope of saving the bulk of the men in the Stalingrad pocket. When Hoth's relief attack had to be called off about Christmas even that hope was lost.

The sector of 2nd Battalion, 64th Panzer Grenadier Regiment, contained something unusual—a snow-covered wheatfield with the ears of grain just about showing above the snow. At night the men would crawl out on their bellies, cut off the wheat-ears, and, then back in their dug-outs, would shake out the grains and boil them with water and horse-flesh to make soup. The horse-flesh was that of their animals, which had either been killed in action or died a natural death and were now lying all over the countryside, frozen rigid under small mounds of snow.

On 8th January Lance-corporal Fischer had just laboriously collected the last handful of wheat-ears and brought them back to his bunker, shaking with cold. Back in the bunker everybody was wildly excited. From Battalion headquarters a report had filtered through that the Russians had made an offer of honourable capitulation. The news spread throughout the pocket like wildfire—heaven only knows by what channels.

It had all happened in the Marinovka area, in the sector of the 3rd Motorized Infantry Division. A Russian captain had appeared under a white flag in front of the foremost positions of the Combat Group Willig. The men sent for their commander, Major Willig. The Russian courteously handed over a letter, addressed "Colonel-General of Panzer Troops Paulus, or representative."

Willig thanked him and allowed the Russian with his white flag to return. Then the telephones started to hum. A courier took the letter to Gumrak. Paulus personally rang through on the telephone with an order that no one was to conduct negotiations for surrender with any Russian officers.

On the following day every trooper could read what Colonel-General Rokossovskiy, the Soviet Commander-in-Chief Don Front, had written to Sixth Army. All over the pocket Russian aircraft dropped leaflets with the text of the Soviet offer of surrender. There it was in black and white, signed by a General from Soviet Supreme Headquarters as well as by Rokossovskiy—all official and sealed:

> To all officers, NCOs, and men who cease resistance we guarantee their lives and safety as well as, at the end of the war, return to Germany or any other country chosen by the prisoner-of-war.
>
> All surrendering Wehrmacht troops will retain their uniforms, badges of rank, and decorations, their personal belongings and valuables. Senior officers may retain their swords and bayonets.
>
> Officers, NCOs, and men who surrender will immediately be issued with normal food rations. All wounded, sick, and frost-bitten men will receive medical attention. We expect your answer in writing on 9th January 1943 at 1500 hours Moscow time by way of a representative personally authorized by you, who will drive in a staff car made clearly recognizable by a white flag along the road from Konnaya passing loop to Kotluban station. Your representative will be met at 1500 hours on 9th January 1943 by duly authorized Soviet officers in Rayon 8, 0·5 km. south-east of passing loop No. 564.
>
> In the event of our call for capitulation being rejected, we hereby inform you that the forces of the Red Army and Red Air Force will be compelled to embark upon the annihilation of the encircled German troops. The responsibility for their annihilation will lie with you.

A leaflet which was dropped simultaneously with the text of the letter, moreover, contained the sinister sentence: "Anyone resisting will be mercilessly wiped out."

Why did not Sixth Army accept this offer of capitulation? Why did it not cease its fruitless struggle before the troops were completely finished physically and mentally? Anybody in reasonable health could expect to survive Russian captivity. It is a question that has been asked continuously to this day.

Paulus continued to declare, even while still in captivity, that he did not surrender on his own initiative because at the beginning of January he could still see a strategic purpose in continued resistance—the tying down of strong Russian forces and hence the protection of the threatened southern wing of the Eastern Front.

The same view is expressed to this day by Field-Marshal von Manstein. He says quite clearly: "Since the beginning of December Sixth Army had been tying down sixty major Soviet formations. The situation of the two Army Groups Don and Caucasus would have taken a disastrous turn if Paulus had surrendered at the beginning of January."

Until not so long ago this thesis might have been dismissed as the arguments of men pleading their own case. To-day this objection no longer applies. The Soviet Marshals Chuykov and Yeremenko in their memoirs both fully confirm Manstein's view. Chuykov attests that in mid-January Paulus was still tying down seven Soviet Armies. Yeremenko makes it clear that the unusual offer made to Paulus on 9th January, the offer of "honourable capitulation," was motivated by the hope of releasing the seven Soviet Armies in order to move them against Rostov with a view to crushing the southern wing of the German Eastern Front. The Sixth Army's fight to the finish foiled this plan. Whether in retrospect this sacrifice makes sense, in a political assessment of the war, is a different question.

But Paulus was confirmed in his attitude also by another circumstance. On 9th January General Hube returned to the pocket from an interview with Hitler and reported what the Fuehrer and also the officers of the Army High Command had told him: a new relief offensive from the west was being planned. Replenished Panzer formations had already been set in motion; they were being concentrated east of Kharkov. Aerial supplies were to be entirely reorganized. Just like the winter crisis at Kharkov in 1941–42, Hube argued, Stalingrad too would yet be turned into a great victory. The prerequisite, of course, was that the southern part of the Eastern Front was re-established and the Army Group Caucasus successfully pulled back. For that reason Sixth Army had to hold out—if need be by progressively reducing the pocket down to the built-up area of Stalingrad. It was just a race against time.

Hube's news agreed with the reports which Major von Below, the chief of operations of 71st Infantry Division, brought back to Stalingrad. Below, now again an officer in the Bundeswehr, had fallen ill in

Stalingrad in September 1942 and had been flown back to Germany; on 9th January he returned to the pocket together with Hube.

Before his return Below had been at Army High Command about the end of December. There he had been extensively questioned both by Major-General Heusinger, the Chief of the Operations Department, and by Zeitzler, the Chief of the General Staff, about the possibilities of attacking from the west, across the Don at Kalach. Below had gained the impression that Army High Command still viewed the situation of Sixth Army optimistically and considered the chances of a renewed relief attack to be favourable. General Zeitzler had dismissed the Major with the words: "True, we've got quite enough General Staff officers in the Stalingrad pocket as it is. But if I don't let you return they'll think we've written them off already."

Considering the strategic situations on the one hand and this glimmer of hope on the other, can one blame Paulus for turning down the Soviet call for capitulation on 9th January?

8. The End

The Soviets' final attack–The road to Pitomnik–The end in the Southern pocket–Paulus goes into captivity– Stracker continues to fight–Last flight over the city–The last bread for Stalingrad.

AS Rokossovskiy had announced in his leaflet, the Soviet full-scale attack against the pocket started on 10th January, twenty-four hours after the rejection of the call for surrender. What happened now has been witnessed in military history on only two occasions—once on the Soviet side and now on the German: starved and ill-equipped troops, cut off from all their communications, resisting a superior enemy with a dogged and furious courage which had few parallels. This had happened before, in the Volkhov pocket, when the Soviet Second Striking Army fought on until it was wiped out. Merciless, just as was that battle in the frosty forests along the Volkhov, was also the final fighting in Stalingrad. Only the rôles had changed. The suffering, the hardship, the gallantry, and the misery were the same.

The Soviet attack against the pocket was launched with tremendous ferocity. The German Luftwaffe flak troops with their 8·8-cm. guns in the foremost lines tried to break the force of the armoured onslaught. They fought to the last round and knocked out a surprising number of tanks. But the infantry were crushed in their positions. The Russians broke through in many places. Casualty figures among the weary German units soared rapidly. So did the number of frost-bite victims. The temperature was 35 degrees below zero, and a blizzard was sweeping across the steppe.

Along the western front the battalions of the divisions employed there were resisting like islands in a sea. One of these divisions was the Austrian 44th Infantry Division, holding the approaches to the vital airfield of Pitomnik. And anyone seeing Stalingrad merely in terms of human suffering, misery, errors, overweening pride, and folly should cast a glance at these battalions. One of these, one among many, was the 1st Battalion, 134th Infantry Regiment.

With its shrunken companies it had dug in its heels outside Baburkin. Its commander, Major Pohl, had received the Knights Cross as recently as mid-December. With it General Paulus had sent him a little package. "Best wishes," it had written on it in Paulus's hand. Inside was a loaf of army bread and a tin of herrings in tomato sauce—a precious prize in Stalingrad at that time, commensurate with the highest decoration for gallantry.

Pohl was in his firing-pit, like all his men, armed with a carbine. Over to the north their last heavy machine-gun was firing belt after belt. "No one's going to shift me from here, Herr Major," the sergeant had said to Pohl a few days before. There was one more burst—and then the machine-gun was silent. The men could see the Russians jumping into the machine-gun party's position. A brief mêlée with rifle-butt and trenching-tool—then it was all over. All through the night the battalion held its position, stiffened by Panzer Jäger Battalion 46 with a few 2-cm. flak guns and three captured Soviet 7·62-cm. guns.

When they had to withdraw on the following morning they had to leave the guns behind: there was no fuel for the captured jeeps to haul them away. Thus each move backward became a Waterloo for the gunners: they had to blow up one gun after another. And even if laboriously they hauled one back with them they would no longer find any shells for it.

The following night Major Pohl drove to Pitomnik to acquaint himself with the situation from his friend Major Freudenfeld, Chief of Luftwaffe Signals in the pocket. It was an eerie journey. In order to mark out the road through the snowy wastes the frozen legs of horses, which had been hacked off the dead animals, had been stuck into the snow, hooves upward—appalling signposts of an appalling battle.

At the airfield itself things looked grim. The Army's vital supply centre was a heap of wreckage. The field was covered with shot-up and damaged aircraft. The two dressing-station tents were crammed full with wounded men. And into this chaos new machines were still being sent, talked down, unloaded, reloaded, and sent off again.

Two days later, on 14th January, Pitomnik fell. That was the end of aerial supplies and the evacuation of the wounded. From that moment onward everything went rapidly downhill. From the pocket fronts the last combat groups were falling back towards the city of Stalingrad. Major Pohl with his men also made this journey through hell. Along the road lay a group of German soldiers who had been hit by a bomb. Those who were still alive had lost some limbs. Their blood had frozen into red ice; no one had bandaged them; no one had moved them off the road. All the columns had moved past them, trudging along in dull apathy, concerned only with themselves. Pohl had the wounded men bandaged and laid alongside each other. He left a medical orderly with them to wait for a lorry to pick up the unfortunate wretches. No lorry came.

That was how tens of thousands experienced the last days of Stalingrad. The fierce hunger and utter defencelessness in the face of the full-scale Soviet offensive led to a rapid decline of fighting strength and morale. Spirits sagged. Casualty figures soared. The crowds at first-aid and dressing stations were colossal. Medical supplies and bandages ran out. Marauders roamed the countryside.

On 24th January at 1645 hours the Sixth Army's chief of operations sent a signal to Manstein, a signal shocking in its unemotional language:

> Attacks in undiminished violence against the entire western front which has been fighting its way back eastward in the Gorodische area since the morning of 24th in order to form hedgehog in the tractor works. In the southern part of Stalingrad the western front along the city outskirts held on to the western and southern edge of Minina until 1600 hours. Local penetrations in that sector. Volga and north-eastern fronts unchanged. Frightful conditions in the city area proper, where about 20,000 unattended wounded are seeking shelter among the ruins. With them are about the same number of starved and frost-bitten men, and stragglers, mostly without weapons which they lost in the fighting. Heavy artillery pounding the whole city area. Last resistance along the city outskirts in the southern part of Stalingrad will be offered on 25.1 under the leadership of energetic generals fighting in the line and of gallant officers around whom a few men still capable of fighting have rallied. Tractor works may possibly hold out a little longer. Chief of operations, Sixth Army Headquarters.

Energetic generals. Gallant officers. A few men still capable of fighting. That was the picture.

On the railway embankment south of the Tsaritsa gorge Lieutenant-General von Hartmann, the commander of the Lower Saxon 71st Infantry Division, fired his carbine at the attacking Russians, standing upright, until mown down by a burst of machine-gun fire.

When Field-Marshal von Manstein read the signal from Sixth Army's chief of operations he realized that there could be no question any longer about any military task being performed by Sixth Army. "Since the Army was no longer able to tie down any appreciable enemy forces," the Field-Marshal reports, "I tried in a long telephone conversation with Hitler on 24th January to obtain his order for surrender—unfortunately in vain. At that moment, but not until that moment, the Army's task of tying down enemy forces was finished. It had saved five German Armies."

What Manstein attempted by telephone Major von Zitzewitz tried to achieve by a personal interview with Hitler.

Zitzewitz had flown out of the pocket at the orders of Army High Command on 20th January. On 23rd January General Zeitzler took him to see Hitler. The meeting was profoundly significant. Here is Zitzewitz's own account of it:

"When we arrived at the Fuehrer's Headquarters General Zeitzler was admitted at once, while I was made to wait in the anteroom. A little while later the door was opened, and I was called in. I reported present. Hitler came to meet me and with both his hands gripped my right hand. 'You've come from a deplorable situation,' he said. The spacious room was only dimly lit. In front of the fireplace was a large circular table, with club chairs round it, and on the right stood a long table, lit from above, with a huge situation map of the entire Eastern Front. In the background sat two stenographers taking down every word. Apart from General Zeitzler only General Schmundt and two personal Army and Luftwaffe ADCs were present. Hitler gestured to me to sit down on a stool by the situation map, and himself sat down facing me. The other gentlemen sat down in the chairs in the dark part of the room. Only the Army ADC stood on the far side of the map table. Hitler was speaking. Time and again he pointed to the map. He spoke of a tentative idea of making a battalion of entirely new tanks, the Panther, attack straight through the enemy towards Stalingrad in order to ferry supplies through in this way and to reinforce Sixth Army by tanks. I was flabbergasted. A single Panzer battalion was to launch a successful attack across several hundred miles of strongly held enemy territory when an entire Panzer Army had been unable to accomplish this feat. I used the first pause which Hitler made in his exposé to describe the hardships of Sixth Army; I quoted examples, I read off figures from a slip of paper I had prepared. I spoke about the hunger, the frost-bite, the inadequate supplies, and the sense of having been written off; I spoke of wounded men and lack of medical supplies. I concluded with the words: 'My Fuehrer, permit me to state that the troops at Stalingrad can no longer be ordered to fight to their last round because they are no longer physically capable of fighting and because they no longer have a last round.' Hitler regarded me in surprise, but I felt that he was looking straight through me.

Then he said, 'Man recovers very quickly.' With these words I was dismissed."

But to Stalingrad Hitler radioed: "Surrender out of the question. Troops will resist to the end."

Bombastic words, however, no longer had any effect. Even the most gallant officers were drained of fighting spirit and of all hope. In the cellar of the OGPU prison regimental commanders, company commanders, and staff officers were lying—filthy, wounded, feverish with ulcerations and dysentery, not knowing what to do. They no longer had any regiments, or battalions, or weapons; they had no bread and often only one round in their pistols—one last round, against the final contingency.

Some of them fired these bullets into their own heads. Headquarters and smaller units blew themselves up with dynamite among the wreckage of their last positions. A few staff officers, airmen and signals troops and a handful of indestructible NCOs took a chance on breaking out and set off into a great uncertain adventure. But most of them simply waited for the end to come—one way or another. The much decorated commander of a famous and frequently cited regiment, Colonel Boje, on 27th January stepped before his men in the OGPU cellar and said, "We've no bread left and no weapons. I propose we surrender." The men nodded. And the colonel, feverish and wounded, led them out of the ruins of the OGPU prison.

The distance to the foremost line along the railway embankment was 50 yards. At the Tsaritsa gorge crossing stood the remnants of Lieutenant-General Edler von Daniels's division. Their commander was with them. None of them had any weapons. They too were ready to surrender. It was a sad procession. Along both sides of the road stood Red Army men, sub-machine-guns at the ready. The men were filmed and photographed, loaded on to lorries, and driven off. The steppe swallowed them up.

Meanwhile units of XI Corps under General Strecker held on to their last positions in the cut-off northern pocket.

And over the air came the worst signal from Stalingrad: "To Army Group Don. Food situation compels suspension of issue of rations to wounded and sick, in order to keep alive fighting personnel. Chief of operations, Sixth Army Headquarters."

In spite of all this, Hitler on 31st January at 0130 hours instructed his Chief of General Staff to send one more signal to Stalingrad: "The Fuehrer asks me to point out that each day the fortress of Stalingrad can continue to hold out is of importance."

Five hours later, in the basement of the department store on Stalingrad's Red Square, a lieutenant of the Headquarters Guard entered the small room of the Commander-in-Chief and reported: The Russians are outside.

During the preceding night Paulus had been appointed a Field-Marshal by a radioed order from Hitler. He had been up and about since 6 A.M., talking to Lieutenant-Colonel von Below, his chief of operations. Paulus was tired and disillusioned, but determined to put an end to it. But he wanted to do it, as he put it, "without fuss"—*i.e.*, without documents of surrender and official ceremonies.

That, presumably, was the reason for the much-discussed and frequently misinterpreted way in which Paulus went into captivity. He stuck to the order not to surrender on behalf of his Army. He went into captivity only with his headquarters staff. The various commanders of the separate sectors made their own arrangements with the Russians about the cessation of hostilities. In Stalingrad Centre everything was over on 31st January.

In the northern pocket, in the notorious tractor works and in the "Red Barricade" ordnance factory, at the very spot where the first shots in the battle of Stalingrad had rung out during the summer, strongpoints of XI Corps were still resisting on 1st February. The battle ended where it had begun.

Although this fighting among the ruins no longer had any strategic significance, Hitler insisted on it in a signal with a threadbare justification. He radioed to General Strecker: "I expect the northern pocket of Stalingrad to hold out to the finish. Every day, every hour, thus gained decisively benefits the remainder of the front."

But XI Corps also died a slow death. During the night of 1st/2nd February Strecker was sitting at the command post of Lieutenant-Colonel Julius Müller's combat group. At daybreak Strecker said, "I must go now." Müller understood. "I shall do my duty," he said. There were no great words. When daylight came the fighting ended also in the northern pocket.

At 0840 hours Strecker radioed to the Fuehrer's Headquarters: "XI Army Corps with its six divisions has done its duty."

Here, too, the starved, hollow-eyed men from famous and much-cited divisions climbed out of their trenches and from among the wreckage, and formed into grey columns. They were led off into the steppe—still an unending procession. How many of them?

The number is being disputed to this day, and a strange juggling with figures is often practised. As if numbers could make any difference to suffering, death, and gallantry. Nevertheless, for the sake of the record, these are the facts. According to the Sixth Army's operation diaries, now in American custody, and the daily reports of the different Corps, the ration strength as of 18th December 1942, given in a return of 22nd December, was 230,000 Germans and German-allied troops in the pocket, including 13,000 Rumanians. In addition, the report speaks of 19,300 Russian prisoners and auxiliaries.

Of these 230,000 officers and men some 42,000 wounded, sick, and specialists were evacuated by air up to 24th January 1943. According to Soviet reports, 16,800 were taken prisoners by the Soviets between 10th and 29th January. During the capitulation between 31st January and 3rd February, 91,000 men surrendered.

Some 80,500 remained on the battlefield of Stalingrad—killed in action or, the greater part, gravely wounded and left without shelter, without attention, and without food during the final days, and not brought to safety after the surrender.

About 6000 men out of 107,800 have returned to their homeland to date.

On 3rd February 1943 Second Lieutenant Herbert Kuntz of 100th Bomber Group was the last German pilot to fly his HE-111 over Stalingrad.

"Have a look to see whether fighting still continues anywhere or whether escaping parties can be seen," Captain Bätcher had said to them. "Then drop your load." The load was bread, chocolate, bandages, and a little ammunition.

Kuntz circled the city at about 6000 feet. Not one flak gun opened up. Dense fog hung over the steppe. Hans Annen, the observer, glanced across to Walter Krebs, the radio operator. Krebs shook his head: "Nothing anywhere."

Kuntz dropped to 300 feet—then to 250. Paske, the flight engineer, kept a sharp look-out. Suddenly the mist parted: they were skimming over the churned-up pitted battlefield, barely 200 feet up. Kuntz snatched the aircraft upward, to a safe altitude, and continued the search. Over there—were those not people behind those shreds of mist? "Load away!" he shouted. And their load dropped earthward. Loaves of bread fell into the snow of Stalingrad—among the dead, the frozen, and the few who were still waiting for death.

Perhaps it would be found by one of the small groups who were trying to fight their way out. Many had set out—staff officers with complete combat groups, such as those of IV Corps headquarters and the 71st Infantry Division. Second Lieutenants and sergeants had marched off with platoons through darkness and fog. Corporals, lance-corporals, riflemen, and gunners had sneaked out of the ruins of the city, in groups of three or four, or even singly. Individual parties were spotted by airmen in the steppe as late as mid-February. Then they were lost. Only one man—Sergeant Nieweg, a sergeant in a flak battery—is known to have got through. Twenty-four hours after his escape he was killed by an unlucky mortar bomb at a dressing station of 11th Panzer Division.

APPENDIX
Acknowledgment

TO describe the battles of a war which was lost and has gone down into history as a criminal act of aggression represents a difficult undertaking, almost too difficult for a chronicler in our decade. There is a great temptation to correct the decisions of the battlefield with the pen, or to wade about in the mire of futility and guilt.

The author intended neither the one nor the other. He wanted to report the military events of Operation Barbarossa, Hitler's campaign of conquest which ended at Stalingrad. He wanted to present a broad canvas for the general reader, a picture based on careful researches, documents, essays, war diaries, accounts of experiences, memoirs, and publications by both sides.

This has been possible only thanks to the help of nearly a thousand voluntary collaborators and a number of experienced advisers. To name them all would take up over twenty pages of print. The list extends from the Colonel-General in command of an Army down to the private soldier, from the Chief of the General Staff down to the signaller, from the divisional commander down to the lance-corporal, and from the chief of supplies down to the rank-and-file medical orderly and horse-groom. To all of these the author expresses his thanks—in particular for the historical documents, originals of orders, sketches, and situation reports which were often saved only with difficulty and at great personal risk from the chaos of the war and the immediate post-war period, and which were made available to the author.

In this way many a controversial question of the history of the war has been elucidated and a number of important new facts brought to light.

Colour photographs: Herbert Adam (1), Color-Dienst Kempter (1), Herbert Kuntz (2), Dr Alfred Ott (10), Asmus Remmer (3), Günther Thien (2), Colonel (rtd) Rudi Wagner (1).

Black-and-white photographs: Bibliothek für Zeitgeschichte (8), Albert Cusian (2), dpa (1), Franz Feigl (1), Walter Hackl (2), Agentur Hecht (1), Carl Henrich (6), Imperial War Museum (4), Hans Klöckner (1), Herbert Krass (1), Ernst A. Paulus (4), Lieutenant-General (rtd) Alfred Reinhardt (1), Dr E. W. Rümmler (1), Hans Schaller (2), Horst Scheibert (2), Otto Sroka (1), Süddeutscher Verlag (10), Gerhard Tietz (4), Gottfried Tornau (1), Ullstein Bilderdienst (2), Archiv (6).

PAUL CARELL

Bibliography

BAILEY, GEOFFREY: *The Conspirators* (Victor Gollancz, London, 1961).
BENARY, ALBERT: *Die Berliner Bären-Division, Geschichte der 257.* (Podzun Verlag, Bad Nauheim, 1955).
BLUMENTRITT, GÜNTHER: "The Battle of Moscow," in *The Fatal Decisions* (Michael Joseph, London, 1956).
BÖHMLER, RUDOLF: *Fallschirmjäger* (Podzun Verlag, Bad Nauheim, 1961).
BRAUN, J.: *Enzian und Edelweiss*, 4.Geb.Div. (Podzun Verlag, Bad Nauheim, 1955).
BREITHAUPT, HANS: *Die Geschichte der 30. Infanterie-Division* (Podzun Verlag, Bad Nauheim, 1955).
BUCHNER, ALEX: *Gebirgsjäger an allen Fronten* (Adolf Sponholz Verlag, Hanover, 1954).
BUXA, WERNER: *11. Infanterie-Division* (Podzun Verlag, Bad Nauheim, 1952).
CARELL, PAUL: *The Foxes of the Desert* (translated by Mervyn Savill; Macdonald, London, 1960).
―――: *Invasion—They're Coming!* (translated by E. Osers; Harrap, London, 1962).
CASSIDY, HENRY CLARENCE: *Moscow Dateline, 1941–1943* (Cassell, London, 1943).
CHALES DE BEAULIEU, WALTER: *Der Vorstoss der Panzergruppe 4 auf Leningrad* (Kurt Vowinckel Verlag, Neckargemünd, 1961).
CHUYKOV, V. I.: *The Beginning of the Road* (Moscow, 1959; in Russian).
CONZE, WERNER: *Die Geschichte der 291. I.D.* (Podzun Verlag, Bad Nauheim, 1953).
DALLIN, ALEXANDER: *Die Sowjetspionage* (Verlag für Politik und Wirtschaft, Cologne, 1956).
DIECKHOFF, G.: *Die 3.I.D. (mot.)* (Erich Börries Druck und Verlag, Göttingen, 1960).
DOERR, HANS: *Der Feldzug nach Stalingrad* (E. S. Mittler & Sohn, Darmstadt, 1955).
VON ERNSTHAUSEN, ADOLF: *Wende im Kaukasus* (Vowinckel Verlag, Neckargemünd, 1958).
ESTEBAN-INFANTES, GENERAL: *Blaue Division* (Druffel Verlag, Leoni, 1958).
FRETTER-PICO, M.: *Missbrauchte Infanterie* (Verlag für Wehrwesen Bernard & Graefe, Frankfurt, 1957).
FULLER, JOHN FREDERICK CHARLES: *The Second World War, 1939–45* (Eyre & Spottiswoode, London, 1948).

GAREIS, MARTIN: *Kampf und Ende der Fränkisch-Sudetendeutschen 98. Division* (Podzun Verlag, Bad Nauheim, 1956).
GARTHOFF, RAYMOND LEONARD: *How Russia makes War* (George Allen & Unwin, London, 1954).
GÖRLITZ, WALTER: *Generalfeldmarschall Keitel, Verbrecher oder Offizier?* (Musterschmidt Verlag, Göttingen, 1961).
——: *Paulus and Stalingrad* (translated by Colonel R. H. Stevens; Methuen, London, 1963).
GRAMS, ROLF: *Die 14. Panzer-Division, 1940–1945* (Podzun Verlag, Bad Nauheim, 1957).
GRASER, G.: *Zwischen Kattegat und Kaukasus, 198.I.D.* (privately published, 1961).
GROSSMANN, H.: *Geschichte der 6.I.D.* (Podzun Verlag, 1958).
——: *Rschew, Eckpfeiler der Ostfront* (Podzun Verlag, 1962).
GSCHÖPF, DR R.: *Mein Weg mit der 45.I.D.* (Oberöster. Landesverlag, Linz, 1955).
GUDERIAN, HEINZ: *Panzer Leader* (translated by Constantine Fitzgibbon; Michael Joseph, London, 1952).
HAUPT, WERNER: *Demjansk, Ein Bollwerk im Osten* (Podzun Verlag, Bad Nauheim, 1961).
HENNECKE, K.: *Die Geschichte der 170.I.D.* (Podzun Verlag, Bad Nauheim, 1952).
HERHUDT VON ROHDEN, H.-D.: *Die Luftwaffe ringt um Stalingrad* (Limes Verlag, Wiesbaden, 1950).
HESS, WILHELM: *Eismeerfront 1941* (Scharnhorst Buchkameradschaft, Heidelberg, 1956).
HOTH, HERMANN: *Panzer-Operationen* (Scharnhorst Buchkameradschaft, Heidelberg, 1956).
HUBATSCH, WALTHER: *61. Infanteriedivision* (Podzun Verlag, Bad Nauheim, 1958).
JACOBSEN, HANS-ADOLF: *1939–1945, Der zweite Weltkrieg in Chroniken und Dokumenten* (Wehr und Wissen Verlagsgesellschaft, Darmstadt, 1959).
JACOBSEN, ROHWER: *Entscheidungsschlachten des zweiten Weltkrieges* (Verlag für Wehrwesen, Bernard & Graefe, Frankfurt-on-Main, 1960).
KALINOW, KYRILL D.: *Sowjetmarschälle haben das Wort* (Hansa Verlag, Hamburg, 1950).
KEILIG, WOLF: *Das deutsche Heer 1939–1945* (Podzun Verlag, Bad Nauheim, n.d.).
KESSELRING, ALBERT: *The Memoirs of Field-Marshal Kesselring* (translated by Lynton Hudson; William Kimber, London, 1953).
KISSEL, HANS: *Angriff einer Infanteriedivision* (Kurt Vowinckel Verlag, Heidelberg, 1958).
KLATT, PAUL: *Die 3. Gebirgsdivision 1939–1945* (Podzun Verlag, Bad Nauheim, 1958).
VON KNOBELSDORFF, O.: *Geschichte der 19.Pz. D.* (Podzun Verlag, Bad Nauheim, 1958).
KONRAD, R.: *Kampf um den Kaukasus* (Copress Verlag, Munich, n.d.).
KRÜGER, HEINZ: *Bildband der 263.I.D.* (Podzun Verlag, Bad Nauheim, 1962).
LANZ, HUBERT: *Gebirgsjäger* (Podzun Verlag, Bad Nauheim, 1954).

LEMELSEN, JOACHIM: 29.I.D.(mot.) (Podzun Verlag, Bad Nauheim, 1960).
LIDDELL HART, BASIL HENRY: *The Soviet Army* (Edited by B. H. L. Hart; Weidenfeld and Nicolson, London, 1956).
LOHSE, GERHARD: *Geschichte der 126.I.D.* (Podzun Verlag, Bad Nauheim, 1957).
LUSAR, RUDOLF: *German Secret Weapons of the Second World War* (translated by P. Heller and M. Schindler; Neville Spearman, London, 1959).
VON MACKENSEN, EBERHARD: *Das III. Panzer-Korps im Feldzug 1941/42 gegen die Sowjetunion*, Mitteilungsblatt der 23.Pz. D. (April 1959).
MANN, MENDEL: *Vor Moskaus Toren* (Verlag Heinrich Scheffler, Frankfurt-on-Main, 1961).
MANNERHEIM, MARSHAL CARL GUSTAV EMIL: *Memoirs of Marshal Mannerheim* (translated by Count Eric Lewenhaupt; Cassell, London, 1953).
VON MANSTEIN, FRITZ ERIC: *Lost Victories* (edited and translated by Anthony G. Powell; Methuen, London, 1958).
VON MELLENTHIN, F. W.: *Panzerschlachten* (Kurt Vowinckel Verlag, Heidelberg, 1963).
MELZER, WALTER: *Geschichte der 252.I.D. 1939–1945* (Podzun Verlag, Bad Nauheim, 1960).
VON METZSCH, F. A.: *Die Geschichte der 22.I.D.* (Podzun Verlag, Bad Nauheim, 1952).
MEYER-DETRING, WILHELM: *Die 137.I.D.* Kameradschaft der Bergmann-Division (1962).
MIDDELDORF, EIKE: *Taktik im Russlandfeldzug* (E.S. Mittler & Sohn, Berlin/Frankfurt, 1957).
MORRISON, SAMUEL ELIOT: *The Battle of the Atlantic*, Vol. I of *The Two Ocean War* (Boston/Toronto, 1963).
MUNZEL, OSKAR: *Panzer-Taktik* (Kurt Vowinckel Verlag, Neckargemünd, 1959).
NEHRING, WALTER K.: *Die 18.Pz.Div. 1941*, in Deutscher Soldatenkalender (1961).
NITZ, GÜNTHER: *Die 292. Infanteriedivision* (Verlag Bernhard & Graefe, Berlin, 1957).
ORLOV, ALEXANDER: *The Secret History of Stalin's Crimes* (Jarrolds, London, 1954).
PANZERMEYER: *Grenadiere* (Schild Verlag, Munich, 1957).
PHILIPPI, ALFRED: *Das Pripjetproblem* (Wehrwissenschaftliche Rundschau, March 1956).
PHILIPPI, ALFRED, and HEIM, FERDINAND: *Der Feldzug gegen Sowjetrussland, 1941 bis 1945* (W. Kohlhammer Verlag, Stuttgart, 1962).
PLATANOV, S. P., PAVLENKO, N. G., PAROTKIN, I. W.: *History of the Second World War, 1939–1945* (three vols.; Moscow, 1958, in Russian).
PLOETZ, A. G.: *Geschichte des zweiten Weltkrieges* (A. G. Ploetz Verlag, Würzburg, 1960).
POHLMAN, H.: *Geschichte der 96.I.D.* (Podzun Verlag, Bad Nauheim, 1959).
RADEK, KARL BERNGARDOVICH: *Leo Schlageter, der Wanderer ins Nichts*, Speech of 20th June 1923 before the Executive of the Communist Internationale.

REINHARDT, HANS: *Der Vorstoss des XXXXI. Panzer-Korps im Sommer 1941 von Ostpreussen bis vor die Tore von Leningrad*, Wehrkunde, Book 3 (March 1956).
REINICKE, ADOLF: *Die 5. Jäger-Division* (Podzun Verlag, Bad Nauheim, 1962).
RÖHRICHT, EDGAR: *Probleme der Kesselschlacht* (Condor Verlag, Karlsruhe, 1958).
ROSKILL, SIMON WENTWORTH: *The War at Sea*, Vol. II (Her Majesty's Stationery Office, London, 1956).
SAMSONOV, ALEKSANDR MIKHAYLOVICH: *The Great Battle before Moscow, 1941–1942* (Moscow, 1956; in Russian).
SAMYALOV, A. S., and KALYADIN, T. Y.: *The Battle of the Caucasus* (Moscow, 1956; in Russian).
SCHEIBERT, HORST: *Nach Stalingrad—48 Kilometer* (Kurt Vowinckel Verlag, Neckargemünd, 1956).
SCHELM, W., and MEHRLE, DR H.: *Von den Kämpfen der 215.I.D.* (privately published, n.d.).
SCHMIDT, G.: *Regimentsgeschichte Pz.A.R.73* (Boettcher Verlag, Bremen, n.d.).
SCHRÖDER, JÜRGEN, and SCHULTZ-NAUMANN, JOACHIM: *Die Geschichte der pommerschen 32. Inf.Div.* (Podzun Verlag, Bad Nauheim, 1956).
SCHRÖTER, HEINZ: *Stalingrad bis zur letzten Patrone* (Kleins Druck- und Verlagsanstalt, Lengerich, n.d.).
SELLE, H.: *Die Tragödie von Stalingrad* (Verlag Das andere Deutschland, Hanover, 1948).
VON SENGER U. ETTERLIN, DR FERDINAND M.: *24. Panzerdivision, vormals 1. Kavalleriedivision* (Kurt Vowinckel Verlag, Neckargemünd, 1962).
SMIRNOV, SERGEY: *In Search of the Heroes of the Fortress of Brest, 1941* (Moscow, 1956; in Russian).
Soviet War News, Vols. I–IV (published by the Press Department of the Soviet Embassy in London, 1941–42).
SPAETER, HELMUTH: *Die Geschichte des Panzerkorps Grossdeutschland* (privately published).
SPEIDEL, HELM: *Reichswehr und Rote Armee* (Vierteljahreshefte für Zeitgeschichte, 1/1953).
STEETS, HANS: *Gebirgsjäger bei Uman* (Kurt Vowinckel Verlag, Neckargemünd, 1955).
———: *Gebirgsjäger in der Nogaischen Steppe* (Kurt Vowinckel Verlag, Neckargemünd, 1956).
———: *Gebirgsjäger zwischen Dnjepr und Don* (Kurt Vowinckel Verlag, Neckargemünd, 1957).
STOVES, ROLF: *Die 1. Panzerdivision* (Podzun Verlag, 1962).
STRAUSS, FRANZ JOSEF: *Geschichte der 2. Pz.Div.* (Kitzingen, 1960).
THORWALD, JÜRGEN: *Wen sie verderben wollen* (Steingrüben Verlag, Stuttgart, 1952).
TIEMANN, R.: *Geschichte der 83. Infanteriedivision* (Podzun Verlag, Bad Nauheim, 1960).
VON TIPPELSKIRSCH, KURT: *Geschichte des zweiten Weltkrieges* (Athenäum Verlag, Bonn, 1951).
TOEPKE, GÜNTER: *Stalingrad wie es wirklich war* (Kogge Verlag, Stade, 1959).

TRESS, KARL, and others: *Das Infanterie- und Sturm-Regiment 14 im zweiten Weltkrieg* (Kameradschaft ehemaliger 114er und 14er, Constance, 1959).
VOYETEKHOV, BORIS: *The Last Days of Sevastopol* (translated by R. Parker and V. M. Genne; Cassell, 1943).
WAGENER, CARL: *Der Vorstoss des XXXX.Pz. Korps von Charkow zum Kaukasus*, in *Wehrwissenschaftliche Rundschau* (September/October 1955).
WERTHEN, WOLFGANG: *Geschichte der 16. Panzerdivision* (Podzun Verlag, Bad Nauheim, 1958).
WICH, RUDOLF: *Baden-Württembergische Divisionen im 2. Weltkrieg* (Verlag G. Braun, Karlsruhe, 1957).
WIEDER, JOACHIM: *Stalingrad* (Nymphenburger Verlagshandlung, Munich, 1962).
YEREMENKO, A. I.: *Towards the West* (Moscow, 1959; in Russian).
————: *Stalingrad* (Moscow, 1961; in Russian).
ZEITZLER, KURT: "The Battle of Stalingrad," in *The Fatal Decisions* (Michael Joseph, London, 1956).
ZHILIN, P. A.: *The Principal Operations of the Great Fatherland War, 1941–1945* (Moscow, 1956; in Russian).
VON ZYDOWITZ, KURT: *Die Geschichte der 58.I.D.* (Podzun Verlag, Bad Nauheim, 1958).

Private Publications

Geschichte der 21. Infanterie-Division (1960).
Geschichte der 24. Infanterie-Division, Study Group of the Division (1956).
Das Buch der 78. Sturm-Division (1956).
Der Weg der 93. Inf. Div., 1939–1945 (1956).
Taten und Schicksale der 197. Infanterie-Division (n.d.).
290. Infanterie-Division (1960).
Geschichte der 56. Infanterie-Division, 1938–1945, Study Group of the Division (n.d.).
Panzerkeil im Osten (Verlag "Die Wehrmacht," Berlin, 1941).

Unpublished Manuscripts, Essays, and Lectures made available to the author by:

Lieutenant-General (rtd) Fritz Bayerlein; Colonel (rtd) Arthur Boje; Air Force General (rtd) Paul Deichmann; Colonel (rtd) Joachim Hesse; Colonel-General (rtd) Hermann Hoth; Major (rtd) Karl Hübner; Major (rtd) Hermann Kandutsch; Colonel of Waffen SS (rtd) Otto Kumm; Major (rtd) F. W. Küppers; General of Panzer Troops (rtd) Hasso von Manteuffel; General of Panzer Troops (rtd) Walter K. Nehring; Colonel (rtd) Heinrich Nolte; Major-General (rtd) Hermann von Oppeln-Bronikowski; General of Flak Artillery (rtd) Wolfgang Pickert; Major (rtd) Eberhart Pohl; Lieutenant-General (rtd) Alfred Reinhardt; Captain (rtd) Helmut K. G. Rönnefarth; General of Panzer Troops (rtd) Ferdinand Schaal; Lieutenant-General Arthur Schmidt; Colonel (rtd) Herbert Selle; Colonel-General (rtd) Strecker; General of Panzer Troops (rtd) Walter Wenck; Major-General (rtd) Heinz-Joachim Werner-Ehrenfeucht; Lieutenant-Colonel (rtd) Coelestin von Zitzewitz.

Index of Names

von Abendroth, First Lieutenant, 13
Adam, Colonel, 600, 603
von Ahlfen, Lieutenant-Colonel, 297
Akhmedov, Major, 64
Alaportsev, Lieutenant, 161 ff.
Aleksandrovskiy, Ambassador, 199
Alicke, Second Lieutenant, 27
Altrichter, General, 395
Andersen, Lale, 180
Annen, Hans, 623
Antonescu, Marshal, 14, 576
von Apell, Major-General Wilhelm, 450
von Arenstorff, 550
von Arnim, General, 54, 135, 269, 407
Arntzen, Major, 257
Aschen, Captain, 383
Auchinleck, Field-Marshal Sir Claude, 355

Baake, Major, 475
Baechle, Lieutenant, 346
Bailey, Geoffrey, 205, 213
Balke, Lieutenant (Waffen SS) Helmut, 278
Barcza, Margarete, 57
Bätcher, Captain, 623
Batov, Lieutenant-General, 575
Batrakov, Colonel, 564
Bauer, Second Lieutenant, 148
Baumann, Second Lieutenant, 155
Bayerlein, Lieutenant-Colonel, 17, 21, 48, 82, 90 ff., 100 ff., 103, 105, 121
Bazna, Elyesa (alias Cicero), 63
Beaverbrook, Lord, 129 ff., 141
Beck, Colonel-General, 59
Becker, Sergeant, 261
Behle, Sergeant, 408
Behrens, Hermann, SS officer, 198, 200
Beisinghof, Lieutenant, 348

Belov, General, 324
von Below, Major, 45, 616, 622
Beneš, President, 196, 199
von Berckefeldt, Captain, 263, 265
Bergener, Lance-corporal, 264 ff.
Berger, Colonel, 269
Beria, Soviet secret-police chief, 44
Bestmann, Major (Waffen SS), 237
Beyle, Lance-corporal, 114, 189
Bieler, Lieutenant-General, 278
Bittlingmeier, Lieutenant, 475
Blaurock, Lieutenant-Colonel, 367 ff., 371
von Blücher, Minister, 220 ff.
Blumenthal, Lieutenant, 489
Blumentritt, Major-General, 163, 164, 193
Blyukher, Marshal, 196, 214
von Bock, Field-Marshal, 15, 74, 89, 90 ff., 93, 99 ff., 122, 131, 152, 162, 164 ff., 229, 302 ff., 324, 331, 447, 455 ff., 483, 488 ff., 494
Bock, Sergeant, 408
Bockholt, medical NCO, 186
Bockmann, Major (Waffen SS), 404
von Boddien, Lieutenant-Colonel, 274, 276, 289
Boehm, Major, 509
Boehringer, Major, 179
von Boineburg-Lengsfeld, Major-General, 477 ff.
Boje, Colonel, 441 ff., 575, 621
Boldin, Lieutenant-General I. V., 190
Boll, Lieutenant, 515
Bollert, Battalion Commander, 378
von Boltenstern, Major-General, 49, 71, 86
Bormann, Martin, 445
Born, Sergeant Willi, 53

Index 631

Born, Colonel, 317
von Borries, Lieutenant-Colonel, 404
von Bose, Lieutenant-Colonel, 148
Bossert, Second Lieutenant, 185
von Both, General of infantry, 394
Bracht, Major, 184
Brämer, Lieutenant Hans, 324, 325, 326
Brandenberger, General, 30, 34
Brandner, Corporal Andreas, 431
Brandt, Second Lieutenant, 88
von Brauchitsch, Field-Marshal, 89, 101, 104, 137, 165 ff., 331, 447
Braun, Lance-corporal, 243
Brehmer, Lieutenant-Colonel, 106
Breith, Major-General, 156 ff.
Breitschuh, Lieutenant, 179
von Bremer, Company Commander, 491
Brennecke, Lieutenant-General, 164
Bried, Captain, 105
Brockdorff-Ahlefeldt, Count, 352 ff., 398 ff., 403, 440, 554
Brockdorff-Rantzau, Count, 207
Brücker, Lieutenant-Colonel, 17
Buchterkirch, Lieutenant, 114 ff.
Buck, Major, 188
Budennyy, Marshal, 83, 116 ff., 119, 121 ff., 139, 212, 214
Buff, Lieutenant, 451
Bukatschek, Sergeant, 72
Bulganin, N. A., 142
Bunzel, Major, 236
Bunzel, Sergeant, 265
Burmeister, Colonel, 511
Burmeister, Lance-corporal Werner, 311, 383, 389
von Burstin, Major, 420
Busch, Colonel-General, 36, 235, 271, 350, 399, 491
Büsing, Lieutenant, 499
Busse, Colonel, 288 ff., 466

CANARIS, ADMIRAL, 34, 57
Caspar, Sergeant, 318
Cassidy, Henry D., Press correspondent, 62
von Chadim, Lieutenant, 472
Châles de Beaulieu, General, 247, 249, 341
von Chappuis, Lieutenant-Colonel, 257

von der Chevallerie, Lieutenant-Colonel, 179, 240, 364
Chiang Kai-shek, 401
Chill, Colonel, 239 ff.
von Choltitz, Colonel, 281, 293, 299, 470
Churchill, Winston, 129 ff., 196
Chuykov, Lieutenant-General, 544, 556, 560 ff., 579 ff., 598, 616
Clausewitz, Carl von, 30, 89, 102, 103, 206, 242, 266, 505
Clausius, Colonel, 595
Colville, John Rupert, Churchill's secretary, 129
Crüwell, Major-General, 117 ff.

DAIJES, LIEUTENANT, 155
Daladier, Edouard, 199
Dallin, David J., 64
Dammann, Second Lieutenant, 300
von Daniels, Lieutenant-General Edler, 621
Darius, Lieutenant, 255, 265, 372
Debois, Colonel, 441
Dechant, Lieutenant, 479
Decker, Lieutenant-Colonel, 173, 177, 376
Dehner, Major-General, 171
Denikin, General, 197, 211
Dettmann, Second Lieutenant, 406, 408
Deutscher, Isaac, 172
Dieckmann, Waffen SS, 496
von Diest, Lieutenant - Commander, 33
Dietl, General, 414 ff., 422 ff., 429
Dietrich, General (Waffen SS), 279
Disselkamp, Major, 339
Doerr, General, 572, 604
Dohrendorf, Corporal, 326
Döll, Second Lieutenant, 73
Domaschk, Captain, 573
Dönitz, Admiral, 129, 434
Dovator, Major-General, 326, 329 ff.
Drabbe, Colonel, 440
Dragalina, Lieutenant-General, 557
Dreger, Sergeant, 132, 443 ff.
Dubyanskiy, Colonel, 565
Dumitrescu, Colonel-General, 577, 599
Duncker, Second Lieutenant, 410
Dürrholz, Lieutenant, 532

632 Index

EBERBACH, COLONEL, 135, 153 ff.
Eberlein, Second Lieutenant, 501
Ebert, Sergeant Fritz, 14
Eckart, Second Lieutenant, 321, 352
Eckinger, Major Joseph, 139, 160, 223, 227, 254
Eder, Major, 281
Eglseer, Major-General, 523
Ehrmann, Second Lieutenant, 155
Eichhorn, Captain, 485
Eichhorn, Lieutenant, 425, 429
Eicke, General, 402
Eikmeier, Lance-corporal, 132
Einbeck, Major, 297 ff.
Eisenhower, General Dwight D., 579
Eismann, Major, 606 ff.
Emmert, Lieutenant, 148
Engel, Major, 416
Engh, Sergeant, 530
von Erdmannsdorff, Colonel, 349 ff.
Esser, Sergeant, 573
Estor, Major, 380
Euler, Second Lieutenant, 528 ff.
Eydemann, Soviet Army leader, 196

FAHRENBERG, SECOND LIEUTENANT, 236
Fahrmbacher, General, 169, 182
Falck, Second Lieutenant, 337
von Falckenberg, Captain, 224
von Falkenhorst, Colonel-General, 418
Fangohr, Colonel, 558
Fechner, Captain Fritz, 511
Fedorenko, Lieutenant-General, 485 ff.
Fege, Sergeant, 242
Fegelein, 356, 372
Feige, General, 423, 436
Feist, Lieutenant, 406
Feldt, General, 85
Felmy, General, 526
Feuer, Sergeant, 348
Feyerabend, Lieutenant-Colonel, 121
Fiebig, Lieutenant-General, 566, 605
Findeisen, Sergeant, 76
Fischer, General, 136, 151
Fischer, Lance-corporal, 615
Fomin, Troop Commissar, 44
Förster, Sergeant Friedel, 472
Förster, General, 335
Förster, Corporal (Waffen SS), 91
Frank, Major, 122, 156
Franz, Lieutenant-Colonel, 71, 477 ff.

Fredebold, Major, 569
Frederick the Great, 240, 332
Fremerey, General, 564
Fretter-Pico, Lieutenant-General Maximilian, 450, 464
Freudenfeld, Major, 618
Frey, Second Lieutenant, 148
Friebe, Colonel, 520
Frinovskiy, Deputy OGPU Chief, 215
Fritsch, Sergeant, 254, 263
Frolov, Lieutenant-General, 435 ff.
Fromme, Second Lieutenant, 224
Fuhn, Corporal, 114
Fuller, General, 157
Fürnschuss, Second Lieutenant, 452
Fuss, Second Lieutenant, 261

GAMARNIK, GENERAL, Deputy Defence Commissar, 200, 209, 214
Gämmeler, Captain, 522
Gavrilov, Petr Mikhaylovich, 44
von Gaza, Lieutenant, 496
Gebhardt, Second Lieutenant, 363
George the Cossack, 525 ff., 527 ff.
Georgi, Corporal, 223
Gerasimenko, Lieutenant-General, 87
Geyer, General, 91, 181
Geyr von Schweppenburg, General Freiherr, 25, 73, 107 ff., 122, 124, 135, 483, 487, 495, 510
Gilbert—see Trepper, Leopold
Gille, Waffen SS, 496, 513
von Gilsa, General, 340, 381, 384
von Glasow, Major, 242
Glinka, 563, 569
Goering, Hermann, Reich Marshal, 52, 300, 431, 445, 482, 591, 593
Goldberg, Sergeant, 514
Golikov, General, 340
Golubyev, Major-General, 67
Goncharov, Lieutenant, 385 ff.
Gordov, Lieutenant-General, 511
Gorin, General, 340
Gorobin, Captain, 47
Gorodnyanskiy, Lieutenant-General, 463
Gottlieb, Lieutenant, 525
Grabert, Lieutenant, 501
Gradl, Major, 135
Grams, Rolf, 570
von Greiffenberg, Major-General, 164, 505

Index 633

Greve, Captain, 451
von Groddeck, Colonel, 450, 452, 453
Grosser, Lieutenant-Colonel, 269
Groth, Captain, 521 ff.
Gruscha, Captain, 168
Guderian, Colonel-General Heinz, 15, 16 ff., 21, 36 ff., 40, 45, 48 ff., 50, 59, 62, 68, 72, 76, 79, 80, 81, 82 ff., 89, 90 ff., 96, 97, 98, 100 ff., 106 ff., 115, 120 ff., 127, 131, 134 ff., 152 ff., 157, 158, 165, 166, 169, 188 ff., 208, 229, 233, 266, 307, 310, 315, 320, 322 ff., 331 ff., 335, 340, 380, 424, 485, 493, 556, 581, 595
Güle, Second Lieutenant, 348
Gulich, Sergeant, 264 ff.
Günther, Dr, medical officer, 345, 348
Gurtyev, Colonel, 573
Guryev, Major-General, 574
Gust, Lieutenant, 460

HACCIUS, COLONEL, 287, 449
Hahne, Colonel, 183
Halder, Colonel-General, 73, 82, 89, 98, 100, 102, 109, 163 ff., 269, 440, 441, 444, 502, 536
von Hammerstein, Lieutenant, 442
von Hammerstein-Equord, Chief of the Army Directorate, 442
Handrick, Major Gotthardt, 285
Hänert, Lieutenant, 85, 94 ff., 154 ff.
Hannig, Major, 88
Hansen, General, 237, 279, 284, 351
Harpe, General, 268
Harriman, Averell, 130, 141
von Hartmann, General, 542, 563, 619
von Hauenschild, Major-General Ritter, 140, 485, 544, 558 ff.
Hauser, Colonel, 317, 318 ff.
Hausser, Lieutenant-General, 92, 140
Hecker, Lieutenant-Colonel, 85
Hederich, Second Lieutenant, 231
von Heigl, Colonel Ritter, 274
Heim, Major-General, 558, 577, 581 ff., 600
Heinrici, General, 157, 331
Heinz, Captain, 410
Heitz, General, 457, 591
Hellmich, Major-General, 87
Hempel, Sergeant, 489

Henrici, Lieutenant-General, 526, 531
Henz, Second Lieutenant, 86
Herhudt von Rohden, Major-General, 606
Herr, Major-General, 496, 513, 538
Herrlein, General, 268, 349
Herzog, General, 29, 256
Hess, Major, 429
Hesse, Colonel, 391
Hesse, Lieutenant-Colonel, 477 ff.
von Heuduck, Major, 104
Heusinger, Major-General, 101, 444, 617
von Heydebrand, Colonel, 150, 159
von Heyden, Major, 563
Heyden-Rynsch, Captain Freiherr von der, 532
Heydrich, Deputy Chief of the Gestapo, 197 ff., 213, 214
Heyeres, Sergeant, 114, 189
Hilger, Second Lieutenant, 527
Hiller, Second Lieutenant, 502
Himer, Lieutenant-General, 297, 299, 301
Himmler, Chief of the Gestapo, 196, 197, 301
Hindemith, Sergeant, 475
Hindenburg, President, 267
Hingst, Captain, 319
Hinkmann, First Lieutenant, 30
von Hirschfeld, Major, 521, 524
Hirthe, Major, 257
Hitler, Adolf, 13 ff., 16, 20, 22 ff., 35 ff., 52, 58 ff., 63, 91, 98 ff., 112, 118 ff., 128, 131, 141, 163 ff., 165 ff., 169, 192 ff., 195, 206 ff., 213, 215, 219, 220, 262, 266 ff., 272 ff., 301, 302 ff., 307, 318, 331 ff., 340 ff., 345, 365, 385, 400, 414, 423, 428, 433, 439, 444 ff., 482 ff., 488 ff., 493 ff., 502 ff., 512 ff., 516, 525, 535 ff., 543, 554, 566, 576 ff., 579 ff., 587, 589 ff., 597 ff., 603, 610, 613 ff., 616, 620 ff.
Hitzfeld, Colonel, 473
Hoepner, Colonel-General, 15, 40, 81, 131, 137, 169, 174, 207, 222, 224, 233 ff., 238, 242, 247, 251, 267, 310, 324, 341, 424, 595
Hoernlein, Colonel, 154
Höfer, Second Lieutenant, 118
Hofer, Second Lieutenant, 378

634 Index

Hoffer, Major, 329
Hofmann, Sergeant, 321
Hofstetter, Second Lieutenant, 407
Hohmeyer, Colonel, 359 ff., 363 ff.
Hollidt, General, 456, 599, 603, 607, 611
Holste, Lieutenant-Colonel, 367
Holzer, Lance-corporal, 23
Holzer, Major (Waffen SS), 379
Hopkins, Harry, 172
Hoppe, Colonel Harry, 252, 258 ff., 392, 394
Hornbogen, Sergeant, 78
Hörst, Lieutenant-Colonel, 602
Hoth, Colonel-General, 15, 40, 48 ff., 65, 72, 81 ff., 86, 89, 91, 93, 97, 103, 131, 137, 150, 207, 324, 484, 486, 506, 544, 547, 554 ff., 560 ff., 566, 579, 584 ff., 592, 599, 603, 604 ff., 609, 611
Höttl, Wilhelm, Dr (alias Walter Hagen), 196
Hube, Major-General, 38, 118, 123, 461, 546, 548 ff., 550 ff., 591, 616
Hubicki, General, 123
Huck, Dr, 407
von Hünersdorff, Colonel, 604

IFFLAND, LIEUTENANT-COLONEL, 324 ff.
Ilgen, Colonel, 401 ff.
Infantes, General, 370
Isenbeck, Second Lieutenant, 77
Israilovich, NKVD representative, 200
Ivanov, 44
Ivashchenkov, Party Secretary, 172
Iwansen, Lieutenant, 599

JAENECKE, GENERAL, 572, 586, 591
Jahnke, member of Heydrich's staff, 197 ff.
Jedermann, Private, 77
Jeschonnek, Goering's chief of staff, 591
Jodl, Colonel-General, 101, 415, 445, 447, 511, 535, 579
Jolasse, Colonel, 383 ff.
Jonasson, Sergeant, 155
Jordan, Second Lieutenant, 39, 442
von Jungenfeldt, Major, 135

KACHALOV, LIEUTENANT-GENERAL, 107

Kalinov, Kyrill, 163, 173
Kambulin, Sergey, 373 ff.
Kandutsch, Captain, 168, 180, 533 ff.
Karabichev, General, 40, 66
Karasov, political officer, 331
Kauffmann, General, 339
Keding, Sergeant, 475
Keitel, Field-Marshal, 101, 194, 333, 445, 482, 536
Keller, Colonel-General, 15
Kempf, General, 123, 131, 484, 487, 557 ff.
Kent—*see* Sokolov, Viktor
Kern, Lance-corporal Walter, 323
Kesselring, Field-Marshal, 15, 43, 60 ff.
Kharitonov, General, 304
Khopka, Major, 565
Khotskilevich, Major-General, 66
Khozyaynov, Sergeant Andrey, 569
Khrushchev, Nikita, 176, 195 ff., 208, 303, 552, 561, 566, 570, 596
Kirchner, Lieutenant-General, 223, 495, 513, 517, 609
Kirponos, Colonel-General, 15, 37 ff., 118, 127
Klaiber, Lance-corporal (Waffen SS), 91
Klatt, Lieutenant-Colonel, 421, 431
Klauke, Lieutenant, 383
Klausing, Second Lieutenant, 363
Klein, Corporal, 77
Kleinjohann, Second Lieutenant, 547
Kleinschmidt, Major, 235
von Kleist, Field-Marshal, 16 ff., 37, 84, 110, 116 ff., 207, 283, 302, 304, 456 ff., 458, 462, 492, 506, 515 ff., 531, 611
Klingenberg, Waffen SS officer, 179
Klossek, Captain, 395
von Kluge, Field-Marshal, 84, 98, 182 ff., 217, 229, 233, 324, 331, 336, 340 ff., 568
Klykov, General, 401
Knaak, First Lieutenant Wolfram, 34 ff., 222
Knaak, Second Lieutenant, 244
Knatchbull-Hugessen, Sir Hughe, 63
Kniess, Lieutenant-General, 394
von Knobelsdorff, Lieutenant-General, 147, 534, 603
Knopf, Lieutenant-Colonel, 318
Koban, Lieutenant, 534 ff.

Koch, Second Lieutenant, 255
Köchling, Lieutenant-General, 395
Köhler, Second Lieutenant, 224
Kolchak, Admiral, 211
Koll, Colonel, 137, 248
Kolpakchi, Major-General, 543, 546
Konev, Colonel-General, 337, 340
Konrad, General, 513, 520
Kopp, Colonel, 318
Kork, General, 196
Koslar, Second Lieutenant, 475
Kozlov, Lieutenant-General D. T., 454
Krakowka, Sergeant, 491
Krämer, Lieutenant, 243
Krauss, Captain, 42
Krauss, Second Lieutenant, 77 ff.
Krebs, Colonel, 372
Krebs, Walter, 623
Kreipe, Colonel, 257, 395
Kreuter, Second Lieutenant, 77
Kreysing, Major-General, 420 ff., 422
Kreyzer, Major-General I. G., 75 ff.
Krivoshein, General, 21
Krüger, Major-General, 223 ff., 228, 255, 317, 367, 369, 371
Krumpen, Colonel, 461, 547, 550, 552
Kruse, Colonel, 369
Krylenko, Chief of Staff, 562
Krylenkov, Ivan Ivanovich, 130
Krylov, General, 564, 569
Kryvolapov, Soviet officer, 172
Kübler, General, 279, 381
von Küchler, Colonel-General, 222, 248, 349, 394
Kühne, Corporal, 77
Kulik, Marshal, 54
Kulikov, Lieutenant, 331
Kumm, Colonel (Waffen SS) Otto, 376, 380
Kümmerle, Sergeant, 523
Kuntz, Second Lieutenant Herbert, 623
Küppers, Captain, 149
Kurochkin, Lieutenant-General, 354
von Kurowski, Colonel, 168
Kutusov, Prince, 166
Kutzner, Major, 452
Kuzmany, Colonel, 383, 384
Kuznetsov, Colonel-General Fedor, 32, 34, 36 ff., 175, 222, 280, 289, 317, 320
Kuznetsov, Lieutenant, 473 ff.

LANDGRAF, MAJOR-GENERAL, 31, 253
von Langermann-Erlenkamp, General Freiherr, 135, 385
Langhammer, Lieutenant, 373
Lanz, Major-General, 462 ff.
Lascar, General Mihail, 581, 599
Lasch, Colonel, 222
Laszar, General, 474
Lattmann, Colonel, 546
Laux, General, 391
von Leeb, Field-Marshal Ritter, 15, 164, 221, 229, 233 ff., 248 ff.
Leibel, Hans, 176
Leliveldt, Second Lieutenant, 260
Lembke, Second Lieutenant, 256 ff.
Lemelsens, General, 131
Lemp, Second Lieutenant, 155
von Lengerke, Colonel, 544, 557
Lenin, 208, 211, 222
von Lewinski, Lieutenant-Colonel, 112
Leyser, Major-General, 584
Liddell Hart, Basil Henry, 58, 65, 192
von Liebenstein, Colonel Freiherr, 17, 48, 82, 90, 103, 381
von der Lieth-Thomsen, Colonel, 205
Lindemann, General, 234, 394, 410
Lindenthal, Captain, 361
Linnika, Major, 330
Lippert, Lieutenant-Colonel Lucien, 524
List, Field-Marshal, 494, 505 ff., 535 ff.
Loerzer, General, 61
Lohmar, Captain, 29
Lohmeyer, Colonel, 29, 32, 33, 391
Löhr, Colonel-General, 15
Lohse, Second Lieutenant, 131 ff., 321
Lopatin, General, 304, 561 ff., 567, 596 ff.
Lübke, Major, 151
Luckner, Corporal, 550
Ludendorff, General, 206, 447
"Ludwig," customs officer, 202 ff.
von Lüttwitz, Colonel, 383
Lvov, General, 298

MACHOLZ, GENERAL, 244
Mack, Major-General, 511
von Mackensen, General, 283, 303 ff., 455, 461 ff., 478 ff., 484, 495, 510, 513, 537
Maier, Lieutenant-Colonel, 325
Mallach, Private Lothar, 243

636 Index

Manitius, Colonel, 405
Mann, Mendel, 144
Mannerheim, Marshal, 59, 219 ff., 262, 267, 423
von Manstein, 15, 29 ff., 35 ff., 45, 64, 175, 223, 225 ff., 229 ff., 248, 271 ff., 279, 282, 285 ff., 292, 294 ff., 299 ff., 302, 440, 448, 453 ff., 464, 469 ff., 472, 475, 495, 503, 567, 599, 601 ff., 606 ff., 616, 619
von Manteuffel, Colonel, 132, 137, 175
Marcard, Lieutenant-Colonel, 105
Maresch, interpreter, 527
Martinek, General, 107, 326, 468
Maslennikov, **Lieutenant-General,** 158 ff., 337
Mastný Czechoslovak Minister, 199
Materna, General, 183
Materne, Major, 397
Matthis, Second Lieutenant, 348
Matussik, Lieutenant, 240, 242
Matzen, Sergeant, 348
Maziol, Sergeant, 361
Meindl, General, 385
Merker, Colonel, 181, 326
Meyer, Private Jan, 284
Meyer, Sergeant, 472
Meyer, Major (Waffen SS), 276
Meyer-Rabingen, Major-General, 105
Michalik, Colonel, 577
Michelsen, Corporal, 168
Model, Lieutenant-General, 25, 73, 112 ff., 115, 122, 126, 127, 150, 159, 366 ff., 371 ff., 375 ff., 379 ff., 567, 594
Moewis, Second Lieutenant, 508
Mölders, Colonel, 61, 276
Molotov, V. M., 52, 130, 142, 220, 386, 494
Moltke, Field-Marshal Count, 45, 132, 206
Momm, Lieutenant-Colonel **Harry,** 477
Montgomery, Field-Marshal, 467
Morozov, General, 349 ff.
Morzik, Colonel, 400
Mues, battalion commander, 546
Mühlenkamp, 496
Müller, Major-General Angelo, 477
Müllard-Gebhard, Lieutenant-General, 474

Müller, Hans, Lance-corporal, 132 ff., 243
Müller, Lieutenant-Colonel, 290, 622
Müller, Sergeant, 362
Mummert, Major, 339, 379
Mundt, Lieutenant, 348
Munzel, Lieutenant-Colonel, 124
Mussolini, Benito, 576

NAGEL, FRITZ, 449, 465 ff.
Napoleon, 45, 73, 86, 89, 90, 139, 167, 506
Naumann, Sergeant, 492
Negendanck, Major-General, 341 ff.
Nehring, General, 25, 48, 72, 133, 380 ff., 384
Neitzel, Lieutenant, 246
Neumann, Captain, 361, 395
Nicholas I, Tsar, 464
von Niedermayer, Professor Ritter Oskar (Neumann), 205
Niemann, Colonel, 467
Nieweg, Sergeant, 623
Nolte, Major, 269 ff.

OCKER, DR, 407
Odinga, Sergeant, 479
Oehrlein, Sergeant, 264 ff.
Ohrloff, Lieutenant, 175, 315
Oktyabrskiy, Vice-Admiral, 473, 476
Olboeter, Captain (Waffen SS), 305 ff.
Oldeboershuis, Private (Waffen SS), 91
von Oppeln-Bronikowski, Colonel, 578, 582, 601
von Oppen, Second Lieutenant, 153 ff.
Ordás, Captain, 348 ff.
Orschler, Captain, 96
Ortlieb, Major, 498 ff., 517 ff.
Osswald, General, 589
Ostarek, Corporal, 132
Ott, General, 514
Ottenbacher, Lieutenant-General, 248, 252
von Oven, Major-General, 26

PAJARI, GENERAL, 221
von Pannwitz, Colonel, 604
Pape, Sergeant, 256 ff.
Pape, Major Günther, 510, 526, 532
Paske, Flight Engineer, 623

Patow, Major, 534
Pätzold, Lieutenant, 369 ff.
Pauli, Second Lieutenant, 261
Paulus, General, 232, 455 ff., 478, 482, 511, 543, 557, 559 ff., 566 ff., 572, 576 ff., 580, 585 ff., 603, 605, 607 ff., 613 ff., 618, 622
Pavlov, Dmitriy, Colonel-General, 47, 51, 54 ff.
Pawendenat, Sergeant, 244
Peitz, Captain, 300, 453
Pepper, Captain, 77 ff.
Pervushin, General, 298
Peschke, Captain, 132
Peter the Great, 259
Petermann, Second Lieutenant, 378
Petrov, General, 286 ff., 476
Pfeiffer, Lieutenant-General, 569
Pflugbeil, General, 486
Pickert, Major-General Wolfgang, 450, 589
Pilsudski, Marshal, 211 ff.
Pingel, medical NCO, 185
Piontek, Lance-corporal, 53
Platanov, Lieutenant-General, 562
Pleyer, Sergeant, 291
Podlas, Lieutenant-General, 463
Pohl, Major, 618 ff.
Pomtow, Major, 125 ff., 508
Poppinga, Colonel, 442
Potapov, Major-General, 119, 127 ff.
Potaturchev, Major-General, 66 ff., 72
Potemkin, Vladimir, 199
Praxa, Captain, 40, 42
Praxa, Corporal Gustav, 361 ff.
Pretz, Major, 290
Prigann, Lieutenant, 329
Pröhl, Captain, 343 ff.
Proske, Lieutenant-Colonel, 363 ff.
von Prott, Lieutenant, 290
Püchler, Colonel, 440, 460
Purkayev, General, 366
Pushkin, Alexander, 221
Putna, General, 213 ff.

QUENTEUX, CORPORAL, 550

RADEK, KARL, 204, 213 ff.
Rado, Alexander, 57 ff., 108, 543
Radu, General, 582
Range, Second Lieutenant, 146

Raus, Colonel, 226, 248, 604
Rehrl, Lieutenant (Waffen SS), 278
Reichel, Major, 479, 490
von Reichenau, Field-Marshal, 26, 37, 116, 301, 302, 307, 595
Reichmann, Major, 178
Reimann, Second Lieutenant, 452
Reinhardt, General, 15, 30 ff., 36, 81, 150, 158, 223, 225, 229 ff., 233, 242, 248, 253, 262, 315, 368
Reinhardt, General Alfred, 497, 515, 520
Reinisch, combat-group leader, 547
Reissner, Lieutenant, 452
Remizov, General, 303
Reuter, Captain, 245 ff.
von Ribbentrop, Joachim, 11, 12
Richter, Major, 373
Richter, Second Lieutenant, 344, 347
von Richthofen, General Freiherr Wolfram, 61, 159, 252, 257, 292, 375, 450, 454, 465, 581, 612
Riebel, Colonel, 485, 545, 557
Riederer, Hans, 425
Ries, Colonel, 113
Rinschen, Lieutenant, 126
Rittmann, Sergeant, 500
Rode, Captain, 85
Rodimtsev, Major-General, 565
Rodt, Colonel, 173 ff., 177, 495
Roettig, Major-General, 38
Rokossovskiy, Colonel-General, 615, 617
Rommel, Erwin, 129, 137, 273, 355, 415, 444, 467, 483, 519, 524, 556, 579
Roosevelt, Franklin D., 172
Roske, General, 598
Rössert, First Lieutenant, 94
Rössler, Rudolf (alias Lucy), 56, 58, 543
Rotter, Captain, 216 ff.
Rowehl, Lieutenant-Colonel, 60
Rüdiger, Major, 391
Ruederer, Lieutenant-Colonel, 149
Rümmler, Dr, 523
von Runstedt, Field-Marshal, 15 ff., 26, 37, 39, 83, 86, 116 ff., 121, 164, 266, 272, 307, 324
Ruoff, General, 173, 492, 506, 513, 516, 519
Rupp, Major-General, 524

SACHS, GENERAL, 440

638 Index

Sailer, Lance-corporal, 430
von Salmuth, General, 280, 281
Samsonov, A. M., 143, 173, 195, 313
Sander, Major-General, 454, 474
Sarge, Sergeant, 11 ff., 78
Sätzler, Captain, 518
Sauerbruch, Captain, 600
Sauvant, Captain, 573, 604
Schaal, Lieutenant-General, 92, 175, 314 ff.
Schaefer, Lieutenant-Colonel, 281
Scharnhorst, General, 447
Schätz, Second Lieutenant, 599
Schaub, Captain, 382 ff.
Schede, Lieutenant-General, 257
von Scheele, Major-General, 381
Scheidt, Sergeant, 451
Schellenberg, Walter, 196
Schenke, Lieutenant-Commander, 33
Scherer, General, 366, 405, 407, 440
Schlageter, Second Lieutenant Albert Leo, 204
Schlemmer, Major-General, 420
Schlieffen, General, 447
Schliep, Second Lieutenant Jürgen, 527, 529 ff.
Schlieper, Major-General, 21
Schlömer, General, 553
Schlösser, Lieutenant Erich, 359
Schlünz, Sergeant, 345
Schmidt, Lieutenant-Colonel Arthur, 392, 482, 586 ff., 607 ff., 610 ff., 614 ff.
Schmidt, Major-General Friedrich, 474
Schmidt, General Rudolf, 259, 268, 323, 380, 599
Schmidt, Captain, 50
Schmidt, Corporal, 243
Schmidt, Sergeant, 550
Schmundt, General, 620
Schneider, Sergeant, 155
Schneider, Lieutenant, 522
von Schobert, Colonel-General Ritter, 38 ff., 116 ff., 271, 276
von Scholz, Waffen SS officer, 513
Schörner, Major-General, 427 ff., 435 ff.
Schrader, Horst, 305
Schröder, Sergeant, 125
Schroeder, Second Lieutenant, 527
Schrottke, Lieutenant, 87
Schubert, General, 337

Schulz, General, 610
Schulze, Major, 524
Schulze-Boysen, Lieutenant, 543
Schulze-Hinrichs, Lieutenant-Commander, 422
Schütze, Corporal, 168
von Schwedler, General, 557
Schwenk, Private (Waffen SS), 92
von Schwerdtner, Captain, 410 ff.
Schwerin, Count, 252, 259, 260
von Seeckt, Colonel-General, 204, 207, 212
Seidinger, Corporal, 133
Seitz, Second Lieutenant, 477
Semenenko, First Lieutenant, 44
von Seydlitz-Kurzbach, General, 402, 456, 512, 541, 547, 554, 560, 563, 591, 596 ff., 605
Shamyakin, Soviet officer, 331
Shaposhnikov, Chief of General Staff, 111, 135, 494
Sheboldayev, Boris, 212
Shilin, Colonel P. A., 182, 287, 448
Shkvor, Czech agent, 64
Shpigelglass, NKVD official, 215
Shuntyayev, 44
Shvetsov, General, 335
Sieckenius, Lieutenant-Colonel, 461, 548
Sierts, Lieutenant, 256
Sievers, Dr, 185
Siilasvuo, Major-General, 423, 436
Silzer, Second Lieutenant, 30
Sinnhuber, Lieutenant-General, 474
Sinzinger, Colonel, 365
Siry, Major-General, 316
Skoblin, General, 197, 213
Smirnov, Sergey, 43 ff.
Smirnov, Lieutenant-General, 283
Smorodinov, General, 46
von Sodenstern, General, 164, 458, 593
Sokirkov, Second Lieutenant, 331
Sokolov, Viktor (alias Kent), 57, 63
Sokolovskiy, Lieutenant-General V. D., 142, 401
Sorge, Richard, 56, 63, 140, 172, 313, 543
Söth, Major, 223
Spang, Lieutenant-General, 600
Specht, Captain, 271
Speer, Minister for Armaments, 194

Index 639

von Sponeck, Lieutenant-General Count, 292, 294 ff.
Sponheimer, General, 238, 240
Stadler, Sergeant, 94
Staedtke, Major, 185
Stahel, Colonel, 600
Stalin, Joseph, 13, 16, 22, 43, 47, 52, 56 ff., 62 ff., 74, 97 ff., 108 ff., 111, 112, 117, 121, 127, 134, 139, 140, 142, 161, 163, 167, 172, 175 ff., 185, 195 ff., 208 ff., 216, 219 ff., 291, 303, 312, 353 ff., 359, 365, 386, 401, 414, 431 ff., 444 ff., 454, 464, 494, 516, 542, 562, 566, 604 ff.
Steiner, General (Waffen SS) Felix, 496, 513
Steinhardt, US Ambassador, 142
Stempel, Second Lieutenant, 574
Stepanchikov, 44
von Stettner, 524
Steves, Sergeant, 347
Stiefvater, Major, 276 ff.
Störck, Second Lieutenant, 114, 188 ff.
Stoves, Second Lieutenant, 254, 263, 317
von Strachwitz, Count Manfred, 25
von Strachwitz, Count, 546, 550 ff.
Strasser, Lieutenant, 395
Strauss, Colonel-General, 98, 335 ff., 338 ff., 367, 372
Strauss, Second Lieutenant, 177
Strecker, General, 575 ff., 591, 622
Strehlke, combat-group leader, 552
Streit, Captain, 13
Ströhlein, Captain, 148
Strucken, Sergeant, 114
von Stülpnagel, General, 27, 37, 116, 303
Stumme, General, 137, 150, 180, 476 ff., 481
Stumpff, Lieutenant-General, 49, 81
von Stünzer, Panzer Company leader, 383
von Le Suire, Lieutenant-Colonel, 417

TANK, LIEUTENANT, 508 ff.
Tarasov, Major-General, 356, 363, 364
Tarorov, Adjutant, 172
Tasya, serving-girl, 563, 569
Tavliyev, Colonel, 330
Teege, Major, 72
Teichert, Major, 72

Teichgräber, 479
Tenning, Private, 492, 514
Teuber, Second Lieutenant, 455, 458 ff.
Teuchler, Lance-corporal, 42, 43
Thilo, Lieutenant-Colonel, 290
Thomale, Lieutenant-Colonel, 146
Thomas, Colonel, 71, 86 ff.
Tietz, Lance-corporal, 322
Timoshenko, Marshal, 51, 55, 73, 82, 83, 85, 99, 108, 111, 142, 304, 442, 456 ff., 461, 489, 490, 493, 503, 511, 516, 542 ff.
von Tippelskirch, Lieutenant, 231
Titulescu, Rumanian Foreign Minister, 213
Tödt, Sergeant, 243 ff.
Tolstoy, 191
Tornau, Lieutenant, 408
Trepper, Leopold (alias Gilbert), 56, 63, 543
von Tresckow, Lieutenant-Colonel, 165
Tripp, Sergeant, 145
Tromm, Lieutenant-Colonel, 408
Trotsky, 46
Tukhachevskiy, Marshal, 196 ff., 208 ff., 216 ff., 401
Tupikov, Lieutenant-General, 127

UBOREVICH, GENERAL, 196
von Uckermann, General, 406
von Usedom, Major, 179

V., SECOND LIEUTENANT, 387
Vassil, Ukrainian, 189
Veiel, Lieutenant-General, 173, 177, 319
von Viebahn, Lieutenant-Colonel, 452
von Vietinghoff, General, 137, 376
Viktor Nikolayevich, 342, 347
Vinogradov, Second Lieutenant, 44
409 ff.
Vlasov, Andrey Andreyevich, 401, 409 ff.
Voelter, Colonel, 481
Vogel, Lance-corporal, 145
Vogel, Sergeant, 133
Vogt, Major, 139, 287
Volskiy, Major-General, 585
Vopel, First Lieutenant, 114 ff., 124
Vormann, Lieutenant, 442
Voroshilov, Marshal, 83, 175, 201, 212, 231 ff., 236 ff., 250, 262, 494
Voss, Second Lieutenant, 348

Vostrukhov, General, 366
Vyshinskiy, General, 213

WAGENER, LIEUTENANT-COLONEL, 534
Wäger, General, 334
Wagner, General, 515
Wagner, Lance-corporal (Waffen SS), 378
Wagner, Lieutenant-Colonel, 145
Waldow, Captain, 408
Wartmann, Lieutenant, 124 ff.
Weber, Sergeant, 77
Weber, Second Lieutenant, 180
von Wedel, Captain Joachim, 464
Wehde, Lieutenant, 106
Weichardts, Major, 105
von Weichs, Colonel-General Freiherr, 98, 121, 484, 494, 531, 558, 559, 586, 589 ff., 593, 596, 600
Weidling, Colonel, 140, 180
Weidner, Second Lieutenant, 11 ff.
Weinrowski, Lieutenant, 30
Weiss, Lieutenant, 28
Weissmeier, Sergeant, 527
Wellmann, Major, 491 ff., 510
Welser, Chief Labour Leader, 418
Wenck, Lieutenant-Colonel, 233, 255, 367, 371, 495, 599 ff., 601, 604
Wendt, Sergeant, 261
von Werthern, Lieutenant, 146
Weseloh, Corporal Heinrich, 284
Westhoven, Colonel, 223 ff., 254, 310, 314, 315, 317 ff.
Westphal, radio operator, 13, 54, 78
Westphal, 278
Wetthauer, 352
Wetzel, General, 517
Weygand, General, 212
Wiebel, Dr, 348
Wichmann, Sergeant, 156
Wiegmann, Kurt, 146
Wieltsch, Second Lieutenant, 41, 42, 45
Wierschin, Sergeant, 25
Wiese, Colonel, 335
Wiesner, Major, 120
von Wietersheim, General, 542, 554
von Wietersheim, Lieutenant-Colonel Wend, 233 ff., 318, 367, 369, 377, 379
Wiktorin, General, 243
Wildhagen, Major, 158
Willich, Sergeant, 348

Willig, Major, 615
Willumeit, Lance-corporal, 239
Winnefeld, Captain, 293
Winter, Lieutenant, 380
Winzen, Captain, 498
Witthöft, Lieutenant-General, 334
Witting, Finnish Foreign Minister, 221
Wittke, Major-General, 39
von Witzleben, combat-group leader, 461
Wöhler, Colonel, 288
Wolff, Major-General, 289, 470, 474
Wolter, battalion commander, 383
von Wrede, Lieutenant-General, 351
Wurm, Second Lieutenant, 532

YAKIR, Soviet Army leader, 196
Yakov, Ukrainian, 189
Yashin, Colonel, 70
Yegorov, Soviet Army leader, 196, 212
Yeremenko, Andrey Ivanovich, 46 ff., 51 ff., 54, 62, 72, 75, 80, 86, 97, 108 ff., 134, 135, 138, 353 ff., 364 ff., 552, 556 ff., 560, 561 ff., 565, 584, 586, 604 ff., 616
Yevstifeyev, Major Ivan, 400
Yezhov, Secret Police Chief, 200
Yukhvin, Colonel, 319
Yushkevich, General, 335

ZAGORODNY, CAPTAIN, 533
Zakharov, General, 250
Zass, regimental commander, 387
Zehender, Colonel (Waffen SS), 372, 375
Zeitzler, General, 594, 596, 600, 613, 617, 620 ff.
Zhdanov, 252 ff., 268
Zhukov, General G. K., 142, 163, 165 ff., 173, 175 ff., 194, 320, 322, 324, 337, 340
von Ziegler, Second Lieutenant, 78, 520, 532
Ziegler, Colonel, 288 ff.
Zimmer, Colonel, 275
Zimmermann, Lieutenant-Colonel, 508
von Zitzewitz, Major Coelestin, 594 ff., 620
Zorn, Major-General, 81, 259, 401 ff.
Zubachev, Captain, 44
Zuckertort, Lieutenant-General, 468
Zumpe, Second Lieutenant, 22 ff., 29